Communications in Computer and Information Science 2491

Rationale

The CCIS series is devoted to the publication of proceedings of computer science conferences. Its aim is to efficiently disseminate original research results in informatics in printed and electronic form. While the focus is on publication of peer-reviewed full papers presenting mature work, inclusion of reviewed short papers reporting on work in progress is welcome, too. Besides globally relevant meetings with internationally representative program committees guaranteeing a strict peer-reviewing and paper selection process, conferences run by societies or of high regional or national relevance are also considered for publication.

Topics

The topical scope of CCIS spans the entire spectrum of informatics ranging from foundational topics in the theory of computing to information and communications science and technology and a broad variety of interdisciplinary application fields.

Information for Volume Editors and Authors

Publication in CCIS is free of charge. No royalties are paid, however, we offer registered conference participants temporary free access to the online version of the conference proceedings on SpringerLink (http://link.springer.com) by means of an http referrer from the conference website and/or a number of complimentary printed copies, as specified in the official acceptance email of the event.

CCIS proceedings can be published in time for distribution at conferences or as post-proceedings, and delivered in the form of printed books and/or electronically as USBs and/or e-content licenses for accessing proceedings at SpringerLink. Furthermore, CCIS proceedings are included in the CCIS electronic book series hosted in the SpringerLink digital library at http://link.springer.com/bookseries/7899. Conferences publishing in CCIS are allowed to use Online Conference Service (OCS) for managing the whole proceedings lifecycle (from submission and reviewing to preparing for publication) free of charge.

Publication process

The language of publication is exclusively English. Authors publishing in CCIS have to sign the Springer CCIS copyright transfer form, however, they are free to use their material published in CCIS for substantially changed, more elaborate subsequent publications elsewhere. For the preparation of the camera-ready papers/files, authors have to strictly adhere to the Springer CCIS Authors' Instructions and are strongly encouraged to use the CCIS LaTeX style files or templates.

Abstracting/Indexing

CCIS is abstracted/indexed in DBLP, Google Scholar, EI-Compendex, Mathematical Reviews, SCImago, Scopus. CCIS volumes are also submitted for the inclusion in ISI Proceedings.

How to start

To start the evaluation of your proposal for inclusion in the CCIS series, please send an e-mail to ccis@springer.com

Deep Gupta · Vipin Kamble · Vishal Satpute ·
Ashwin Kothari
Editors

Paradigm Shifts in Communication, Embedded Systems, Machine Learning, and Signal Processing

Third International Conference, PCEMS 2024
Nagpur, India, November 11–12, 2024
Revised Selected Papers, Part II

Springer

Editors
Deep Gupta
VNIT Nagpur
Nagpur, Maharashtra, India

Vipin Kamble
VNIT Nagpur
Nagpur, Maharashtra, India

Vishal Satpute
VNIT Nagpur
Nagpur, Maharashtra, India

Ashwin Kothari
VNIT Nagpur
Nagpur, Maharashtra, India

ISSN 1865-0929 ISSN 1865-0937 (electronic)
Communications in Computer and Information Science
ISBN 978-3-031-90576-6 ISBN 978-3-031-90577-3 (eBook)
https://doi.org/10.1007/978-3-031-90577-3

This Springer imprint is published by the registered company Springer Nature Switzerland AG
The registered company address is: Gewerbestrasse 11, 6330 Cham, Switzerland

Preface

The 3rd International Conference on Paradigm Shifts in Communication, Embedded Systems, Machine Learning and Signal Processing (PCEMS 2024) was organized by Visvesvaraya National Institute of Technology Nagpur, Maharashtra, India on 11th–12th November 2024. PCEMS is a premier flagship conference of the Department of Electronics and Communication Engineering, VNIT Nagpur, focused on different applications of communication, embedded systems, machine learning and signal & image processing. The conference witnessed extraordinary success with publications in multiple domains of Computer Science, Electronics and Communication Engineering and other engineering as well.

PCEMS 2024 received 352 regular paper submissions, which went through a rigorous review process by more than 200 reviewers from different national and international renowned institutes and universities. The technical program chairs coordinated the overall review process which resulted in the acceptance of 123 research articles. Each paper received at least 3, and several received up to 5 single-blind reviews. The papers were evaluated on their relevance to the conference track or topics, original contributions, scientific correctness and clarity of presentation. The selection was based on these reviews and further discussion of each paper among the involved program committee members. Out of them, 73 papers were registered and presented for oral presentation in 15 technical sessions around the different research areas based on communication, antennas, computer vision, medical image analysis, deep learning, AI-based systems and applications, classification problems, embedded systems and IoT, etc. Moreover, 4 tutorials on machine learning, Signal Processing, IoT, and the healthcare domain and 2 industry workshops by MathWorks and Designtech Systems were scheduled along with an Industry Expo.

Keynote Talks: PCEMS 2024 was organized with seven keynote talk sessions planned for the two-day conference. PCEMS 2024 commenced with a keynote talk on "Artificial Intelligence: Evolution and Current Trends" by Sanghamitra Bandyopadhyay (Director, ISI Kolkata, India), followed by keynote talks by Petia Ivanova Radeva (University of Barcelona, Spain), R. Venkatesh Babu, (Indian Institute of Science, Bangalore, India) and Ujjwal Maulik (Jadavpur University, Kolkata, India). On the last day of the conference, an informative keynote talk on Advancements and Opportunities in Next-Generation MIMO Wireless Communication for 6G was given by Robert W. Heath (University of California, San Diego, USA), followed by a keynote talk by M. V. Kartikeyan (Director, IIITDM-Kancheepuram, INDIA) finally Sajal Das (Missouri University of Science and Technology, USA) shared deep insights with the audience on Smart Sensing to Smart Living.

Awards: PCEMS 2024 presented high-quality research works with innovative ideas. All the session chairs were invited to vote for the different categories of awards: Visvesvaraya Best Paper Award, PCEMS 2024 and Best Paper Award, PCEMS 2024 Best Student Paper Award, and Best Paper Awards in the AI, Communication, Signal processing,

IoT and Embedded systems tracks sponsored by various industries. All the awards were announced in the valedictory ceremony of PCEMS 2024 with the announcement of the next edition of PCEMS 2025.

Our heartfelt gratitude goes to our academic sponsor institution for their cooperation, support and assistance: Visvesvaraya National Institute of Technology, Nagpur, India. We also thank our financial sponsors: Science and Engineering Research Board (SERB), Anusandhan National Research Foundation (ANRF), Board of Research in Nuclear Sciences (BRNS), Indian Space Research Organisation (ISRO), Data Maven Pvt. Ltd., Anysys, Intellipaat, DigiTech Systems, Scientech and MathWorks. Most importantly, we thank the authors for submitting and trusting their work to the conference.

Organizing Chair

PCEMS 2024

Organization

PCEMS 2024 Organizing Committee

Patron

Prem Lal Patel	VNIT Nagpur, India

General Chairs

Ashwin Kothari	VNIT Nagpur, India
Petia Ivanova Radeva	University of Barcelona, Spain
Vikram M. Gadre	IIT Bombay, India

General Co-chairs

Abhay Gandhi	VNIT Nagpur, India
Kishor Bhurchandi	VNIT Nagpur, India
Vishal Satpute	VNIT Nagpur, India

Conference Chairs

Deep Gupta	VNIT Nagpur, India
Ashish Tiwari	VNIT Nagpur, India

Technical Program Chairs

Prabhat Sharma	VNIT Nagpur, India
Saugata Sinha	VNIT Nagpur, India
Himanshu Padole	IIT Bhubaneshwar, India
Ankit Bhurane	VNIT Nagpur, India
Jayendra Kumar	NIT Jamshedpur, India
Shyam Lal	NIT Surathkal, India
Pradnya Ghare	VNIT Nagpur, India

Ganesh Patil	VNIT Nagpur, India
Pankaj Pratap Singh	CIT Kokrajhar, India
Deepak Mishra	IIST Trivandrum, India
Sudhir Kumar	IIT Patna, India
Radhika Gour	IIIT Allahabad, India
Raghavendra Ramachandra	NTNU, Norway
Chhavi Dhiman	DTU Delhi, India
Vivek Raghuwanshi	IIT Bhilai, India
Anurag Singh	NIT Delhi, India
Chiranjeev Kumar	IIT-ISM Dhanbad, India
Harsh K. Verma	NIT Jalandhar, India
Mayank Pandey	MNNIT Allahabad, India

Website Chairs

Praveen Pawar	VNIT Nagpur, India
Vipin Kamble	VNIT Nagpur, India
Ravindra Keskar	VNIT Nagpur, India
Poonam Sharma	VNIT Nagpur, India

Publication Chairs

Neeraj Rao	VNIT Nagpur, India
Anamika Singh	VNIT Nagpur, India

Publicity Chairs

Arvind Kumar	VNIT Nagpur, India
Amit Agarwal	VNIT Nagpur, India

Local Organizing Chairs

Joydeep Sengupta	VNIT Nagpur, India
K. Surender	VNIT Nagpur, India
Punit Bhavsar	VNIT Nagpur, India

Industry Forum and Exhibition Chair

Saugata Sinha	VNIT Nagpur, India

PCEMS'24 International Advisory Committee

Sebastiano Battiat	Università di Catania, Italy
Jonathan Wu	University of Windsor, Canada
Satish Viswanath	Case Western Reserve University, USA
Fabio Dell'Acqua	University of Pavia, Italy
A. G. Ramakrishnan	IISc Bangalore, India
Ram Bilas Pachori	IIT Indore, India
Jayanta Mukhopadhyay	IIT Kharagpur, India
R. S. Anand	IIT Roorkee, India
Dharmendra Singh	IIT Roorkee, India
Phalguni Gupta	IIT Kanpur, India
Ravibabu Mulaveesala	IIT Delhi, India
Ganapati Panda	IIT Bhubaneswar, India
Ujjawal Maulik	Jadavpur University, India
Yegnanarayana B.	IIT Hyderabad, India
Mahesh P. Abegaonkar	IIT Delhi, India
Rajeev Srivastava	IIT BHU, India

PCEMS 2024

List of Reviewers

A. S. Gandhi	Visvesvaraya National Institute of Technology, Nagpur, India
Abhijeet M. Pimpale	Visvesvaraya National Institute of Technology, Nagpur, India
Abhijit Devidas Bijwe	Priyadarshini College of Engineering, India
Abhijit Titarmare	G H Raisoni College of Engineering, India
Abhilash Mandloi	SVNIT, Surat, India
Abhishek Pahuja	Koneru Lakshmaiah Education Foundation, India
Abhishek Basu	RCCIIT. West Bengal, India
Adesh Kumar	UPES Dehradun, India
Aditi Jain	Visvesvaraya National Institute of Technology, Nagpur, India

Ajay K. Vyas	Adani University, India
Aleefia Khurshid	Shri Ramdeobaba College of Engineering and Management, India
Amarana Prabhakara Rao	Vishnu Institute of Technology, Andhra Pradesh, India
Amarana Prabhakara Rao	KLEF, Bowrampet, Hyderabad, India
Ambigavathi Munusamy	Indian Institute of Information Technology Una, India
Amit Agarwal	Visvesvaraya National Institute of Technology, Nagpur, India
Amit Kumar	NIT Delhi, India
Amit R. Shewale	IIIT Nagpur, India
Amit Kumar Chanchal	School of Computing, MIT Vishwaprayag University, Solapur, India
Amol P. Bhopale	Indian Institute of Information Technology Nagpur, India
Amruta Sanjay Lambe	Visvesvaraya National Institute of Technology, Nagpur, India
Anagha Prashant Rathkanthiwar	Priyadarshini College of Engineering, India
Anamika Singh	Visvesvaraya National Institute of Technology, Nagpur, India
Anand Kumar M.	National Institute of Technology Karnataka, India
Anish Kumar Gupta	Visvesvaraya National Institute of Technology, Nagpur, India
Anish Kumar Vishwakarma	UPES, Dehradun, India
Anjali Bhalchandran	GCOE, Aurangabad, India
Ankit Ashokrao Bhurane	Visvesvaraya National Institute of Technology Nagpur, India
Ankush Jamthikar	Visvesvaraya National Institute of Technology, Nagpur, India
Anupama D. Sakhare	R T M Nagpur University, India
Anveshkumar Nella	VIT Bhopal University, India
Apurva Naik	MIT Art Design & Technology University, Pune, India
Archana R. Tiwari	RCOEM, Nagpur, India
Archana S. Nanoty	NIT Hamirpur, India
Arun Balodi	Dayananda Sagar University, Bangalore, India
Arun Rasika Karunakaran	Tata Consultancy Services, India
Arvind Kumar	Visvesvaraya National Institute of Technology, Nagpur, India
Asha C. S.	Manipal Institute of Technology, Manipal, India
Ashish Chandelkar	NIT Nagpur, India

Joydeep Sengupta	VNIT Nagpur, India
K. Surender	Visvesvaraya National Institute of Technology, Nagpur, India
K. M. Bhurchandi	VNIT Nagpur, India
Kanchan S Tiwari	MESCOE, India
Khushboo Jain	IIIT, Nagpur, India
Kiran Kumar Gurrala	National Institute of Technology Andhra Pradesh, India
Kishore Puli Kumar	National Institute of Technology Andhra Pradesh, India
Komal Baburao Umare	VNIT Nagpur, India
Krishna Chaitanya A	NIT Andhra Pradesh, India
Krishnamaraju Suraparaju	VNIT Nagpur, India
Kuldeep Baderia	Jaypee Institute of Information Technology
Laxmi Kumre	MANIT Bhopal, India
Lekshmi R. R	Amrita School of Engineering, India
Madhav Chandane	VJTI Mumbai, India
Madhuri A. Tayal	GHRCEM, Nagpur, India
Madhurima Bandyopadhyay	Continental Automotive Components (India) Pvt. Ltd., India
Mahantesh P. Mattada	PES Institute of Technology and Management, Shivamogga, India
Mahendra A. Gaikwad	G H Raisoni College of Engineering Nagpur, India
Mahendra Shukla	ABV-IIITM, Gwalior, India
Mahesh M. Shirole	VJTI, Mumbai, India
Manish Kurhekar	VNIT Nagpur, India
Manish Mandloi	PDEU Gandhinagar, India
Manisha Manohar Ingle	Visvesvaraya National Institute of Technology, Nagpur, India
Manisha D. Raut	GHRCEM, Nagpur, India
Manisha Sawant	Visvesvaraya National Institute of Technology, Nagpur, India
Mayur Katwe	National Institute of Technology Raipur, India
Mayur R. Parate	IIIT Nagpur, India
Mayuri A. Digalwar	IIIT, Nagpur, India
Megha B. Chakole	YCCE Nagpur, India
Megha M. Kolhekar	Fr. C. Rodrigues Institute of Technology, India
Milind R. Penurkar	IIIT Nagpur, India
Milind S. Shah	Agnel Charities' Fr. C. Rodrigues Institute of Technology, India
Mohammad Farukh Hashmi	National Institute of Technology, India

Mridula Korde	Shri Ramdeobaba College of Engineering and Management, India
Naveen Cheggoju	Indian Institute of Information Technology Una, India
Neeraj Rao	Visvesvaraya National Institute of Technology, Nagpur, India
Neeraj Rao	VNIT Nagpur, India
Neha Bhusari	Visvesvaraya National Institute of Technology, Nagpur, India
Neha Ravindra Kasture	Indian Institute of Information Technology Nagpur, India
Nikhil Agrawal	Indian Institute of Information Technology Nagpur, India
Nikhil Dhengre	Visvesvaraya National Institute of Technology, Nagpur, India
Nikhil Bhagwatrao Gaikwad	Visvesvaraya National Institute of Technology, Nagpur, India
Nikhil Deep Gupta	Visvesvaraya National Institute of Technology, Nagpur, India
Nilesh Bhaskarrao Bahadure	GSFC University, Vadodara, Gujarat, India
Nitesh K. Bharadwaj	National Institute of Technology Raipur, India
Nitin B. Chakole	VNIT, Nagpur, India
Nrusingha Charan Pradhan	Reykjavik University, Iceland
Omkar Suresh Vaidya	Sandip Institute of Technology & Research Centre, Nashik, India
P. H. Ghare	VNIT-Nagpur, India
Pallavi K. Parlewar	Shri Ramdeobaba College of Engineering and Management, Nagpur, India
Pankaj Pratap Singh	Central Institute of Technology (CIT) Kokrajhar, India
Pankaj Chandankhede	G H Raisoni College of Engineering, India
Parasuram Yadav Palla	National Institute of Technology Karnataka, India
Paritosh D. Peshwe	IIIT, Nagpur, India
Parul Sahare	Indian Institute of Information Technology, Nagpur, India
Prabhat Kumar Sharma	VNIT-Nagpur, India
Prachi Palsodkar	YCCE, Nagpur, India
Pradeep Gupta	PSIT Kanpur, India
Pradip Selokar	Shri Ramdeobaba College of Engineering and Management Nagpur, India
Pradnya Zode	Yeshwantrao Chavan College of Engineering, Nagpur

Sambhudutta Nanda	VIT AP University, India
Sampada Wazalwar	G H Raisoni College of Engineering, Nagpur, India
Sandeep Kumar Singh	National Sun Yat-sen University, Taiwan
Sanjay Dorle	G H Raisoni College of Engineering, Nagpur, India
Sanjit Maitra	Indian Statistical Institute North-East Centre, India
Sarita Patil	GHRCE Nagpur, India
Sashikala Mishra	Symbiosis International University, India
Satish Kumar Thakur	VNIT, Nagpur, India
Saugata Sinha	VNIT Nagpur, India
Saurabh Kumar	NIT Hamirpur, India
Shankar Aenagandula	Visvesvaraya National Institute of Technology Nagpur, India
Shankar Bhattacharjee	IIIT Nagpur, India
Shankargoda M. Patil	VIT Vellore, India
Shilpa Suresh	Manipal Institute of Technology, India
Shyam Lal	NIT Karnataka, India
Sourabh Tiwari	Chandigarh University, India
Sravya N.	National Institute of Technology Karnataka, India
Sudeep Baudha	BITS Pilani, K K Birla Goa Campus, India
Sudhir Kumar	IIT Patna, India
Sudhir Kumar Singh	VNIT Nagpur, India
Sumalatha S Saleti	SRM University AP, India
Suman Talla	Visvesvaraya National Institute of Technology Nagpur, India
Sumithra P.	Indian Institute of Technology Madras, India
Suraj K. Bhatnagar	GHRCE Nagpur, India
Suresh S. Kurumbanshi	MPSTME NMIMS Shirpur Campus, India
Syed Taqi Ali	Visvesvaraya National Institute of Technology Nagpur, India
Tapan K. Jain	IIIT NAGPUR, India
Tejaswini G. Panse	YCCE Maharashtra, India
Tushar Muratkar	VNIT Nagpur, India
Ujwala S. Ghodeswar	YCCE Maharashtra, India
Ujwalla Gawande	Yeshwantrao Chavan College of Engineering, Nagpur, India
Vaidehi Padmawar	VNIT, Nagpur, India
Vaijayanti H. Panse	Indian Institute of Information Technology, Nagpur, India
Varsha R. Ratnaparkhe	Government College of Engineering Aurangabad, India

Venkata Sainath Gupta Thadikemalla	V R Siddhartha Engineering College, India
Vibha Kevala	NIT Surathkal, India
Vidyashree R. Kanabur	National Institute of Technology Karnataka, India
Vijay Kumar Bohat	NSUT, Delhi, India
Vinay Gautam	Chitkara University, Punjab, India
Vipin Milind Kamble	Visvesvaraya National Institute of Technology, Nagpur, India
Vishal Satpute	VNIT, Nagpur, India
Vivek E. Khetade	RKNEC, Nagpur, India
Vivek Raghuwanshi	IIT Bhilai, India
Vrinda Yadav	IIIT Nagpur, India
Yashika A. Gaidhani	Y.C.C.E Nagpur, India
Yogesh Shriram Solunke	VNIT Nagpur, India
Yogita Dubey	Y.C.C.E Nagpur, India
Yuvaraj Sivasubramanian	National Institute of Technology Andhra Pradesh, India

Contents – Part II

Contents – Part I

Web-Based Application for Detection of Gingival Phenotype in Preventive Dentistry Using Artificial Intelligence

Vibha Bora[1](✉), Piyush Shahu[2], Surekha Rathod[3], Supriya Kaule[3], and Prateek Dutta[4]

[1] Department of Electronics Engineering, G H Raisoni College of Engineering, Nagpur, India
vibha.bora@raisoni.net

[2] Department of Electronics and Telecommunication Engineering, G H Raisoni College of Engineering, Nagpur, India

[3] Department of Periodontics and Implant Dentistry, Ranjeet Deshmukh Dental College and Research Centre, Nagpur, India

[4] Department of Artificial Intelligence Engineering, G H Raisoni College of Engineering, Nagpur, India
dutta_prateek.ai@ghrce.raisoni.net

Abstract. Detecting the Gingival Phenotype (GP) is a crucial component in dental treatment planning, as it helps evaluate the thickness and quality of gingival tissues. Conventional methods, such as direct assessments and probe transparency, are commonly employed in preventive dentistry. However, these methods are invasive and can cause discomfort. This research examines the application of deep learning as a non-invasive alternative for identifying the Gingival Phenotype from intraoral images, offering a more patient-friendly approach in the dental field.

A dataset comprising 412 intraoral images of gingiva was gathered from Ranjeet Deshmukh College of Dental Sciences and Research. The dataset was augmented to increase the total number of images to 1,236. Three deep learning models—Sequential CNN, VGG-16, and ResNet50—were trained using transfer learning techniques on this gingival phenotype dataset. The Sequential CNN, VGG-16, and ResNet50 models achieved accuracy rates of 64.51%, 89.51%, and 93.54%, respectively. Given these outcomes, ResNet50 was identified as the most effective model for detecting Gingival Phenotype and was chosen for implementation in a web-based application for real-time usage. This web-based application serves as a non-invasive, highly accurate tool for oral health professionals, allowing for more precise diagnosis and treatment of gingival diseases This application could significantly improve the diagnosis and management of gingival disorders, leading to better patient outcome.

Keywords: Preventive Dentistry · Gingival Phenotype · Web Based Application · Artificial Intelligence · Intelligent Systems

D. Gupta et al. (Eds.): PCEMS 2024, CCIS 2491, pp. 1–11, 2025.
https://doi.org/10.1007/978-3-031-90577-3_1

1 Introduction

The term "gingival phenotype" (GP) describes the noticeable differences in the thickness and width of the facial keratinized tissue observed clinically [1]. Assessing the GP is essential in treatment planning and risk evaluation before any dental procedures, particularly those involving soft tissue manipulation in preventive dentistry. These characteristics are useful for identifying potential gum problems, such as inflammation and periodontal disease, and can also help assess the risk of developing these conditions in the future. Additionally, gingival phenotypes play a key role in diagnosing and planning treatment for conditions like gingival hyperplasia [2]. The conventional methods that are being used by dentists are:

Visual examination: The dentist or periodontist will assess the gums by visually inspecting their color, shape, and texture, while also checking for any indications of inflammation or gum recession.

Probe transparency-method:
In this technique, a thin GP is recognized when the outline of the probe can be seen through the gingival margin from within the sulcus [3] as shown in Fig. 1(a)

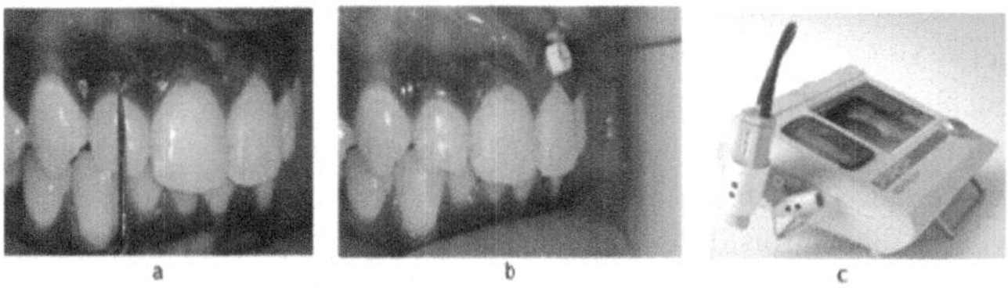

Fig. 1. (a) Probe Transparency method (b) Direct Method (c) Pirop Ultrasonic Sensor

Direct Method: The direct method can be somewhat uncomfortable for patients, as it involves the use of anesthesia before employing an endodontic probe to assess tissue thickness. A rubber stopper is used to mark the measurement, and a digital caliper ensures accurate readings as shown in Fig. 1(b). A thickness of ± 1.5 mm and above is classified as a thick phenotype, while thicknesses less than 1.5 mm are categorized as thin tissue phenotypes [4].

Ultrasonic-probing:
During the initial consultation, gingival thickness was assessed non-invasively using a Pirop ultrasonic biometer equipped with an A-scan probe as shown in Fig. 1(c). The probe has a diameter of approximately 1.7 mm, operates at a frequency of 20 MHz, and has an ultrasonic pulse speed of 1540 m/s, ensuring accurate measurements. A chlorhexidine gel is applied to the probe tip before it is gently positioned at the application site. Ten automatic measurements are taken, and their average is calculated, with a standard deviation not exceeding 0.05 mm. The results are then displayed on the device screen [5].

These traditional methods can be painful, costly, and time-consuming. Recent studies have utilized AI, particularly deep learning models like Convolution Neural-Networks

(CNNs) [6, 7], for oral diseases detection [8], Mammogram classification [9, 10], Cytological image classification [11], Heart disease prediction [12], Skin disease classification [13], Osteoporosis detection from Xray images [14, 15] and so on. Deep learning approaches have been also effectively applied for the detection of gingivitis through image analysis [16]." AI-based photographic tools have shown high accuracy in identifying gingivitis, offering faster and more objective diagnostics compared to traditional methods. These advancements demonstrate AI's growing role in enhancing periodontal disease detection and treatment [17]. This study proposes a non-invasive approach that utilizes deep learning architecture to extract gingival features and classify them as either Thick or Thin.

2 Material and Methods

2.1 Dataset

In this study, 412 intraoral images were supplied by Ranjeet Deshmukh College of Dental Sciences and Research, India. The images were categorized into two classes: Thick and Thin, as illustrated in Fig. 2. Data collection involved a manual method, where gingival thickness was measured at teeth numbers 11, 12, 21, and 22. Based on the average thickness obtained, each image was classified as either Thick or Thin. Consequently, each image was labeled with its corresponding gingival phenotype, which was also recorded in an Excel file provided by the hospital.

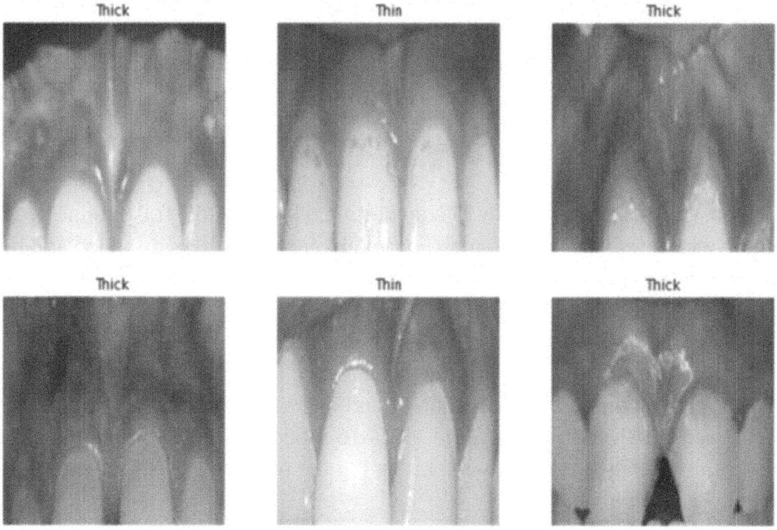

Fig. 2. Sample images in dataset as Thick and Thin Gingiva Phenotype.

2.2 Preprocessing of Data

The dataset was resized to a consistent input dimension of (224, 224, 3) and normalized by applying a rescale factor of 1/255. To further enhance the dataset, image augmentation was performed using the ImageDataGenerator. This included rotating the images by $\pm 15^\circ$ and applying horizontal flips. After augmentation, the dataset grew to a total of 1,236 images.

2.3 Deep Learning Methods

In this study, the training set comprised 80% of the dataset, containing 988 images, while both the validation and testing sets each represented 10%, consisting of 124 images each. The model was initialized and trained using the specified data split, employing three different architectures: Sequential CNN, VGG-16, and ResNet50 [18]. A categorical class mode was used, with a batch size of 62 chosen for training over 15 epochs.

2.3.1 Sequential CNN

In Sequential CNNs, the layers are organized in a linear sequence, enabling the network to gradually acquire more intricate features from the input data. A Sequential CNN consists of four key components: convolutional-layers, activation-functions, pooling-layers, and fully connected layers. Convolutional layers capture features from input images, while activation-functions add non-linearity to the model [19].

This study implemented a 16-layer Sequential CNN using Keras/TensorFlow, consisting of 6 convolutional layers, 6 max pooling layers, 1 flatten layer, and 2 dense layers. The model used the 'adam' optimizer and Sparse Categorical Cross Entropy for the loss function. Performance was evaluated based on accuracy, precision, recall, and F1 score. ReLU was applied to the hidden layers, while Softmax was used in the output layer.

2.3.2 VGG-16

VGG-16 consists of 16 layers, with 13 convolutional layers and 3 fully connected layers. Its distinctive feature is the use of small 3×3 convolutional filters in a deep architecture, allowing the network to capture intricate image features. It also employs pooling to reduce spatial resolution and help prevent overfitting.

VGG-16 has demonstrated exceptional performance on large datasets and serves as an effective feature extractor for transfer learning. Despite its impressive results, it requires significant computational resources and memory, which can hinder deployment on some devices [10]. In this study, the VGG-16 model was modified into a sequential format by removing its last two layers, including the output layer and five performance parameters were calculated.

2.3.3 ResNet-50

ResNet is a widely utilized CNN architecture in computer vision, recognized for its strong performance. ResNet-50 consists of 48 convolutional layers, along with one MaxPool and one average pool layer. It utilizes residual blocks with shortcut connections, which add the original input to the output of the final convolutional layer, effectively tackling the vanishing gradient problem and enhancing convergence. The bottleneck design reduces parameters and matrix multiplications, enabling faster training. However, ResNet's impressive capabilities come at the cost of high computational requirements and significant memory usage during training [20].

ResNet-50 incorporates bottleneck blocks and skip connections, allowing the network to learn complex features while reducing overfitting. It also features batch normalization layers to normalize activations and mitigate internal covariate shift.[8]In this study, a dense layer and a flatten layer were added to the ResNet-50 model to enhance its performance.

3 Performance Metric

Evaluation metrics play a vital role in measuring the performance of deep learning models for specific tasks [21]. The choice of metrics is influenced by the nature of the data. Commonly used evaluation metrics in deep learning include:

Confusion Matrix: A confusion matrix is a useful tool for evaluating model performance, providing insights into true positives, true negatives, false positives, and false negatives. In the case of a binary classification problem, it generally consists of two rows and two columns.

Accuracy: Accuracy is a fundamental evaluation metric that measures the proportion of correct predictions made by a model. It is widely used in classification tasks to assess performance.

$$Accuracy\ (A) = \frac{TP + TN}{TP + TN + FP + FN} \tag{1}$$

Precision: Precision calculates the ratio of true positive predictions to the total number of positive predictions made by the model. This metric is commonly used to assess a model's effectiveness in reducing false positives.

$$Precision(P) = \frac{TP}{TP + FP} \tag{2}$$

Recall: Recall, or sensitivity, is an evaluation metric that calculates the percentage of true positive instances correctly identified by the model.

$$Recall(R) = \frac{TP}{TP + FN} \tag{3}$$

F1 score: The F1 score is a widely used metric for assessing classification models, balancing precision and recall. A higher F1 score indicates the model accurately detects positive instances while reducing false positives.

$$F1 - score = 2 * \frac{P * R}{P + R} \tag{4}$$

4 Result

In this study, three distinct CNN architectures were employed for gingival phenotype detection. The dataset comprised 1,236 augmented images, labeled into two classes: thin (class '1') and thick(class '0'). Total 988 images (80%) were used for training, 124 images (10%) for validation, and 124 images (10%) for testing. Table 1 offers a in depth-analysis of the models' performance across various classes, presenting a detailed breakdown of key performance metrics for the CNN, VGG-16, and ResNet-50 models. The ResNet50 architecture yielded optimal parameters, achieving an accuracy of 93.54%, precision of 90.90%, recall of 96.77%, and an F1 score of 93.74.

Table 1. (a) Comparison of performance metrics of Sequential CNN, VGG-16 and ResNet-50 on testing dataset of 124 intraoral images.

Models	Total Thin Images	Total Thick Images	Thin obtained as Thin (TP)	Thick obtained as Thick (TN)	Thin obtained as Thick (FN)	Thick obtained as Thin (FP)	Accuracy	Precision	Recall
Sequential CNN	62	62	42	38	20	24	64.51%	63.63%	67.74%
VGG-16	62	62	53	58	9	4	89.51%	92.98%	85.48%
ResNet-50	62	62	60	56	2	6	93.54%	90.90%	96.77%

Figure 3 shows plots of the accuracy and loss function on the training data Vs epochs for CNN, VGG-16 and ResNet-50 model. Similarly, Fig. 4 displays the confusion matrices for the test dataset generated by the CNN, VGG-16, and ResNet-50 models, providing insights into their classification performance.

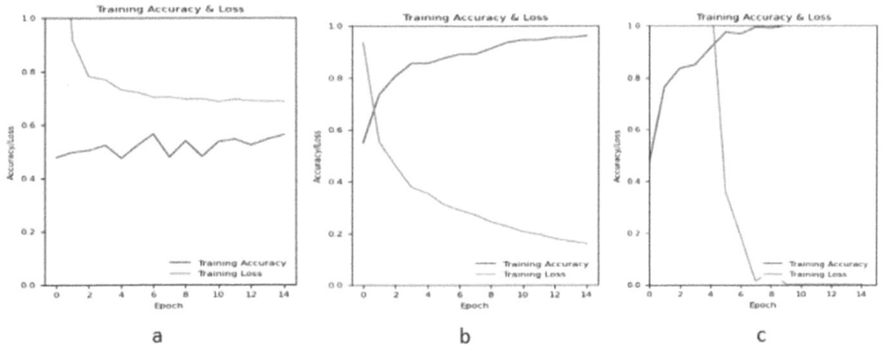

Fig. 3. Plots of Training accuracy & Loss function Vs Epoch (a) Sequential CNN (b) VGG-16 (c) ResNet-50

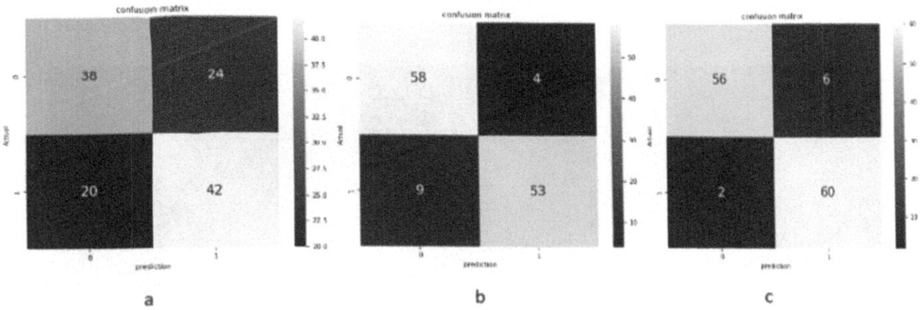

Fig. 4. Confusion matrices for testing data—comparative analysis of (a) Sequential CNN (b) VGG-16 (c) ResNet-50 for Gingiva Phenotype classification.

Figure 5 presents a visual comparison of the ROC curves for the three models. These curves illustrate the balance between the true positive rate and the false positive rate across different classification thresholds.

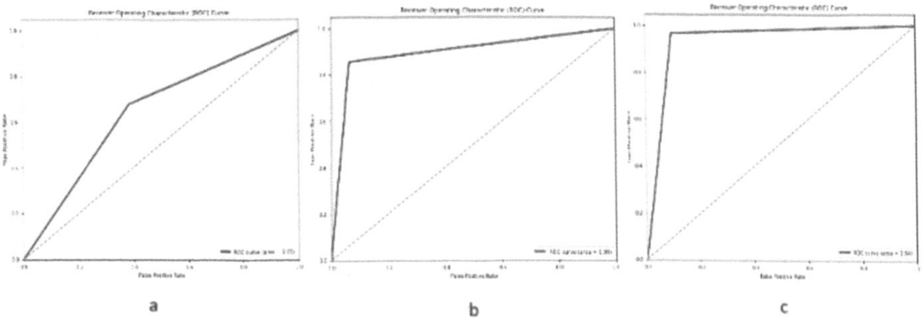

Fig. 5. ROC curve for testing data—comparative analysis of (a) CNN (b) VGG-16 (c) ResNet-50 for Gingiva Phenotype classification

4.1 Graphical User Interface

In this study, a web-based graphical user interface (GUI) was created to allow users to upload intraoral gingival images for feature extraction and phenotype detection using the ResNet-50 deep learning model. The interface consists of two sections: a homepage and a results page. On the homepage, users can upload the intraoral image, which undergoes preprocessing after submission. The image is resized to (224, 224, 3) before being processed by the deep learning model. The detected gingival phenotype is then displayed on the results page, as shown in Fig. 6.

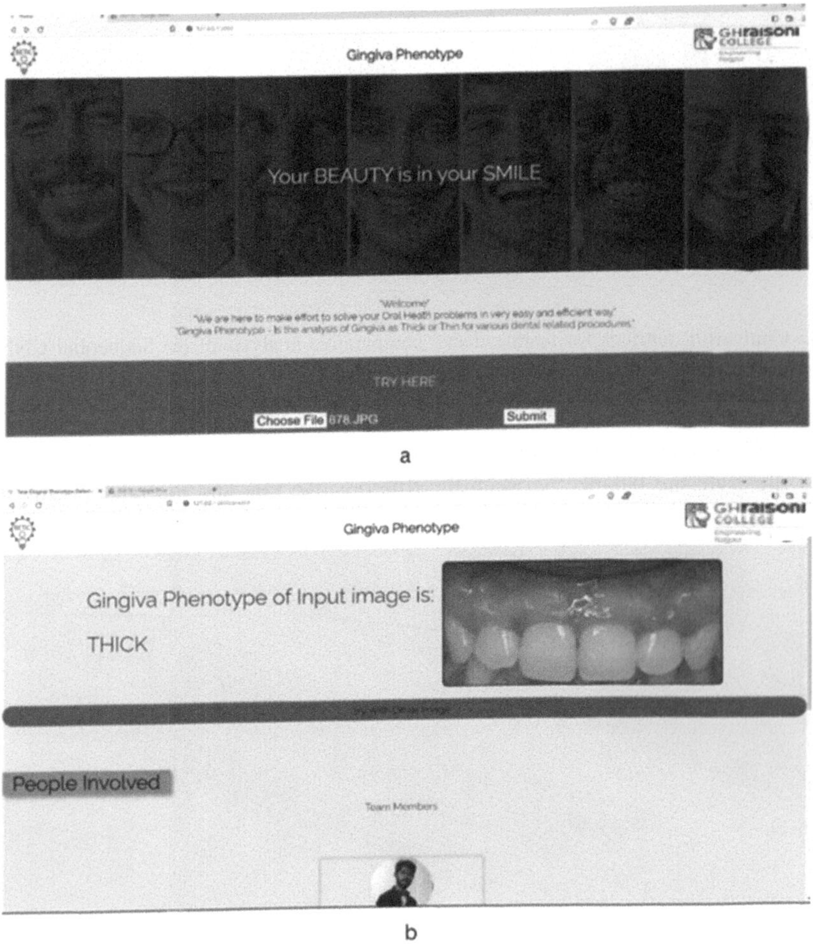

Fig. 6. (a) Homepage of the GUI for Web-application to detect Gingiva Phenotype (b) Result page that shows the input image and its Gingival Phenotype

5 Discussion

Three deep learning algorithms were evaluated in this study: Sequential CNN, VGG-16, and ResNet-50, using 15 epochs and a batch size of 62. Each model was tested on 124 images, equally split between thin (62 images) and thick (62 images) gingival phenotypes.

For the Sequential CNN model, 42 thin images were correctly classified as thin (True Positive), while 38 thick images were correctly identified as thick (True Negative). However, 20 thin images were misclassified as thick (False Negative), and 24 thick images were incorrectly classified as thin (False Positive). The Sequential CNN achieved an accuracy of 64.52%, precision of 63.63%, recall of 67.74%, and an F1-score of 65.62%.

For the VGG-16 model, 53 thin images were correctly identified as thin (True Positive), and 58 thick images were accurately classified as thick (True Negative). However, 9 thin images were misclassified as thick (False Negative), and 4 thick images were incorrectly labeled as thin (False Positive). The VGG-16 model achieved an accuracy of 89.51%, precision of 92.98%, recall of 85.48%, and an F1-score of 89.07%.

The ResNet50 model outperformed the other two models, correctly classifying 60 out of 62 thin images (True Positive) and 56 out of 62 thick images (True Negative). It demonstrated a low false negative rate, with only 2 thin images misclassified as thick, and 6 thick images misclassified as thin (False Positive). As a result, the ResNet50 model achieved an accuracy of 93.54%, precision of 90.90%, recall of 96.77%, and an F1-score of 93.74% (Refer to Table 1).

In this study, the ResNet-50 model outperformed the other models in detecting gingival phenotype from intraoral images, demonstrating higher accuracy and recall. The ResNet50 architecture was selected for transfer learning because of its strong performance. Furthermore, the trained ResNet50 model was integrated into a web-based application for real-time detection of gingival phenotype.

6 Conclusion

Detecting gingival phenotype is crucial in preventive dentistry as it helps assess the thickness and quality of gingival tissue, which directly impacts the success of dental treatments like veneers, crowns, and implant restorations. Gingival health is also essential for periodontal well-being, as thin or fragile gums are more susceptible to inflammation and recession. This study aimed to develop a non-invasive and efficient method for detecting gingival phenotype using intraoral images combined with transfer learning via ResNet-50. Our approach demonstrated superior performance compared to traditional detection techniques, such as the Probe Transparency Method and direct assessment methods. The trained model was integrated into a web-based application, achieving an accuracy of 93.54% after image preprocessing. Compared to sequential conventional methods our technique demonstrated significant improvements in classification accuracy. This algorithm has the potential to transform gingival phenotype detection, offering a cost-effective, non-invasive alternative. Future research could expand the model's capabilities to detect various gingival diseases, and further improvements could be made by incorporating larger datasets, enhancing accuracy, precision, and recall.

References

1. Shah, H.K., Sharma, S., Shrestha, S.: Gingival biotype classification, assessment and clinical importance: A review. J. Nepal. Soc. PerioOral. Implantol. **4**(8), 83–88 (2020)
2. Vlachodimou, E., Fragkioudakis, I., Vouros, I.: Is there an association between the gingival phenotype and the width of keratinized gingiva? A systematic review. Dent. J. (Basel). **9**(3), 34 (2021). https://doi.org/10.3390/dj9030034
3. Manjunath, R.G., Rana, A., Sarkar, A.: Gingival biotype assessment in a healthy periodontium: Transgingival probing method. J. Clin. Diagn. Res. **9**(5), 66–69 (2015)

4. De Rouck, T., Eghbali, R., Collys, K., De Bruyn, H., Cosyn, J.: The gingival biotype revisited: transparency of the periodontal probe through the gingival margin as a method to discriminate thin from thick gingiva. J. Clin. Periodontol. **36**(5), 428–433 (2009). https://doi.org/10.1111/j.1600-051X.2009.01398.x
5. Seba Abraham, K.T., Deepak, R., Ambili, C., Preeja, V.A.: Gingival biotype and its clinical significance – A review. Saudi J. Dent. Res. **5**(1), 3–7 (2014). https://doi.org/10.1016/j.ksujds.2013.06.003
6. Agrawal, R., Bajaj, P.: Design of EEG Based Classification of Brain States Using STFT by Deep Neural Network. In: Advances in Systems, Control and Automations Lecture Notes in Electrical Engineering, vol. 708. Springer, Singapore (2021). https://doi.org/10.1007/978-981-15-8685-9_66
7. Sapra, S.J., Thakur, S.A., Kapse, A.S.: Constructive approach for text summarization using advanced techniques of deep learning. Intelligent data communication technologies and internet of things. In: Lecture Notes on Data Engineering and Communications Technologies, vol. 101. Springer, Singapore (2022). https://doi.org/10.1007/978-981-16-7610-9_65
8. Kumar Yadav, R., Ujjainkar, P., Moriwal, R.: Oral cancer detection using deep learning approach. In: IEEE International Students' Conference on Electrical, Electronics and Computer Science (SCEECS), pp. 1–7, Bhopal, India (2023). https://doi.org/10.1109/SCEECS57921.2023.10062993
9. Bora, V.B., Kothari, A.G., Keskar, A.G.: Mammogram segmentation using rough k-means and mass lesion classification with artificial neural network. In: Hassanien, A.E., Salem, A.B.M., Ramadan, R., Kim, T. (eds.) Advanced Machine Learning Technologies and Applications. AMLTA, Communications in Computer and Information Science, vol. 322. Springer (2012). https://doi.org/10.1007/978-3-642-35326-0_7
10. Sarode, A., Bora, V.: Breast cancer detection using deep learning and thermal images. In: Recent Evolutions in Energy, Drives and e-Vehicles. REEDEV 2022 Lecture Notes in Electrical Engineering, vol. 1162. Springer (2024). https://doi.org/10.1007/978-981-97-0763-8_11
11. Ailawar, V., Bora, V. Comparison of cell nuclei classification in cytological breast images using machine learning algorithms. Proceedings of the International Conference on Cognitive and Intelligent Computing. Cognitive Science and Technology. Springer (2023) https://doi.org/10.1007/978-981-19-2358-6_54
12. Gaikwad, M.J., Asole, P.S., Bitla, L.S.: Effective Study of Machine Learning Algorithms for Heart Disease Prediction. 2nd International Conference on Power Electronics & IoT Applications in Renewable Energy and its Control (PARC), 1–6 (2022). https://doi.org/10.1109/PARC52418.2022.9726613
13. Kshirsagar, P.R., Manoharan, H., Shitharth, S., Alshareef, A.M., Albishry, N., Balachandran, P.K.: Deep learning approaches for prognosis of automated skin disease. Life. **12**(3), 426 (2022). https://doi.org/10.3390/life12030426
14. Kawade, V., Naikwade, V., Bora, C.S.: A comparative analysis of deep learning models and conventional approaches for osteoporosis detection in hip X-ray images. World Conference on Communication & Computing (WCONF), 1–7 (2023). https://doi.org/10.1109/WCONF58270.2023.10235129
15. Kale, R.D., Khandelwal, S.: A review on: Deep learning and computer intelligent techniques using X-ray imaging for the early detection of knee osteoarthritis. In: Machine Learning, Image Processing, Network Security and Data Sciences. MIND. Communications in Computer and Information Science, vol. 1762. Springer (2022). https://doi.org/10.1007/978-3-031-24352-3_8
16. Reinhard Chun, W.C., Guan-H, L., In, M.T., Khaing, M.T., Colman, M., Wai-L, L., Wing-Kuen, L., Richard Tai-Chiu, H., Walter Yu, H.L.: Accuracy of Artificial Intelligence-Based

Photographic Detection of Gingivitis. International Dental Journal. **73**(5), 724–730 (2023). https://doi.org/10.1016/j.identj.2023.03.007
17. He, K., Zhang, X., Ren, S., Sun, J. Deep Residual Learning for Image Recognition. IEEE Conference on Computer Vision and Pattern Recognition (CVPR).pp. 770-778.(2016)
18. Alalharith, D.M., Alharthi, H.M., Alghamdi, W.M., Alsenbel, Y.M., Aslam, N., Khan, I.U., Shahin, S.Y., Dianišková, S., Alhareky, M.S., Barouch, K.K.: A deep learning-based approach for the detection of early signs of gingivitis in orthodontic patients using faster region-based convolutional neural networks. Int. J. Environ. Res. Public Health. **17** (2020). https://doi.org/10.3390/ijerph17228447
19. Patil, S., et al.: Artificial intelligence in the diagnosis of oral diseases: applications and pitfalls. Diagnostics (Basel) **12**(5) (2022). https://doi.org/10.3390/diagnostics12051029.
20. Gehring, J., Auli, M., Grangier, D., Yarats, D., Dauphin, Y.N.: Convolutional sequence to sequence learning. In Proceedings of the 34th International Conference on Machine Learning (ICML). **70**, 1243–1252 (2017)
21. Dalianis, H.: Evaluation metrics and evaluation. In: Clinical Text Mining. Springer (2018). https://doi.org/10.1007/978-3-319-78503-5_6

AutoBill: An Integrated AI-Powered Autonomous Checkout System

Mr Patil Chetan Sunil(✉), Mr Tadge Atharva Somnath, Ms Joby Nikita Maria, and Prof Ghatte Najib Nisar

Department of Electronics Engineering, Sardar Patel Institute of Technology, Mumbai, Maharashtra, India
{chetan.patil2,atharva.tadge,nikita.joby,najib_ghatte}@spit.ac.in

Abstract. AutoBill represents a paradigm shift in retail checkout systems, introducing an AI-powered autonomous solution that leverages computer vision and machine learning for a seamless shopping experience. Emphasizing speed and safety, AutoBill minimizes human interactions through real-time object detection, weight measurement, and instant bill generation. Developed on Edge Impulse with Raspberry Pi integration, the system demonstrates precision with a 93.75% accuracy rate, offering a glimpse into the future of retail technology with potential enhancements such as digital receipts, transaction history, and multi-store integration.

AutoBill, an innovative retail checkout system, blends cutting-edge technology with practical design. Using computer vision, load cells, and Edge Impulse's machine learning capabilities, the system achieves instant, contact-free checkout. The integration of a WS2812B RGB LED strip adds a dynamic visual element, while calibration ensures accurate weight measurement. AutoBill's success paves the way for future retail technology, envisioning features like digital receipts and multi-store analytics, signaling a transformative evolution in the retail landscape.

Keywords: Autonomous Checkout System · AI-Powered Retail · Contact-Free Shopping

1 Introduction

In the current global scenario, the demand for contact-free solutions is growing rapidly. AutoBill addresses this need by combining cutting-edge technologies to create a seamless and efficient checkout process. This project utilizes a Raspberry Pi 3 Model B as the core computing platform, along with components such as a Seeed Studio Weight Sensor, SparkFun Load Cell Amplifier (HX711), REES52 5 Megapixel Camera, WS2812B RGB Led Strip, and a customized cabinetry for housing the system.

D. Gupta et al. (Eds.): PCEMS 2024, CCIS 2491, pp. 12–23, 2025.
https://doi.org/10.1007/978-3-031-90577-3_2

2 Related Work

This section provides an overview of existing auto billing applications and their limitations. It examines various systems, including traditional POS setups, RFID-based solutions, and emerging IoT platforms, to identify areas requiring improvement, such as accuracy, efficiency, and adaptability to diverse operational environments.

Traditional POS systems have been the backbone of retail transactions for decades. These systems involve manual scanning of barcodes by cashiers and are generally efficient in environments with low to moderate customer volumes. However, as identified in the work by Sumak et al. [7], POS systems suffer from scalability issues during peak hours, resulting in longer wait times and increased potential for human error. The reliance on manual labor not only slows down the checkout process but also introduces a risk of inaccuracies in pricing and transaction recording.

RFID technology was introduced as a means to automate and accelerate the checkout process by allowing multiple items to be scanned simultaneously without manual interaction. Perez et al. [2] discussed the integration of RFID into autonomous systems, emphasizing the reduced need for human intervention. However, while RFID-based systems offer improved efficiency over traditional POS setups, they are costly to implement, particularly in stores with a large inventory, as noted by Bostjan et al. [7]. Moreover, the reliability of RFID can be compromised due to interference from metals or liquids, and the potential for security vulnerabilities remains a concern.

The use of computer vision and machine learning (ML) in checkout systems has gained significant attention in recent years. Huang et al. [4] demonstrated a real-time object detection system using Raspberry Pi, which shows great promise for retail automation. This technology enables automated identification of products without relying on barcodes or RFID tags, improving both speed and accuracy. However, as noted by Ishiyama and Ushiama [5], there are challenges related to object occlusion, lighting variability, and the need for continuous model training to ensure accuracy in dynamic retail environments.

In the realm of computer vision and object detection, Huang et al. [4] proposed a real-time object detection system based on Raspberry Pi, integrating LiDAR and image processing algorithms. Their work showcases the advancement of Raspberry Pi-based systems in complex applications such as autonomous navigation and surveillance.

Although the aforementioned technologies offer potential improvements to the retail checkout process, each has its limitations. POS and barcode systems are labor-intensive and slow, RFID systems are costly and prone to interference, and computer vision technologies struggle with real-world complexities like overlapping objects and poor lighting. Furthermore, sensor-based systems alone cannot account for all the nuances of product differentiation. This leaves a gap in the market for an affordable, scalable, and accurate billing system that can operate efficiently in diverse retail settings without requiring significant infrastructure investments.

AutoBill addresses these gaps by integrating computer vision, machine learning, and weight sensors into a single, cohesive system. By leveraging the accuracy of object detection techniques like those described by Redmon et al. [8] and the precision of load cells as demonstrated by Yahaya et al. [1], AutoBill offers a robust solution that reduces human intervention, increases checkout speed, and ensures reliable billing, even in high-volume retail environments (Table 1).

Table 1. Customer's total waiting time.

	Scenario 1 Avg. (s.d.)	Scenario 2 Avg. (s.d.)	Scenario 3 Avg. (s.d.)	Scenario 4 Avg. (s.d.)
Shoppers in Store	106.85	106.15	106.15	106.15
Time Shopping	53.63	53.48	53.48	53.48
Items per Customer	89.29	88.99	89.00	88.96
Payment Time	1.38	1.39	1.39	1.39
Checkout Time	7.04	7.00	7.00	6.99
Bagging Time	1.86	1.85	1.85	1.85
#In Store	140.10	135.85	134.71	139.20
Time in store	70.30	68.27	67.68	69.77
Total wait time	6.37	5.92	5.33	7.46

3 System Architecture

3.1 Object Detection and Classification

AutoBill's system architecture begins with a robust process of object detection and classification. The high-resolution REES52 5 Megapixel Camera captures detailed images of products on the checkout counter. Leveraging state-of-the-art computer vision algorithms, this stage involves the identification and classification of each item, ensuring accuracy in recognizing a diverse range of products. The seamless integration of the OpenCV library facilitates real-time image processing, allowing AutoBill to efficiently handle varying lighting conditions and product orientations (Fig. 1).

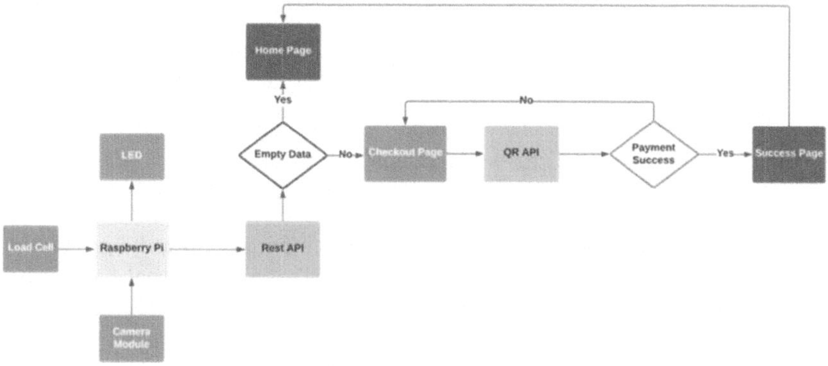

Fig. 1. Flowchart

3.2 Edge Impulse Model Training

The labeled data obtained from the object detection phase is fed into the Edge Impulse platform for model training. This pivotal stage involves the creation and refinement of a neural network specifically tailored for precise product recognition. AutoBill utilizes the NVIDIA TAO Toolkit, built on TensorFlow and PyTorch, to streamline model training through transfer learning. By adapting pre-trained models, AutoBill minimizes the need for large datasets and complex AI expertise.

For object detection, the system employs YOLOv5 via Edge Impulse, optimized for the Renesas DRP-AI platform. YOLOv5 offers fast and accurate real-time detection of multiple products, even in challenging conditions, ensuring efficient and precise checkout operations.

YOLOv5 operates by dividing an image into grids and predicting bounding boxes and class probabilities for each object. YOLOv5 is a single-stage detector, meaning it performs both object localization and classification in one pass, making it fast and efficient for real-time applications.

The NVIDIA TAO Toolkit simplifies the model development process by utilizing transfer learning. Pre-trained models are fine-tuned on AutoBills dataset, allowing the system to recognize retail products with minimal data. This approach accelerates training and ensures that the model can adapt to new products or variations quickly.

Through multiple iterations, the Edge Impulse model achieves an impressive 93.75% precision which was achieved through rigorous testing using Edge Impulse. During data collection, the dataset was split into training and testing sets to ensure reliable model evaluation. The model dataset consisted of 60 samples for each product was trained exclusively on the training set, while the testing set was used to evaluate its performance on unseen data, preventing overfitting, making AutoBill adept at accurately identifying products in a fast-paced retail environment. The model is trained to adapt to changes in product packaging, ensuring continuous improvement and adaptability.

The trained YOLOv5 model is deployed on the Raspberry Pi 3B for real-time object detection. This setup enables AutoBill to handle product recognition swiftly, enhancing the checkout process with high inference speed and accuracy.

3.3 Real-Time Processing with Raspberry Pi 3B

AutoBill's agility and efficiency are owed to the integration of the Raspberry Pi 3B as the central processing unit. Once the Edge Impulse model is trained, it is seamlessly integrated into the Raspberry Pi 3B, enabling real-time processing of the captured images. This compact yet powerful hardware ensures swift decision-making during the checkout process, contributing to the system's overall speed and reliability. The integration of the Edge Impulse Linux runner with the Raspberry Pi 3B allows for seamless communication between the model and the hardware, creating a cohesive and responsive autonomous checkout system.

4 Hardware Components

The hardware components of this are carefully selected to ensure optimal performance and reliability. The Raspberry Pi 3B serves as the central processing unit, orchestrating the entire checkout process. The Seeed Studio Weight Sensor and SparkFun Load Cell Amplifier provide accurate weight measurements, essential for determining product presence and quantity. The REES52 5 Megapixel Camera captures high-resolution images for object detection, while the WS2812B RGB LED strip adds a visual element to the user interface, enhancing the overall shopping experience.

4.1 Image Acquisition and Processing

AutoBill employs the high-performance REES52 5 Megapixel Camera as a key component for capturing detailed images of products during the checkout process. This camera, equipped with advanced optics, ensures optimal image quality even in varying lighting conditions. The captured images serve as the input for the object detection and classification phase, enabling the system to accurately identify and categorize a diverse range of products.

4.2 Precision Weight Measurement

The Seeed Studio Weight Sensor, in conjunction with the SparkFun Load Cell Amplifier, constitutes the precision weight measurement system of AutoBill. This hardware combination enables the system to not only identify the presence of items on the checkout counter but also accurately measure their weight. The load cell amplifier processes the weight data from the sensor, providing precise measurements that contribute to the overall reliability of the autonomous checkout system.

4.3 Real-Time Processing Unit - Raspberry Pi 3B

Raspberry Pi 3B employs a quad-core ARM Cortex-A53 processor, ensuring high-performance real-time processing for complex tasks like object recognition. With built-in Wi-Fi and Bluetooth, it seamlessly communicates with other devices and networks, facilitating data transfer and updates for the AutoBill system.

The VideoCore IV GPU enhances graphics processing and accelerates machine learning tasks, optimizing the performance of object detection algorithms. Its compact form factor, GPIO pins for sensor integration, and low power consumption make Raspberry Pi 3B an ideal choice, ensuring efficient and sustainable operation in the AutoBill checkout system (Fig. 2).

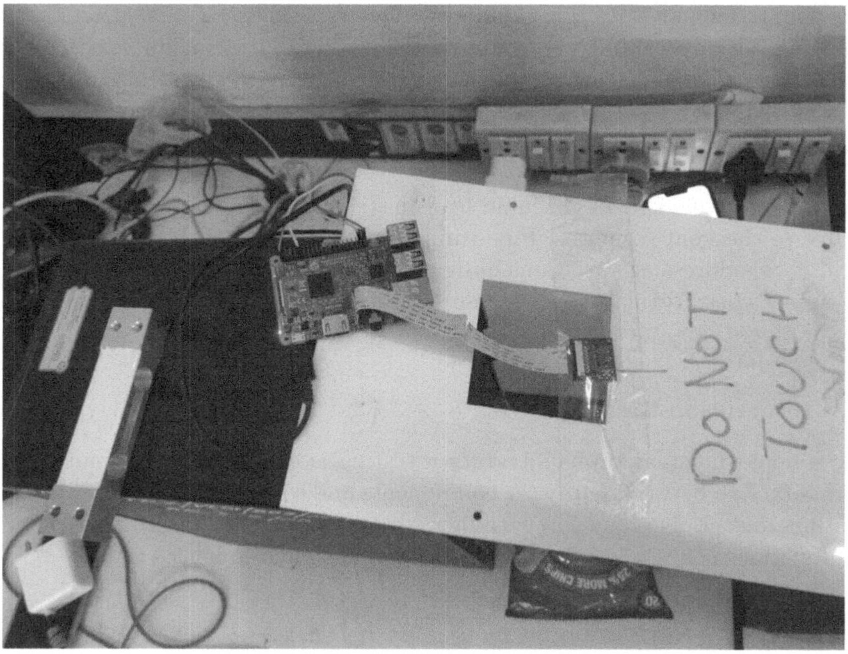

Fig. 2. Hardware Integration

4.4 Intuitive User Interface - WS2812B RGB LED Strip

AutoBill prioritizes user experience with the integration of the WS2812B RGB LED strip, creating an intuitive visual interface. This LED strip serves not only as an aesthetic element but also as a functional indicator during the checkout process. The strip communicates information such as successful product identification and transaction completion through a combination of colors and patterns, providing users with a clear and user-friendly interaction.

5 Calibration and Deployment

Ensuring the accuracy of weight measurements is critical for AutoBill's functionality. The calibration process involves meticulous adjustments to the load cell, guaranteeing precise weight readings. Setting up the Raspberry Pi 3B involves installing the necessary software, including the Edge Impulse Linux runner, creating a user-friendly and accessible checkout system. The local deployment capability of AutoBill enables retailers to implement this innovative solution seamlessly in their stores, minimizing disruption and maximizing efficiency.

5.1 Sensor Calibration

To guarantee precise object detection and recognition, calibration of sensors is paramount. This involves fine-tuning parameters such as focal length, exposure time, and sensor sensitivity. Automated calibration routines, supported by computer vision algorithms, streamline this process, enhancing system accuracy.

5.2 Algorithm Fine-Tuning

Continuous refinement of machine learning algorithms is crucial for adapting to diverse checkout scenarios. Fine-tuning involves training the system on real-world data, addressing variations in lighting conditions, object orientations, and packaging types. Regular updates ensure the system remains adept at recognizing new products.

5.3 Integration Testing

Before deployment, rigorous integration testing is conducted to validate seamless interaction between hardware components and software modules. This phase identifies and resolves any compatibility issues, ensuring the entire AutoBill system functions cohesively in a real-world checkout environment.

5.4 Deployment Strategies

The deployment strategy encompasses considerations such as system scalability, ease of installation, and adaptability to different retail settings. Whether deploying in a small convenience store or a large supermarket, a well-defined deployment plan ensures minimal disruption to existing checkout processes.

5.5 User Training and Support

Adequate training for store personnel is essential for a smooth transition to the AutoBill system. This includes familiarizing staff with the system interface, troubleshooting common issues, and understanding maintenance procedures. A robust support system is established to address any challenges during the initial deployment phase.

5.6 Performance Monitoring

Post-deployment, continuous performance monitoring is implemented to track the system's accuracy, speed, and reliability. Analytics tools provide insights into system usage, enabling proactive maintenance and future enhancements based on user feedback and evolving retail dynamics.

6 User Interface and Future Enhancements

AutoBill's user interface is designed with simplicity and functionality in mind. Combining HTML and JavaScript for the front end, and a NodeJS-Express backend API, the interface provides a smooth and intuitive checkout experience. Looking ahead, future enhancements for AutoBill include the implementation of digital receipts, transaction history tracking, promotional displays, and the potential to integrate with multiple retail outlets. These enhancements aim to elevate the autonomy and intelligence of the system, creating a comprehensive solution for modern retail environments.

7 Experimental Results

In this section, we delve into the outcomes of rigorous experimentation conducted to validate the functionality and performance of the AutoBill checkout system. The experiments were designed to assess various aspects of the system's capabilities, providing valuable insights into its efficiency and reliability.

7.1 Object Detection Accuracy

Our first set of experiments focused on evaluating the accuracy of the multi-object detection feature in diverse scenarios. We considered varying lighting conditions, object orientations, and distances to ensure a comprehensive assessment of the system's robustness. The results highlight the system's exceptional accuracy in identifying and delineating objects accurately, even in challenging real-world conditions (Fig. 3).

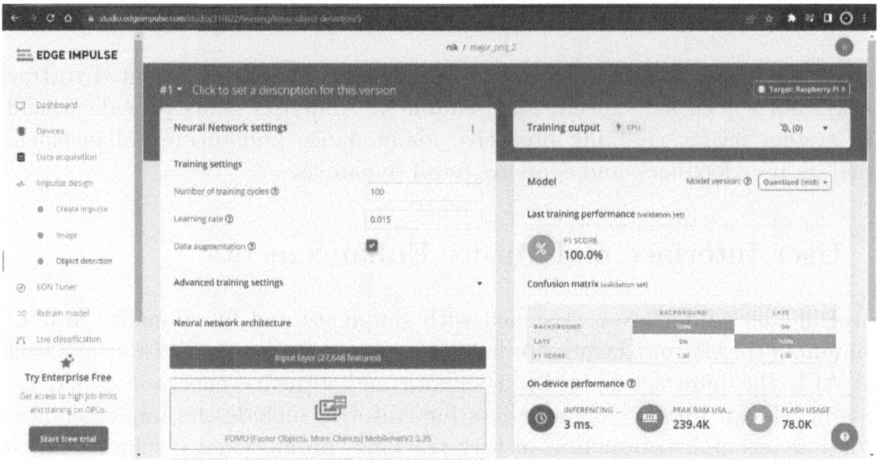

Fig. 3. ML Model Traning/Accuracy

7.2 Real-Time Processing Performance

A crucial aspect of the AutoBill system is its ability to process transactions in real-time. We conducted experiments to measure the processing speed and efficiency, analyzing how quickly the system detects and prices multiple items on the checkout counter. The results demonstrate the system's impressive real-time processing capabilities, meeting the demands of high-speed retail environments (Fig. 4).

Fig. 4. Object Detection

The billing processing time is calculated from the moment an item is placed on the checkout counter to when the bill is generated. This includes the time required for object detection, weight measurement, and price calculation. The total time for each transaction was measured and averaged across multiple test scenarios, considering various factors such as the number of items and lighting conditions. To gather customer feedback, several performance metrics were considered, ensuring that the system met user expectations in both speed and accuracy. The following metrics were evaluated:

- Billing Speed: Measured by the average time taken to complete a transaction, ensuring that the system provides a fast and seamless experience.
- Detection Accuracy: The accuracy of the system in correctly identifying and classifying products, as validated by the model's performance on the testing dataset.
- User Satisfaction: This metric was derived from customer surveys, where users were asked to rate their experience based on ease of use, transaction speed, and accuracy.
- Error Rate: Any misidentification or incorrect price calculations were tracked, providing insights into areas for improvement.

7.3 User Interaction and Interface Responsiveness

Beyond technical performance, user experience is paramount. We conducted experiments to evaluate the system's user interface responsiveness and overall interaction. User feedback, response times, and intuitiveness were considered, providing insights into the system's user-friendliness and its potential to enhance the overall checkout experience (Fig. 5).

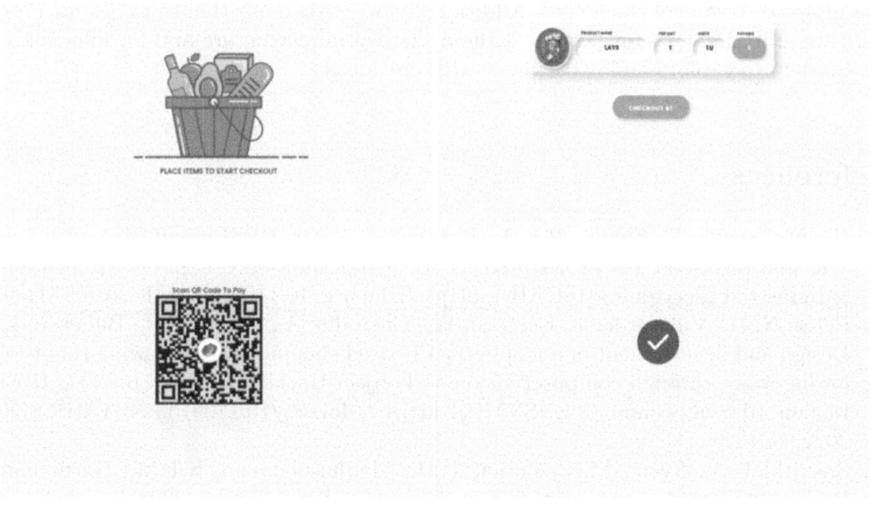

Fig. 5. UI/UX of Website

7.4 Scalability and Load Handling

Assessing the system's scalability is crucial for its application in diverse retail environments. Our experiments included scenarios with varying transaction volumes and loads. The results elucidate the system's ability to scale efficiently, ensuring consistent performance even during peak transaction periods.

8 Conclusion

The development and deployment of AutoBill mark a significant advancement in autonomous retail checkout systems. The integration of computer vision and machine learning, specifically through the use of NVIDIA TAO and YOLOv5, has enabled the system to achieve an impressive accuracy of 93.25% in real-time product detection. Through careful optimization, AutoBill also achieves efficient billing processing times, ensuring seamless and fast transactions for customers.

Additionally, AutoBill demonstrates scalability by effectively utilizing Edge Impulse to adapt pre-trained models with minimal training data. This reduces the computational burden and makes the system adaptable to various retail environments. Customer feedback indicates high satisfaction due to the systems speed, accuracy, and ease of use, with a minimal error rate observed during testing. In summary, AutoBill has successfully addressed key challenges in retail automation by providing a cost-effective, scalable, and accurate solution, paving the way for further enhancements in digital retail infrastructure.

Acknowledgments. We are deeply grateful to our mentor, Prof. Najib Ghatte, for his unwavering support. His valuable guidance and insights throughout this project have been instrumental in helping us develop and implement our ideas effectively. His encouragement allowed us to reach our full potential, unbounded by any preconceived limitations or technical challenges. Additionally, we extend our thanks to Sardar Patel Institute of Technology for providing the necessary infrastructure and for allowing us the freedom to explore and experiment with our ideas.

References

1. HanumYahaya, F., Shauri, R.L.A., Bakar, S.A.: Dorper BSI monitoring with load cells and raspberry PI. In: 2019 IEEE 9th Symposium on Computer Applications & Industrial Electronics (ISCAIE). https://doi.org/10.1109/ISCAIE.2019.8743693
2. Perez, N. D., Villaverde, J., Cereneo, W., Custodio, A., Conde, S.B., Bulawan, L.: Design and development of a raspberry Pi-based shopping cart following robot following robot through computer vision and object tracking method. In: 2022 IEEE Region 10 Symposium (TENSYMP). https://doi.org/10.1109/TENSYMP54529.2022.9864520
3. Sreejithlal, A., Syam, M.N., Letha, T.M., Madhusoodanan, K.P.M., Harikumar, R.G.: Automated sensor test system using Raspberry Pi. In: 2022 International Conference on Connected Systems & Intelligence (CSI). https://doi.org/10.1109/CSI54720.2022.9923977

4. Huang, C.J., Chen, T.W., Fan, Y.C.: Chip and system design of real time object detection based on LS-R-YOLO network. In: 2020 IEEE 9th Global Conference on Consumer Electronics (GCCE). https://doi.org/10.1109/GCCE50665.2020.9291775
5. Ishiyama, K., Ushiama, T.: Machine learning model to evaluate the appropriateness of layout for automatic generation of graphic design works. In: 2023 17th International Conference on Ubiquitous Information Management and Communication (IMCOM). https://doi.org/10.1109/IMCOM56909.2023.10035646
6. Nuryanti, Rifa'i, A.F., Rivaldo, M.: Design of multipoint weighing in three-dimensional object center of mass measurement system based on raspberry Pi (2022). https://doi.org/10.1109/ICEVT55516.2022.9924788
7. Bostjan, S., Pusnik, M., Hericko, M.: An empirical study of factors affecting the adoption of self-service checkout termi-nals in Slovenia. In: Information and Communication Technology Electronics and Microelectronics (MIPRO),: 37th International Convention on. IEEE **2014**,(2014). https://doi.org/10.1109/MIPRO.2014.6859610
8. Redmon, J., Divvala, S., Girshick, R., Farhadi, A.: You only look once: unified real-time object detection (2015). https://doi.org/10.1109/CVPR.2016.91
9. Yangqing, J., et al.: Caffe: convolutional architecture for fast feature embedding. In: Proceedings of the ACM International Conference on Multimedia (2014). https://doi.org/10.1145/2647868.2654889
10. Alex, K., Sutskever, I., Hinton, G.E.: Imagenet classification with deep convolutional neural networks. Adv. Neural. Inf. Process. Syst. (2012). https://doi.org/10.1145/3065386
11. Girshick, R., Donahue, J., Darrell, T., Malik, J.: Rich feature hierarchies for accurate object detection and semantic segmentation. In: Computer Vision and Pattern Recognition (CVPR) 2014 IEEE Conference on, pp. 580–587 (2014). https://doi.org/10.1109/CVPR.2014.81
12. Girshick, R.: Fast R-CNN. In: Computer Vision 2015 IEEE International Conference on, pp. 1440–1448 (2015). https://doi.org/10.1109/ICCV.2015.169
13. Lienhart, R., Maydt, J.: An extended set of haar-like features for rapid object detection. In: Image Processing. 2002. Proceedings. 2002 International Conference on, vol. 1 (2002). https://doi.org/10.1109/ICIP.2002.1038171
14. Viola, P., Jones, M.J.: Robust real-time object detection. Int. J. Comput. Vision **4** (2001). https://doi.org/10.1109/ICCV.2001.937709

Multiple Controls Wheelchair for Elderly People & Lower Limb Disabilities for Hilly Region

Sudipta Chatterjee[1(✉)] and Sahadev Roy[2]

[1] Department of ECE, Brainware University, Kolkata, India
[2] Department of ECE, National Institute of Technology, Jote, Arunachal Pradesh, India
sahadevroy@nitap.ac.in

Abstract. This paper introduces a multi-control wheelchair customized to support elderly people or lower-limb disabilities person for hilly region using advanced assistive technologies. It presents a novel approach for a assistive wheelchair with a unique emphasis on object identification for navigation. The MCW offers versatile control options, including joystick operation, internet connectivity, or a combination of both. By employing a network of sensors, the system evaluates terrain conditions and translates user commands into appropriate control actions. The proposed control system, powered by a Raspberry Pi, facilitates precise maneuvering and control of the wheelchair. By integrating Raspberry Pi with camera, sensors, and other control components with the Android system using wifi technologies for supervisory control system. This integration results in multi-layered system architecture with specialized control functions, leveraging using Android devices achieve remote capture of surrounding environment videos and for automatic control. The study's results demonstrate that users can execute essential tasks with improved control, thus enhancing the functionality of traditional wheelchairs and potentially civilizing the overall class of life for individuals with lower-limb disabilities.

Keywords: Deep neural network · Distributed control · Far-end control · Object detection · Opencv · Raspberry Pi

1 Introduction

In recent era, there has been an emergent focus among researchers on rehabilitation, particularly in the domain of assistive technology. This current area of study explores how certain robotic devices can enhance the capabilities of individuals with disabilities or older adults, facilitating their engagement in daily living activities. Among these innovative devices, intelligent wheelchairs have emerged as a tangible solution for individuals facing mobility challenges [1]. These wheelchairs have gained widespread acceptance as prominent indoor navigation tools, owing to their user-friendly controls, seamless mobility, and specialized human-machine interface tailored for specific applications. The contemporary focus on assistive wheelchairs has garnered extensive attention in

D. Gupta et al. (Eds.): PCEMS 2024, CCIS 2491, pp. 24–34, 2025.
https://doi.org/10.1007/978-3-031-90577-3_3

both academic research and industry worldwide. Despite this, only a limited number of these studies have effectively addressed real-life customer needs and translated into successful market outcomes. Back in 1595, an anonymous maker crafted a wheelchair for "Philips II" in Spain, marking an early historical instance. "In 1655, Stephen Farfer, a watchmaker" with a penchant for innovative design, constructed a "three-wheeled self-propelled chair" inspired by this concept, later recognized as a wheelchair. In the twenty-first century, wheelchair design has assumed heightened significance, particularly in the integration of emerging technologies.

Numerous research studies have indicated that assistive wheelchairs can radically enhance the superiority of life for people with lower limb disabilities and the elderly. There is now a diverse array of wheelchairs, including tilt-in-space [2], elevation, and stair-climbing models [3]. These wheelchairs can be operated using methods such as a joystick [4], head gestures [5], or other innovative technologies. Several assistive wheelchair technologies feature a human-machine HMI control interface, allowing for varied control methods. Examples encompass voice command control [6], eye-tracking for follow-up movements [7], and tongue movement tracking [8]. In one study by Sahoo et al., [9] the author explores a system for wheelchair movement control through voice commands. However, the solutions proposed in their work are notably expensive. Users are required to undergo voice recognition within the system before utilizing this technology. Alternatively, another research endeavor [10] introduces a system enabling wheelchair control through head movements, facilitated by a mounted camera capturing and tracking the user's head motions. Additionally, an alternative control interface [11] is discussed, wherein wheelchair operation is dictated by monitoring the electrical motion of the user's muscles via EMG signals. Nevertheless, the utilization of electromagnetic signals poses potential health risks to the user. In a different context researchers have developed a control mechanism allowing users to steer the wheelchair using brain activity, measured through EEG signals [12]. However, such systems have many limitations. The present study introduces a novel multi-control protocol and hardware machine capable of interfacing with actuators, sensors and additional peripherals in an electric wheelchair using image processing techniques. Users have the option to operate this system via joystick or image processing mode. Furthermore, caregivers have the ability to remotely control and monitor this system from anywhere in the world using IoT technology. Still, concerns remain about the safety and risk management of these modern technology wheelchairs. Studies have highlighted the increasing number of accidents and associated risks, especially among older users, associated with the proliferation of such devices. Therefore, ensuring safe operation is paramount, where users have control over exposure to potential hazards.

To address this scenario, we've designed the architecture of the wheelchair to ensure stability across various environments and facilitate easy control. The key contributions of this study are as follows: (a) Design of a smart wheelchair with multiple control capabilities, featuring an object detection algorithm utilizing MPU 6050 Gyro-accelerometer, Pi-noir camera, ultrasonic sensor integrated with Raspberry Pi and Arduino for motor control system. (b) Implementation of automatic object detection tailored to user preferences, enabling subsequent adjustments to wheelchair navigation.

The execution of experimental wheelchair navigation serves to authenticate the algorithm, control functions, and performance in accordance with user preferences. This involves conducting practical tests and trials wherein the wheelchair equipped with the proposed control system is operated in real-world scenarios or simulated environments. During these experiments, various aspects such as the accuracy of object detection, responsiveness of control commands, stability of wheelchair movement, and overall user experience are evaluated and analyzed. User preferences may include factors such as speed of movement, sensitivity of controls, ease of navigation, and reliability of object detection. By conducting experiments tailored to these preferences, researchers can assess the effectiveness and efficiency of the proposed system in meeting the needs of wheelchair users with lower-limb disabilities.

The rest of the paper is structured as follows: Section 2: Comprehensive overview of the wheelchair architecture, detailing all hardware components and software elements. Section 3: The methodology employs in this proposed wheelchair including specifications, requirements, and operational processes. Section 4: Discussion of results obtained and their implications, along with mention of specific desired output datasets. Section 5: Conclusion drawn from this research.

2 Structural Design

The intention of this study is to navigate and present assistance to the user through decision-making based on identified objects linked to the user's preferences. A Pi camera is integrated into the system. Whenever the user moves their head and directs attention to certain objects, the system autonomously recognizes those objects and navigates to them. Control signals for wheelchair movement are determined by the wheelchair's immediate surroundings (provided by sensor outputs) shown in Fig. 1

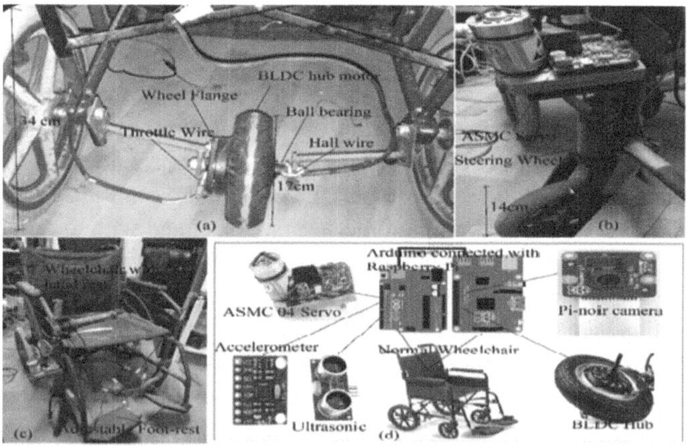

Fig. 1. Architectural configuration of the proposed Smart Wheelchair for Hilly Region

The direction and speed of wheelchair movement are then indicated by the object detection results. The main aspect of the proposed work lies in the automatic selection of

Table 1. Experimental result of detection probability

Tag	Probability
Door	23%
Cycle	19%
Chair	54%
Bottle	3%

objects. Figure 1 (a), likely depicts a visual representation or schematic diagram of the wheelchair, highlighting the dimensions of the BLDC hub motor (17 cm in diameter) and the back wheels (34 cm in diameter). This illustration helps visualize how these components are integrated into the wheelchair design scheme. Overall, this wheelchair design appears to prioritize functionality, comfort, and safety, incorporating key elements such as an electric motor for propulsion, suspension for smoother rides, and a robust chassis for structural support. In Fig. 1(b), illustration depicts the steering wheel assembly, which is an essential component for controlling the wheelchair's direction. The steering wheel has a diameter of 14 cm, indicating its size. It is connected via the ASMC-04 servo motor. A servo motor like the ASMC-04 is commonly used in applications where precise control of angular position is required, making it suitable for steering mechanisms. Figure 1(c), likely shows a depiction of a standard wheelchair, serving as a reference point for comparison with the proposed system concept presented in Fig. 1(d). The standard wheelchair may lack some of the advanced features or modifications proposed in the design scheme. Figure 1(d), illustration provides an overview of the proposed system concept. It likely integrates all the components mentioned earlier, such as the motor wheel assemblage, suspension device, chassis, steering wheel assembly, and other relevant features. This figure serves to visualize how all the components come together in the proposed wheelchair design scheme. Overall, Fig. 1 provides a comprehensive representation of the various components and concepts involved in the wheelchair design scheme, allowing for a better understanding of its features and functionalities.

The proposed design incorporates multiple modules, utilizing a Raspberry Pi 4 single-board computer to segregate computationally intensive tasks from control tasks, which are managed by Arduino. Figure 2, illustrates the block diagram, demonstrating the interconnection of all devices with the Raspberry Pi.

2.1 Steering Wheel Assembly

In manual wheelchairs, turning is typically achieved by exerting pressure on the hand rims attached to the wheels. To turn left, the user pulls the right hand rim backward while pushing the left hand rim forward, and conversely for a right turn. To turn the wheelchair in desired direction an ASMC 04 power servo motor [13] is installed in an either front wheel. As soon as the system detects an object and needs to navigate to a specific location, the steering wheel, positioned at the front of the wheelchair, rotates according to instructions provided by the 3-axis accelerometer.

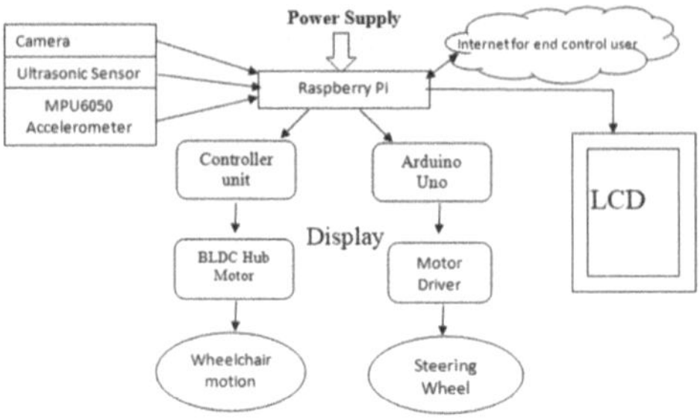

Fig. 2. Block diagram Smart Wheelchair for Hilly Region.

The Fig. 3, shows a 24-volt servo motor, model ASMC 04, connected to the front wheel of a vehicle, serving as the steering mechanism. The motor is mounted alongside a control circuit board, which likely manages the steering commands. The setup demonstrates how the servo motor is used to provide precise steering control, integral to the vehicle's movement and direction. This configuration enables accurate adjustments and responsiveness necessary for effective steering functionality.

Fig. 3. ASMC 04 servo is connected in a front wheel serves as steering wheel.

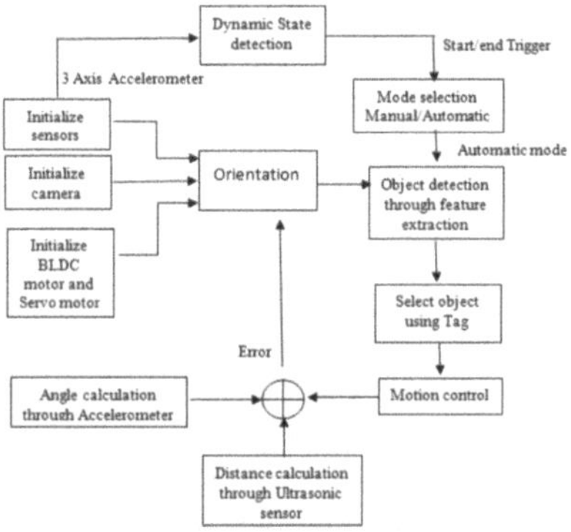

Fig. 3. Control methodology of the proposed Smart Wheelchair for Hilly Region.

The Fig. 3. outlines a system workflow for object detection and motion control involving various sensors and components. It begins with initializing sensors, including a 3-Axis Accelerometer, camera, and motors. The system determines initial orientation using these inputs. Dynamic state detection is performed via the accelerometer. A start/end trigger initiates the process, followed by mode selection (manual or automatic). In automatic mode, objects are detected through feature extraction and tagged for processing. Motion control adjusts based on orientation, angle (calculated by the accelerometer), and distance (measured by an ultrasonic sensor). Errors are fed back to correct orientation, creating a feedback loop for precise motion control and object handling.

2.2 Driving Mechanism

BLDC motors have gained significant popularity in motor control technology due to their versatility and efficiency across various applications. They offer advantages over traditional brushed DC motors, including better variable speed management, increased efficiency, and improved heat dissipation capabilities. BLDC motors contribute to the advancement of digital control systems, providing reliability and cost-effectiveness. The integration of BLDC hub motors into the system is depicted in Fig. 4(a), likely showing how these motors are incorporated into the wheelchair design scheme. BLDC hub motors are commonly used in electric vehicles and devices where compact size and high efficiency are desirable. Fig. 4(b) illustrates the connection of brake and throttle wires to the controller for manual operation. This suggests that the wheelchair design allows for both electric and manual modes of operation.

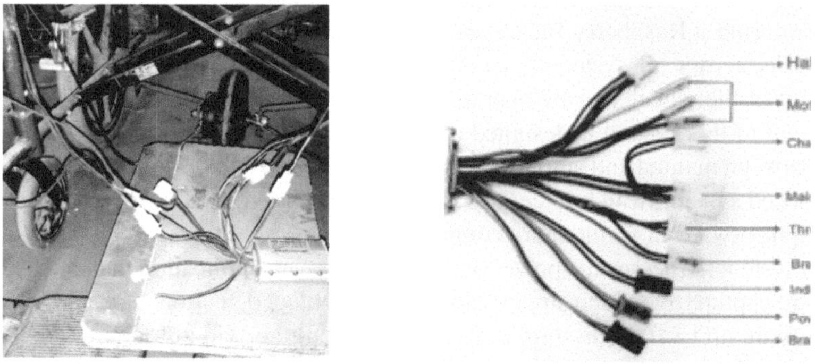

(a) (b)

Fig. 4. Assembly for wheel drive (a) BLDC hub motor interfacing with controller (b) Wire description of controller unit.

BLDC motors operate as permanent magnet synchronous electric motors powered by direct current (DC). They achieve motion by modifying the phase flow, which induces motor movement and generates rotational torque through an electronically controlled commutation mechanism. This mechanism allows for precise control of motor speed and direction.

2.3 Interfacing Sensors

A MPU 6050 sensor is employed to ascertain the correct angle along the X, Y, and Z axes, alongside rotation velocity. It comprises an x-y-z axis Gyroscope capable of discerning orientation and rotational polarity. Within this system, it is affixed alongside a Pi camera.

A camera module and an ultrasonic sensor are also interface into the framework enabling live video footage capture and obstacle avoidance capabilities.

2.4 Software Development

The principal algorithm for the proposed system is developed using Python, general-purpose programming language known for its interpretive, object-centric nature. An object detection technique is also developed into the Raspberry Pi bord using OpenCv. The OpenCV library boasts over two thousand optimized algorithms for tasks like facial recognition, object detection, human action categorization from videos, object tracking, and 3D model extraction.

3 Methodology

The essential components of a wheelchair and the necessary adjustments required to accommodate the installation of the intended control board are outlined. The proposed board, featuring a Raspberry Pi, serves as the central control unit responsible for executing tasks across all systems. A motor control driver is employed to regulate the movement of the wheelchair by operating the motor-wheel assembly [14]. The central control unit of this system is designed to offer dual operation modes, allowing users to switch between manual and automatic control as needed. A novel approach is adopted in our proposed system, which integrates Raspberry Pi, proving to be more cost-effective and straightforward in architecture compared to previous systems.

This system harnesses the power of OpenCV [15], an application that enables Raspberry Pi to capture live video streams and perform object detection. OpenCV is capable of identifying up to 1000 dissimilar objects in a picture or video frame. This functionality empowers the system to detect and recognize various objects or individuals in its surroundings, enhancing the overall capabilities and versatility of the wheelchair's control system. Tensorflow is utilized to identify different objects and output is provided in Fig. 5, demonstrating the detection of two distinct objects.

Image classification stands out as a key capability, entailing the prediction of an image's contents. Such a model proves invaluable in discerning image contents, empowering users to adapt any model to their specific needs. When identifying an object, the TensorFlow algorithm initially generates numerous bounding boxes. The algorithm categorizes the objects present within the bounding boxes. Following the object detection

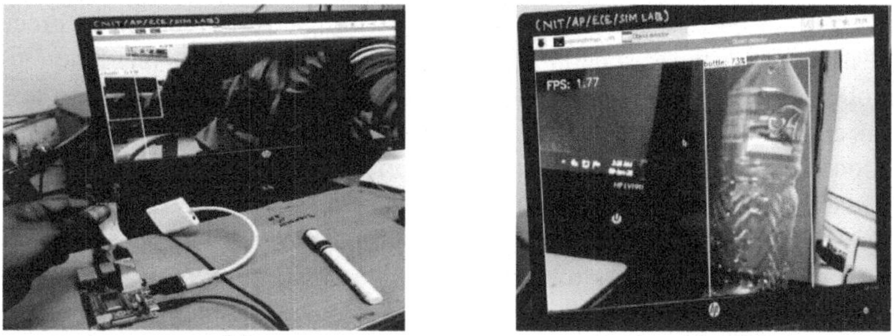

(a) (b)

Fig. 5. Some example of object detection (a) Object detection (b) system is identifying a bottle.

process, the algorithm proceeds to verify all detected bounding boxes against its pre-trained model for matching. Upon finding a match, the system identifies the object and displays its name above the corresponding bounding box. Although the system is trained to recognize thousands of objects, the algorithm is specifically designed to initiate navigation only once the pi-camera focuses on particular objects identified by predefined tags. This approach ensures that the wheelchair's navigation is triggered only when specific objects of interest are detected, aligning with the user's preferences and intended functionality of the system. Upon detecting an object, an ultrasonic sensor emits a sound wave towards it, which then reflects back to the sensor, allowing the system to calculate the object's distance. This technology enables the proposed wheelchair to maintain a safe distance from obstacles and accurately measure distances [16]. Additionally, the accelerometer is utilized to compute the proper angle for steering the wheel.

The system's performance is tested by using a Pi camera, which captures videos and extracts frames for classification. For instance, the system is trained to identify five specific types of objects: Door, Cycle, chairs, bottles, and Cup.

The Raspberry Pi is equipped with an IR camera, which is directed towards tracking the user's focal point. Upon detecting the object, the Raspberry Pi sends instructions to the motor driver to adjust the object's position accordingly. Programming for real-time computer vision utilizes OpenCV, followed by the installation of TensorFlow [17]. This framework encompasses numerous models relevant to everyday items like TVs, computers, people, bicycles, boards, laptops, glasses, tables, chairs, and more. Additionally, it allows for the inclusion of custom models for further object detection training within the library. Through this approach, programming tags are employed to exclusively select objects from a predetermined library, which are then relayed to the GPIO pin of the Raspberry Pi during live video streaming. Utilizing this GPIO pin, the Raspberry Pi transmits instructions to the motor drive circuit, facilitated by an Arduino. These instructions prompt the motor driver circuit to maneuver the wheelchair towards the specified object. For instance, within a room containing various objects such as TVs, refrigerators, tables, chairs, cupboards, beds, and computers, our system is capable of detecting each of these items. To establish communication between the Raspberry Pi and Arduino, a connection setup is implemented, enabling swift execution of tasks. While solely relying on the Raspberry Pi could eliminate the necessity for Arduino and motor drive circuits, the inherent delay poses a significant challenge. To mitigate this issue, image processing tasks are delegated to the Raspberry Pi, while the Arduino manages other functions such as controlling the ultrasonic sensor and adjusting the motor driver for precise steering.

Users can control the wheelchair system using an Android mobile application. This application likely communicates with the Raspberry Pi to send commands for controlling various functions of the wheelchair, such as movement and possibly other features using VNC Viewer. The VNC Viewer application is utilized to remotely access and control the Raspberry Pi from the Android mobile device. VNC stands for Virtual Network Computing, and it allows users to interact with a remote computer or device's desktop interface. By installing VNC Viewer on the Android device, users can remotely control the Raspberry Pi as if they were directly interacting with its interface. This is typically achieved by connecting both devices to a Wi-Fi hotspot. Ensuring they are on the same network is essential for communication between the devices. This IP address allows the VNC Viewer app to establish a connection with the Raspberry Pi over the network. The IP address serves as the unique identifier for the Raspberry Pi within the network. The IP Scanner application can assist in identifying the IP address of the Raspberry Pi. By scanning the network, the IP Scanner displays the IP addresses of all devices connected to the network, including the Raspberry Pi. This helps users locate the Raspberry Pi's IP address for input into the VNC Viewer app. This setup enables remote control of the wheelchair system from an Android mobile device, providing flexibility and convenience for users to operate the system from a distance. In scenarios where an individual is unable to personally attend to their elderly parents or patients due to being situated far away, such as at their workplace or during specific time constraints, this proposed system can prove invaluable. The caregiver can offer timely assistance by remotely managing tasks via their mobile device; ensuring essential needs such as medication or water are attended to promptly.

This process involves assigning labels to model images, with each label representing a distinct category, thereby facilitating model identification (Fig. 6). An image is inputted into the model for estimation, yielding an array of probabilities ranging from 0 to 100. Subsequently, each image output undergoes amalgamation with previously trained labels, initiating a compression process. The system then assigns the label corresponding to the most probable model. Notably, the sum of all values within the array always equates to 100.

Fig. 6. Experimental setup (a) Object chair is detected (b) navigate towards the desire object.

4 Result and Discussion

In the laboratory setting, various objects such as tables, chairs, various fruits, bicycle, computer, laptop, etc. are positioned in front of the system for testing purposes. Simultaneously, an accelerometer and ultrasonic sensor identifies the necessary angle and distance of the targeted object. Based on the data provided by these sensors, the wheelchair autonomously navigates to its intended destination according to a predefined algorithm. If an unforeseen obstacle obstructs the system's path, it halts its movement. In manual mode, the system maneuvers smoothly using the throttle. External configuration of the proposed system is achievable via internet connectivity facilitated by Wi-Fi, utilizing VNC and Putty applications on an Android mobile device. Furthermore, control over the system can be exerted using the same means. The operational characteristics of this assistive wheelchair were documented during its control, as delineated in Table 2.

Table 2. Testing features

Features	Estimate
Rotation of radius	1.4 feet
The space essential to stop the system	2 to 2.5 feet
Utmost velocity achieved by wheelchair	5 km/hr
Frame-capture speed for pi noir camera	30 fps

Following the testing of our proposed system in the laboratory, it yielded satisfactory results in terms of object detection at a frame rate of 30 fps. When steering towards an object, it can rotate up to a maximum of 1.4 feet. Additionally, the observed maximum speed of navigation for the proposed wheelchair is 5 km/hour, with the capability to halt precisely 2 to 2.5 feet before the intended object.

5 Conclusion

We have developed and tested wheelchair for hilly region it effectively recognizes diverse objects and navigates towards specified targets. Through SSH and VNC, it seamlessly integrates into the Internet of Things (IoT), while its manual functionality is facilitated by throttle control linked to a BLDC motor. Various peripherals, including microcontrollers, sensors, and cameras, are procured from local markets at a significantly reduced cost, rendering it much more affordable compared to traditional wheelchairs. This technology holds immense potential to enhance the lives of individuals with elderly or disabilities, aiming to bolster their self-reliance. The proposed wheelchair could serve as a viable alternative to commercially imported models, offering support to elderly, vulnerable, and disabled individuals globally. Future endeavors will focus on refining aspects such as mapping, localization, trajectory planning, and tracking to enhance the navigation process, potentially allowing caregivers to visualize the wheelchair's real-time position through animated representations.

References

1. Stroh, A., Desai, J.: Hand gesture-based artificial neural network trained hybrid human–machine interface system to navigate a powered wheelchair. J. Bionic Eng. **18**(5), 1045–1058 (2021)
2. Koda, H., Okada, Y., Fukumoto, T., Morioka, S.: Effect of tilt-in-space and reclining angles of wheelchairs on normal force and shear force in the gluteal region. Int. J. Environ. Res. Public Health. **19**(9), 5299 (2022)
3. Amini, E., Etemadi, M., Shahabi, S., Barth, C.A., Honarmandi, F., Karami Rad, M., Lankarani, K.B.: Barriers and enabling factors for utilizing physical rehabilitation services by Afghan immigrants and refugees with disabilities in Iran: a qualitative study. BMC Public Health. **24**(1), 1–15 (2024)
4. Krishna, H., Emani, S., Manapaka, R., Rao, S.P.V.S.: Voice controlled and joystick based wheel chair for differently abled people. Int. Res. J. Modern. Eng. Technol. Sci. **3**(6), 3111–3114 (2021)
5. Pathan, S.M.K., Ahmed, W., Rana, M.M., Tasin, M.S., Islam, F., Sultana, A.: Wireless head gesture controlled robotic wheel chair for physically disable persons. J. Sens. Technol. **10**(4), 47–59 (2020)
6. Mnassri, A., Nasri, S., Boussif, M., Cherif, A.: Real-time voice-controlled human machine interface system for wheelchairs implementation using Raspberry Pi. Int. J. Veh. Inf. Commun. Syst. **9**(1), 81–102 (2024)
7. Fischer-Janzen, A., Wendt, T.M., Van Laerhoven, K.: A scoping review of gaze and eye tracking-based control methods for assistive robotic arms. Front. Robot. AI. **11**, 1326670 (2024)
8. Assaf, M.H., Kumar, R., Sharma, K., Sharma, B.: Optimized tongue driven system using artificial intelligence. Comput. Methods Biomech. Biomed. Eng. Imaging Vis. **11**(3), 696–710 (2023)
9. Sahoo, S., Choudhury, B.: Voice-activated wheelchair: An affordable solution for individuals with physical disabilities. Manag. Sci. Lett.. **13**(3), 175–192 (2023)
10. Miftahussalam, I., Julian, E.S., Prawiroredjo, K., Djuana, E.: Wheelchair control system with hand movement using accelerometer sensor. Microelectron. Eng. **278**, 112018 (2023)
11. Abdal, Y.M., Ayoub, M.G., Farhan, M.N., Abdulla, H.A.: Human–machine interaction for motorized wheelchair based on single-channel electroencephalogram headband. Bull. Electr. Eng. Inform. **12**(2), 902–910 (2023)
12. Artanto, D., Pranowo, I.D., Wicaksono, M.B., Siswoyo, A.: Combining Mindwave, MPU6050, internet of things for reliable safe monitored wheelchair control system. Indones. J. Electr. Eng. Comput. Sci. **32**(2), 742–751 (2023)
13. Xu, X., Liu, P., Lu, S., Wang, F., Yang, J., Xiao, G.: Iterative learning with adaptive sliding mode control for trajectory tracking of fast tool servo systems. Appl. Sci. **14**(9), 3586 (2024)
14. Dutta, A., De Sarkar, M., Roy, S.: A review on performance of distributed embedded system. Social Networking and Computational Intelligence: Proceedings of SCI-2018, 605–614 (2020)
15. Kabir, M.F., Roy, S.: Development of an accident prediction model for freeways systems. Int. J. Crit. Infrastruct. **19**(3), 274–289 (2023)
16. Rietveld, T., Vegter, R.J., der Woude, L.H., de Groot, S.: The interaction between wheelchair configuration and wheeling performance in wheelchair tennis: a narrative review. Sports Biomech. **23**(3), 370–391 (2024)
17. Mondal, A., Roy, S.: Design and development of dynamic gesture recognition system based on deep neural network for driver assistive devices. Int. J. Syst. Syst. Eng. **13**(3), 271–283 (2023)

Web-Based Intelligent Chatbot for Depression Therapy

Anupam Agrawal(✉), Saloni Doshi, Anjali Sahu, Kavita, and Himanshu Mishra

IIIT Allahabad, Prayagraj, India
{anupam,iit2020059,iit2020063,iit2020252,mrm2022005}@iiita.ac.in

Abstract. Due to the increase in depression cases and people's reluctance to seek medical therapy, it has been an alarming situation to develop a chatbot that offers therapy in an accessible and easy manner so that individuals suffering from depression do not feel uncomfortable during social interactions with medical professionals. This researchâĂŹs main objective is to present a web-based intelligent chatbot that offers therapy to users depending on the emotions shown in their conversations. The emotion detection part utilizes models such as BERT and RoBERTa coupled with CNNs to correctly categorize emotion in the user's message, and the chatbot is integrated with Generative AI for giving personalized responses to the user. For testing the model we chose the ISEAR dataset that included 7666 sentences categorized into 7 major emotions that are joy, guilt, anger, disgust, fear, sadness, and shame.

Keywords: Depression therapy · Intelligent Chatbot · ISEAR Dataset · RoBERTa · Emotion Detection

1 Introduction

A deep feeling of sadness that lasts for a long time and never goes away, along with a noticeable degradation in cognitive and physical abilities, is the definition of depression, a mental illness that affects a large number of people. Based on estimations, depression is likely to be the primary root of disease worldwide in 2030, ranking among the top 10 leading root of illness according to the WHO (World Health Organization) [1]. As such, the field of web-based chatbots for depression has pushed research and development to a critical position [2]. When it comes to providing help, these chatbots are a very appealing option because of their ability to effectively combine anonymity, immediacy, and availability.

Depression is a common issue, affecting an estimated 3.8% of people globally. This covers 5% of adults, with women making up a higher percentage (6%) than men (4%). The disease also affects (5.7%) of those over 60 [3]. Across the globe, an estimated 280 million people suffer from anxiety/depression as shown in Fig. 1.

There is an enormous variation in the frequency of the disorder across genders, with rates about 50% higher in women than in men. Furthermore, about

D. Gupta et al. (Eds.): PCEMS 2024, CCIS 2491, pp. 35–47, 2025.
https://doi.org/10.1007/978-3-031-90577-3_4

10% of women experience depression either during or after giving birth [4]. For those between the ages of 15 and 29, suicide is disappointingly among the top four causes of death. More than 700,000 people commit suicide yearly.

Even though there are excellent therapies for mental illnesses, countries with low or middle incomes have very limited access to these treatments. In these areas, more than 75% of people do not receive any kind of care for mental health issues [5].

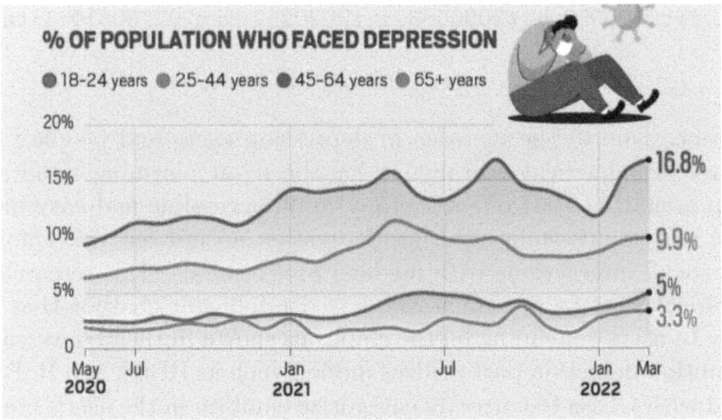

Fig. 1. Depression Statistics [6]

Due to an increasing shortage of mental health experts, it is becoming more and more challenging to provide the widely accepted gold standard of care traditional individual therapy sessions. This situation requires the investigation of further, more effective methods for therapy. An interesting idea is the use of therapeutic chatbots. These AI-powered solutions can provide people with a platform to manage their depression that is friendly, easy to use, and judgment-free. These chatbots may enhance traditional treatment in addition to providing round-the-clock assistance, increasing accessibility, and reducing stigma around mental health care.

Using these concepts, the purpose of this study is to provide an intelligent chatbot for the web that may help users with therapy by analyzing the emotions they express in their conversations. This system uses Convolutional Neural Networks (CNNs) in combination with advanced models like BERT and RoBERTa for emotion detection. For Chatbot, when a user sends a message to the chatbot, the tool first preprocesses it, augmenting it with additional context that is the detected emotions. This updated message is then passed to the OpenAI ChatCompletion API. Here, the model analyzes the input message to generate a logical and personalized response.

Our paper is organized as follows: this combination guarantees the right categorization of the user's emotional state from their communications, allowing the chatbot to reply accordingly. Section 2 examines earlier studies conducted in the field, Sect. 3 presents the dataset that was used in this study, and Sect. 4

describes our Emotion detection model for Depression therapy and explains how it works with text and audio inputs. Section 5 gives our findings, observations, and comparison with previous research; and Sect. 6 highlights the conclusion and directions for future work.

2 Background and Related Work

Text and audio inputs are important to human communication because they provide important information. They are simple to use and have proven value for evaluating patients' mental and emotional health [7]. We will thus talk about some current research in this field that utilizes chatbots to treat depression.

Patel et al. [8] proposed a sophisticated social therapeutic chatbot that categorizes user text into specific emotion labels, including Joy, Anger, Shame, Sadness, Disgust, Fear, and Guilt. These emotion labels are used by Chatbot to further analyze their mental state using the user's chat data, such as if they are experiencing depression or anxiety. CNN (Convolutional Neural Network), RNN (Recurrent Neural Network), and HAN (Hierarchical Attention Network) are three standard deep learning classifiers that the authors used to achieve accurate identification of emotions. To test their proposed model, the authors used the ISEAR Dataset. With excellent stability over 15 epochs, CNN has achieved accuracy as high as 75%. 15 epochs of up to 70% accuracy have been obtained by RNN and HAN respectively. The significance of a social Curative chatbot, particularly for students, was highlighted in this paper, although the results were not sufficiently consistent across all groups.

Biswas et al. [9] constructed a Web Application that integrates a Virtual Voice and Chat Assistant. The building of a specific voice and text/chatbot is the key goal of this study. The application is meant mainly to support people who are depressed by motivating them to interact with others without bias, which can be calming. With a metaphorical reference to a cat purring, the application, called "App," aims to comfort its users in a way similar to the peace a cat feels when its mother is around. Dialog Flow will be used in the application's design, and artificial intelligence's machine learning component will be used for NLP (natural language processing). This method offers a simple way to put Machine Learning libraries into practice. A database will be included in the application's backend to cache the history of conversations between the chatbot and the user. According to the present results, the correctness of the training dataset is 84.14%, the correctness of the cross-validation dataset is 83.32%, and the accuracy of the test dataset is 60.17%.

Adoma et al. [10] conducted a detailed evaluation of the performance of used pre-trained transformer models such as BERT, DistilBERT, RoBERTa, and XLNet in the task of textual data emotion recognition. The process includes analyzing each model's result to that of the others. These models are capable of differentiating between seven main categories of emotions: fear, disgust, anger, joy, sorrow, guilt, and shame using the ISEAR dataset. The models rank in descending order based on accuracy in emotion recognition: DistilBERT, BERT,

XLNet, RoBERTa. The corresponding recognition accuracies of these are 0.7431, 0.7299, 0.7009, and 0.6693. The models were successful in identifying emotions from the userâĂŹs text; RoBERTa achieved the highest accuracy.

3 Dataset

This research study has been carried out using the International Survey on Emotion Antecedents and Reactions (ISEAR) dataset [11] which is a collection of emotional textual data that was established for research purposes in the field of affective computing and emotion analysis. It has 7666 sentences that are categorized into 7 major emotions that are joy, guilt, disgust, anger, sadness, fear, and shame as mentioned in Table I. The dataset was collected from around 3000 respondents including both psychologists and non-psychologists from 37 countries across 5 continents were asked to express situations in which they had experienced these major 7 emotions Table 1.

Table 1. The ISEAR DatasetâĂŹs Sentiment Distribution

Emotion Label	Quantity
Joy	1094
Guilt	1093
Disgust	1096
Anger	1096
Sadness	1096
Fear	1095
Shame	1096
Total	**7666**

4 Proposed Work

In this section, we will present a comprehensive overview of our approach. Following that, we have provided a detailed discussion of the emotion detection model, followed by a thorough examination of the chatbotâĂŹs functionalities and design principles.

4.1 Overview of the Approach

As shown in Fig. 2, in the initial phase we collected a dataset specifically tailored for emotion detection i.e. the ISEAR dataset. This dataset was carefully cleaned and preprocessed to ensure high-quality training data for our deep learning models. Leveraging advanced architectures such as BERT and RoBERTa coupled

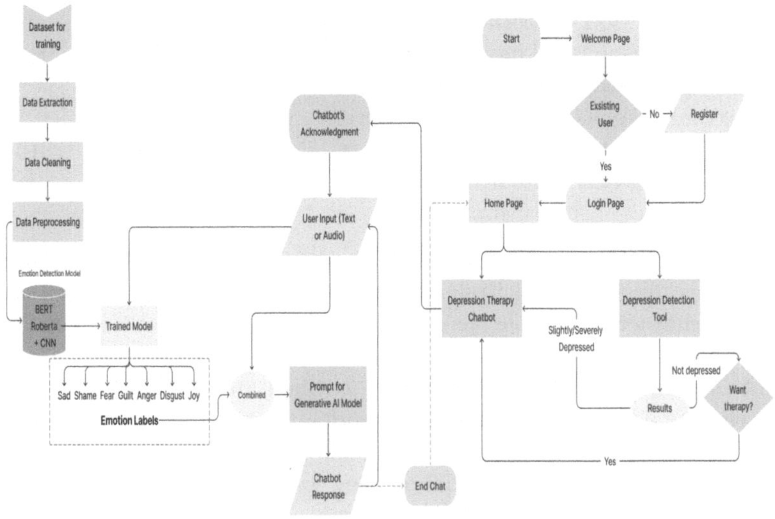

Fig. 2. An outline of each step involved in our Proposed work

with Convolutional Neural Networks (CNNs), we proposed a model capable of accurately identifying emotions within textual data.

Subsequently, we moved into the development of a web-based chatbot interface, allowing users to interact using both text and audio messages. By utilizing the previously trained emotion detection model, the chatbot efficiently analyzes user messages and assigns corresponding emotion labels to them.

When a user sends a message to the chatbot, the tool first preprocesses it, augmenting it with additional context that is the detected emotions. This updated message is then passed to the OpenAI ChatCompletion API. Here, the model analyzes the input message to generate a logical and personalized response. For instance, the chatbot can offer valuable mental health support or suggest mood-boosting activities catering to individuals who lack a support network for discussing personal challenges or may lack access to conventional therapy services.

4.2 Emotion Detection Model

This section aims to analyze our proposed methodology for the Emotion Detection model. We will discuss both the BERT approach and a hybrid approach combining RoBERTa (a robustly optimized BERT pretraining approach) with CNN.

Data Extraction. The ISEAR dataset that we have collected includes several columns, capturing various components of the questionnaire responses. However, our primary interest lies in two specific columns: one containing individuals' textual responses, and the other indicating the corresponding emotion labels assigned to each response. These two columns were extracted as they are crucial for the subsequent stages of our work.

The textual responses provide the context and content from which we aim to derive emotional states, while the emotion labels serve as ground truth annotations for training and evaluating our emotion detection model.

Data Cleaning. After gathering the data, the next step involves cleaning the data to remove mistakes and irrelevant information. This includes the removal of irrelevant characters, double spacing, URLs, or special symbols from the text data. Additionally, columns lacking textual responses (such as "[No response]" text) but containing emotion labels were identified and removed. As a result of this process, the dataset size was reduced from an initial count of 7,666 entries to 7,564 entries, ensuring a more focused and relevant dataset for subsequent analyses. These adjustments were crucial as they were providing a negative impact on the recognition performance of the system.

Data Preprocessing. To prepare the data for model compatibility, a mapping of sentiment labels to numerical values was performed, such as mapping "anger" to 0, "joy" to 1, and so on. After this, the text data underwent tokenization utilizing the BERT tokenizer for the BERT approach and the RoBERTa tokenizer for a hybrid approach from the Hugging Face Transformers library, text data underwent tokenization, converting textual inputs into numerical tokens suitable for deep learning models. These tokens were then encoded into input IDs, and attention masks were generated to highlight relevant tokens during model training.

An additional step involved identifying the token corresponding to the word "not" within the tokenized sequences. The attention mask was then modified to assign a higher weight specifically to the "not" token. This modification was made to highlight the significance of negation introduced by "not" in emotion detection tasks

For instance, in a tokenized sequence comprising [CLS] (Classification Token), "I", "am", "not", "happy", [SEP] (Separator Token), the attention mask was adjusted to increase the attention weight of the "not" token, ensuring that the model prioritizes comprehending negation and its influence on emotion rather than focusing primarily on tokens following the "not" token. Following this, the preprocessed dataset underwent division using a classified split method to create training and testing data with an 80-20 ratio, to maintain a balanced class distribution across both sets.

Model Architecture

BERT Approach. After preprocessing, the BERT model (bert-base-uncased) as depicted in Fig. 3 has been used where the information was subjected through 768 hidden layers and twelve levels of self-attention yielding about one hundred ten million parameters to obtain the contextual vector representation. The aggregated sequence representation which was utilized for classification was derived from the transformed vectors representing prediction probabilities in the output of the 12th transformer layer.

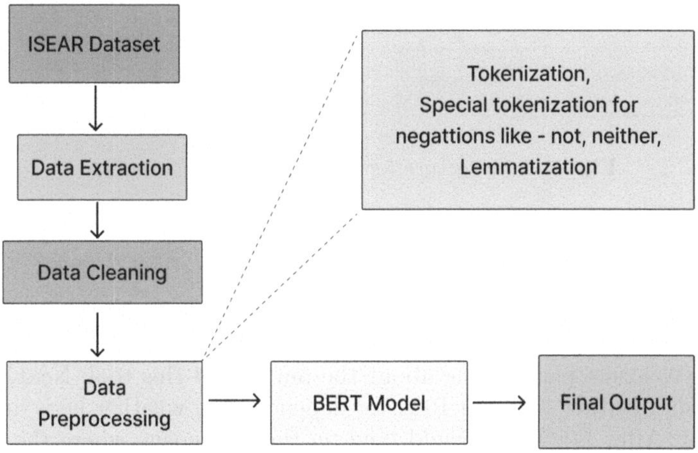

Fig. 3. Architecture for BERT model

RoBERTa + CNN Hybrid Approach The preprocessed data i.e. input_ids and attention_masks are fed into the RoBERTa model to obtain hidden states. The hidden states represent contextualized embeddings for each token in the input text. The hidden states from RoBERTa are passed through multiple CNN blocks, each with different filter sizes. CNN blocks use the 1D convolutional layer, batch normalization, Dropout, and MaxPooling layers to extract features from the token embeddings. Each CNN block processes the hidden states independently to capture different types of features based on filter sizes. The output of each CNN block (after max-pooling) is concatenated into a single vector. The CLS token representation from RoBERTa is also extracted. Both representations are then concatenated to combine information from RoBERTa and CNN as depicted in Fig. 4. The concatenated representation is passed through fully connected layers for sentiment classification. Finally, logits from RoBERTa and CNN are combined to get the final output logits for classification.

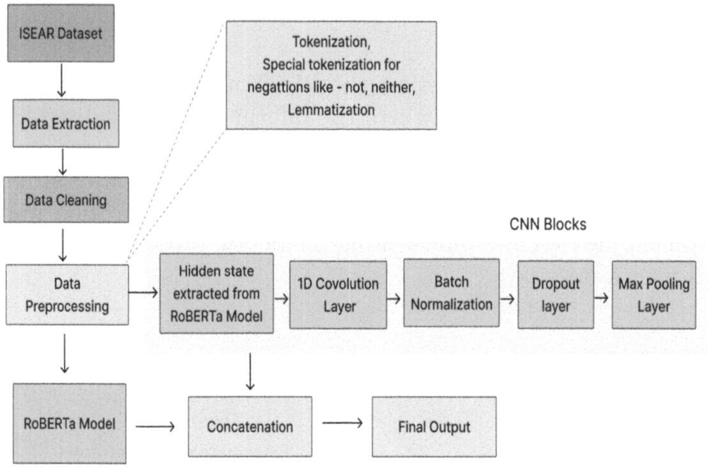

Fig. 4. Architecture for RoBERTa + CNN model

4.3 Chat-Bot Implementation

Our web-based intelligent chatbot is implemented using ReactJS for making the front-end and Python for handling the back-end. As we open the website, it shows the Welcome page telling about the purpose of this tool. Next, the individual needs to either Login or Register depending on whether he is an existing user or not. After Login, he would land on the Homepage where the user gets options like taking a depression detection test, opting for chatbot therapy, or visiting his profile. When the user opts for chatbot therapy, he will be directed to a new page, where he can communicate with the chatbot via either text or audio messages as shown in Fig. 5. As the user sends a message to the chatbot, first with the help of our proposed model, we extract the user's emotion from the message. The chatbot leverages generative AI, specifically the GPT-3.5-turbo model, for generating human-like responses because of its advanced natural language processing capabilities. When a user sends a message to the chatbot, the tool first preprocesses it, augmenting it with additional context that is the detected emotions. This updated message is then passed to the OpenAI ChatCompletion API. Here, the model analyzes the input message to generate a logical and personalized response. By implementing this approach, the chatbot can create high-quality responses that closely resemble human conversation. This not only increases user engagement but also leads to more meaningful conversations. Moreover, the chatbot would stay up to date with the current trends and patterns because of OpenAI's dedication to ongoing learning and development. Thus this chatbot provides a more personalized reply by understanding the user's feelings and hence providing them with therapy to overcome depression. The user can continue the chat for as long as he wants or exit the page after completion.

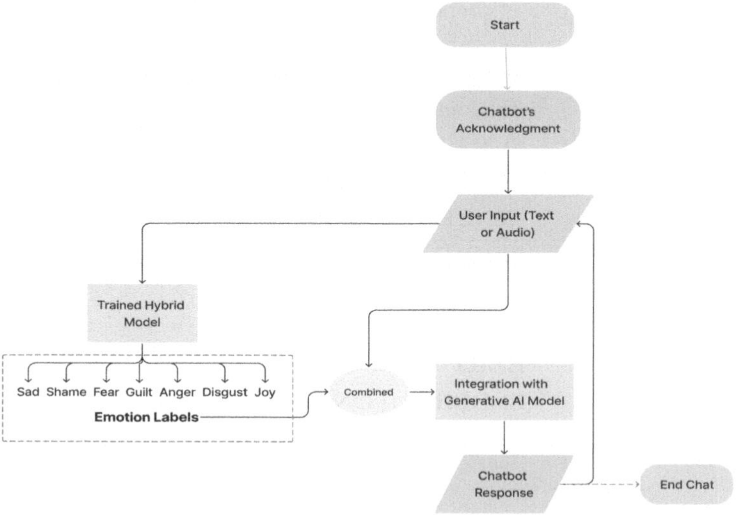

Fig. 5. Chatbot Implementation

5 Results and Analysis

This section presents a comprehensive summary of the results achieved for the proposed model by providing a comparison with existing baseline models. Notably, our hybrid model performed well, outshining the baseline models with an F1-score, Recall, and Precision of 0.78, 0.78, and 0.77 respectively. Table 2 and Fig. 6 present the results of our proposed emotion detection model compared with existing approaches using the ISEAR dataset, where our hybrid model leveraging RoBERTa coupled with CNN has shown substantial improvement in the F1-score from baseline models.

Also, as mentioned in Sect, 4, we created a web- interface for our intelligent chatbot. Thus Figs. 7, 8, 9, and 10 show a collection of images that provide an understanding of the same. âĄďâĄďThe images display the welcome page, login page, home page, and the interface for chatbot therapy.

Table 2. Comparison of proposed model's F1-score with existing models

Model	F1-score
Random Forest [12]	0.49
SVM [12]	0.55
BiLSTM+CNN [12]	0.63
XLNET [10]	0.73
BERT [10]	0.70
DistilBERT [10]	0.69
RoBERTa [10]	0.74
BERT	**0.73**
RoBERTa +CNN	**0.78**

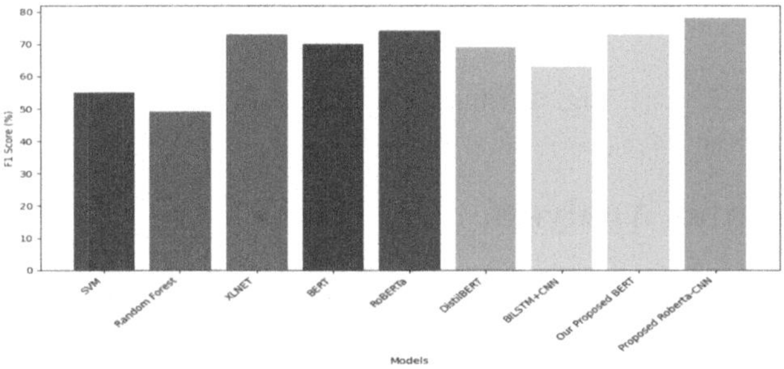

Fig. 6. Comparing the F1-Score of our proposed model with a number of previously suggested models

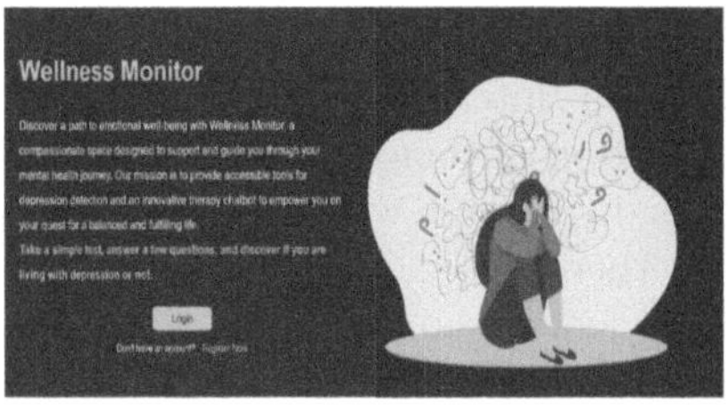

Fig. 7. Welcome Page

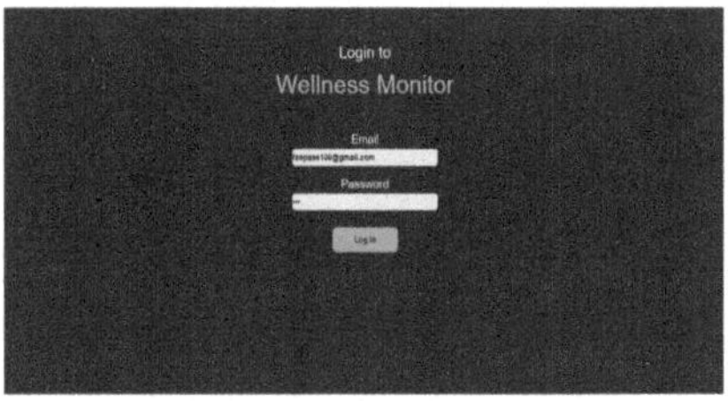

Fig. 8. Login Page

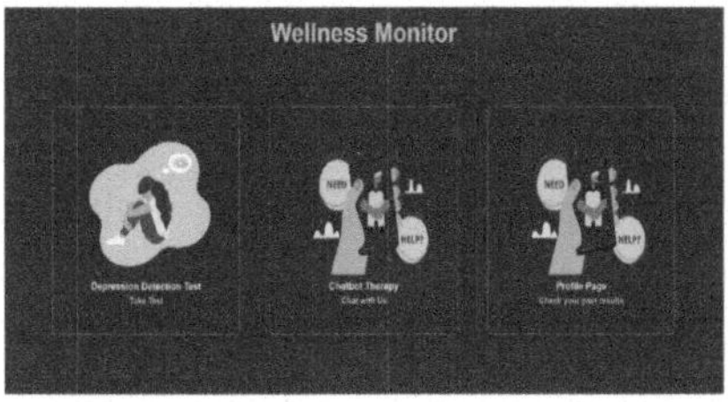

Fig. 9. Home Page

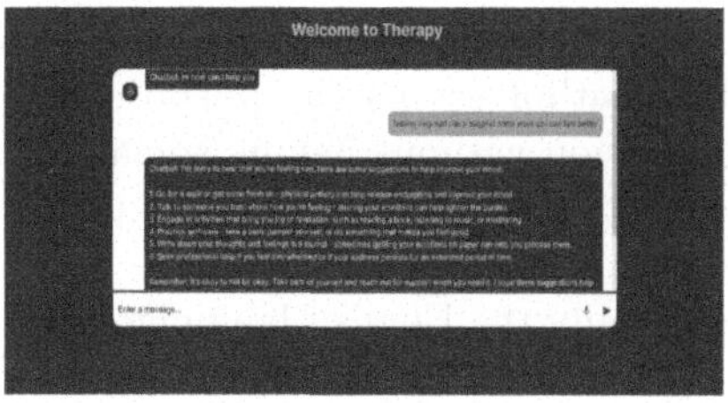

Fig. 10. Interface for chatbot therapy

Additionally, Table 3 presents a comparison between our chatbot and other existing apps using various factors such as its goal, pricing structure (free or in-app purchase), media type used in conversations, and whether the conversations are bounded or flexible, etc.

Table 3. Comparison of features of existing chatbot-based apps with our proposed chatbot

App Name	Goal	Addressed Problems	Conversation Flow	Mode of Media Used	Price	Country
Serenity [13]	Conversational companion	Anxiety, relationship issues, self-care, and sleep disorder	Bounded	Text and emoji	Free	US
Youper [14]	Virtual therapist	Self-care	Bounded	Text	In-app purchases $70/year	US
Woebot [15]	Digital coach	Loneliness, self-care, stress, sleep disorder, anxiety, and loneliness	Semi-bounded	GIF, text, audio, emoji	Free	US
Wysa [16]	Digital coach	Sleep disorder, anxiety, stress, depression, and self-care	Flexible	GIF, text, audio, emoji	In-app purchases $300/year	India
Our Chatbot	**Virtual therapist + Conversational companion**	**Depression, Stress, anxiety, self-care, loneliness, anger issues**	**Flexible**	**Text, emoji, and audio**	**Free**	**India**

6 Conclusion and Future Scope

This study makes significant progress in the field of depression therapy by utilizing a web-based intelligent chatbot. This chatbot takes input from users in the form of both text and audio. Further, by leveraging models such as BERT and RoBERTa coupled with CNNs which acquires F1-score of 0.78, it detects the emotion of the user input, which is further combined with the user message and sent to generative AI to create a personalized response for the user. Thus, this chatbot serves the purpose of providing therapy to any individual facing depression without having social interactions.

In the future, this work could be further expanded by introducing multiple input languages such as Hindi, Telugu, etc. for our chatbot as currently it supports only the English language thus benefiting a bigger audience. Additionally, we believe consulting domain experts and incorporating their feedback could significantly enhance this tool because of their expertise in effectively understanding human emotions and mental health.

Acknowledgment. We express our sincere thanks to the authors of the ISEAR dataset, which served as the basis for our research. Furthermore, we extend our heartfelt gratitude to the Indian Institute of Information Technology, Allahabad, Uttar Pradesh, India, for providing an environment that supports research activities and helped us to enhance the quality of this research and complete it successfully.

References

1. Depression Statistics. https://www.who.int/news-room/fact-sheets/detail/depression
2. Deepa, B., Madhul, M., Kuzhali, R., Senthamilselvi, P.: Depression healing chatbot using AI and ML. Int. J. Sci. Eng. Technol. **12**(2), 1–8 (2024)
3. Institute of Health Metrics and Evaluation. Global Health Data Exchange (GHDx)). https://vizhub.healthdata.org/gbd-results/ Accessed 4 Mar 2023
4. Woody, C., Ferrari, A., Siskind, D., Whiteford, H., Harris, M.: A systematic review and meta-regression of the prevalence and incidence of perinatal depression. J. Affect. Disord. **219**, 86–92 (2017)
5. Evans-Lacko, S., Aguilar-Gaxiola, S., Al-Hamzawi, A., et al.: Socio-economic variations in the mental health treatment gap for people with anxiety, mood, and substance use disorders: results from the WHO World Mental Health (WMH) surveys. Psychol. Med. **48**(9), 1560–1571 (2018)
6. https://www.indiatoday.in/diu/story/youth-women-suffered-depression-anxiety-india-during-pandemic-who-data-shows-2362947-2023-04-21
7. Uddin, M.A., Joolee, J.B., Sohn, K.-A.: Deep multi-modal network based automated depression severity estimation. IEEE Trans. Affect. Comput. **14**(3), 2153–2167 (2023)
8. Patel, F., Thakore, R., Nandwani, I. and Bharti, S.K.: Combating depression in students using an intelligent chatBot: a cognitive behavioral therapy. In: 2019 IEEE 16th India Council International Conference (INDICON), pp. 1–4. IEEE (2019)
9. Biswas, A., Kaushik, A.: Depression therapy using chatbot. Int. J. Modern Trends Sci. Technol. **6**(12), 323–327 (2020)
10. Adoma, A.F., Henry, N.M., Chen, W.: Comparative analyses of bert, roberta, distilbert, and xlnet for text-based emotion recognition. In 2020 17th International Computer Conference on Wavelet Active Media Technology and Information Processing (ICCWAMTIP), pp. 117–121. IEEE (2020)
11. https://www.unige.ch/cisa/research/materials-and-online-research/research-material/

12. Polignano, M., Gemmis, M., Basile, P., Semeraro, G.: A comparison of word-embeddings in emotion detection from text using bilstm, cnn and self-attention. In: Adjunct Publication of the 27th Conf. on User Modeling, Adaptation and Personalization, Laranca Cyprus, pp. 63–68 (2019)
13. Serenity. https://apps.apple.com/us/app/serenity-guided-meditation/id1414802569
14. Youper. https://apps.apple.com/us/app/youper-cbt-therapy-chatbot/id1060691513
15. Woebot. https://apps.apple.com/us/app/woebot-the-mental-health-ally/id1305375832
16. Wysa. https://apps.apple.com/us/app/wysa-mental-health-support/id1166585565

Design of RAM Cell Using Quantum Dot Cellular Automata

Pamidi Mohammad Ashraf(✉) and Vobulapuram Ramesh Kumar

Department of ECE, Indian Institute of Information Technology Sri City, Chittoor, Andhra Pradesh, India
{ashraf.p21,rameshkumar.v}@iiits.in

Abstract. This paper presents the design and analysis of a Random Access Memory (RAM) cell using Quantum Dot Cellular Automata (QCA) technology. QCA offers a promising alternative to conventional transistor-based approaches, leveraging Coulombic interaction between cells for computation. The proposed RAM cell design incorporates QCA's unique properties, such as majority logic gates and clocking to achieve high-speed operation, minimal power consumption, and fault tolerance. Through comprehensive analysis including power efficiency, fault tolerance, and polarization effects, the paper highlights the importance of optimizing RAM cell design parameters for reliable performance in QCA-based digital systems. The results show that the proposed RAM circuit is more efficient. An improvement of 17.80% in cell count, 17.78% in cost function and used the coplanar approach that reduced the complexity of the design over the other efficient RAM.

Keywords: Quantum Dot Cellular Automata (QCA) · RAM cell design · digital circuits · power efficiency · fault tolerance · polarization effects · nanotechnology

1 Introduction

In today's tech landscape, the dominance of Complementary Metal Oxide Semiconductor (CMOS) technology in microprocessors and integrated circuits is undeniable [1, 2]. However, as transistor sizes shrink towards the nanometer scale, quantum phenomena like tunnelling start to impose performance limitations [3, 4]. Moreover, while device scaling progresses, interconnection scalability lags due to issues with wire resistance and capacitance [5, 6].

Quantum Dot Cellular Automata (QCA), a promising alternative to CMOS technologies [7, 8]. QCA operates on the principle of Coulombic interaction between cells, bypassing the need for charge transport [9, 10]. Its advantages, including low power consumption and high density, make it an attractive choice for various applications, from general-purpose computing to embedded systems [11, 12]. QCA relies on cells housing multiple quantum dots, exhibiting intriguing bistable behavior owing to Coulombic forces [13, 14]. These cells interact with one another to perform various logic functions, with each cell's state influenced by its neighbors' states during the previous clock cycle

D. Gupta et al. (Eds.): PCEMS 2024, CCIS 2491, pp. 48–61, 2025.
https://doi.org/10.1007/978-3-031-90577-3_5

[15, 16]. The necessity arises from the growing demand for faster, more energy-efficient computing systems [17, 18]. Traditional RAM designs may struggle to meet the stringent requirements of future digital systems, necessitating innovative solutions that mitigate power consumption, enhance speed, and ensure fault tolerance [19, 20].

To address these challenges, the research proposes a novel RAM cell design tailored for QCA technology. By incorporating QCA's intrinsic features such as majority logic gates and clocking mechanisms, the proposed design aims to achieve high-speed operation, minimal power consumption, and robust fault tolerance [6, 7]. Through rigorous analysis and simulation, the paper evaluates the performance of the proposed RAM cell, highlighting its superiority over existing designs and contributing to the evolution of next-generation digital systems.

Additionally, it investigates further applications of QCA technology, elucidating its potential for revolutionizing digital circuit design. Moreover, a comprehensive analysis has been conducted on various aspects of QCA technology, including power consumption, fault tolerance, and polarization effects [8].

In the subsequent Sect. 2 provides the brief review of related works, highlighting existing RAM cell designs and their key features. Section 3 details the proposed RAM cell design using QCA technology, including its architecture, operation, and analysis. Simulation results and analysis are presented in Sect. 4, then power efficiency, fault tolerance, and polarization effects are thoroughly examined, followed by a comparative study of physical properties of QCA-based RAM cells in Sect. 5. Finally, Sect. 6 concludes the paper by summarizing the key findings.

2 Related Works

A design technique for a RAM cell without set/reset control using a D latch module and a loop-based structure was described by Singh and Sharma (2020) [1]. The RAM cell is a three-layer structure made up of two inverters and six majority voters. Applications such as "In," "Enable," and "W/R'"are applied at Layer 1, and at Layer 3 are obtained the D latch's "Qn" output and the RAM cell's "Out" output. Eq. (1) provides the output function of the suggested RAM, which makes use of two inverters and six majority gates. To perform write and read operations, set Enable to "1," where "1″ denotes a write operation (W/R' = 1) and "0″ denotes a read operation. The multilayer RAM cell is shown in Fig. 1 [1].

$$Out = M\left(\overline{Enable}, -1, M\left(Qn, -1, \frac{W}{R'}\right)\right) \tag{1}$$

[2] Loop-based structures are widely used in QCA technology for RAM cell design, where data storage is accomplished by circulating a small amount of data among QCA cells in a closed-loop configuration. An essential part of these loop-based RAM cell designs is the D-latch. The RAM cell design makes use of a D-latch that was designed using a novel technique that was proposed by Mubarakali et al. (2019). Using a self-designed 2 × 1 multiplexer, the input of the first multiplexer is connected to the output signal to produce the D-latch. Figure 2 (a) and 3 (b) show the schematics of the D-latch that is based on the 2 × 1 multiplexer.

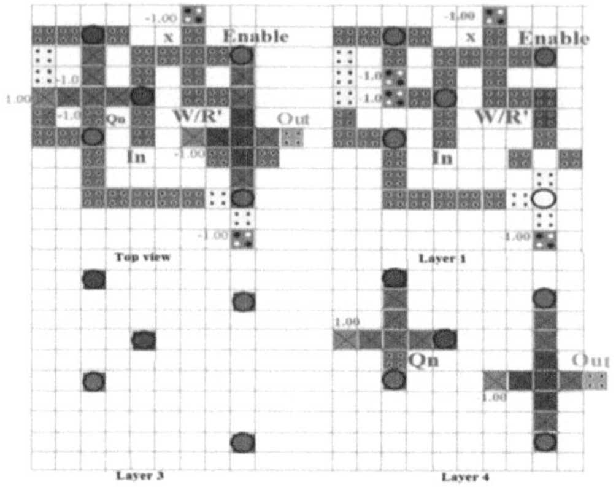

Fig. 1. Multilayer RAM cell (a) top view (b) Layer 1 (c) Layer 2 (d) Layer 3 [1]

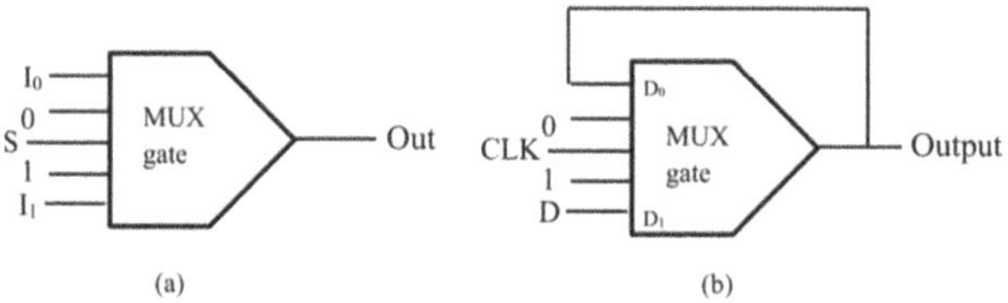

Fig. 2. Schematic of 2 × 1 multiplexer (a) D-latch using 2 × 1 multiplexer (b)

In Feedback is applied in the proposed D-latch from the multiplexer's (D0) input to its output. The input is sent to the output when CLK is 1. On the other hand, by setting the clock signal to zero, the feedback mechanism preserves the value of the output that has been stored. Figure 3 illustrates the suggested D-latch configuration.

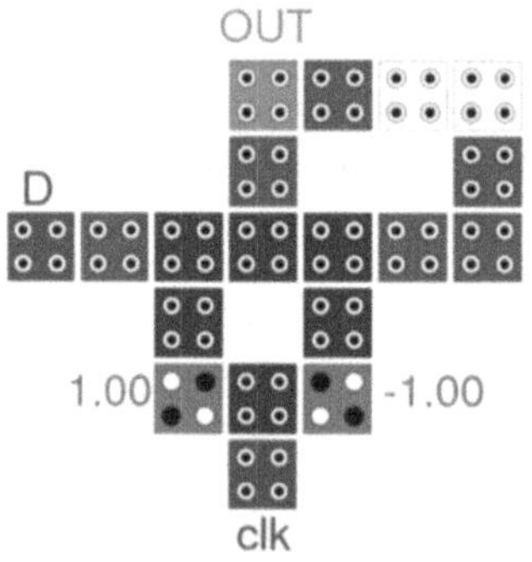

Fig. 3. D-latch layout [2]

Mubarakali et al. (2019) further introduced a new design for RAM cell with structural optimization considerations. The RAM cell features two inputs, one output, and an additional input called "Select," which deter-mines whether the RAM cell is enabled or disabled. When Select input is set to 0, both the input and outputs are deactivated, rendering the RAM cell disabled. Conversely, setting the Select input to 1 enables read/write operations. The RAM cell layout is depicted in Fig. 4 [2]. The MUX gate has three inputs labelled as a, s, and b, with the output given by Q_t ($Q_t = sa' + sb$). Table 1 outlines the functionality of the proposed RAM cell, showing the behavior of the output (Q_{t+1}) based on the input (Q_t) and the control signals (R/W, S, a, b).

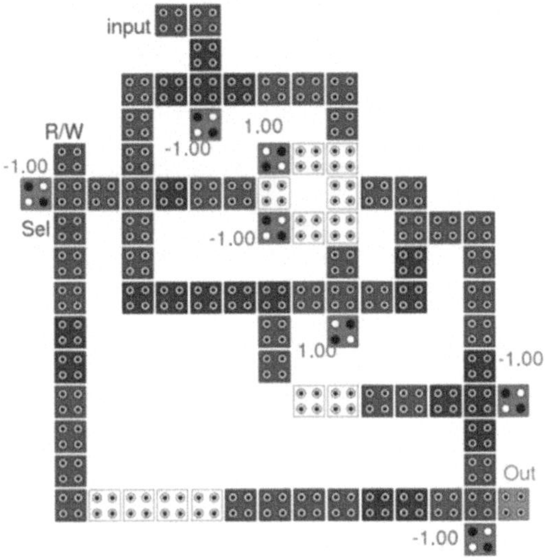

Fig. 4. QCA layout of the RAM cell [2]

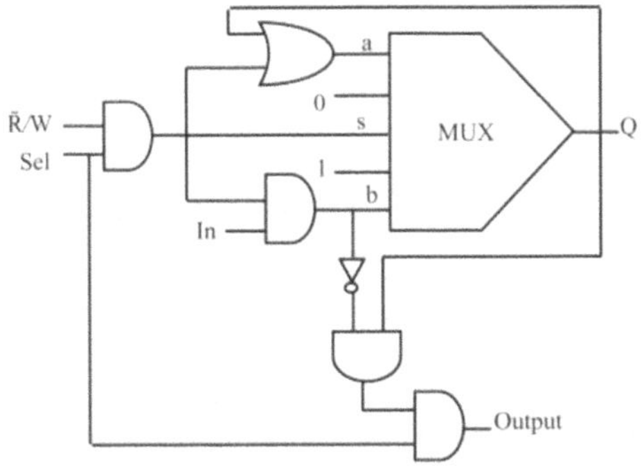

Fig. 5. RAM cell circuit diagram

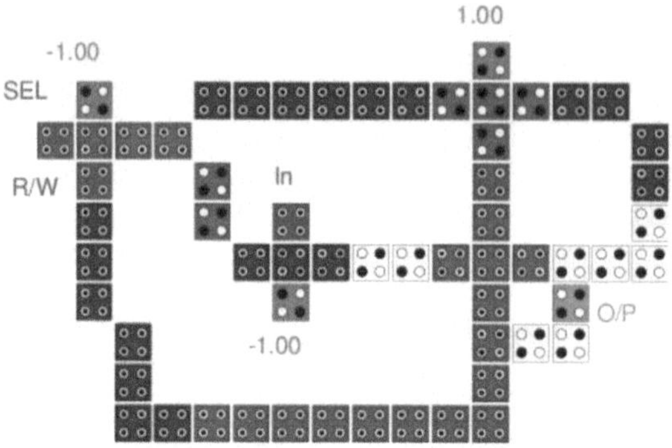

Fig. 6. Proposed RAM QCA layout

Table 1. Truth table of RAM cell

SEL	R/W	Input	Q_t	Q_{t+1}
0	X	X	X	0
1	0	X	0	0
1	0	X	1	1
1	1	0	X	0
1	1	1	X	1

Table 2. Analysis table due to external cell addition

Expected Output	0 0 0 0 0 1 1 0
Affecting cells	
Cell number 2	1 1 1 1 1 0 0 0
Cell number 11	0 0 0 0 0 0 0 0

3 Proposed RAM Cell

The proposed RAM cell design operates on QCA technology, utilizing the unique properties of QCA devices to achieve high-speed and energy-efficient memory storage. The RAM cell consists of interconnected QCA cells arranged in a specific topology. At its core, the RAM cell incorporates a modified D latch structure [2] implemented with a 2x1 multiplexer, allowing for efficient read and write operations.

The design aims to achieve high-speed operation and minimal power consumption while maintaining reliability and fault tolerance. The RAM cell's operation begins with the input signals, including the R/W signal and the data input. During a read operation (R/W = 0), the R/W signal directs the flow of signals within the cell to prioritize data retrieval. The stored data is circulated within the QCA cell structure, guided by clock signals and majority logic gates, eventually reaching the output stage. The output signal reflects the stored data, allowing external systems to read the information without altering it.

Conversely, during a write operation (R/W = 1), the R/W signal initiates the flow of new data into the RAM cell. The incoming data is processed through the input stage and directed to the appropriate QCA cells for storage. Depending on the input data (0 or 1), the polarization of the QCA cells is adjusted accordingly, either clearing the contents or up-dating the stored information. This process ensures that the RAM cell accurately reflects the input data.

When the R/W input is set to 0, indicating a read operation, the RAM cell maintains its current state. The stored data, represented as Q_t, remains unchanged, and the output Q_{t+1} provides access to the stored information. This read operation involves the circulation of signals within the QCA cell structure, enabling rapid access to stored data without altering its contents.

When the R/W input is set to 1, the RAM cell updates its state based on the input data. If the input data is logic 0, the RAM cell clears its contents, setting the output Q_{t+1} to 0. This process involves manipulating the polarization of QCA cells to store or erase information selectively.

Overall, the RAM cell design leverages the integral capabilities of QCA technology, to achieve efficient read and write operations while minimizing power consumption and ensuring reliability in digital systems.

4 Simulation Results

The simulation results of the RAM cell utilizing the D-latch is shown in Fig. 7. These results validate the proper functioning of the RAM cell, aligning with the specified values in Table 1. During the write mode, denoted by the R/W signal set to 1, new input data is transmitted to the output after approximately 1.75 clock cycles.

Conversely, in the read mode with the R/W = 0, the previous output data is latched within the loop, rendering the output is independent of input. Additionally, if Select = 0, the RAM cell is effectively deactivated. This detailed analysis confirms the reliability and operational modes of the RAM cell design, crucial for its integration into digital systems.

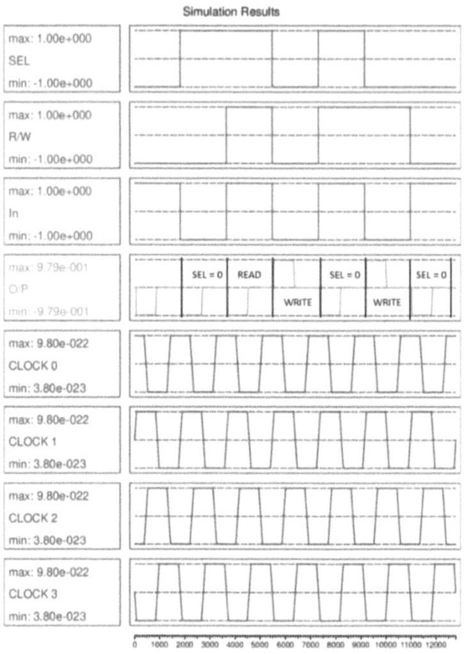

Fig. 7. Simulation waveform of proposed RAM cell.

Fig. 8. Layout of Power dissipation

4.1 Power Analysis

Power consumption is a critical factor in digital circuit design, especially in energy-constrained applications. The RAM cell design is analyzed for power efficiency using QCA Pro Designer, considering factors such as switching activity, signal propagation delay, and energy dissipation in QCA devices. The results of the power analysis provide in-sights into optimizing the RAM cell design for low-power operation.

Table 3. Analysis table due to omission of cell

Expected Output	Cell number													
	1	2	3	4	5	6	7	9	10	11	12	13	14	20
0	1	1	1	1	1	1	1	0	0	0	0	0	0	1
0	0	1	1	1	1	1	1	0	0	0	0	0	0	1
0	0	0	1	1	1	1	1	0	0	0	0	0	0	0
0	1	1	1	1	1	1	1	0	0	0	0	0	0	1
0	1	1	1	1	1	1	1	0	0	0	0	0	0	0
1	1	1	1	1	1	1	1	0	0	0	0	0	1	0
1	1	1	1	1	1	1	1	0	0	0	0	0	0	0
0	1	1	1	1	1	1	1	0	0	0	0	0	0	1

Fig. 9. Layout of Polarization

Table 4. Analysis table due to Displacement of cell

Direction	Cell number																	
	2	3	4	5	6	7	9	10	12	13	14	50	51	54	56	57	58	59	60
Horizontal (+)		12									13	15	10	3	8		3	48	5
Horizontal (−)			10									15	10	5	7	44	5		5
Vertical (+)	15	9	6	5	6	8	6	4	4	6	6			2	24	9		34	
Vertical (−)	15	8	8	5	6	8	6	3	3	6	7.5					9			

4.2 Fault Tolerance

Fault tolerance is a critical aspect of circuit design, ensuring the proper functioning of the circuit. In this section, we conduct a comprehensive analysis of fault tolerance in our QCA RAM cell design, focusing on three key types of faults: addition of cell, missing cell, and displacement of cell.

4.2.1 Addition of Cell

Addition of an extra cell within the QCA RAM cell layout can occur due to fabrication errors or defects. The additional cell may introduce parasitic effects, altering the circuit's functionality and performance. We evaluate the percentage error induced by the presence of the additional cell through simulation studies. This involves measuring the deviation in circuit behavior from the ideal operation.

The analysis reveals the extent of performance degradation caused by the extra cell and provides insights into mitigating strategies to minimize its adverse effects.

4.2.2 Omission of Cell

Omission of a cell in the QCA RAM cell layout can result from fabrication defects or environ-mental factors. The absence of a cell disrupts the circuit's structure, potentially leading to non-functionality or erratic behavior. We investigate the consequences of a missing cell on the RAM cell's operation through rigorous simulations. This involves examining changes in circuit behavior and performance metrics.

The analysis highlights the criticality of each cell in the QCA RAM cell design and quantifies the loss in circuit functionality due to the absence of a cell. Recommendations for enhancing fault tolerance in the presence of missing cells are discussed.

4.2.3 Displacement of Cell

Cell displacement refers to the unintended shifting of cells from their designated positions in the QCA layout. Misplacement of cells can disrupt the circuit's topology, affecting its functionality and reliability. We simulate various scenarios of cell displacement and observe the resulting changes in circuit behavior. This includes assessing the circuit's resilience to cell misplacement and identifying critical regions prone to displacement.

The analysis elucidates the circuit's vulnerability to cell displacement and provides insights into design modifications or layout optimization techniques to enhance fault tolerance.

The values listed above serve as threshold values, indicating the point at which the output remains constant regard-less of any displacement of the cells. Empty cells denote scenarios where the output remains constant despite potential cell displacement.

4.3 Polarization Check

Polarization effects in QCA devices can lead to errors in signal propagation and logic operations, affecting the overall functionality of digital circuits. A polarization check is performed to assess the polarization state of QCA cells in the RAM design and to identify potential issues such as polarization-induced errors and signal distortion. (Fig. 3.7 represents the polarization check w.r.t temperature).

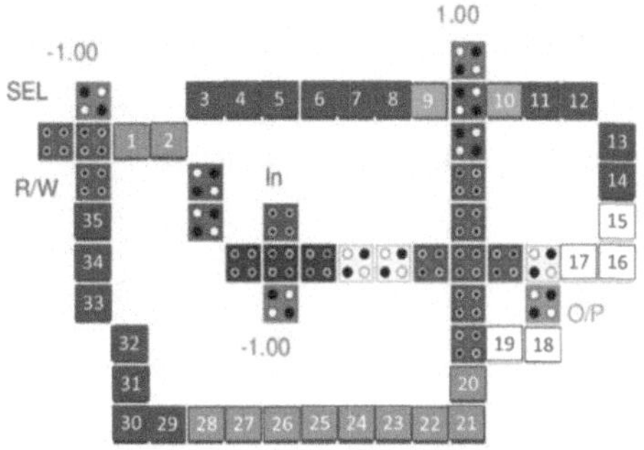

Fig. 10. Addition of external cell to each of the numbered cells layout

Table 5. Complete fault tolerance analysis of the proposed cell

Fault Tolerance analysis	Percentage of Error (%)
Addition of Cell	3.93
Omission of Cell	15.36
Displacement of Cell	31.67

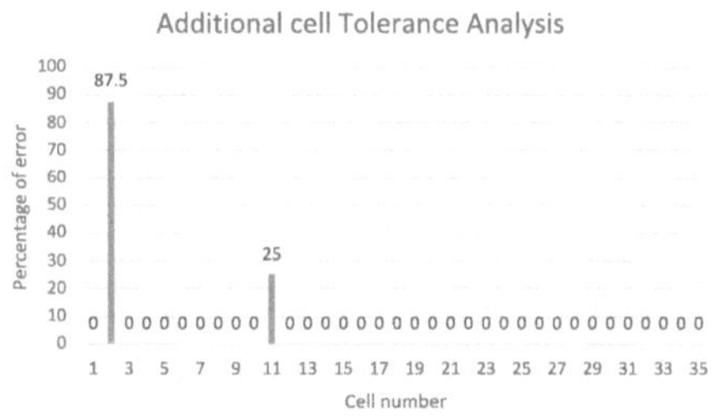

Fig. 11. Graphical Representation of the percentage error

5 Analysis

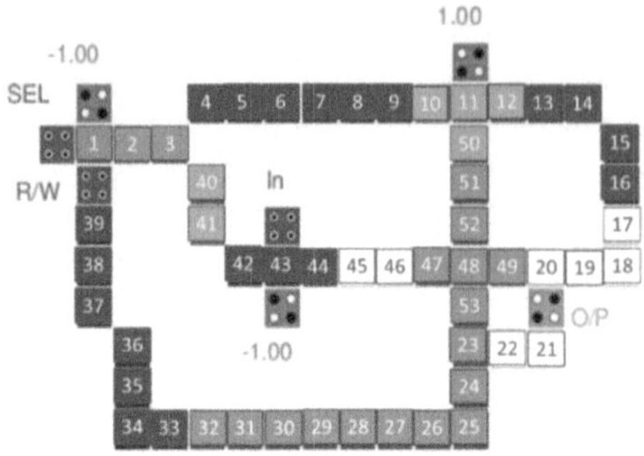

Fig. 12. Omission of numbered cells layout

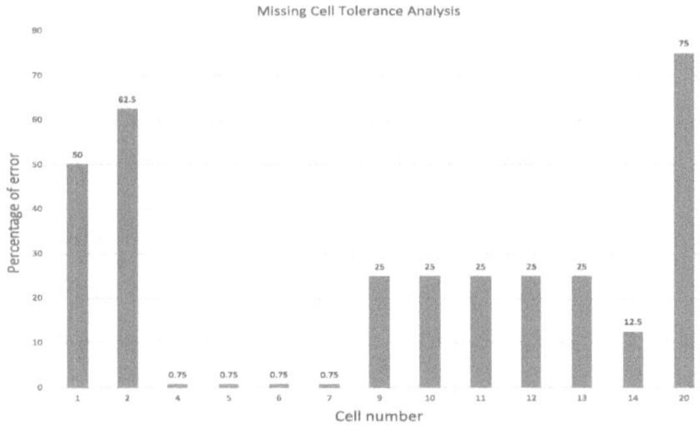

Fig. 13. Graphical Representation of the percentage error

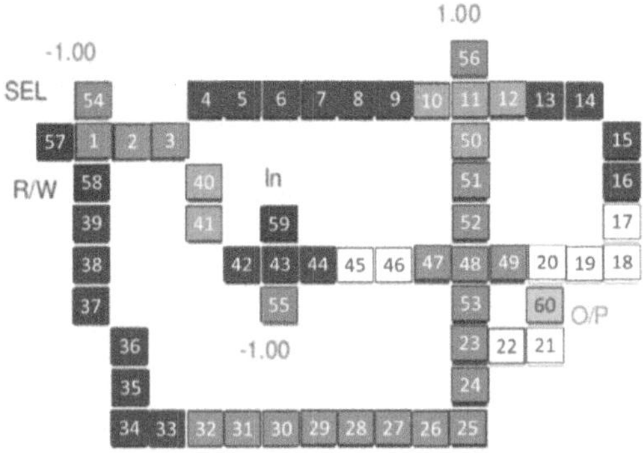

Fig. 14. Displacement of each cell layout

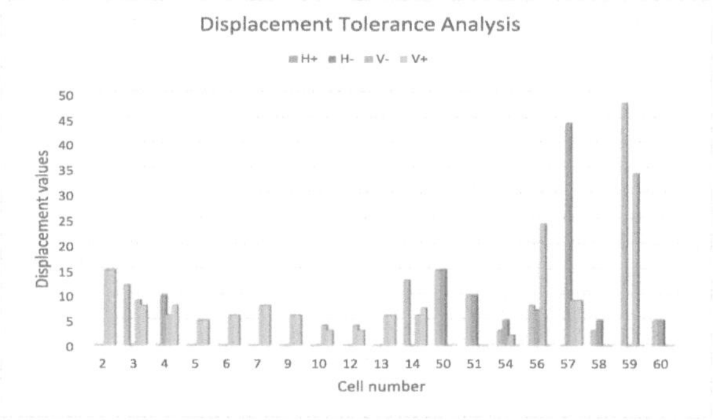

Fig. 15. Graphical Representation of the displacement values

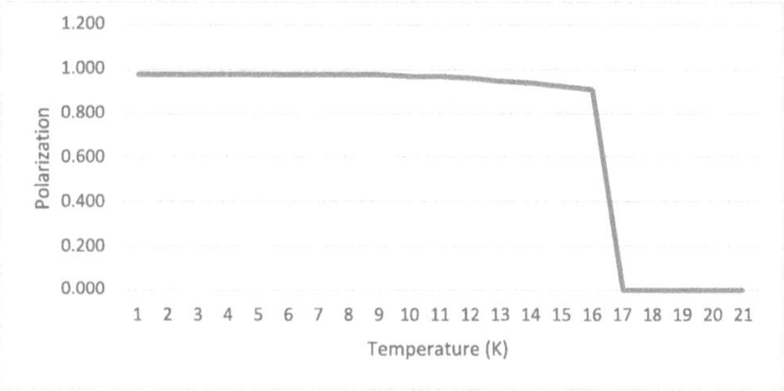

Fig. 16. Polarization vs Temperature analysis

Table 6. Complete properties of the proposed cell

Factors	Proposed memory cell
Cell count	60
Area (μm2)	0.07
Type	Coplanar
Total Energy dissipation (eV)	2.97×10^{-2}
Avg. Energy dissipation per cycle (eV)	2.70×10^{-3}

6 Conclusions

This study proposes an optimized single-layer RAM cell using an existing but modified D latch circuit, comprising only 60 cells. The suggested designs are more efficient than the best of the prior designs, according to the results of the comparisons between the proposed and previous multiplexer designs. For instance, when compared to the best RAM cell among the prior designs, the suggested RAM cell shown improvements of 17.80% and 17.78%, respectively, in terms of the cell count and cost function. The

Table 7. Comparative results of RAM cells

Memory Cell	Cell count	Area (μm^2)	Cost function	Latency	Type
Presented Memory Cell [1]	73	0.07	8.94	1.75	Multilayer
Presented Memory Cell [2]	87	0.12	15.66	1.50	Coplanar
Presented Memory Cell [3]	178	0.38	202.92	3	Multilayer
Presented Memory Cell [4]	55	0.06	8.25	2.5	Multilayer
Presented Memory Cell [5]	63	0.07	8.82	2	Coplanar
Presented Memory Cell [6]	100	0.11	35.75	3.25	Multilayer
Presented Memory Cell [7]	109	0.13	24.79	1.75	Coplanar
Presented Memory Cell [8]	158	0.16	50.56	2	Multilayer
Proposed Memory Cell	60	0.07	7.35	1.75	Coplanar

suggested design is more effective than the earlier ones in terms of cell counts, occupied area, latency, cost function, and complexity.

References

1. Singh, R., Sharma, D.K.: Design of efficient multilayer RAM cell in QCA framework. Circuit World. **47**(1), 31–41 (2020). https://doi.org/10.1108/cw-10-2019-0138
2. Mubarakali, A., Ramakrishnan, J., Mavaluru, D., Elsir, A., Elsier, O., Wakil, K.: A new efficient design for random access memory based on quantum dot cellular automata nanotechnology. Nano Commun. Netw. **21**, 100252 (2019). https://doi.org/10.1016/j.nancom.2019.100252
3. Liolis, O., Mardiris, V.A., Sirakoulis, G.C., Karafyllidis, I.G.: Quantum-dot cellular automata RAM design using crossbar architecture. J. Circuits Syst. Comput. (2018b). https://doi.org/10.1145/3232195.3232216
4. Fam, S.R., Navimipour, N.J.: Design of a loop-based random access memory based on the nanoscale quantum dot cellular automata. Photon Netw. Commun. **37**(1), 120–130 (2018). https://doi.org/10.1007/s11107-018-0801-9
5. Dehkordi, M.A., Shamsabadi, A.S., Ghahfarokhi, B.S., Vafaei, A.: Novel RAM cell designs based on inherent capabilities of quantum-dot cellular automata. Microelectron. J. **42**(5), 701–708 (2011). https://doi.org/10.1016/j.mejo.2011.02.006
6. Shamsabadi, A.S., Ghahfarokhi, B.S., Zamanifar, K., Movahedinia, N.: Applying inherent capabilities of quantum-dot cellular automata to design: D flip-flop case study. J. Syst. Archit. **55**(3), 180–187 (2008). https://doi.org/10.1016/j.sysarc.2008.11.001
7. Hashemi, S., Navi, K.: New robust QCA D flip flop and memory structures. Microelectron. J. **43**(12), 929–940 (2012). https://doi.org/10.1016/j.mejo.2012.10.007
8. Gadim, M.R., Navimipour, N.J.: Quantum-dot cellular automata in designing the arithmetic and logic unit: Systematic literature review, classification and current trends. J. Circuits Syst. Comput. **27**(10), 1830005 (2018d). https://doi.org/10.1142/s0218126618300052
9. Hanckel, B., Petticrew, M., Thomas, J., Green, J.: The use of qualitative comparative analysis (QCA) to address causality in complex systems: a systematic review of research on public health interventions. BMC Public Health. **21**(1) (2021). https://doi.org/10.1186/s12889-021-10926-2

10. Sridharan, K., Pudi, V.: Design of Arithmetic Circuits in Quantum Dot Cellular Automata Nanotechnology. (2015b). https://doi.org/10.1007/978-3-319-16688-9
11. Asfestani, M.N., Heikalabad, S.R.: A unique structure for the multiplexer in quantum-dot cellular automata to create a revolution in design of nanostructures. Phys. B Condens. Matter. **512**, 91–99 (2017). https://doi.org/10.1016/j.physb.2017.02.028
12. Thiem, A.: Qualitative comparative analysis (QCA), pp. 607–628. Edward Elgar Publishing eBooks (2022). https://doi.org/10.4337/9781839101014.00044
13. Abedinia, O., Amjady, N., Ghadimi, N.: Solar energy forecasting based on hybrid neural network and improved metaheuristic algorithm. Comput. Intell. **34**(1), 241–260 (2017). https://doi.org/10.1111/coin.12145
14. Aghajani, G., Ghadimi, N.: Multi-objective energy management in a micro-grid. Energy Rep. **4**, 218–225 (2018). https://doi.org/10.1016/j.egyr.2017.10.002
15. Ahmadian, I., Abedinia, O., Ghadimi, N.: Fuzzy stochastic long-term model with consideration of uncertainties for deployment of distributed energy resources using interactive honey bee mating optimization. Front. Energy. **8**(4), 412–425 (2014). https://doi.org/10.1007/s11708-014-0315-9
16. Nouri, A., Khodaei, H., Darvishan, A., Sharifian, S., Ghadimi, N.: RETRACTED: Optimal performance of fuel cell-CHP-battery based micro-grid under real-time energy management: An epsilon constraint method and fuzzy satisfying approach. Energy. **159**, 121–133 (2018). https://doi.org/10.1016/j.energy.2018.06.141
17. Moharrami, E., Navimipour, N.J.: Designing nanoscale counter using reversible gate based on quantum-dot cellular automata. Int. J. Theor. Phys. **57**(4), 1060–1081 (2017). https://doi.org/10.1007/s10773-017-3638-6
18. Angizi, S., Sarmadi, S., Sayedsalehi, S., Navi, K.: Design and evaluation of new majority gate-based RAM cell in quantum-dot cellular automata. Microelectron. J. **46**(1), 43–51 (2014). https://doi.org/10.1016/j.mejo.2014.10.003
19. Gadim, M.R., Navimipour, N.J.: Correction to: A new three-level fault tolerance arithmetic and logic unit based on quantum dot cellular automata. Microsyst. Technol. **24**(2), 1307 (2018e). https://doi.org/10.1007/s00542-018-3716-6
20. Vaish, U., Abraham, J.S., Kumar, V.R.: A fault-tolerant QCA-based convolution encoder for robust data transmission. Nano Communication Networks: 100507. (2024). https://doi.org/10.1016/j.nancom.2024.100507

Voice Controlled Wheelchair for Blind with Hearing Assistant

P. A. Anjana(✉), Deepraj Dinanath Govenkar, Jayton Denilson Rodrigues, Ish Sinai, Jayalaxmi Devate, and Satish Gangavati

Department of Electronics and Telecommunication Engineering, Padre Conceicao College of Engineering, Verna 403722, Goa, India
anjanapa4602@gmail.com, deeprajgovenkar2@gmail.com, jaytonrodrigues2002@gmail.com, ishrameshsinai2003@gmail.com, jayaa_2002@yahoo.com, satishgangavati9@gmail.com

Abstract. This paper helps enhance the quality of life of the visually impaired through better mobility. These comprise of sound processing, LiDAR, ultrasonic real-time obstacle detection, and webcam-based environment capture. Voice control lets users use voice commands to rotate the wheelchair while audio guidance helps the operator. LiDAR and ultrasonic sensors can either avoid an object or let the user know that the object is present through the speakers. It captures facial expressions, communicates them to users through the sound system, and alerts them on interferences. The integration of voice control, intelligent sensing, and auditory feedback thus extends independence, safety, and quality of life, for users who are with visual impairments.

Keywords: Voice controlled wheelchair · object detection · hearing assistance

1 Introduction

Voice Controlled Wheelchair: Tan Kian Hou [1] used an ATmega328P microcontroller and developed a voice recognition system that works on a laptop. Both devices are connected via two Bluetooth modules. However, the voice recognition system is more limited, and the microcontroller lacks multitasking capabilities, causing delays when receiving signals from the laptop. Mohammad Ilyas Malik [2] used an Arduino UNO to control the wheelchair using voice commands from a phone. The communication between the phone and the Arduino UNO is easy via Bluetooth. While building the small prototype, he encountered a delay in the transmission of signals to the Arduino uno. Dr. K. Sravan Kumar [3], the microcontroller used is an Arduino UNO and for voice recognition, a phone was used, and these were connected through the Bluetooth module. However, some challenges were observed in the course of implementing the project; the following are the disadvantage: high cost of production since individual parts had to be produced, the Arduino UNO had restricted functionality in receiving signals from the phone thus slowing down the operation. R. Jamuna [4] the author used the PIN Microcontroller and HM-2007 Voice IC in terms of voice recognition in their work.

D. Gupta et al. (Eds.): PCEMS 2024, CCIS 2491, pp. 62–73, 2025.
https://doi.org/10.1007/978-3-031-90577-3_6

However, they suffer from too much delay as they use low-power microcontrollers and the HM-2007 voice IC. This increases its error margin to 56% on average and brings about a considered delay in translating speech into words. K. Sangeetha [5] used the Node MCU as a microcontroller and phone for voice recognition and a relay module for motor control. However, the Node MCU, being a low-power microcontroller, lacks multitasking capabilities, resulting in a significant delay in the signal received from the phone. R. Bhambare [6], which employs voice command along with hand gestures to operate the wheelchair with a microcontroller, an accelerometer, and RF modules. On the brighter side, it makes communication easier; however, it has the following problems: dependence on certain speakers, limited word recognition, reliability issues in noisy environments, more power consumption, and environmental constraints. A. V. Mohan [7] invented the smart wheelchair, which employs an AVR microcontroller for voice commands, YOLO for real-time object identification, and incorporated power train control and sensor systems for mobility. There are also some downside that may be associated with articulating arm lights: the initial cost may be relatively high, their design is quite complex, these lamps can consume a lot of energy, and they are very sensitive to the environment, for instance, inadequate light. R. Neema [8], voice processing via Arduino and Bluetooth, and YOLO-based real-time object recognition using the COCO dataset have been combined into a web and Android application for wheelchairs with intelligence. However, there are several disadvantages, including limited battery life, high cost, complexity, and exposure to environmental factors that affect reliability.

Object Detection: Abdul Muhsin M. [9] presents a YOLO-based real-time object search algorithm using OpenCV and DNN on a Raspberry Pi to assist the visually impaired. Advantages include speed and accuracy, but disadvantages include complexity, lower potential accuracy of particles or overlaps, and higher hardware costs. M. Thulasi [10], YOLO method for real-time search divides an image into a grid and predicts bounding boxes and probabilities simultaneously. Advantages include speed and efficiency. However, it can struggle with clusters of small objects nearby and can have localization errors. Marek Vajgl [11] and M. Y. Babu [12] the methodology involves YOLO (You Look Once) object search using a combination of the DNN module and OpenCV library for real-time search and remote computation. Disadvantages include greater computational complexity and potentially reduced accuracy under environmental conditions. Thariq Hassan et al. [13] proposed an object detection and recognition system called YOLO that can be used to aid blind people. However, as for its drawbacks, YOLO has some problems. For example, it is difficult for YOLO to detect small objects; it has some limitations in complex environments; it consumes a large amount of computing resources; and it is sensitive to lighting conditions. Pooja Maid [14] this project incorporates IoT and ML with TensorFlow and RCNN on Raspberry Pi 3 to identify objects within images and videos to improve the experiences of the blind. The system also has flexibility issues, cost issues, and environmental issues, yet it has the capability to offer real-time visualization. Pradnya Kasture [15] performed real-time object detection for identifying various objects by using the YOLO algorithm, the COCO 2014 dataset, and LabelImg for labeling. Although the system yields accurate and quick results, the need to access large sets of data, high computational demand, and environmental factors present considerable issues.

The proposed project pushes assistive technology forward specifically for the visually impaired through the use of mechanically complex, motorized wheelchairs along with deep learning recognition engines. This system avoids obstacles in real time and reacts promptly to voice queries, offering continuous verbal prompts. Testimonies taken from the users show that they like it because it is easy to use and helps transform the way of navigation for visually impaired people. The concept of using voice commands and assistive hearing can be used to enhance the lives of good people.

2 Methodology

Figure 1 shows the general system layout of this project and represents an overall framework of how the different parts of the Voice-Controlled Wheelchair for Blind with Hearing Assistant work together to allow for voice-controlled mobility and obstacle avoidance.

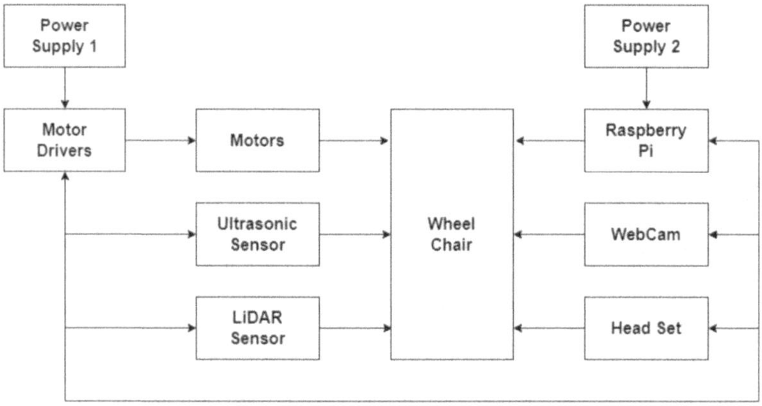

Fig. 1. Block diagram of the connections

Headphone (Input and Output): In the case of the headphones, the participant was both the record (input) and playback (output) system. One of the inputs is the microphone incorporated in the headphones, which record the user's vocal directions, including forward, left, right, or stop. All these voice inputs are communicated to the Raspberry Pi for processing.

The output produced by the headphones is sound that is fed back to the user through a built-in speaker. It also reacts to the voice command, which may include 'Moving forward' and sends out a message alert such as 'There is an obstacle ahead' whenever the sensors on the wheelchair detect an object on the way.

Raspberry Pi (Central Processing Unit): By processing speech data received through an integrated microphone in a headphone, the Raspberry Pi functions as the core of the system, while speech recognition is done through the SpeechRecognition Python library. Its output words are in text form, which are then relayed to the motor driver after the Raspberry Pi processes the commands to control the movement of the wheelchair.

It also controls data from the LiDAR and ultrasonic sensors, which detect obstacles in front of the wheelchair and halt the wheelchair or change direction to avoid those obstacles.

Motor Driver and DC Motors: On controlling the wheels of the wheelchair, the motor driver receives signals from the Raspberry Pi and drives the DC motors. This enables the wheelchair to perform maneuverability by voice commands (for instance, forward, left, right, and stop) and speed by voice commands (slow, faster, slower, with speeds adjustable by ± 2 km/h).

Camera: When an object is within the range of 80 cm–200 cm, the camera module is triggered to capture video of the environment in real-time. This serves to help identify objects that need to be referred for further elaborative analysis using Deep learning systems, which act in parallel to the sensing systems.

Thus conclusion, Fig. 1 illustrates the ways whereby the headphone serves as an input: the microphone and the output: the headphone. As for the real-time control of the wheelchair, the Raspberry Pi acts as the control center for the voice prompt and the sensor data; the motor driver and sensors make the wheelchair promptly react to obstacles, and the camera plays a part in the whole decision.

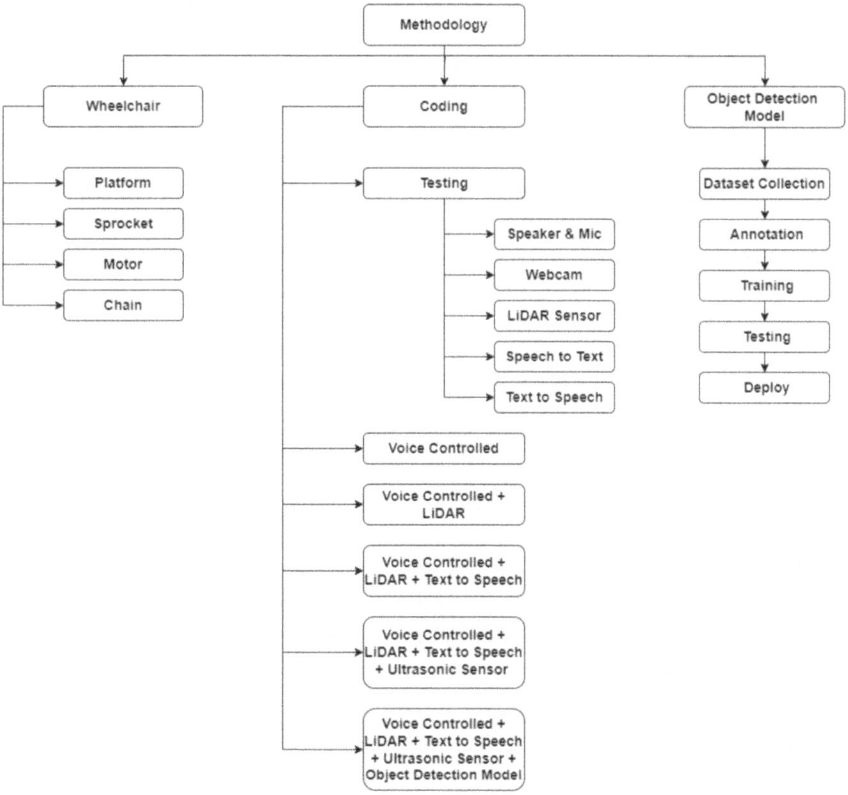

Fig. 2. Methodology

We have divided our work into three distinct phases, each focusing on important aspects of the development process:
1) Wheelchair Assembly
The process began with careful planning and planning though specified the details and dimensions of the wheelchair platform. Choosing the right mechanism based on torque and speed requirements was critical, followed by careful alignment of the wheelchair frames to ensure proper alignment and mounting to transfer energy so well. The addition of a sprocket and chain drive system simplified wheelchair wheel maintenance. Proper testing of the motorized wheelchair improved its efficiency and usability. Fine-tuning of motor settings and drive system parameters was performed to optimize performance. Practical tests were conducted to determine stiffness, load-carrying capacity, and overall performance.
2) Object Detection Model
The data collection process involved the collection of photographs containing various objects that could interfere with the wheelchair. Manual registration dataset of objects of interest, such as pedestrians, obstacles, and furniture, for training. For model training and optimization, a suitable deep learning algorithm was selected, and a convolutional neural network (CNN) architecture, such as YOLOv8, was used for real-time feature recognition. Training the model with an annotated dataset enabled it to identify and classify features within the wheelchair. A well-trained model was developed to run on Raspberry Pi, facilitating onboard inference. The integration of the pattern recognition system into the wheelchair system enabled real-time detection and response to obstacles in the environment.
3) Coding/Interfacing
Feature development includes text-to-speech (TTS) and speech recognition services using libraries such as Google Text-to-Speech API and Speech Recognition. Control algorithms were developed to interpret the commands of the operator and come to the wheelchair based on visual cues. The system integration process required the integration of the object recognition model with the control software to enable autonomous navigation and obstacle avoidance. Extensive testing and debugging ensured the reliability and robustness of the entire system. In addition, safety features including fault-handling devices were implemented to prevent accidents and ensure the safety of users during operations.

We choose the speed of 5 km/h based on user safety criteria and standard walking speeds. This normally varies between 3 and 7 km/h. This speed ensures that the wheelchair can move efficiently in both indoor and outdoor environments. Without compromising user safety This allows sufficient response time when LiDAR and ultrasonic sensors detect obstacles. While maintaining a stable and controlled movement, the speed can also be adjusted up to ±2 km/h using the engine driver. This provides the flexibility to adapt to the needs of different users and environments.

1. Convert speed to meters per second:

$$\begin{aligned}\text{Speed } &= 5 \text{ km/h} \times 1000 \text{ m3600 s} \\ \text{Speed } &\approx 1.39 \text{ m/s}\end{aligned}$$

2. Assuming a time interval for covering the distance of 1 meter:

$$\begin{aligned}\text{Acceleration} &\approx \frac{1.39\ \text{m/s}}{1\ \text{s}}\\ \text{Acceleration} &\approx 1.39\ \text{m/s}^2\end{aligned}$$

3. Use Force = Mass × Acceleration to find the force:

$$\begin{aligned}\text{Force} &= 70\ \text{kg} \times 1.39\ \text{m/s}^2\\ \text{Force} &\approx 97.3\ \text{N}\end{aligned}$$

4. Finally, use the torque formula:

$$\text{Torque}(\tau) = \frac{\text{Force} \times \text{Radius}}{\text{Angular velocity}}$$

To find ω, we can use

$$\begin{aligned}\omega &= \frac{\text{Speed}}{\text{Radius}}\\ \omega &\approx \frac{1.39\frac{\text{m}}{\text{s}}}{0.3\ \text{m}}\\ \omega &\approx 4.63\,\text{rad/s}\end{aligned}$$

Now, substitute this into the torque formula:

$$\begin{aligned}\text{Torque} &\approx \frac{97.3\ \text{N} \times 0.3\ \text{m}}{4.63\,\text{rad/s}}\\ \text{Torque} &\approx 6.3\,\text{Nm}\end{aligned}$$

3 Working

The Voice-Controlled Wheelchair for Blind with Hearing Assistant operates using two primary interfaces: voice recognition, voice command control, and object recognition, all controlled by a Raspberry Pi.

Voice Command Interface: The voice command control system is activated when the user says forward, left, right, or stop while wearing an integrated microphone headset. These spoken words are translated in real-time by the SpeechRecognition Python library into texts, which instantly reach the control system of the wheelchair to perform the corresponding motion. Once a command is received and processed, the system discharges auditory feedback to the user through the headset, for example, moving forward or moving left, thus enabling the user to record the wheelchair's actions. This system also incorporates the SpeechRecognition library to transform speech to text and combines it with the Google Text-to-Speech API for better sound yield in noisy conditions. This ensures that in case of distractions or even loud environments, the recognition of commands is successful. It has been realized that the system has undergone various

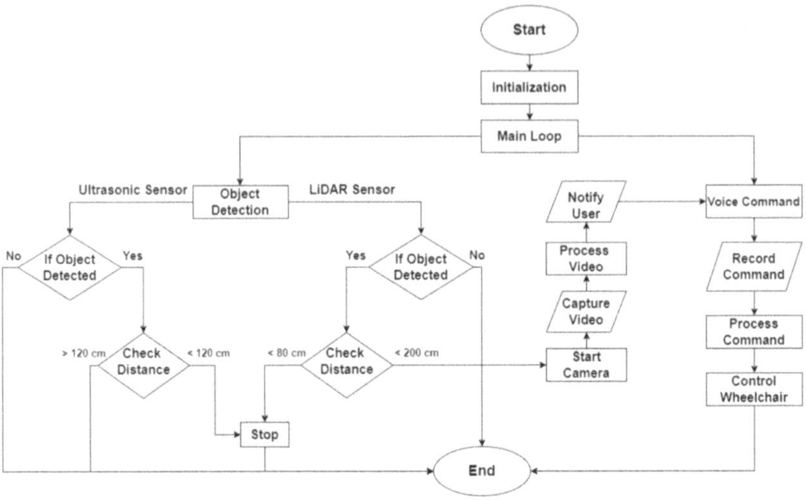

Fig. 3. Control algorithm

tests in various environments with an actual recognition accuracy of 90% at the optimal test environments and about 85% at suboptimal test environments. Also, it is possible to design the set of commands that this software uses for controlling the wheelchair and change them accordingly to the user's needs and accents making the experience of working with the wheelchair control system more comfortable and effective.

Object Detection Interface: Object detection is achieved using ultrasonic sensors and LiDAR sensors, and the control also uses ultrasonic sensors to detect surrounding objects. The distance of the ultrasonic sensor is 120 cm. If the object is within this distance, the wheelchair will stop immediately. The additional LiDAR sensor has a range of up to 200 cm and allows for the detection of obstacles at a farther distance than those provided by the rest of the sensors. If an object is found in the range of a distance of seventy-eight cm and below, the wheelchair halts and an audible message, such as "There is an object ahead," is given. If the object is detected between 80 cm and 200 cm, the camera turns on, and the object is recognized by using real-time video to draw a complete picture of the object.

Control Algorithm: In addition, the two parts, namely the voice command and the object detection, are perfectly synchronized within the control algorithm of the system. For example, when the user wants to make the wheelchair move forward, it does that, but at the same time, it constantly searches for any obstacles using its sensors. When an object is sensed within that range, the chair stops and announces the presence of an object, thereby giving the subject a chance to serve in the right direction. Once the obstacle is addressed, the system returns back to the voice command loop to wait for the next command of the user.

To summarize, the wheelchair can be controlled for an independent journey when responding to spoken commands and including obstacle recognition for constant and secure navigation. The combination of voice control and the sensors that allow the detection of obstacles ensure the safety of the journey and do not require the hands to control the mobility scooter, which will be especially useful for a person with low vision.

4 Implementation

Fig. 4. Final wheelchair design

Fig. 5. Circuit connections

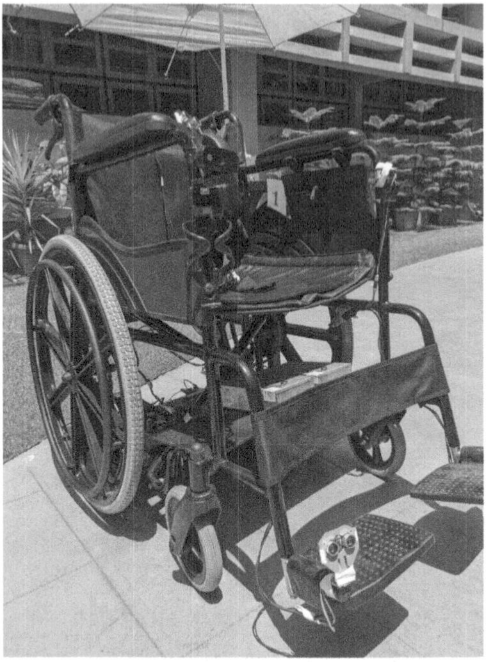

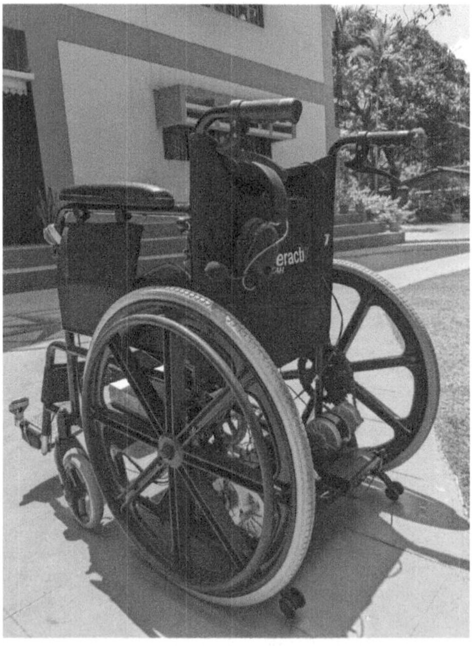

5 Comparison

5.1 Voice Control vs. Eyeball Tracking

It is an effectively developed technology for users that control a wheelchair by eyes as a unique kind of mobility for patients with severe motor diseases and those who lost their voice. Nevertheless, this system is accurate and needs calibration; it does not work well in practice owing to lighting or when the person has to rotate his head rapidly, and it may execute an undesirable action. As opposed to that, our voice-controlled system means that a user is able to control a wheelcair, move ahead, turn left, or stop, for instance, simply by speaking to their wheelchair. This system is easier and more comfortable for users who have no motor disabilities but visual impairments. Further, our system also includes a hearing aid as we give the user constant verbal instructions and feedback, which an eyeball tracking system does not provide.

5.2 Voice Control vs. Joystick Control

Powered mobility devices can be operated with a joystick and are among the most commonly used. They offer greater precision as well as ease of handling for people with functional limb movements but limited mobility. Nevertheless, for the visually impaired, joystick-based systems may not be ideal because they rely on the sight of the subject to avoid the barriers. Our voice-controlled system is developed exclusively for visually impaired users. It is non-tactile and includes real-time obstacle sensing through LiDAR and ultrasonic sensors with verbal directions to the user. Although joystick systems are perfect for people with perfect vision and fine motor skills, our approach is designed for people requiring a natural, speech-controlled interface.

5.3 Voice Control vs. Gesture-Based Control

Other developing technologies in wheelchair control include gestures of hand or body movements through sensors, such as touchless technologies of controls that include buttonless quadric controls. These operation styles are useful for visually impaired users and also as a variation to voice control since they are hand-operated. Nonetheless, gesture systems are very sensitive to the motor skills of the user and have problems in noisy or cluttered environments since gestures might be interpreted wrong and might be safety hazardous.

This is in contrast to other systems based on gestures, to which our voice-controlled wheelchair does not fall prey and can successfully function in different conditions of noise or visual interference. That means the headset microphone gets the user's voice command properly, and noise filtering makes sure that the command is understood well. This also makes it easier to use for people with less control over their motor function or muscles or for those just wanting to use their voice instead of their hands.

5.4 Summary

Compared with other typical wheelchair control systems, including eyeball tracking, a joystick, and gesture control, our Voice-Controlled Wheelchair for Blind with Hearing Assistant is more creative and friendly to use. Together with actual recognition of obstacles, the ability to activate a voice, and other functions, and in addition, the constant auditory feedback, it makes it a possibly useful mechanism of control both with voice and independently of vision. This combination makes it even more suitable for the group of visually impaired end-users who require constant feedback and secure and independent mobility in different terrains.

6 Limitations

Although this project have achieved great progress. But two main limitations need to be addressed:

6.1 Dependence on Main Power Source

A major disadvantage of this project is that the wheelchair can only be powered if it is connected directly to the main source of power. Lack of capital and battery mass means that, for the present, the system does not employ a portable battery power supply. This greatly restricts the freedom and utility of the wheel; the chair cannot run by itself using the battery power; it has to be charged. In general, the lack of a battery decreases its function ability for use in some cases, such as in the distance or outdoor ones.

6.2 Speech-to-Text Processing Delay

There is a definite latency between the time a user speaks a command and the wheelchair responds to the verbal command with movement. This delay comes from the speech-to-text conversion, processing time to determine what word has been said before activating motor control in responding to the command.

7 Conclusion

The "Voice-Controlled Wheelchair for the Blind with Hearing Assistant" project depicts how technology can be of significant importance in improving the quality of life of the physically challenged. Using voice commands and sensors, the system provides the most secure and autonomous solution for the desired navigation. The ultrasonic and LiDAR sensors are used to detect obstacles in the vicinity, while voice control helps in controlling the wheelchair. The system is powered by a Raspberry Pi, through which interaction between the user and the environment is facilitated.

8 Future Scope

New improvements in battery systems and speech recognition will enhance the capabilities of electric wheelchairs in the future. The integration of a better battery connection will reduce the need for direct wiring, which leads to a very neat design and minimal maintenance. In addition, the integration of advanced speech and text recognition will enhance the user's ability to command the wheelchair through voice without much struggle. Overall, these new features will enhance the functionality, capabilities, and usability of power wheelchairs.

References

1. Tan, H., et al.: Arduino based voice controlled wheelchair. J. Phys. Conf. Ser. **1432**, 012064 (2020)
2. Malik, I., Bashir, T., Khan, O.F.: Voice controlled wheel chair system. IJCSMC. **6**(6), 411–419 (2017)
3. Kumar, K.S., Deepika, P., Farzana, M., Kaveri, O., Preethi, K., Sailaja, P.B.S.: A realtime voice controlled wheelchair for specially abled people. TURCOMAT. **14**(02), 742–751 (2023)
4. Jamuna, R., Bharath Kumar, K., Karthikeyan, P.: Wheelchair control using voice recognition. Int. J. Adv. Res. Biol. Ecol. Sci. Technol. **1**(1) (2015)
5. Sangeetha, K., Arun, K., Goutham, K.K., Karthick, J.: Voice controlled wheelchair for physically disabled people. Int. J. Health Res. **5**(2), 254–225 (2020) ISSN: 2455-7587
6. Bhambare, R.R., Ghaiwate, A.P., Jadhav, K.B., Patil, H.K.: Smart wheelchair control using voice and gesture control. IJARIIE. **3**(2), 4150 (2017) ISSN (Online): 2395-4396
7. Mohan, A.V., Chandran, A.R., Sooraj, S.S., Syedali, S.N., Nampoothiri, V.N.V.: Voice controlled autonomous wheelchair with inbuilt GSM: A review. Int. J. Adv. Eng. Mgmt. **3**(4), 256–260 (2021., ISSN: 2395-5252). https://doi.org/10.35629/5252-0304256260
8. Neema, R., Rajkumar, R.: Voice controlled smart wheelchair. J. Manag. Inf. Decis. Sci. **24**(3), 1–9 (2021) ISSN: 1532-5806-24-S3-113
9. Muhsin, M.A., Alkhalid, F.F., Oleiwi, B.K.: Online blind assistive system using object recognition. Int. Res. J. Mod. Eng. Technol. Sci. **5**(6), 47–51 (2019)
10. Thulasi, A., et al.: The object recognition voice assistant for visually impaired people. Int. Res. J. Modern. Eng. Technol. Sci. **05**(06) (2023) e-ISSN: 2582–5208
11. Vajgl, P.H., Nejezchleba, T.: Dist-YOLO: Fast object detection with distance estimation. Appl. Sci. **12**, 1354 (2022)
12. Babu, M.Y., Jatavath, A., Reddy, G.Y.K., Kumar, P.A.: Object detection system with voice alert for blind. Int. J. Res. Appl. Sci. Eng. Technol. **11**(II), 707 (Feb. 2023) ISSN: 2321-9653
13. Thariq Hussan, M.I., Saidulu, D., Anitha, P.T., Manikandan, A., Naresh, P.: Object Detection and Recognition in Real Time Using Deep Learning for Visually Impaired People. Int. J. Electr. Electron. Res. **10**(2), 80–86 (2025) e-ISSN: 2347-470X
14. Maid, P., Thorat, O., Deshpande, S.: Object Detection for Blind Users. Int. Res. J. Eng. Technol. **7**(6), 3089 (June 2020) e-ISSN: 2395-0056, p-ISSN: 2395-0072
15. Kasture, P., Tangade, A., Pole, A., Kumkar, A., Jagtap, Y.: Real-Time Object Detection Using Yolo Algorithm for Blind People. Int. Adv. Res. Sci. Commun. Technol. **5**(2), 152 (2021., ISSN (Online): 2581-9429,). https://doi.org/10.48175/IJARSCT-1191

An IoT-Based Soil Analysis for Smart Agriculture Using Machine Learning

Shabnam Chandrakar and Manju Pandey(✉)

Department of Computer Applications, NIT, Raipur, C.G., India
{shabnam.phd2022.mca,mpandey.mca}@nitrr.ac.in

Abstract. As the Internet of Things (IoT) has advanced, life has altered. The majority of countries rely heavily on agriculture, which needs to become "smart". Recently, a lot of progress has been made, mostly in the fields of machine learning and data analysis, which help farmers make better decisions at every stage of the production process. Based on the analysis, the nutrients are organized and the best soil conditions for various regions are identified. The classification of the micronutrients present in soil is the primary objective of this study. This study thoroughly examines a range of soil micronutrients, such as phosphorus, nitrogen, sulfur, potassium, copper, boron, and others, to enhance crop recommendation systems. It looks at the several databases, algorithms, and soil analysis components used in precision farming. Therefore, this approach uses an examination of soil health and machine learning to classify micronutrients for crop suggestion.

Keywords: IoT · Smart agriculture · Machine learning · Soil Micro Nutrients

1 Introduction

Internet of Things (IoT) is an advanced technology for allows devices to be monitored and controlled from anywhere in the world. It is capable of establishing connections between living beings and machines. The Internet of Things is becoming widely used in many different sectors. These days, the adaptable nature of IoT has changed and is accessible to regular users. One of the primary sources that are vital to the functioning of an entire economy is agriculture. Agriculture is the foundation of most nations' economies because it is always necessary for it to function. As the world's population increases to 10 billion by 2050, so will the need for agricultural goods, particularly food (Zul Azlan et al., 2024). A major pillar of the Indian economy, agriculture employs two-thirds of the country's workforce directly and remains so today (Motwani A et al., 2022). Farmers require quick and accurate access to data on soil, pesticide, fertilizer, and weather conditions an order to make informed decisions about the kinds of crops to plant and to ensure a successful harvest. An appraisal of the suitability of the land is a prerequisite to crop cultivation and maximum production. Conventional data collection methods will be replaced by Internet of Things (IoT)-based sensors as the solution (Vincent DR et al., 2019). The values in the dataset are utilized to train (V. Vimal et al., 2024) and validate the model, after which the outcomes are evaluated and calculated. When

D. Gupta et al. (Eds.): PCEMS 2024, CCIS 2491, pp. 74–82, 2025.
https://doi.org/10.1007/978-3-031-90577-3_7

developing the model, this is done by classifying the dataset into training and testing data. After the soil has been examined, several machine learning algorithms are applied to classify the soil's fertility according to its chemical characteristics and micronutrient content. Crop yield and sowing are also analyzed using ML models. The categorization and prediction problems are being successfully addressed by a (MS Arefin et al., 2021) number of machine learning (ML) models and algorithms in recent times. The emergence of machine learning techniques has also played a major role in helping professionals in the agricultural area solve their issues. Regression and classification algorithms can be used to predict agrarian productivity effectively. These models are useful for estimating, mapping yield, matching supply and demand, and general crop management. Machine learning algorithms are highly beneficial for agricultural and farming applications, such as crop failure detection, water management, soil management, and livestock prediction. Machine learning techniques like support vector machines (SVM), decision trees, ANNs, linear regression, and Naïve Bayes etc. are used in the current systems. Authors have employed machine learning algorithms to identify the quality of soil used in a recent study by expanding the soil. The research focuses on using the Internet of Things (IoT), machine learning, and data analytics to transform agriculture into "smart farming." As IoT evolves, these technologies are increasingly being used in precision farming, allowing farmers to make smarter decisions throughout the agricultural process.

1.1 IoT in Agriculture

The advancement of IoT allows for the real-time collection of data related to soil health, environmental conditions, and crop status. By integrating sensors, connected devices, and cloud platforms, IoT facilitates a more automated, data-driven approach to agriculture. A select few Internet of Things technologies, like robots, drones, and remote sensors, have improved and eased people's lives. Farmers can increase their productivity by using IoT -based crop yield prediction. Smart farming systems based on the Internet of Things (IoT) are typically deployed in agricultural fields to monitor the crop field through the use of sensors like NPK, and DHT11 (temperature and humidity sensor). With this setup, farmers can keep an eye on the field conditions from anywhere. The prediction engine predicts the outcomes and provides information to the notification server. Gateways collect and route data from the crop area to the storage unit. Crop production prediction by hand is a difficult undertaking because of various factors. The idea of "smart farming" is still relatively new because of (Pugazhenthi K et al., 2023) Internet of Things (IoT) devices that can provide data on agricultural regions. The paper aims to recommend the best crops for the specific field based on an investigation of the soil's NPK value means a machine learning algorithm. A monitoring system powered by the Internet of Things (IoT) that makes use of several sensors to monitor temperature, humidity, and other variables including soil moisture. The device also features a control mechanism that allows irrigation and other settings to be changed remotely. According to the authors, their strategy can reduce water waste and increase crop output. The study focuses on leveraging the Internet of Things (IoT), machine -learning, and data analysis to transform agriculture into "smart farming (Fig. 1).

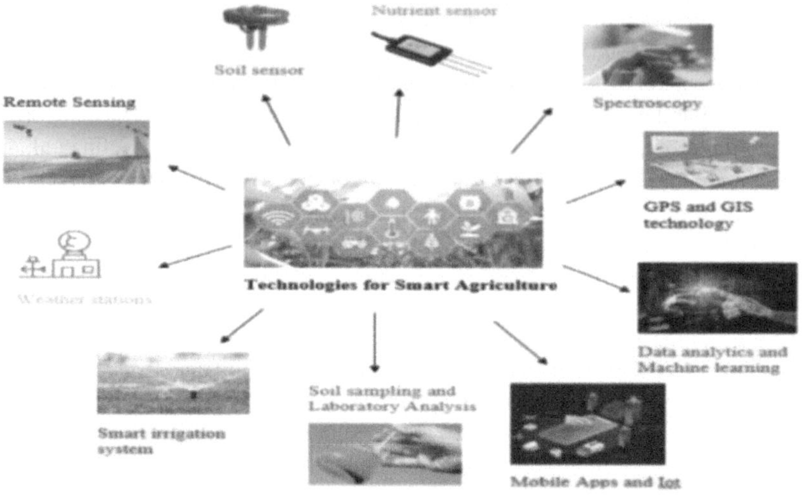

Fig. 1. Component of IoT for Smart Agriculture. (P. Abinaya et al., 2024)

1.2 Machine Learning Concepts for Agriculture

Machine learning models are applied to analyze data collected from the field, helping to predict and classify various soil properties and micronutrient levels. These insights help farmers optimize crop selection, planting schedules, and input usage, ultimately increasing yield and reducing resource waste. The primary objective of this study is the classification of soil micronutrients, such as phosphorus, nitrogen, sulfur, potassium, copper, boron, and others. Understanding the distribution and concentration of these micronutrients is crucial for determining the best-suited crops for specific soil conditions. With the use of more accurate algorithms, the growing field of "smart agriculture" increases agricultural productivity. An Internet of Things (IoT) based smart agriculture system can continually learn the properties of the soil by leveraging discrete data acquired from sensors at regular intervals. Computer data can be transferred to a base station via cellular communications in the Machine-to-machine (M2M) scenario. Cellular M2M is an interesting technology in agriculture because wired networks have drawbacks and strengths in terms of cost, performance, usability, and other areas. (Bhanu KN et al., 2020) though numerous vector quantification techniques have been created for the coding of signal processing, compression, and telecommunications, the term "machine learning" actually has its roots in computer science. The author proposes a communication channel that allows two or more (SW Lee et al., 2021) items to communicate with each other without direct human intervention in M2M interactions. A few wireless technologies that function on particular WiFi, Bluetooth, WAN, and ZigBee mobile and cellular forms have been suggested; these are frequently utilized in Internet of Things (IoT) applications related to agriculture. Wireless technology uses M2M communication, and as IoT-based smart agriculture improves, waste will be reduced and awareness raised. Real-time micro-controller data are fed into a device that assesses target-trained data sets using data mining techniques to produce efficient crop selection. The KNN classifier is given the sensor readings, and the Euclidean distance is calculated.

The two categories of machine learning algorithms that the author discussed are supervised learning and unsupervised learning. Naive Bayes, logistic regression, K-nearest neighbor (KNN), Support (Bhanu KN et al., 2020), Vector Machine (SVM), and linear regressions are discussed in supervised learning. Clustering and vector quantization are studied in the context of unsupervised learning. Agronomic management systems apply machine learning to sensor data to create artificial intelligence systems that offer the greatest guidance and overview for future decisions and actions that can ultimately increase performance.

2 Literature Survey

This section analyzes and discusses some of the materials that is currently available for soil analysis utilizing machine learning models and crop yield generation. Machine learning, deep learning, and neural networks are examples of decision-making models that can be used to address problems that can be modeled using data and numerical parameters. Employ a dataset based on the Soil Resources Development Institute (SRDI) to develop KNN, SVM, and bagged tree models. This dataset comprises 495 samples divided into 11 classes. The proposed ML model can be used to predict the soil series and recommend an appropriate crop yield for every given soil. In this paper, SVM classification yields the highest accuracy in soil categorization. (P. A. Harlianto et al., 2017), employ SVM, Naïve Bayes, decision trees, neural networks, and various other machine learning methods. The system for automating soil type classification achieves around 70% accuracy.

(Prakash et al.), present an advanced machine learning-based soil moisture prediction model. Numerous machine learning approaches, including multiple linear regression, support vector regression, and recurrent neural networks, are used to forecast soil. The models were implemented and numerous databases that were sourced from different web repositories were utilized. The prediction performance is assessed using methods such as mean squared error (MSE) and coefficient of determination (R2). The multiple regression models, which are better than the other models, yields the values 0.975 and 0.14 for R2 and MSE, respectively, based on the comparison and findings. Machine learning algorithms are utilized to forecast soil moisture content, soil type, and soil nutrient content. To classify the soil nutrient levels and fertility indices village-wise, a variety of classifiers are employed, including neural networks, (U. Shruthi et al., 2019) SVM, AdaBoost, RF, and bagging. The soil is assigned labels corresponding to low, middle, and high scales based on the numerical values. By using several regression models and numerical data to generate pseudo transfer functions, it is possible to predict the fertility indices of the soil village-wise directly. Extreme Learning Machines (ELMs), which are widely employed because to their very rapid learning speed and enhanced generalization performance over the past few years, are a promising machine learning technique that introduced.

In this work (T. Blesslin Sheeba et al., 2022), the parameter soil response (pH) and the test report to various aspects with fertility indices for phosphorus (P), potassium (K), boron (B), organic carbon (OC), and accessible boron (B) are used to classify the soil. For the purpose of cross-validation, ten steps are employed, with 10% of the data used for validation and the remaining 90% for training. The data is used to train the extreme learning method (ELM), a quick classification technique that finds the micronutrients in the soil. To optimize the process, activation functions such sine-squared, hyperbolic tangent, hard limit, triangular basis, and Gaussian radial basis are employed. The irrigation system that the authors have suggested helps to reduce water waste and mechanize the water system structure for large areas of crops. The system uses the behavior of soil moisture, humidity, and air temperature to determine how much water the crop needs. Using a machine learning technique, the framework compares perceived values obtained from sensors with a limit value supplied to the machine learning for additional analysis. Following this process, the machine learning system compares the obtained results with the forecast and determines whether or not water supply should be implemented.

In (K. S. Pratyush Reddy et al., 2020), order to reduce water waste and mechanize the water system structure for large areas of agricultural, the authors have suggested an irrigation system. The system uses the behavior of soil moisture, humidity, and air temperature to determine how much water the crop needs. By using a machine learning technique, the framework compares sensed values obtained from sensors with a limit value that is supplied to the machine learning for additional analysis. Following this process, the machine learning system compares the obtained result with the forecast and determines whether or not water supply should be implemented.

Farmers can now more easily employ a variety of Precision Agriculture techniques in conjunction with mobile computing (MC), wireless (G. S. Prasanna et al., 2023) sensor networks (WSN), and intelligent agricultural equipment according to the Internet of Things. An irrigation control system is created over two years using data from 22 soil sensors. Numerous regression and classification methods were applied to the dataset. Out of the ML algorithms selected, Gradient Boost Regression produced predictions with 93% accuracy, while Boosted Tree Classifier produced the best results with 95% accuracy when forecasting the need for irrigation. Sensors and an indirect technique of estimating the crop's water requirements were used to determine how much irrigation would be needed.

3 Methodology

The study investigates the classification and analysis of soil fertility using machine learning techniques, focusing on key parameters such as soil pH and the fertility indices for various nutrients like phosphorus (P), nitrogen (N), potassium (K), boron (B), copper (Cu), manganese (Mn), sulfur (S), iron (Fe), and zinc (Zn) are utilized in the current study to differentiate the soil. The soil dataset used in this study is obtained from Kaggle repository (www.kaggle.com), containing various features that represent the soil's chemical properties, such as pH and micronutrient levels. The dataset is cleaned by handling missing values, removing outliers, and normalizing/standardizing the data. Any irrelevant features are dropped, and data is scaled to ensure the uniform contribution of all parameters during model training. The relevant soil parameters (pH, P, N, K, B, Cu, Mn, S, Fe, Zn) are selected as input features for the model. These features are essential for soil fertility assessment and are used for training machine learning algorithms to classify soil data. When developing the model, the dataset is divided into training and testing sets for this reason. The dataset is divided into 80% training and 20% testing sets using a standard ratio. The training set is used to develop the model, while the testing set is used for validating its performance on the data. After training, the models are evaluated on the testing dataset to assess their generalization performance. After the soil is examined, its fertility is categorized using a variety of machine learning algorithms that take into account the chemical characteristics and micronutrients that the soil contains. Crop yield and sowing are also analyzed using ML models. Machine learning techniques like logistic regression, random forest, Naïve Bayes, Gradient Boosting, KNN, AdaBoost, and support vector machine (SVM) are used in the current systems. This study highlights the importance of using cutting-edge technologies such as IoT and machine learning to promote smarter, more sustainable farming practices, while simultaneously improving the quality and productivity of agricultural soils.

4 Results

This study explored the effectiveness of different machine-learning techniques in fitting models for soil nutrient classification and crop recommendation. A variety of elements are included in smart agriculture for soil nutrient analysis, and they all function together to maximize crop productivity and soil health. In Fig. 2 shows the relationships between features in the dataset, especially when selecting features for machine learning models and displaying a heatmap with correlation values for various soil parameters (N, P, K, pH, EC, S, Cu, Fe, Mn, Zn, B).The heatmap is a powerful way to visualize which soil nutrients are related, helping in soil analysis and prediction tasks in smart agriculture. The heatmap uses color intensity to visualize the strength of correlations between these elements. The positive correlations are closer to 1, indicating a strong positive relationship between those parameters. Negative correlations are closer to -1, indicating a strong negative relationship. Naïve Bayes, Random Forest and Gradient Boosting were found to deliver the highest accuracy in classifying soil fertility based on nutrient content (e.g., phosphorus, nitrogen, potassium, and other micronutrients). These models were highly effective in predicting optimal conditions for soil management and crop yield improvement. In Table 1, we can find various algorithms that were utilized in the proposed model along with the corresponding accuracies, precision, f1-score and recall.

Table 1. Various algorithm corresponding accuracies obtained.

Algorithm	Accuracy	Precision	Recall	F1-score
Random Forest	95.16	95.41	95.16	95.19
SVM	93.55	93.8	93.55	93.58
KNN	93.55	93.75	93.55	93.58
Logistic Regression	93.55	93.75	93.55	93.47
Gradient Boosting	95.16	95.44	95.16	95.21
AdaBoost	52.42	56.29	52.42	42.38
Naive Bayes	95.97	96.27	95.97	96

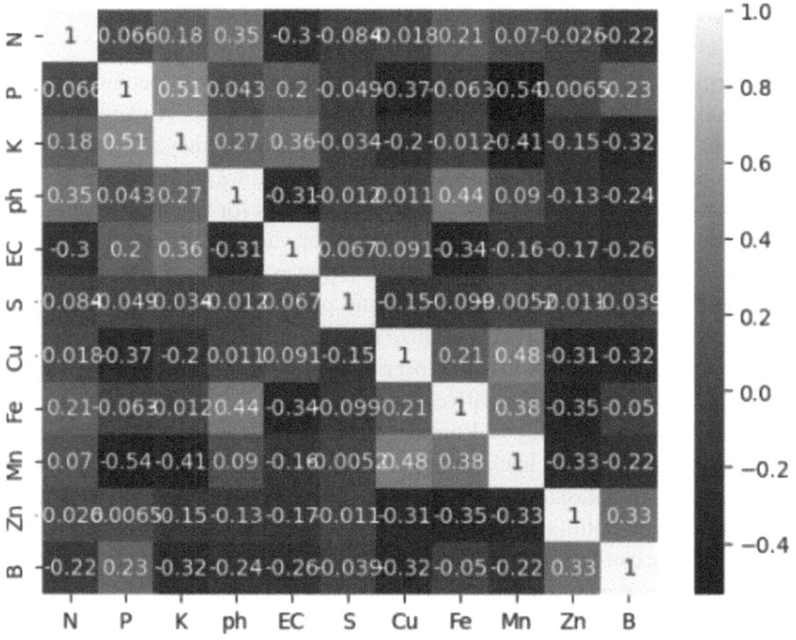

Fig. 2. Analysis of correlation matrix.

5 Conclusion

This study highlights the significant progress being made in smart farming, driven by IoT and machine learning. By focusing on the classification of soil micronutrients and analyzing their availability, this approach enhances crop recommendation systems, enabling farmers to make more informed decisions. By using a combination of machine learning techniques like SVM, Random Forest, and Decision Trees, this study provides a framework for soil fertility classification based on micronutrient analysis. Ultimately, the integration of IoT, machine learning, and soil health analysis leads to smarter, more sustainable agriculture, helping countries that heavily depend on farming transition into more efficient and productive agricultural systems. In future work is expanding the model by incorporating deep learning approaches and exploring unsupervised learning techniques to detect hidden patterns in soil data.

References

Zul Azlan, Z.H., Junaini, S.N., Bolhassan, N.A., Wahi, R., Arip, M.A.: Harvesting a sustainable future: an overview of smart agriculture's role in social, economic, and environmental sustainability. J. Clean. Prod. **434**, 140338 (2024). https://doi.org/10.1016/j.jclepro.2023.140338

Motwani, A., Patil, P., Nagaria, V., Verma, S., Ghane, S.: Soil analysis and crop recommendation using machine learning. In: 2022 International Conference for Advancement in Technology (ICONAT): IEEE, Goa, India, pp. 1–7, January 2022. https://doi.org/10.1109/ICONAT53423.2022.9725901

Vincent, D.R., Deepa, N., Elavarasan, D., Srinivasan, K., Chauhdary, S.H., Iwendi, C.: Sensors driven AI-based agriculture recommendation model for assessing land suitability. Sensors **19**(17), 3667 (2019). https://doi.org/10.3390/s19173667

Bhanu, K.N., Jasmine, H.J., Mahadevaswamy, H.S.: Machine learning Implementation in IoT based Intelligent system for agriculture. In: 2020 International Conference for Emerging Technology (INCET), Belgaum, India, IEEE, pp. 1–5, June 2020. https://doi.org/10.1109/INCET49848.2020.9153978

Abinaya, P., Faaiza, A., Tharshni, S., Vishali, G., Aravindan, M., Santhi, G.: Survey on soil micronutrients analysis and crop recommendation system in smart agriculture. Int. Res. J. Adv. Eng. Hub **2**(03), 457–463 (2024). https://doi.org/10.47392/IRJAEH.2024.0066

Harlianto, P.A., Adji, T.B., Setiawan, N.A.: Comparison of machine learning algorithms for soil type classification. IEEE Xplore, 01 July 2017. https://ieeexplore.ieee.org/abstract/document/8011843. Accessed 30 June 2023

Pugazhenthi, K., Sengamalam, A., Ganesan, B.: Milk quality monitoring system using IoT. In: 2023 International Conference on Sustainable Computing and Smart Systems (ICSCSS), Coimbatore, India, pp. 1096–1098, 2023. https://doi.org/10.1109/ICSCSS57650.2023.10169262

Arefin, M.S., Kaiser, M.S., Bandyopadhyay, Md.A.R., Ahad, K.R. (Eds.): Proceedings of the International Conference on Big Data, IoT, and Machine Learning: BIM 2021, vol. 95. In: Lecture Notes on Data Engineering and Communications Technologies, vol. 95. Springer, Singapore (2022). https://doi.org/10.1007/978-981-16-6636-0

Shruthi, U., Nagaveni, V., Raghavendra, B.K.: A review on machine learning classification techniques for plant disease detection. In: 2019 5th International Conference on Advanced Computing & Communication Systems (ICACCS), Coimbatore, IEEE, India, pp. 281–284, March 2019. https://doi.org/10.1109/ICACCS.2019.8728415

Lee, S.W., Singh, I., Mohammadian, M. (eds.): Blockchain technology for IoT applications. In: Blockchain technologies. Springer Nature, Singapore (2021)

Vimal, V., Perikos, I., Mukherjee, A., Piuri, V. (eds.): Multi-Strategy Learning Environment: Proceedings of ICMSLE 2024. In: Algorithms for Intelligent Systems. Springer, Singapore (2024). https://doi.org/10.1007/978-981-97-1488-9

Pratyush Reddy, K.S., Roopa, Y.M., Rajeev, K.L.N., Nandan, N.S.: IoT based smart agriculture using machine learning. In: 2020 Second International Conference on Inventive Research in Computing Applications (ICIRCA), July 2020. https://doi.org/10.1109/icirca48905.2020.9183373

Blesslin Sheeba, T., et al.: Machine learning algorithm for soil analysis and classification of micronutrients in IoT-enabled automated farms. J. Nanomater. **2022**, e5343965 (2022). https://doi.org/10.1155/2022/5343965

Prasanna Lakshmi, G.S., Asha, P.N., Sandhya, G., Vivek Sharma, S., Shilpashree, S., Subramanya, S.G.: An intelligent IOT sensor coupled precision irrigation model for agriculture. Meas. Sens. **25**, 100608 (2023). https://doi.org/10.1016/j.measen.2022.100608

https://www.kaggle.com/code/ortalsava/crop-recommender-with-ensemblemodeling/data

Multiband Staircase Shaped MIMO Antenna for Defence Applications

Sneha Moghe(✉) and Leeladhar Malviya

Shri.G.S Institute of Technology and Science, Indore, M.P, India
snehamoghe07@gmail.com

Abstract. Wireless communication is an emerging field for THz frequency band specially for space,imaging, defence and biomedical applications. In this paper a staircase shaped THz MIMO antenna of size $54 \times 40 \times 9\ \mu m^3$ for 8.61–9.66 THz operating band is designed on silicon dielectric substrate with Graphene patch. The maximum gain for the proposed design is 7.5 dBi with radiation efficiency 79%. Enevelope correlation coefficient and Channel capacity loss both are less than 0.5 and isolation greater than 10 dB for all over the frequency range. The phase combinations of 180°, 180° and 120°, 180° excitation signals, shows the similarity with the measured S_{11} parameter. Designed antenna placed in line of sight configuration and results are analysed in the form of S-parameters.

Keywords: THz · MIMO · Graphene

1 Introduction

THz frequency band posses so many attractive features. Its one of the main feature is that it can easily penetrate through fabrics and plastics. So it is mostly used for searching of bombs and concealed weapons which makes it very useful for defence related applications. Because of its unique properties this band is used for diagnosis of various deceases like cancer. Various designs are available in research articles. Some of them are discussed here. A MIMO array was designed using Rogers RT/Duroid 6010 as a substrate and distance between the nodes is equal to $\frac{\lambda}{4}$. At the resonance frequency gain in the operating frequency range was 8.2 dBi by keeping isolation between antennas more than -22 dB [1]. A novel MIMO antenna of size $37 \times 44\ \mu m^2$ was designed with silicon dioxide (Sio_2) substrate. The gain of the antenna was 6.5 dB for frequency rage 4.8-8 THz [2]. An elliptical-shaped microstrip feed MIMO antenna of size $1000 \times 1400 \times 101.29 \mu m^3$ and gain of 19 dB was introduced was proposed for high-speed terahertz (THz) applications [3]. A highly efficient multiband optimized slotted pentagonal terahertz patch antenna and its 2 × 2 MIMO was proposed for multiple terahertz (THz) applications like high speed indoor communications, explosive detections, arms detection, and industrial inspections etc. [4]. A MIMO array antenna is designed for 1.6-1.87 THz with RT5880 and Polyimide subatrates having more

D. Gupta et al. (Eds.): PCEMS 2024, CCIS 2491, pp. 83–97, 2025.
https://doi.org/10.1007/978-3-031-90577-3_8

than 92 % radiation efficiency and gain of 12.05 dBi [5]. A MIMO array designed with four elements having size 2348.18 × 1200 × 80.8 μm^3 with Polyimide substrate. Gain of the antenna were 9.7 dBi with efficiencies 90.2 % 0.281 THz and 89.1 % at 0.282 THz [6].

1.1 Antenna Design

The designed staircase shaped MIMO antenna for THz spectrum is designed on the silicon substrate with Graphene patch. At high frequency gain and efficiency are affected by dielectric losses and metallic losses. For substrates having high dielectric constant antenna size is less which increases fabrication complexity while substrates having low dielectric permittivity gain of the antenna get reduced and cross polarization increases. Graphene($\epsilon_r = 4$ and $tan\delta = 0.36$) has special surface plasmon polaritron property (SPP) property which makes it the most appropriate material used for fabricating THz antennas [7]. The surface conductivity of Graphene is defined by intraband and interband conductivities given by Eqs. (1) and (2) [8].

$$\sigma_{intra} = j\frac{e^2 k_B T}{\pi h^2(w + j\tau^{-1})}\left[\frac{\mu_c}{k_B T} + 2\ln(e^{\frac{-\mu_c}{k_B T}} + 1)\right] \tag{1}$$

$$\sigma_{inter} \cong \frac{je^2}{4\pi h}\ln\left(\frac{2|\mu_c| - w_h}{2|\mu_c| + w_h}\right), \tag{2}$$

where, e shows electron charge, w is the radian frequency, h is plank's constant, k_B is boltzmann's constant, T is temperature, μ_c is the chemical potential and τ is the momentum relaxation time.

Silicon ($\epsilon_r = 11.68$ and $tan\delta < 4 \times 10^{-5}$ at 1 THz) is the most widely used dielectric material as it is less lossy, can be etched easily and stable over large range of frequency. Its absorption coefficient is less than 0.05 cm^{-1} in the THz range. For dielectric materials having low values of loss tangent and high values of relative permittivity at THz frequency, maximum radiation efficiency is achieved. Sio_2 and quartz are preferred substrates for large bandwidth applications due to their low relative permittivities [9,10].

2 Evolution of Staircase MIMO Antenna Design

An offsetted MIMO antenna consists of four design steps shown in Fig. 1. In first step, a single port with offsetted Graphene patch is designed with conventional full ground. The single port in this step covers 2:1 VSWR in three different frequency bands of 8.53–8.67 THz, 8.74–8.87 THz and 9.29–9.64 THz having 0.14 THz, 0.13 THz and 0.35 THz bandwidth. In second step, same patch is designed at a distance 'a' from the left corner of the first step. In second step, this single port antenna covers 2:1 VSWR band of 8.55–8.90 THz and 9.32–9.67 THz, band having 0.35 THz bandwidth. In step three, one more patch of same size is designed on the patch of step two. This in turn covers 2:1 VSWR in two frequency bands of 8.58–8.91 THz and 9.35–9.68 THz having 0.43 THz and 0.33 THz bandwidth. Similarly, in step four, one more patch of same size is designed on the patch of step three. This in turn covers 2:1 VSWR frequency band of 8.59–8.93 THz and 9.35–9.71 THz GHz band having 0.34 THz and 0.36 THz bandwidth.

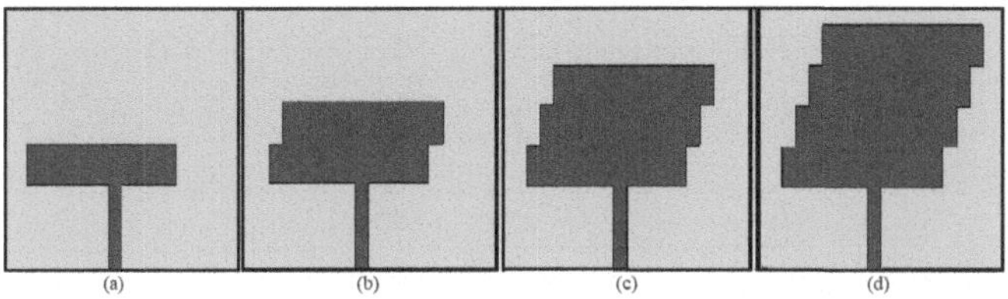

Fig. 1. Evolution stages of designed staircase antenna: (1) step 1 (b) step 2 (c) step 3 (d) step 4

The resonant frequency of designed staircase MIMO antenna is given by Eq. (3) [11].

$$f_r = \frac{7.2}{\sqrt{\epsilon_{eff}}(l + r + h)} \tag{3}$$

where, l and r are length and radius of patch, h is the length of the gap between the patch and ground plane and ϵ_{eff} is the effective permittivity.

The effective dielectric constant is given by Eq. (4).

$$\epsilon_{eff} = \frac{\epsilon_r + 1}{2} + \frac{\epsilon_r - 1}{2}\left[1 + 12\frac{h}{W}\right]^{-0.5} \tag{4}$$

where, ϵ_r is the dielectric constant, h is the dielectric thickness and W is width of radiating element.

The front and back view of the single patch antenna is is separately shown in Fig. 2 along with the CST MWS PSO optimized dimension parameters. Return loss is improved by converting into MIMO antenna where the distance between two antennas is taken as W µm as shown in Fig. 3. The dimensions of all the parameters is shown in Table 1.

3 Results and Discussion

3.1 Resonant Frequency

Staircase MIMO antenna is designed with respect to the 2:1 VSWR band (or -10 dB return loss). The return loss for each design step can be obtained here. The maximum return loss S_{11} is obtained in fourth stage is -33.47 dB while -31.97 dB in third stage and for second and first stage its value is -23.45 dB and -21.86 dB respectively shown in Fig. 4.

Table 1. Dimensions of the proposed staircase MIMO antenna

Parameter	value (μm)
$SW1$	27
SW	54
Sl	40
Sh	3
FW	3
Fl	13
h	6
L	35
a	3
gh	3
$Graphene\ thickness$	0.003
$Graphene\ temperature$	293
$Graphene\ potential$	0.2
$Graphene\ relaxtime$	1

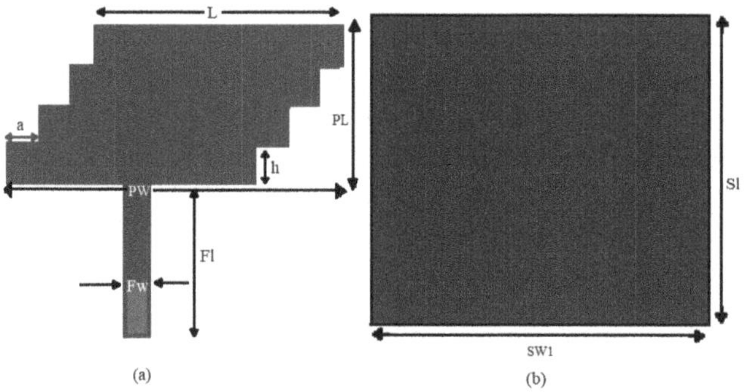

Fig. 2. Schematic view: a) front view b) back view

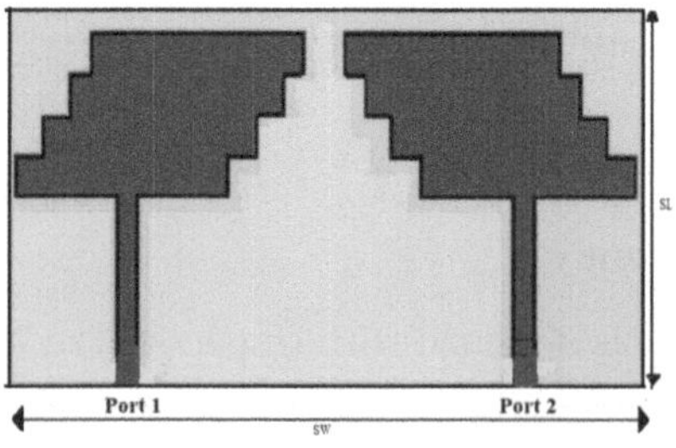

Fig. 3. Designed staircase MIMO antenna

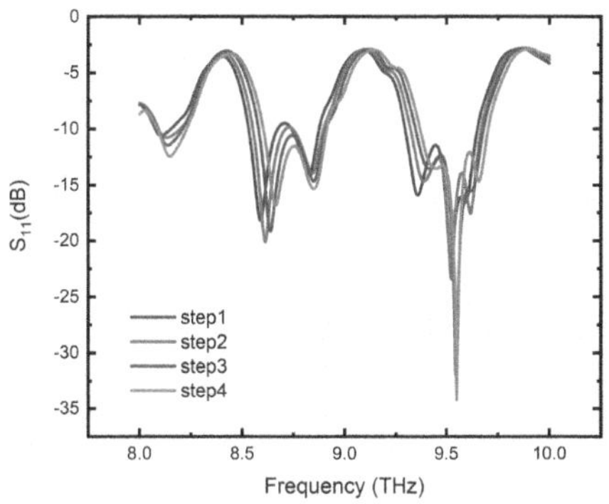

Fig. 4. S_{11} of four steps of designed staircase antenna: (a) step 1 (b) step 2 (c) step 3 (d) step 4

Proposed MIMO antenna designed for the frequency band 8.61–9.66 THz and resonates for three different bands. Out of three bands frequency range for the first band is from 8.61–8.74 THz,second band is from 8.8–8.93 THz and third band lies in the frequency range from 9.38–9.66 THz. S_{11} in these bands are -14.67 dB at 8.66 THz, -12.88 dB at 8.86 THz and -26.97 dB at 9.61 THz respectively. The isolation is more than 10 dB and voltage standing wave ratio (VSWR) is below two for all over the frequency. Variation in S_{11}, S_{12} and VSWR with frequency is shown in Fig. 5 and Fig. 6.

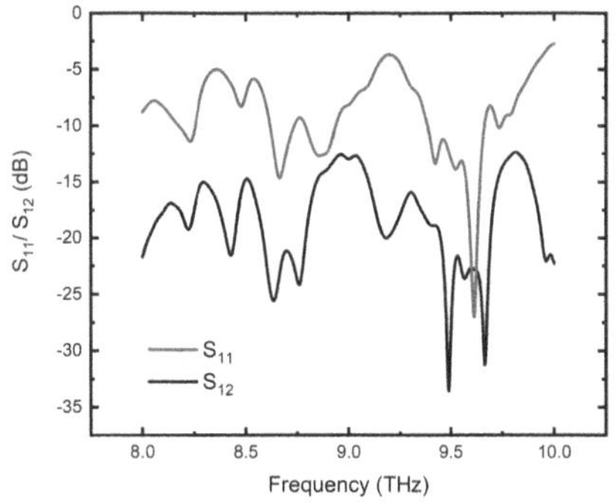

Fig. 5. S_{11} and S_{12} of designed staircase MIMO antenna

3.2 Farfield Parameters of the Proposed Staircase MIMO Antenna

The maximum gain of the proposed staircase MIMO antenna is 7.5 dBi obtained at 9.4 THz. The maximum radiation efficiency and total efficiency of the MIMO

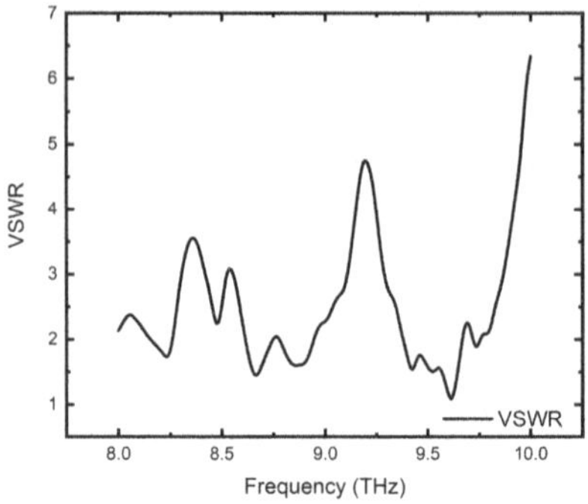

Fig. 6. VSWR of designed staircase MIMO antenna

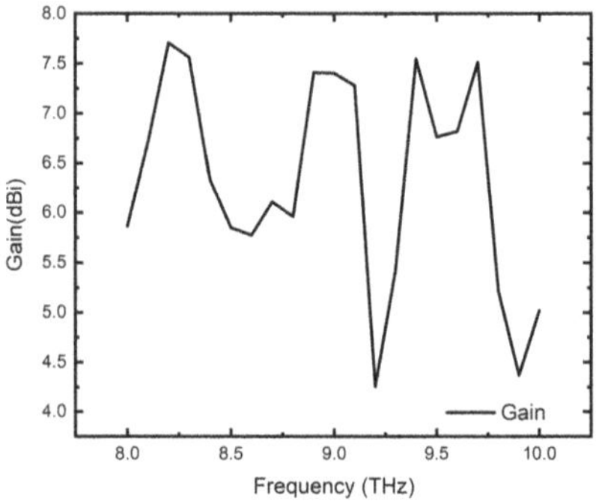

Fig. 7. Gain of designed staircase MIMO antenna

is antenna is 79% and 74.9%. Variation in gain and efficiencies with frequency are shown in Fig. 7 and Fig. 8.

The main lobe directions of E-field and H-field radiation pattern at 8.66 THz for the two radiators are 4° and 21°. The angular 3 dB beam widths of each radiator in E-field and H-field pattern are 29.9^o and 37.7^o. Similarly, E and H field radiation pattern at 8.86 THz for the two radiators are 57° and 46° and angular 3 dB beam widths of each radiator in E-field and H-field pattern are 95.5^o and 22.5^o. E and H field radiation pattern at 9.61 THz have main lobe directions as 30° and 33°. The angular 3 dB beam widths of each radiator in E-field and H-field pattern are 34.2^o and 19.2^o. E-field and H-field radiation patterns for MIMO antenna with homogeneous substrate at three different resonant frequencies for two ports are shown in Fig. 9, Fig. 10 and Fig. 11.

3.3 Surface Current Distribution of the Staircase MIMO Antenna

The information regarding the direction of flow of current is determined by the radiation pattern. Staircase MIMO antenna resonates at three different resonant frequencies which are 8.66 THz, 8.86 THz and 9.61 THz. The maximum value of

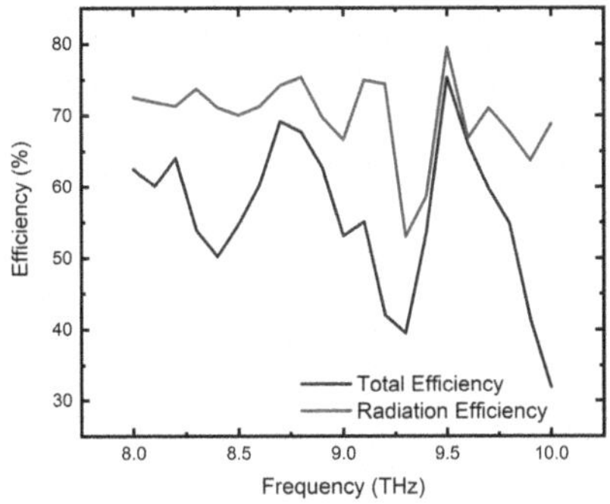

Fig. 8. Efficiency of designed staircase MIMO antenna

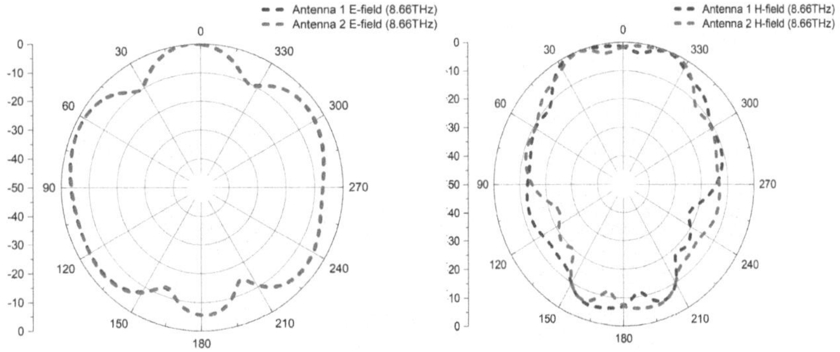

(a) (b)

Fig. 9. Radiation pattern for designed staircase MIMO antenna at 8.66 THz : a) E-field b) H-field

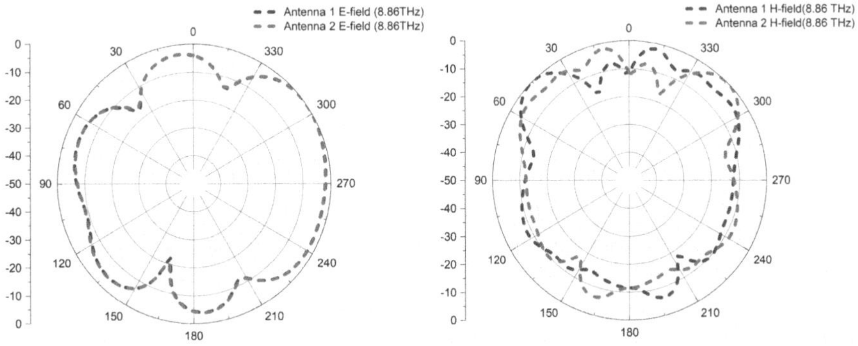

(a) (b)

Fig. 10. Radiation pattern for designed staircase MIMO antenna at 8.86 THz : a) E-field b) H-field

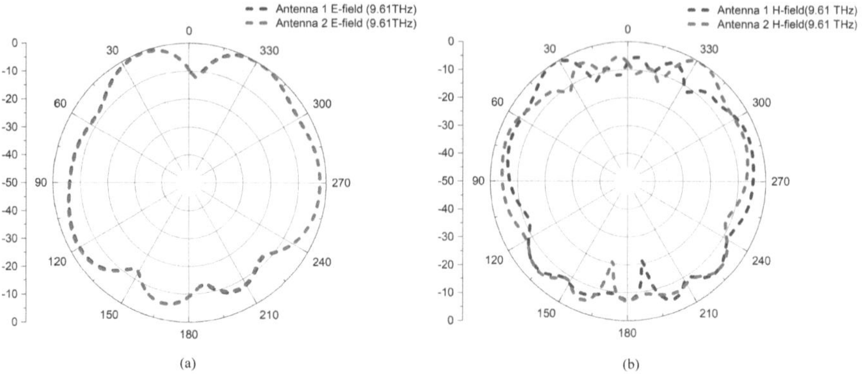

Fig. 11. Radiation pattern for designed staircase MIMO antenna at 9.61 THz : a) E-field b) H-field

surface current for proposed MIMO antenna is 89.2 88.5 and 90.1 at 8.66 THz, 8.86 THz and 9.61 THz respectively shown in the Fig. 12, Fig. 13 and Fig. 14 given below.

Fig. 12. Surface current distribution of designed staircase MIMO antenna at 8.66 THz

3.4 ECC of Designed Staircase MIMO Antenna

Correlation between the radiation patterns of different antennas is determined by ECC. Its value is zero for antennas radiates in opposite directions to each other. For MIMO antenna its value is always less than 0.5 [12–14]. ECC can be determined by far field characteristics of the radiating elements. Let the $\tilde{F}_1(\theta,\phi)$ and $\tilde{F}_2(\theta,\phi)$ are the field patterns of two radiating elements with respect to θ and ϕ components then ECC can be written in Eq. 5 [15].

$$\rho_e = \frac{\left| \int_{4\pi} \left[\tilde{F}_1(\theta,\phi) \bullet \tilde{F}_2(\theta,\phi) \right] \right|^2}{\int \int_{4\pi} \left| \tilde{F}_1(\theta,\phi) d\Omega \right|^2 \left| \tilde{F}_2(\theta,\phi) \right|^2 d\Omega} \tag{5}$$

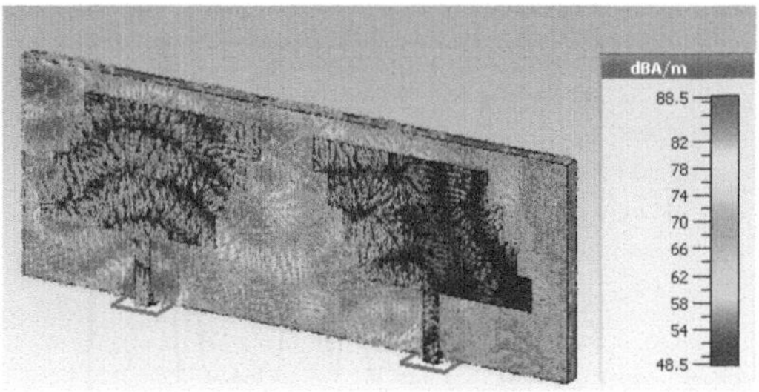

Fig. 13. Surface current distribution of designed staircase MIMO antenna at 8.86 THz

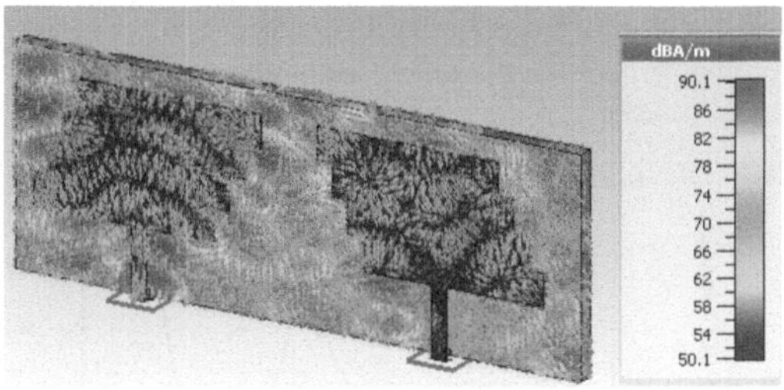

Fig. 14. Surface current distribution of designed staircase MIMO antenna at 9.61 THz

where • is the Hermitian product operator and Ω is the solid angle.

ECC is calculated for the staircase MIMO antenna and its values are 0.011, 0.013 and 0.04 respectively for the three resonating frequencies 8.66 THz, 8.86 THz and 9.61 THz as shown in Fig. 15.

3.5 CCL and TARC of Designed Staircase MIMO Antenna

Channel capacity loss (CCL) defines the capacity of the channel upto which signal can be transmitted successfully. The value of CCL should be less than 0.5 bits/sec/Hz. It is calculated by Eq. (6).

$$CCL = -log_2 Det(\varnothing^R), \tag{6}$$

where, $\varnothing^R$ is calculated as

$$C = \begin{bmatrix} \varnothing_{11} & \varnothing_{12} \\ \varnothing_{21} & \varnothing_{22} \end{bmatrix} \tag{7}$$

CCL is calculated for the designed staircase MIMO antenna and its value is 0.11 bits/s/Hz, 0.21 bps/Hz and 0.05 bps/Hz at frequencies 8.66 THz , 8.86 THz and 9.61 THz respectively means it is less than 0.4 bps/Hz for the frequency bands as shown by a line in Fig. 16.

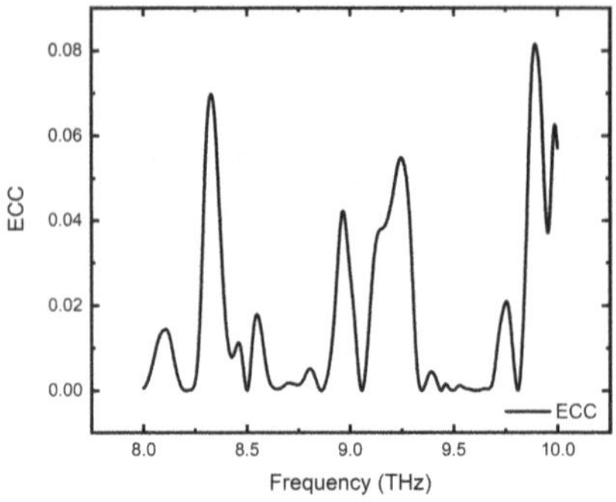

Fig. 15. Variation of ECC with frequency for designed staircase MIMO antenna

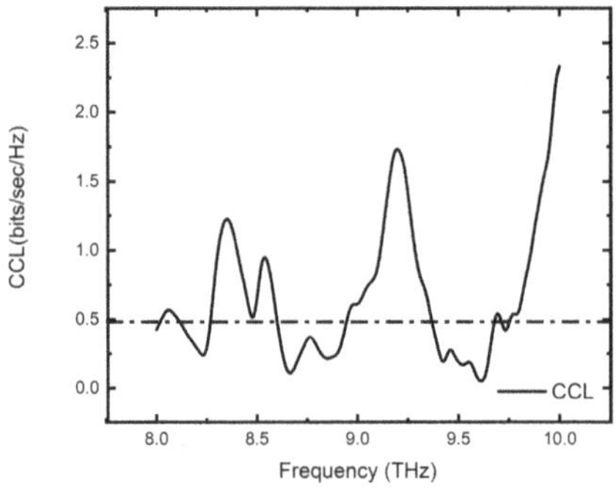

Fig. 16. Variation of CCL with frequency for designed staircase MIMO antenna

Similarly, the total active reflection coefficient (TARC) is also one of the parameter, which is determined in terms of on the return loss and correlation. TARC computes the radiation performance of designed MIMO antenna by involving random signals and their phase angles [16]. It analyzes behaviour of the S-parameters like S_{12}, S_{13}, and S_{14}. Its value should be between zero or one where zero shows no power and one shows that maximum power is radiated given by Eq. (8) [17].

$$\Gamma_a^t = \frac{\sqrt{(|S_{ii} + S_{ij}e^j\theta|^2)(|S_{ji} + S_{jj}e^j\theta|^2)}}{\sqrt{N}} \tag{8}$$

where, the phase angle between any of the two ports is given by θ. S_{ii} and S_{ij} are the input reflection coefficient and isolation.

The proposed MIMO antenna with 180°, 180° and 120°, 180° of phase combinations of excitation signals are similar with S_{11} parameter obtained from simulation while other values of phase angles show deviation from the actual S_{11} parameter. The calculated TARC for the designed staircase MIMO antenna is shown in Fig. 17 at changing values and same values of θ. TARC calculated is less than 0 for all values of θ . Therefore the designed staircase MIMO antenna is better for all applications as it has least amount of unwanted reflected power.

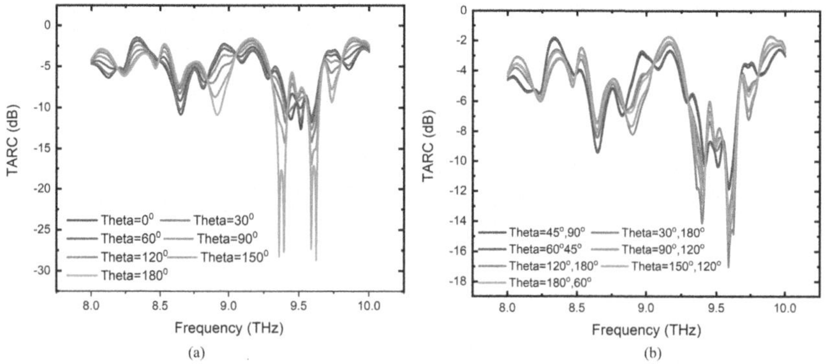

Fig. 17. TARC of designed staircase MIMO antenna a) same angle theta b) different angle theta

4 Variation of Gain with Variation in PW and FW

In order to show the effectiveness of the patch width on S parameters, parametric sweep is done. As change in patch width causes changes in far-field parameters like gain. The considered value of patch width is 35 µm. Gain of the proposed staircase MIMO antenna varies with frequency for five different samples of patch width from 32 µm - 37 µm. The The best results are evident for the patch width of 32 µm for which gain is 7.66 dBi shown in Fig. 18.

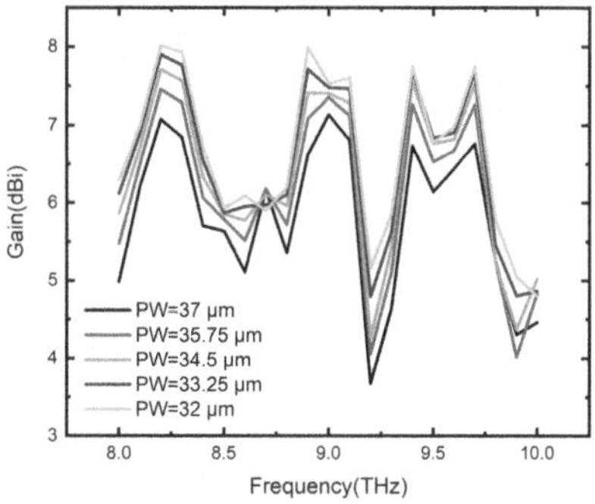

Fig. 18. Variation in gain of staircase MIMO antenna with patch width

Similarly S parameters are determined by changing the width of feed. As variation in feed width leads to change in port impedance which affects far-field MIMO antenna parameters. The gain of the MIMO antenna varies with frequency for different values of feed width from 1 µm - 5 µm. Gain is determined for five samples. The maximum gain as 7.95 dBi is obtained for feed width equal to 5 µm shown in Fig. 19.

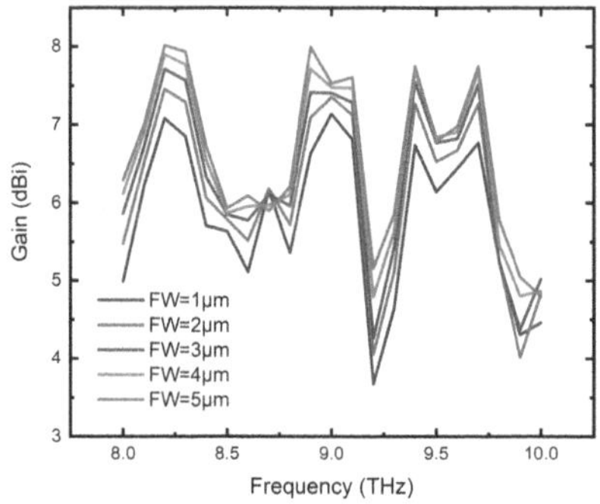

Fig. 19. Variation in gain of staircase MIMO antenna with feed width

5 Proposed Staircase MIMO Antenna Placed in Face to Face Configuration

MIMO antennas are mostly used for communication. As signals travelled from different channels travelled through different paths and arrived at the receiver at different time intervals. In line of sight communication two MIMO antennas are placed face to face at some distance expressed by the variable dd (in µm) from each other shown in Fig. 20. The distance between the two MIMO's is varied from 5 µm to 25 µm. Five samples are taken and simulation results are calculated in terms of four S-parameters. There are total sixteen S-parameters are obtained where S_{11} is same as S_{22}, S_{33} and S_{44}. Similarly S_{12} is same as S_{21}, S_{34} and S_{43} and S_{13} is same as S_{31}, S_{24} and S_{42}. S_{14} is similar to S_{41}, S_{23} and S_{32}, So only four out of sixteen S parameters are shown in the Fig. 21 which are S_{11}, S_{12}, S_{13} and S_{14}.

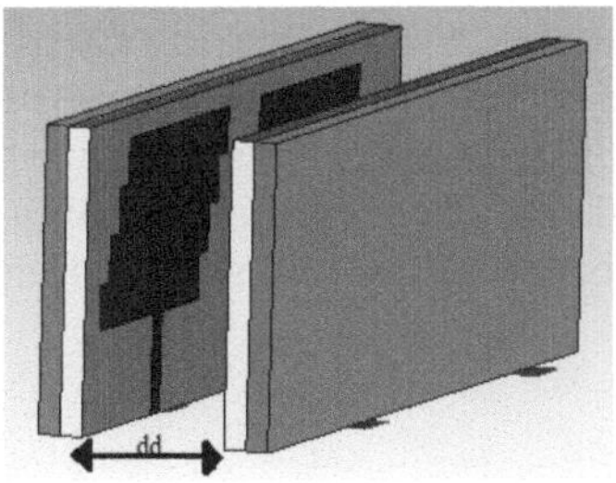

Fig. 20. Schematic view of staircase MIMO antennas

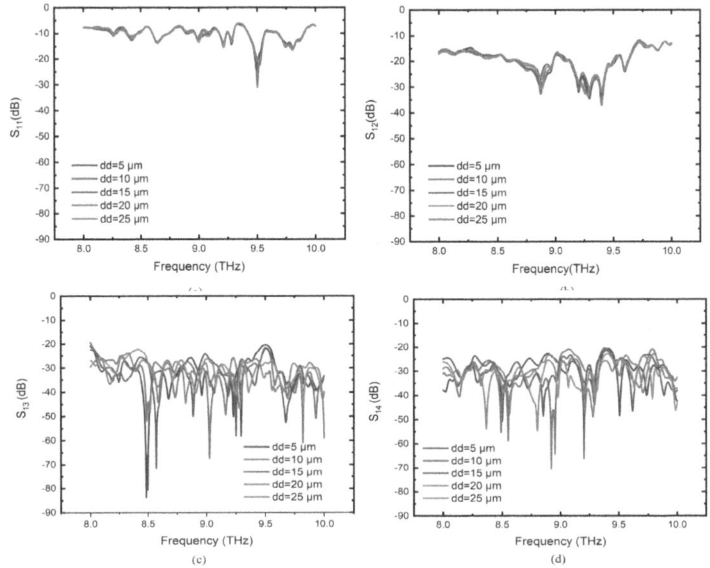

Fig. 21. Proposed staircase MIMO antennas placed face to face: (a) S_{11} (b) S_{12} (c) S_{13} (d) S_{14}

Table 2. Comparison of the proposed staircase MIMO antenna with the previous antennas

Ref.	Frequency band (THz)	Antenna size (μm^2)	Design	Substrate	Isolation (dB)	ECC	Gain (dBi)
[1]	8-9	35×35	MIMO array	RTDuroid 6010	-22.26	0.0005	8.2
[2]	4.8-8	37×44	MIMO	SiO_2	-10	less than 0.5	6.5
[3]	0.33-10	1000×1400	MIMO	RT5880	-25	0.0015	19
[4]	2-11	90×90	MIMO Array	Polyimide	less than -10	less than 0.5	13.92
Proposed work	8.61-9.66	54×40	MIMO	Silicon	-15	0.011, 0.013 and 0.04	7.5

6 Conclusion

Staircase multiband MIMO antenna achieves gain of 7.5 dBi obtained for the band 8.61–9.66 THz. Four strips one over the other in the designed antenna improved the return loss and it was further improved by using MIMO version. Comparison of different MIMO antennas on the basis of gain, bandwidth and size is given in Table 2. Variation in gain for different values of width of patch and feed is calculated and analysed. Proposed antenna is appropriate for space,defence and wireless communication because of its simple design and high efficiency.

References

1. Rubani, Q., Hak Gupta, S., Rajawat, A.: A compact MIMO antenna for WBAN operating at Terahertz. Int. J. Light Electron. Optics. 1–9 (2020)
2. Rasheed, F.M., Abdulnabi, H.A.: Multiband graphene-based Mimo antenna in terahertz antenna regime. In: International Conference on Multimodal Interaction, pp. 1–6 (2022)
3. Saxena, G., Awasthi, Y.K., Jain, P.: High isolation and high gain super-wideband (0.33-10 THz)MIMO antenna for THz applications. Int. J. Light Electron. (2020)
4. Singh, M., Singh, S.: Design and performance investigation of miniaturized multi-wideband patch antenna for multiple terahertz applications. Photonics and Nanostructers - Fundam. Appl. 1–10 (2021)
5. Moghe, S., Malviya, L.: Inverted C shaped THz array antenna for wireless body area network applications. In: Antennas and propagation Conference MAPCON Dec, Microwave, pp. 1–6 (2022)
6. Moghe, S., Yadav, R., Malviya, L.: International Conference on Communication System and Network Technologies, Sub-THz MIMO Array Antenna for Future Wireless Applications, pp. 49–53 (2022)
7. Stephen, J.P., Hemanth, D.J. : An investigation on specific absorption rate reduction materials with human tissue cube for biomedical applications. Int. J. RF Micro. Comput. Aided Eng. 1–19 (2019)
8. Dash, S., Patnaik, A.: Dual band reconfigurable plasmonic antenna copper dipole antenna using bilayer Graphene. In: IEEE International Symposium on Antennas and Propagation AP-S. pp. 921–922 (2017)
9. Zou, L., Withayachumnankul, W., Shah, C.M., Mitchell, A., Bhaskaran, M., Sriram, S., Fumeaux, C: Dielectric resonator nanoantennas at visible frequencies. Optics Express. 1344–1352 (2013)
10. Zou, L., et al.: Spectral and angular characteristics of dielectric resonator metasurface at optical frequencies. Appl. Phys. Lett. 1–4 (2014)
11. Balanis,C.A.: Antenna Theory: Analysis and Design (John Wiley and Sons), 4 th edition, ISBN: 978-1-118-64206-1 (2016)
12. Malviya, L., Kartikeyan, M.V., Panigrahi, R.K.: A 2×2 Dual-Band MIMO Antenna with Polarization Diversity for Wireless Applications. Progress In Electromagnetics Research C. 91–103 (2016)
13. Malviya, L., Panigrahi, R.K., Kartikeyan, M.V. : MIMO Antennas for Wireless Communication. 1st Edition,Publisher: CRC Press ISBN: 9781003080275 (2020)
14. Malviya, L., Panigrahi, R.K., Kartikeyan, M.V. : Four element planar mimo antenna design for long-term evolution operation. IETE J. Res. 367–373 (2017)
15. Yang, D.G., Kim, D.O., Kim, C.Y.: Design of a dual band MIMO monopole antenna with high isolation using slotted CSRR for WLAN. Microwave Optical Technology Letters **56**(10), 2252–2257 (2014)

16. Malviya, L., Kartikeyan, M.V., Panigrahi, R.K.: Offset planar MIMO antenna for omnidirectional radiation patterns. Int. J. RF Microwave. 1–10 (2018)
17. Malviya, L., Panigrahi, R.K., Kartikeyan, M.V.: MIMO antennas with diversity and mutual coupling reduction techniques: a review. Int. J. Microw. Wirel. Technol. 1763–1780 (2017)

Design of a Fifth-Order Wideband Bandstop Filter Using Stepped Impedance Resonators

Selvakumar Arumugam[1], Kanaparthi V. Phani Kumar[1(✉)], and Vamsi Krishna Velidi[2]

[1] Department of Electronics and Communication Engineering, Faculty of Engineering and Technology, SRM Institute of Science and Technology, Kattankulathur, Tamil Nadu 603203, India
kvphanikumar264@gmail.com

[2] Communication Systems Group, U R Rao Satellite Center, Indian Space Research Organization, Bangalore, KA, India

Abstract. In this paper, a fifth-order wideband bandstop filter employing a quarter wavelength series transmission line and stepped impedance resonators is proposed. The wideband bandstop filter consists of a dual-stepped impedance resonator at one end of the series transmission line and tri-stepped impedance resonator at the other end. This combination produces five transmission zeros in the stopband. The frequency of the transmission zeros can be controlled by varying the impedance values. The proposed filter is designed on AD250C substrate using Ansys HFSS full-wave simulator. The stopband response is centered at 1 GHz and the cut-off frequencies are 0.32 GHz to 1.69 GHz. The simulated 3 dB and 20 dB fractional bandwidths are 136.3% and 118.36%, respectively.

Keywords: Wideband · Bandstop filter · Stepped impedance resonator · quarter wavelength transmission line

1 Introduction

In RF communication systems, distortion suppression using band stop filters (BSFs) plays a major role in avoiding signal interference and unwanted signal attenuation for RF transmitters, Power amplifiers, mixers, and high-frequency devices. The band stop filter is generally designed as shunt open circuit stubs to provide low pass and band stop responses with low and high impedance lines. The transmission line connected with stepped impedance resonators, spilt ring resonators, lumped resonators, and coupled line section proposes bandstop filters with wide frequency responses reported from [1–12]. BSF realizes with two parallel coupled lines and two open circuit stubs have a low 20 dB fractional bandwidth (FBW) of 29.44% and poor insertion loss [1]. A coupled line with a half wavelength transmission line-based filter proposes [2] a moderate rejection level at 24 dB and the FBW of 64% at 17 dB level. High impedance phase shift section and coupled line with short stub model are realized in [3] and [8] to provide direct control rejection of stopband signals over the FBW about 64% and 83%, respectively.

Sharp rejection wideband BSF is proposed in [4] based on shunt open stubs and the transmission line generates five transmission zeros and a higher 20 dB FBW of about

D. Gupta et al. (Eds.): PCEMS 2024, CCIS 2491, pp. 98–104, 2025.
https://doi.org/10.1007/978-3-031-90577-3_9

145%. The compact size wideband BSF is reported in [5] using stepped split ring resonators and its circuit size is $0.078\lambda_g{}^2$. Unequal width three coupled line (UWTCL) based filter is demonstrated in [6] which has a higher rejection level and FBW as 112% with a design frequency of 3GHz. Sharp rejection characteristics bandstop filter realized using coupled line and open stub has FBW of 94% and sharp rejection at 50 dB level. Electrically coupled open stub type wideband BSF proposed in [9] has a rejection bandwidth of 52% and adding a spur line increases the rejection bandwidth by about 125%. T-shaped SIR and open stub model realized for wideband signal rejection from 3.45 GHz to 15.84 GHz [11]. Interdigital resonators with stepped impedance resonators are used to design [12] a wideband BSF that has a bandwidth of about 4 to 10 GHz and FBW greater than 80%.

This paper proposes a fifth-order WBBSF filter that generates five transmission zeros and a wide stopband using a quarter wavelength transmission line and stepped impedance resonators at both ends. These type of filter provides greater selectivity and wider bandwidth control as compared to lower order filters.The wideband stopband response obtained by properly selecting the impedances of resonators and the circuit simulated result compared with Full wave simulation and the results are more identical with each other. A prototype WBBSF filter operating at center frequency 1 GHz provides wide stopband from 0.32 GHz to 1.69 GHz and the 3 dB FBW of about 136.3% is modeled on Rogers - AD250C substrate with dielectric constant $\varepsilon_r = 2.5$, thickness $h = 0.76$ mm and loss tangent $\delta = 0.0014$.

2 WBBSF Filter Design and Analysis

The proposed WBBSF circuit model is shown in Fig. 1(a), designed with a quarter wavelength transmission line impedance of (Z_1), and the stepped impedances are (Z_2, Z_3) connected in one end and (Z_2, Z_3, Z_4) in another end. The electrical length of the transmission line and stepped impedance resonators equal to θ for the stop band center frequency.

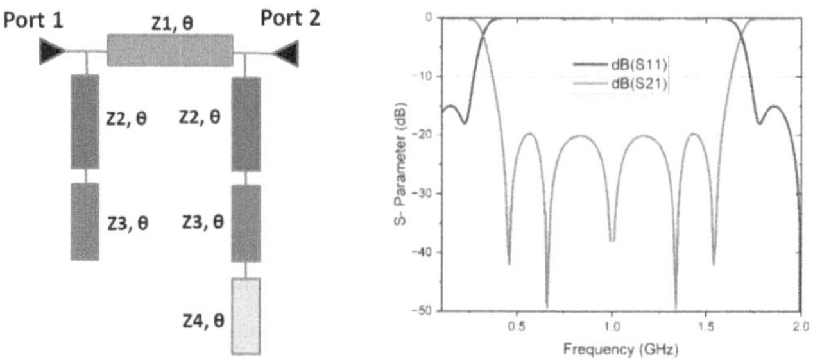

(a) (b)

Fig. 1. (a) Ideal circuit of the WBBSF. (b) Circuit simulated result of WBBSF

Fig. 2. (a) Equivalent circuit Z_S (b) Equivalent circuit Z_T

The equivalent circuit analysis, based on transmission line theory is shown in Fig. 2 (a) and (b). The frequency response of WBBSF is characterized five transmission zeros (TZs) and a wide stopband derived from two port network analysis model for different impedance configurations. The stepped impedance resonators provides flexibility in control of bandwidth by adjusting the impedance ratios and electrical lengths of the transmission line sections. To achieve a stopband response, the open ended input impedance resonators Z_S, Z_T and impedance ratio $R_1 = Z_3/Z2 < 1$, where $Z3 > Z2$ are are properly selected. The resonant response is given by Eqs. (6) and (11).

$$Z_{OC} = -jZ_4 \cot\theta \tag{1}$$

$$Z_T = Z_2\left[\frac{Z_{S1} + jZ_2 \tan\theta}{Z_2 + jZ_{S1} \tan\theta}\right] \tag{2}$$

$$Z_{S1} = jZ_3\left[\frac{jZ_3 \tan\theta - jZ_4 \cot\theta}{Z_3 + Z_4 \cot\theta \tan\theta}\right] \tag{3}$$

$$Z_{S1} = \left[\frac{\tan\theta - R_1 \cot\theta}{1 + R_1}\right] \tag{4}$$

$$\tan^2\theta - R_1 + jZ_2 \tan^2\theta(1 + R_1) = 0 \tag{5}$$

$$\tan^2\theta_T = \frac{Z_3Z_4}{Z_3^2 + Z_2(Z_3 + Z_4)} \tag{6}$$

$$\theta_T = \tan^{-1}\left[\sqrt{\frac{Z_3Z_4}{Z_3^2 + Z_2(Z_3 + Z_4)}}\right] \tag{7}$$

$$Z_S = jZ_2\left[\frac{jZ_2 \tan\theta - jZ_3 \cot\theta}{Z_2 + Z_3 \cot\theta \tan\theta}\right] \tag{8}$$

$$Z_S = \left[\frac{\tan\theta - R_2 \cot\theta}{1 + R_2}\right] \tag{9}$$

$$\tan^2\theta - R_2 = 0 \tag{10}$$

$$\tan^2\theta_S = \frac{Z_3}{Z_2}, \theta_S = \tan^{-1}\left[\sqrt{\frac{Z_3}{Z_2}}\right] \tag{11}$$

The transmiision zeros at the centre frequency (f_o) produced by uniform impedance resonators are f_{z1} and f_{z2}. The electrical length θ = 90° and the relation between θ and frequency is $f_{TZ} = (\theta/\theta_0) f_0$ where the θ_0 is the resonant frequency of WBBSF. The transmission zero frequencies are calculated as $f_{z1} = 0.46$ GHz and $f_{z2} = 0.66$ GHz, $f_0 = 1$ GHz and two inherent transmission zero frequencies are $f_{z4} = 1.33$ GHz, $f_{z5} = 1.54$ GHz considering with $Z_0 = 50\ \Omega$, $Z_1 = 140\Omega$, $Z_2 = 31\Omega$, $Z_3 = 90\Omega$, $Z_4 = 124\Omega$ and $f_0 = 1$ GHz.

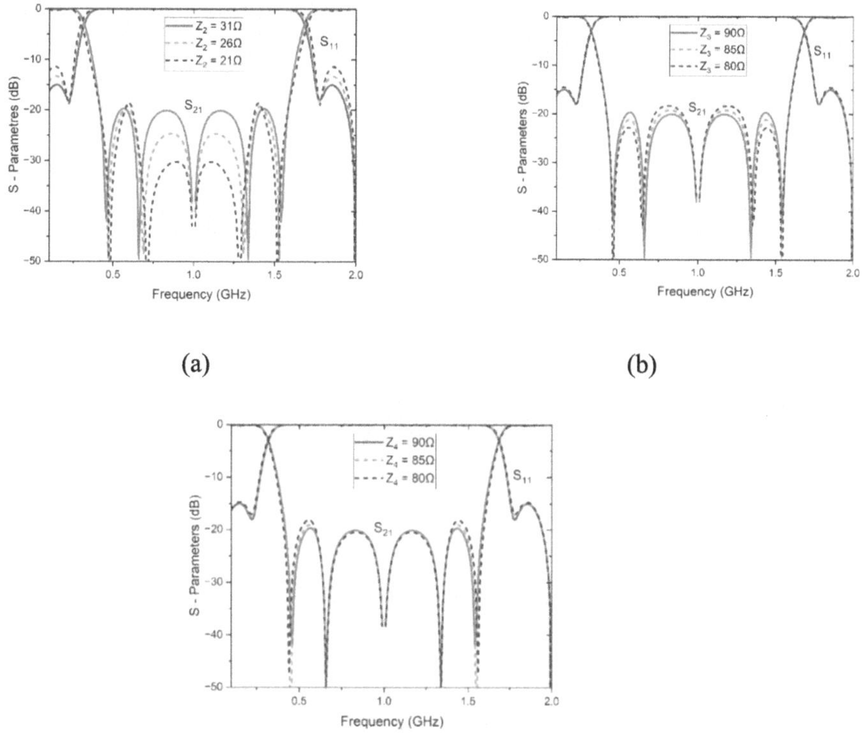

(a) (b)

(c)

Fig. 3. S-Parameter response with different impedance values (a) Z_2 (b) Z_3 (c) Z_4

The parametric analysis of proposed WBBSF for various impedance values Z_2, Z_3, Z_4 and changes in S-parameter responses are shown in Fig. 3 and compared with the optimized impedance values. The low impedance resonator Z_2 changes with increasing the stopband insertion loss, decreases the return loss and shift in transmission zeros to be noted. For the different values of high impedance resonators Z_3 and Z_4 produces small variations IL and RL and the locations of TZs are not changed.

3 WBBSF Filter Design Using HFSS

The proposed WBBSF filter 3D model top view and layout are shown in Fig. 4 (a) and (b). Based on the theoretical analysis, the circuit design parameters are optimized with $Z_0 = 50\ \Omega$, $Z_1 = 150\ \Omega$, $Z_2 = 170\ \Omega$, $Z_3 = 100\ \Omega$, $Z_4 = 100\ \Omega$, design frequency f_0 = 1GHz and wideband cut-off frequencies are $f_1 = 0.32$ GHz and $f_2 = 1.69$ GHz. The WBBSF is designed on Rogers- AD250C substrate with dielectric substrate with $\varepsilon r = 2.5$, thickness h = 0.76 mm, and loss tangent $\delta = 0.0014$ and simulated using HFSS V 2022.

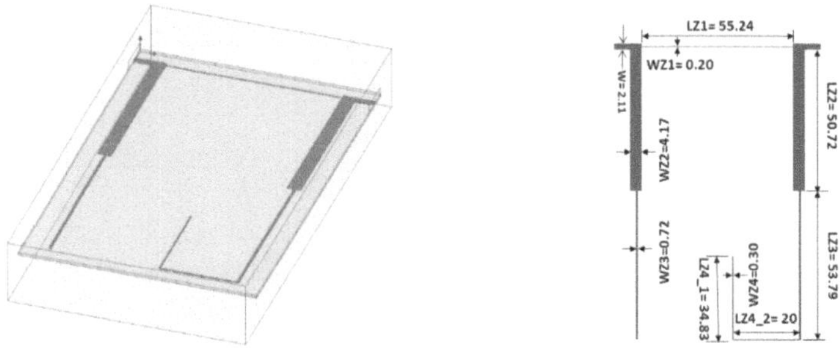

(a) (b)

Fig. 4. (a) Wide-band bandstop filter 3-D model. (b) Layout

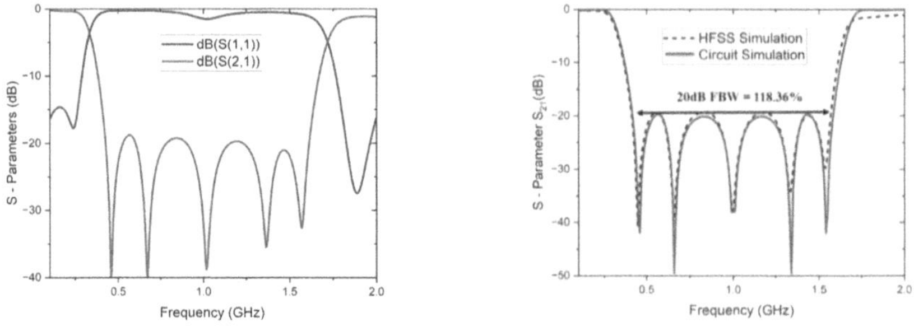

(a) (b)

Fig. 5. (a) Full-wave simulated S-parameters response(b) S_{21}- Comparison

The full-wave simulated S-parameter responses are shown in Fig. 5(a). The cut-off frequencies of the wideband response are located at 0.32 and 1.69 GHz with 3-dB FBW of 136.3% and 20-dB FBW of about 118.36% at the frequencies are 0.40 and 1.56 GHz The five transmission zeros are located at 0.45, 0.65, 0.99, 1.33 and 1.53 GHz and the stopband rejection level up to 42 dB. The passband insertion loss of less than 1 dB at the wide band frequencies from 0.40 and 1.69 GHz. The compact size of Wide-band BSF is 105 mm × 56 mm.

The simulated result obtained from WBBSF is compared with other related works listed in Table 2, which shows the improved stopband response of about 20 dB FBW 118.36% and the number of TZs. Figure 5(b) shows the comparison of circuit and full wave simulation response agrees good matching between circuit and full wave simulated frequencies and the transmission zero frequencies are listed in Table 1.

Table 1. TZs frequency comparison with Circuit and Full wave simulation.

TZ Frequency (GHz)	*Theoretical Calculation*	*Circuit Simulated*	*HFSS Simulated*
f_{z1}	0.455	0.46	0.45
f_{z2}	0.662	0.66	0.65
$f_{z3} = f_0$	1	1.01	0.99
f_{z4}	1.337	1.34	1.33
f_{z4}	1.544	1.54	1.53

Table 2. Proposed WBBSF comparison with other reported filters.

Ref.	f_c *(GHz)*	*20 dB FBW %*	f_{c1} *(GHz)*	f_{c2} *(GHz)*	*IL (dB)*	*No. of Tz*	*Circuit size (mm)*
[1]	1	29.44	0.71	1.16	<1.5	–	120 × 12
[2]	1	64	0.66	1.3	<1	5	140 × 49
[3]	1.5	122.5	0.57	2.37	<1.8	5	28 × 22
[4]	1	145.8	0.51	2.54	<1	5	114 × 65
[5]	2.5	39.76	2.03	3.05	<1	–	32 × 30
[6]	3	112	1.38	4.5	<1	–	0.57 × 0.35
[7]	3	106	1.43	4.57	<1	–	54 × 32
[8]	3	83	1.44	4.67	<1	5	64 × 28
[9]	5	52	4.3	6	<1	–	44 × 16
[10]	2	66.2	0.5	1.5	<1		78 × 25
[11]	9	59.1	3.45	15.8	<1	3	12.8 × 12.8
[12]	7	80	4	10	<0.5	–	–
This work	**1**	**118.36**	**0.32**	**1.69**	**<1**	**5**	**105 × 56**

4 Conclusion

In this paper, a fifth-order WBBSF filter employing a quarter wavelength transmission line and stepped impedance resonators has been presented. The proposed filter characteristics are analyzed through transmission line theory-based equations are derived and the performance of the designed filter frequency responses is compared with simulated results. The wide stop band frequencies are at 0.40 and 1.56 GHz at 20 dB fractional bandwidth of about 118.36% indicating sharp rejection over the stop band bandwidth and improved insertion loss. The proposed filter is applicable for wireless communication systems with wide frequency suppression needed.

References

1. Rao, M., Yongle, W., Wang, W., Liu, Y.: Design of compact Wideband high-selectivity band-stop filter based on coupled lines. Progress Electromagn. Res. C **53**, 55–66 (2014)
2. Kong, M., Yongle, W., Zhuang, Z., Liu, Y.: Wideband bandstop filter with extreme sharp skirt selectivity. IEEE Microw. Wirel. Compon. Lett. **28**(12), 1104–1106 (2018)
3. Liu, L., Liang, X., Fan, H., Jin, R., Bai, X., Geng, J.: Compact wideband bandstop filter with directly controlled rejection. IEEE Trans. Circuits Syst. II Express Briefs **68**(7), 2282–2286 (2021)
4. Chen, H., Wang, X., Geyu, L.: Analysis and design of a sharp-rejection wideband bandstop filter with multi-transmission zeros and reflection zeros. IET Microw. Antennas Propag. **15**(9), 995–1008 (2021)
5. Munirathinam, S., Balachandran, A.: Compact wideband band-stop filter using stepped complementary split ring resonators. Analog Integr. Circ. Sig. Process **102**, 363–367 (2020)
6. Zheng, Y., Wang, W., Wu, Y.: Synthesis of high-selectivity ultra-wideband reflective/absorptive bandpass filters with unequal-width three-coupled line sections. IEEE Trans. Circuits Syst. II: Express Briefs (2023)
7. Liang, L., et al.: A novel wide-stopband bandstop filter with sharp-rejection characteristic and analytical theory. Progress Electromagn. Res. C **40**, 143–158 (2013)
8. Qiu, L.-L., Zhu, L.: Wideband bandstop filters based on wideband 180° phase shifters. IET Microw. Antennas Propag. **14**(13), 1662–1670 (2020)
9. Lee, H.-S., Lim, W.-G., Kim, C.-Y.: Compact wideband bandstop filters with electrically coupled open stubs. Microw. Opt. Technol. Lett. **55**(11), 2701–2703 (2013)
10. Yang, C., Chengfang, F.: Analysis and design of a new ring bandstop filter using lumped equivalent circuit. Progress Electromagn. Res. Lett. **93**, 81–87 (2020)
11. Lee, T.-H., et al.: A band-stop filter with wide stop band using T-shaped stepped impedance resonator and L-shaped open stub. In: 2017 International Conference on Emerging Trends & Innovation in ICT (ICEI). IEEE, 2017
12. Singh, K., Soni, B.K., Nirmal, A.V.: Wideband microstrip band stop filters realization with interdigital topologies. Int. J. Inf. Technol. **11**, 479–483 (2019)

Performance Analysis of 16 Bit RISC Processor Using Vedic Multiplier

Mithilesh Kavade(✉), Prasad Deshmukh, Sankalp Mhatre, Mayank Patil, and Akhil Masurkar

Vidyalankar Institute of Technology, Mumbai, India
mithileshkavade007@gmail.com

Abstract. In this work a comprehensive performance analysis of a 16-bit RISC processor integrated with two different multiplier architectures namely Vedic multiplier and shift and add multiplier. The objective was to evaluate and compare the efficiency and effectiveness of these multiplier designs in terms of performance and power consumption. The designs were implemented and synthesized using Verilog in Xilinx Vivado. The tested architecture was practically implemented on a Spartan 7 FPGA board and simulated on spartan 7, basys 3 and nexys A7 100T FPGA boards. The above two multipliers were implemented inside a 16-bit RISC processor and the processor performance was measured. The performance was measured using parameters like WNS (Worst Negative Slack), TNS (Total Negative Slack), WHS (Worst Hold Slack), THS (Total Hold Slack), TPWS (Total Pulse Width Slack), Total Power, Number of LUT's used and Number of Flip Flops used. It is observed that the performance of RISC processor implemented on Spartan 7 FPGA board improves by around 34.89% when a Vedic multiplier is used instead of Shift and Add multiplier.

Keywords: 16-bit RISC · Vedic multiplier · shift and add multiplier · Xilinx Vivado · FPGA · Spartan 7

1 Introduction

Multiplication is a key operation in an ALU and significantly impacts the execution time of DSP algorithms, making high-performance multipliers essential. The time required for multiplication influences the instruction cycle duration of a DSP chip. As demand for computing power in battery-powered mobile devices increases, design focus has shifted to minimizing power dissipation while maintaining high performance. This paper evaluates a 16-bit (RISC) processor with two multiplier architectures: the Vedic multiplier and the shift-and-add multiplier. These designs are compared based on area, power consumption, and speed, implemented on an FPGA board using Verilog in Xilinx Vivado.

Vedic mathematics simplifies calculations through formulas that align with natural human thought processes. Its Urdhva Tiryakbhyam Sutra, or "Vertically and Crosswise," is a universal multiplication method. It multiplies digits at the line ends and adds their product to the preceding carry. For multiple lines, products are summed with the previous carry. The least significant digit becomes part of the result, while the remaining digits serve as the carry for the next step, initially assumed to be zero.

D. Gupta et al. (Eds.): PCEMS 2024, CCIS 2491, pp. 105–119, 2025.
https://doi.org/10.1007/978-3-031-90577-3_10

2 Literature Review

2.1 Previous Related Work

The foundation of Vedic multipliers is rooted in the ancient mathematical principles described by Swami Bharati Krshna Tirthaji [4] in his book Vedic Mathematics. This work introduced the key sutras, such as the Urdhva Tiryakbhyam Sutra, which have been adapted to modern digital circuit design. Yadav et al. [2] demonstrated that implementing a Vedic-based multiplier in a 16-bit RISC processor leads to a 44% reduction in power for the MAC and 12% for the ALU, along with a significant reduction in delay. Similarly, Dhillon and Mitra [5] and Pushpalata and Mehta [6] highlighted the efficiency of Vedic multipliers, especially the Urdhva Tiryakbhyam Sutra, in high-speed operations. The simplicity and speed of Vedic multipliers make them suitable for DSP and communication systems, as supported by Bandi [8], who compared different partial product addition methods and found that combining Vedic methods with half and full adders reduced power dissipation and area usage.

Further, Vedic multipliers have demonstrated advantages in advanced applications like Direct Digital Synthesizers, as shown by Shanthi et al. [9], where latency and power were significantly reduced. Agarwal et al. [10] emphasized their use in digital signal processing, while Shaik et al. [11] and Sridevi and Sathiya [12] explored energy-efficient Vedic multiplier designs using CMOS, FinFET, and the Nikhilam Sutra. These designs showcased substantial reductions in power consumption and operational delay, making Vedic multipliers highly suitable for applications in VLSI, IoT, and biomedical systems where power efficiency is crucial. Overall, the literature highlights the consistent benefits of Vedic multipliers in enhancing the performance and energy efficiency of digital systems across various fields.

2.2 16-Bit RISC Processor

The minimalist architecture of the 16-bit RISC processor prioritizes speed and efficiency over complexity. In the paper titled "Design and 16-bit RISC processor" [7] the authors describe the development and implementation of a custom 16-bit RISC processor optimized for embedded applications. This processor features a reduced instruction set, efficient pipelining, and hardware support for essential operations, showcasing improvements in performance and area utilization compared to traditional designs.

A 16-bit Reduced Instruction Set Computing (RISC) processor emphasizes a simplified instruction set to enhance efficiency and performance. It operates on 16-bit data units with fixed-length instructions designed to execute in a single clock cycle, streamlining the instruction decoding process. With a small number of general-purpose registers (GPRs) and a load-store architecture, RISC processors separate data access from arithmetic operations, resulting in efficient data processing and reduced memory access latency.

A 16-bit RISC (Reduced Instruction Set Computer) processor operates efficiently by utilizing a streamlined instruction cycle. This cycle begins with the program counter (PC) fetching the next instruction from memory. Each instruction, being 16 bits in size, is then decoded to identify the operation and its operands. The subsequent stage involves fetching the necessary operands from registers or memory. With the operands available, the processor proceeds to the execute stage, where the Arithmetic Logic Unit (ALU) performs the specified operation, which can include arithmetic, logical, or control functions.

2.3 Sutras Used in Vedic Mathematics to Perform Mathematical Operations

Vedic Mathematics is a collection of techniques and sutras designed to solve arithmetic problems quickly and easily. It comprises 16 Sutras (formulas) and 13 sub-sutras (sub-formulas) that can be applied to arithmetic, algebra, geometry, calculus, and conics. These Sutras and Sub-Sutras are commonly used for division, multiplication, calculating squares, calculating square roots, calculating cubes, calculating cube roots. In Vedic mathematics, multiplication is carried out using general or specific Techniques. The Specific Techniques are Nikhilam Sutra, Anurupyena Sutra, Ekayunena Purvena, Antyaordaske'pi. The general technique is known as Urdhva Tiryagbhyam Sutra. In the proposed work Urdhva Tiryagbhyam Sutra is used.

2.4 Urdhva Triyagbyham Sutra

This general method allows for the quick multiplication of any type of numbers and can be easily applied to multiply 3-digit, 4-digit, or even larger numbers. A 3-digit number can be multiplied using the Urdhva Tiryak Sutra as follows:

$$\text{Number 1} = \text{abc} = \left(a \times 10^2 + b \times 10^1 + c \times 10^0\right)$$

$$\text{Number 2} = \text{def} = \left(d \times 10^2 + e \times 10^1 + f \times 10^0\right)$$

$$\left(a \times 10^2 + b \times 10^1 + c \times 10^0\right) \times \left(d \times 10^2 + e \times 10^1 + f \times 10^0\right) = ad \times 10^4 + (ae + bd) \times 10^3$$
$$+(af + be + cd) \times 10^2 + (bf + ce) \times 10^1 + cf \times 10^0$$

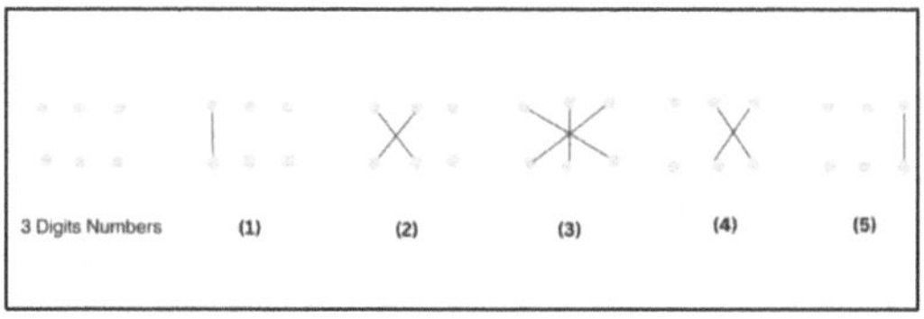

Fig. 1 Multiplication of 3-digit numbers using Urdhva Triyagbhyam Sutra

The steps to multiply two three-digit numbers are given in Fig. 1.

1. Multiply the first digits of the two numbers vertically.
2. Add the result of crosswise multiplication of the first two digits of both numbers.
3. Add the result of crosswise multiplication of all three digits of both numbers.
4. Add the result of crosswise multiplication of the last two digits of both numbers.
5. Multiply the last digits of the two numbers vertically.
6. Ensure that each compartment, except the first one, contains only one digit. If there are more digits, carry forward the initial digits to the previous compartment (Refer to the examples below for clarity).

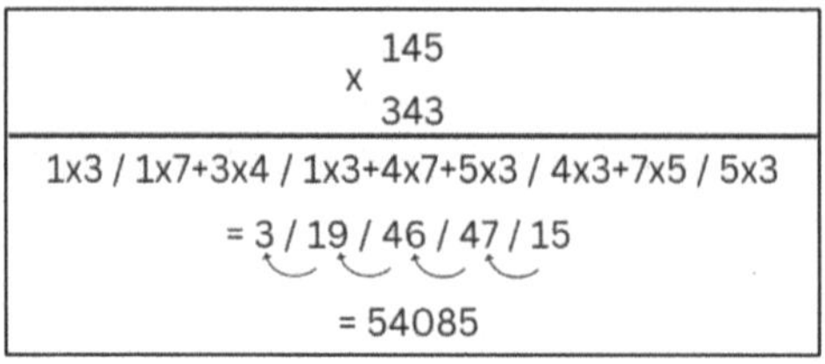

Fig. 2. 3-Digit Multiplication using Urdhva Triyagbhyam Sutra.

2.5 8 × 8 Vedic Multiplier

The 8 × 8 Vedic multiplier is constructed by employing four 4 × 4 Vedic multipliers. Initially, the two 8-bit inputs are divided into two parts, each 4 bits wide. These 4-bit inputs are subsequently fed into the four 4 × 4 multipliers. The products obtained from these multipliers are then forwarded to three adders, where the data is combined through addition operations. The result of these additions yields a final 16-bit output, representing the product of the two 8-bit inputs. This approach enables the multiplication of two 8-bit numbers by leveraging parallel processing and structured addition operations, ensuring efficient computation while maintaining simplicity and speed.

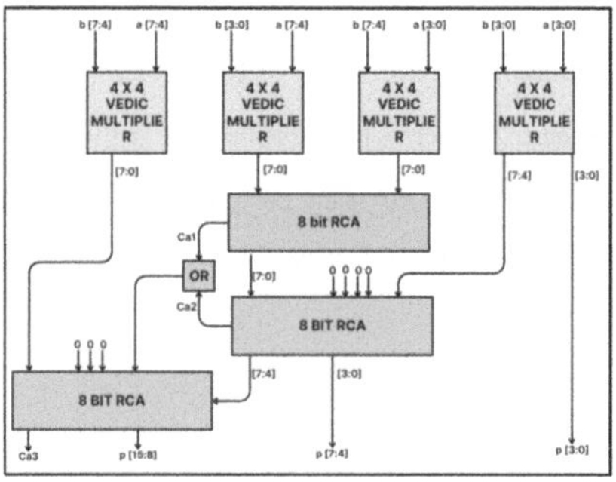

Fig. 3. 8 × 8 Vedic multiplier.

2.6 Shift and ADD Multipler

The shift and add multiplication algorithm, also known as the multiply-and-accumulate method, is a fundamental approach used to perform multiplication operations using only basic arithmetic operations such as addition and shifting (rightward). This algorithm is particularly useful for implementing multiplication in hardware circuits where complex multiplication units may not be feasible or efficient. First step is to initialize Accumulator (AC) as 0 and two operands, the multiplier (QR) and the multiplicand (BR), Also, initialize carry bit (C) to 0 and the LSB of QR to be Qn. Initialize counter SC to be equal to the number of bits (n) of both the multiplier and the multiplicand which will also be the number of iterations need. If the least significant bit (rightmost bit) of the multiplier QR (Qn) is equal to 1 then add the multiplicand (BR) to the accumulator (AC). After addition perform logical shift right on the carry bit, accumulator and the multiplier QR. After logical shift right, decrement the SC count and check if it is zero, if yes then the 2n bit result will be stored in the accumulator (AC) and the multiplier (QR). If no then repeat from step 2. If the least significant bit (rightmost bit) of the multiplier QR (Qn) is equal to then implement logical shift right on the carry bit, accumulator and the multiplier QR. After logical shift right, decrement the SC count and check if it is zero, if yes then the 2n bit result will be stored in the accumulator (AC) and the multiplier (QR). If no, then repeat from step 2.

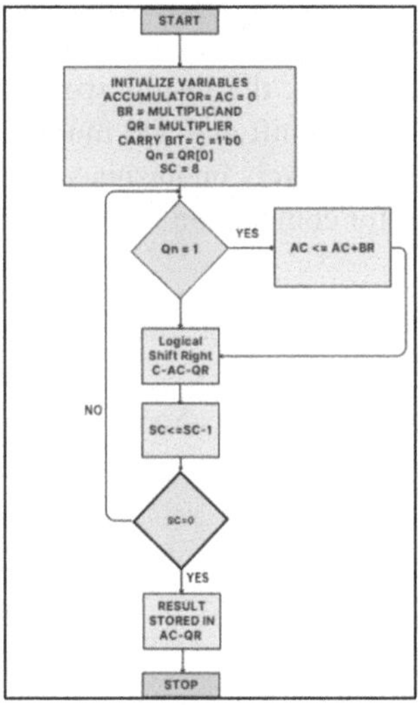

Fig. 4. 8 × 8 Shift and add multiplier.

2.7 Comparative of Standalone Vedic multiplier and Booth multiplier

Kumawat et al., (2014), [3] has presented a comparison between standalone Vedic Multiplier and Booth Multiplier using Urdhva Triyagbhyam Sutra, with an emphasis on minimal delay in multiplication operations across various numerical types and magnitudes. They have illustrated the practical applicability and performance of this method through its implementation using Verilog HDL for 8x8 bit multiplication, followed by FPGA synthesis utilizing Xilinx tools on the Spartan 3E kit. The synthesis results presented by the authors reveal a notable computation time of 15.05 ns for computing the product of 8x8 bits, underscoring the speed advantage of the Urdhva Triyagbhyam method over conventional multiplier architectures. Additionally, the study introduces innovative modifications to ripple carry adders, optimizing hardware utilization and further enhancing computation speeds.

Anju et al. (2013), [1] has presented a performance comparison of Vedic Multiplier and Booth Multiplier which emphasizes the importance of reducing power consumption and delay in processors to enhance overall performance. The research highlights the need for optimizing multiplier architecture to achieve efficient processor design. The methodology involves implementing the Vedic Multiplier and Booth Multiplier using VHDL in Modelsim 10.1b, followed by synthesizing the designs in Xilinx 8.2i ISE, this closely aligns with our paper's approach. Table 3 in their study compares the area, delay, and power consumption of both Booth and Urdhava Tiryakbhyam (Vedic) multipliers. The results indicate that the Urdhava Tiryakbhyam multiplier outperforms the Booth multiplier in terms of area, delay, and power consumption.

Regarding the choice of the shift-and-add multiplier over other common multipliers, such as Booth's multiplier. The shift-and-add multiplier was selected for this study because the Vedic multiplier we designed is specifically capable of multiplying only unsigned integers. If we were to compare the Vedic multiplier with Booth's multiplier, which supports signed multiplication, the results would likely be biased in favor of the Vedic multiplier. Therefore, the shift-and-add multiplier, which closely resembles Booth's algorithm but focuses exclusively on unsigned multiplication, provides a more appropriate and unbiased basis for comparison in our research.

Table 1. Performance of Standalone Vedic Multiplier and Booths Multiplier in [1]

FPGA Device Package (XC2550 TQ144)	Area in LUT's	Delay (ns)	Speed (MHz)	Memory (Kb)	Power (mW)
Booth's Multiplier	9	11.176	89.477	156,440	11.30
Urdhava Multiplier	86	8.460	118.203	166,744	9.29

3 Methodology

3.1 RISC Processor

The 16-bit RISC processor is driven by a clock signal and a system reset input. The clock signal provides timing synchronization for all processor operations, while the system reset (active-low) initializes the processor state. When the clock signal is active and the system reset is low, the processor activates. The Program Counter (PC) determines the address of the next instruction in the program memory. Instructions stored in the program memory are fetched and loaded into the Instruction Register (IR). The IR holds the current instruction being executed.

The instruction stored in the IR is decoded by the Control Unit. The Control Unit interprets the instruction opcode to determine the corresponding operation to be executed. Based on the decoded instruction, the Control Unit directs the processor to execute specific arithmetic or logical operations using data stored in the General-Purpose Registers (GPRs) or memory. Within the set of instructions handled by the Control Unit, there

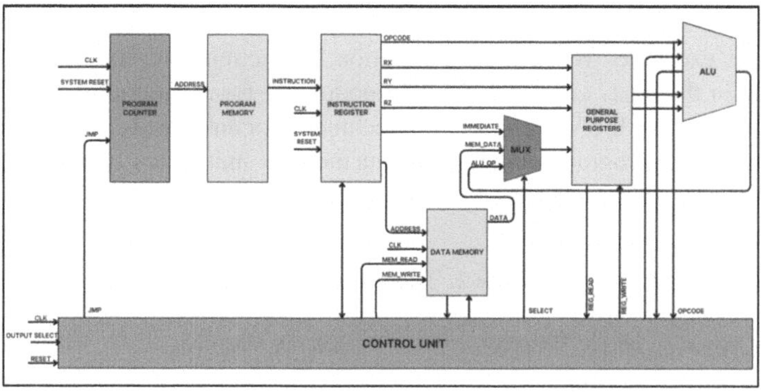

Fig. 5. Block diagram of 16-bit RISC processor.

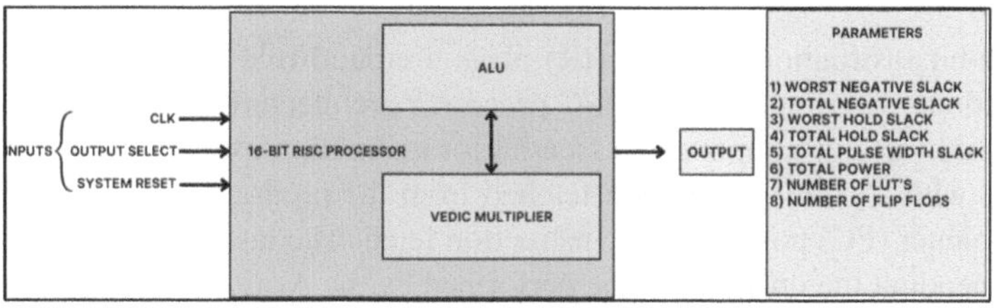

Fig. 6. Block Diagram of 16-Bit RISC Processor with 8 × 8 Vedic multiplier, Inputs, Outputs, and Comparison Parameters.

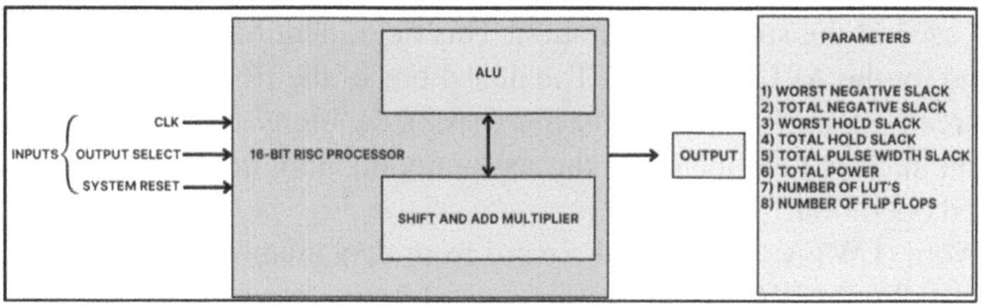

Fig. 7. Block Diagram of 16-Bit RISC Processor with Shift and Add multiplier, Inputs, Outputs, and Comparison Parameters.

exists a specific instruction for multiplication (mul). When this instruction is encountered during execution, it triggers the Vedic multiplier module / Shift and Add multiplier module. The Vedic multiplier module / Shift and Add multiplier module performs multiplication operations on the specified operands, leveraging its unique design for efficient and rapid computation.

The Program memory unit can store 16 instructions, each 16 bits wide, serving as the primary storage for program instructions. 16-bit wide instruction register holds the currently fetched instruction from the program memory for decoding and execution. The ALU performs a range of arithmetic and logical operations (14 operations in total) on data from the register bank based on control signals from the control unit. Notably, the

ALU includes a specialized multiplier module that is instantiated when a multiplication instruction is executed. When a multiplication instruction is decoded, the multiplier module within the ALU is activated. This module performs multiplication operations efficiently, leveraging the Vedic multiplier technique for enhanced computational speed and efficiency. Like program memory, the data memory unit stores 16 lines of data, each 16 bits wide, providing storage for variables and data utilized during program execution.

Table 2. Instruction Register

16-BIT INSTRUCTION REGISTER			
15–12 bits	11–8 bits	7–4 bits	3–0 bits
Operation Type	Destination Register	Source Register 1	Source Register 2

The 16-bit Instruction Register (IR) plays a crucial role in decoding and executing instructions within the 16-bit RISC processor architecture. The IR is structured to accommodate various components essential for instruction execution. Specifically, the IR is filled with a 16-bit instruction fetched from the program memory, to which the program counter (PC) points during instruction fetch. The last 4 bits of the IR are dedicated to encoding the operation to be performed by the Arithmetic Logic Unit (ALU). These bits select one of the 14 different operations supported by the ALU. The subsequent 4 bits of the IR specify the address of the destination General Purpose Register where the result of the operation will be stored after execution. The next 4 bits of the IR indicate the address of the first source General Purpose Register 1(rsrc1), representing Operand 1 used in the specified operation. This field identifies the register holding the first operand for the ALU operation. The final 4 bits of the IR specify the address of the second source General Purpose Register 2 (rsrc2), denoting Operand 2 required for the ALU operation. This field identifies the register containing the second operand used in the specified operation.

Load Word (LW) is used to load a word from data memory into a general-purpose register. Store Word (SW)is used to store a word from a general-purpose register into a data memory. Addition (ADD) performs addition of two operands, Subtraction (SUB) performs subtraction of two operands, Multiplication (MUL) initiates multiplication of two operands. Logical and (AND) performs bitwise AND operation between two operands, Logical or (OR) performs bitwise OR operation between two operands, Logical not (NOT) performs bitwise NOT operation on a single operand, Logical ex-or (XOR) performs bitwise XOR (exclusive OR) operation between two operands, Logical nand (NAND) performs bitwise NAND operation between two operands, Logical nor (NOR) performs bitwise NOR operation between two operands and Logical ex-nor (XNOR) performs bitwise XNOR (exclusive NOR) operation between two operands. Jump (JUMP) changes the program counter to jump to a specified memory address and Halt (HALT) stops the processor's operation, effectively halting further execution. These opcode assignments enable the control unit of your processor to decode instructions fetched from memory (stored in the Instruction Register) and determine the appropriate operation to be executed by the ALU or other processor components.

Table 3. Instruction Set

OPERATION TYPE	OPCODE
LOAD WORD (LW)	0000
STORE WORD (SW)	0001
ADDITION (ADD)	0010
SUBTRACTION (SUB)	0011
MULTIPLICATION (MUL)	0100
LOGICAL AND (AND)	0101
LOGICAL OR (OR)	0110
LOGICAL NOT (NOT)	0111
LOGICAL EX-OR (XOR)	1000
LOGICAL NAND (NAND)	1001
LOGICAL NOR (NOR)	1010
LOGICAL EX-NOR (XNOR)	1011
JUMP (JUMP)	1100
HALT (HALT)	1101

GPR (0) is loaded with 16-bit hexadecimal 0002 from data memory (0). GPR (1) is loaded with 16-bit hexadecimal 0003 from data memory (1). GPR (15) is loaded with 16-bit hexadecimal 00AB from data memory (15). GPR (14) is loaded with 16-bit hexadecimal 00CD from data memory (14). GPR (0) and GPR (1) are added, and the result is stored in GPR (2). The result from GPR (2) is then stored in data memory (2). GPR (15) and GPR (14) are multiplied using the Vedic and shift-and-add multipliers, with the result stored in GPR (13) and then in data memory (13). GPR (14) and GPR (15) are subtracted, and the result is stored in GPR (3) and then in data memory (3). A bitwise AND operation is performed on GPR (14) and GPR (15), with the result stored in GPR (4) and then in data memory (4). A bitwise OR operation on GPR (14) and GPR (15) stores the result in GPR (5) and then in data memory (5). A bitwise XOR operation on GPR (14) and GPR (15) stores the result in GPR (6). Finally, a halt operation indicates the completion of all instructions.

3.2 Implementation of Vedic Multiplier

When the system reset (sys_rst) signal is low and the multiplication operation (mul_op) signal is high, it indicates a specific condition where the system is ready to perform a multiplication this scenario triggers the following actions.

The processor module detects the high state of mul_op during a normal operational state (after reset). This condition signals the processor to initiate a multiplication instruction. Inside the processor module, the mul operation case in the decode_inst task is activated. This prepares the necessary operands (temp1 and temp2) by assigning them

Table 4. Data Given to Program Memory

Sr. No	INSTRUCTION	HEXADECIMAL FORM
1	0000_0000_0000_0000	0000
2	0000_0001_0001_0000	0110
3	0000_1111_1111_0000	0FF0
4	0000_1110_1110_0000	0EE0
5	0010_0010_0000_0001	2201
6	0001_0010_0010_0000	1220
7	0100_1101_1111_1110	4DFE
8	0001_1101_1101_0000	1DD0
9	0011_0011_1110_1111	33EF
10	0001_0011_0011_0000	1330
11	0101_0100_1110_1111	54EF
12	0001_0100_0100_0000	1440
13	0110_0101_1110_1111	65EF
14	0001_0101_0101_0000	1550
15	1000_0110_1110_1111	86EF
16	1101_0000_0000_0000	D000

the values from the source registers (GPR [rsrc1] and GPR [rsrc2]). The vedic multiplier module begins the multiplication operation using the provided operands (ip1 and ip2). The result of the multiplication (op) is calculated and made available once the operation.

3.3 Implementation of Shift and Add Multiplier

When the system reset (sys_rst) signal is low and the multiplication operation (mul_op) signal is high, it indicates a specific condition where the system is ready to perform a multiplication this scenario triggers the following actions.

The processor module detects the high state of mul_op during a normal operational state (after reset). This condition signals the processor to initiate a multiplication instruction. Inside the processor module, the mul operation case in the decode_inst task is activated. This prepares the necessary operands (temp1 and temp2) by assigning them the values from the source registers (GPR[rsrc1] and GPR[rsrc2]). Additionally, the t_start signal is asserted (t_start <= 1'b1), which serves as a trigger to start the multiplication process in the shift and add multiplier module (shift_add_mul_new). With t_start set high, the shift and add multiplier module begins the multiplication operation using the provided operands (ip1 and ip2). The result of the multiplication (op) will be calculated and made available once the operation completes.

3.4 Parameters Used for Comparison

Worst Negative Slack (WNS), Total Negative Slack (TNS), Worst Hold Slack (WHS), Total Hold Slack (THS), and Total Pulse Width Slack are critical timing metrics that directly impact the performance and reliability of processors and multipliers. WNS indicates the largest timing violation in signal arrival at flip-flops, with smaller values suggesting greater risks of functional errors. TNS aggregates all negative slack values, providing an overview of timing issues that may require redesign. WHS measures the minimum time difference for data to be latched correctly under hold conditions, while THS sums all hold slack values to highlight potential violations that could affect data integrity. Total Pulse Width Slack assesses the minimum width needed for signals to be correctly recognized, ensuring reliable transitions. Together, these metrics help evaluate timing performance.

The operands required for the multiplication operation are selected from the GPRs or memory locations based on the instruction's addressing mode. Upon completion of instruction execution, the results of arithmetic or logical operations can be accessed based on the output select input provided to the processor. Post-synthesis and implementation using Xilinx Vivado, the processor's performance is evaluated through various parameters:

- Worst Negative Slack (WNS): Measures the critical path delay, indicating the longest delay in meeting timing requirements.

WNS = Data Required Time – Data Arrival Time.

Register to Register Setup Slack
Arrival Time = Tck- > q + Combo Delay
Required Time = Clock Period - Setup Time
Setup Slack = Required Time - Arrival Time
Setup Slack = Clock Period - Setup Time - Tck- > q – Combo Delay

- Total Negative Slack (TNS): Sum of all negative slack values, reflecting overall timing constraints.
- Worst Hold Slack (WHS): Represents the minimum hold time margin for timing stability.

WHS = Data Arrival Time – Data Required Time.

Register to Register Hold Slack:
Required Time = Hold Time
Arrival Time = Tck- > q + Combo Delay
Hold Slack = Arrival Time - Required Time
Hold Slack = Tck- > q + Combo Delay - Hold Time

- Total Hold Slack (THS): Aggregate hold time margin across all timing paths.
- Total Path Worst Slack (TPWS): Quantifies the most critical timing path within the design.
- Total Power Consumption: Reflects the dynamic and static power consumption of the synthesized design.
- Number of Look-Up Tables (LUTs) and Flip-Flops (FFs): Indicate the hardware resources utilized in the design.

3.5 Hardware Used

The Spartan 7 series provides a balanced mix of affordability and computational power, making it suitable for academic projects with budget constraints. Additionally, these FPGAs offer low power consumption without sacrificing performance, which is beneficial for applications targeting low-power devices.

The Spartan-7 XC7S50, offers several key features that align well with our research needs. It features up to 52,160 logic cells and 2700 Kbits of block RAM, providing ample space for implementing complex digital circuits. The FPGA also includes up to 120 DSP slices, crucial for efficient arithmetic operations in the Vedic multiplier. Advanced clock management capabilities, with up to 6 clock management tiles, allow precise control of clock signals essential for synchronous designs. To diversify the simulation results gained from implementation purpose we have also used spartan 7, basys 3 and nexys a7 100t FPGA boards.

4 Results

Table 5. Results for Processor integrated with 8 × 8 Vedic multiplier.

PARAMETERS	Units	8 × 8 VEDIC MULTIPLIER			8 × 8 SHIFT AND ADD MULTIPLIER		
		SPARTAN 7	BASYS 3	NEXYS A7 100T	SPARTAN 7	BASYS 3	NEXYS A7 100T
Worst negative slack (WNS)	ns	8.161	8.728	8.743	6.050	7.249	7.240
Total negative slack (TNS)	ns	0	0	0	0	0	0
Worst hold slack (WHS)	ns	0.197	0.215	0.201	0.182	0.174	0.113
Total hold slack (THS)	ns	0	0	0	0	0	0
Total pulse width slack (TPWS)	ns	0	0	0	0	0	0
Total power consumption	W	0.044	0.097	0.123	0.045	0.097	0.112
Number of LUT's	–	373	207	208	307	307	307
Number of FF's	–	27	27	27	65	65	65

The integration of the processor with the Vedic multiplier results in a higher Worst Negative Slack (WNS) compared to the shift and add multiplier, indicating better setup

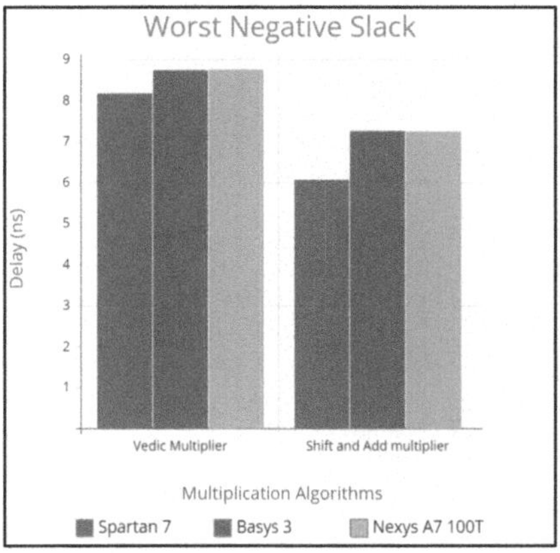

Fig. 8. Bar Graph Comparing Worst Negative Slack of FPGA Boards

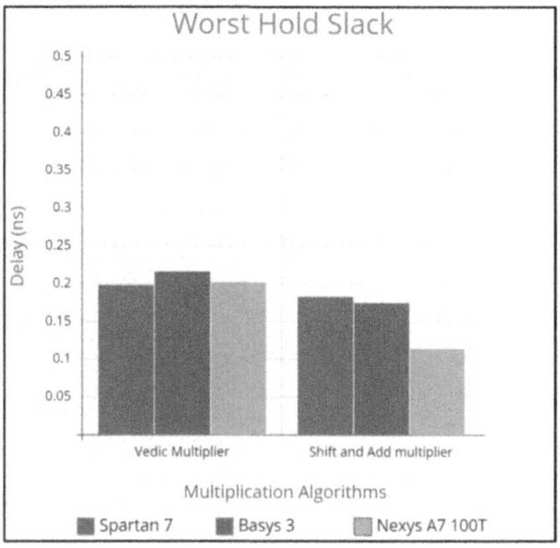

Fig. 9. Bar Graph Comparing Worst Hold Slack of FPGA Boards

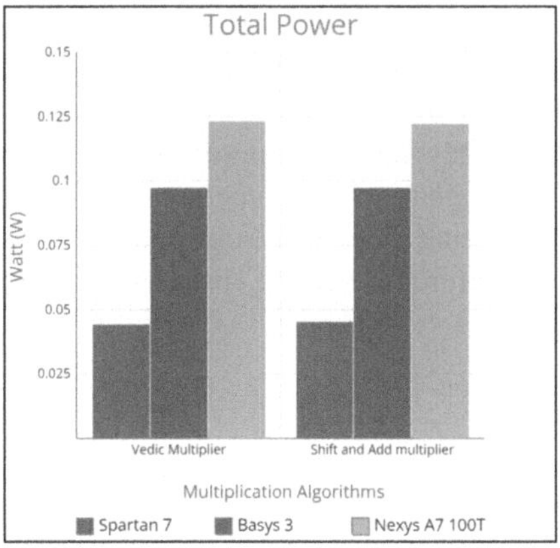

Fig. 10. Bar Graph Comparing Total Power of FPGA Boards

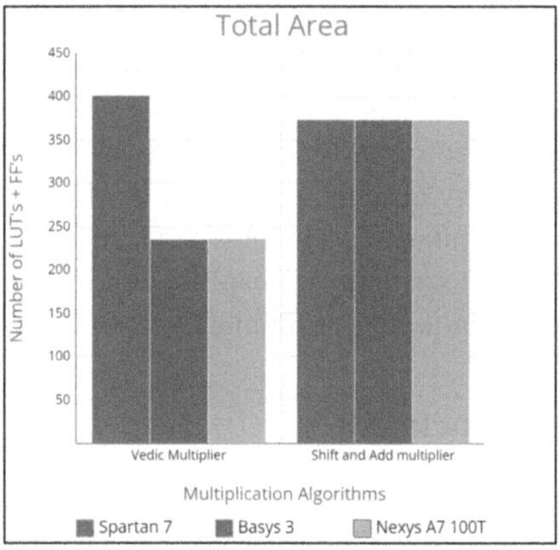

Fig. 11. Bar Graph Comparing Total Area of FPGA Boards

delay performance but at the cost of increased area and power consumption. This improved WNS is due to the shorter critical path and reduced number of flip-flops in the Vedic design. In contrast, the shift and add multiplier shows a lower WNS, posing potential timing challenges, but it offers advantages in power efficiency (0.045 W) and reduced area utilization, using fewer Look-Up Tables (LUTs) and flip-flops. Both designs meet their timing requirements with a Total Negative Slack (TNS) of 0 ns, and have positive Worst Hold Slack (WHS), Total Hold Slack (THS), and Total Pulse Width Slack (TPWS), ensuring proper data capture and signal integrity throughout the system. Finally, the design demonstrates successful multiplication of two hexadecimal numbers, producing the result 88EF on the output LEDs as shown in Fig. 12.

Fig. 12. Implementation on FPGA Board.

5 Future Scope

We acknowledge the current study's focus on metrics like Worst Negative Slack (WNS), Total Negative Slack (TNS), Worst Hold Slack (WHS), and Total Power Consumption. While these are critical for understanding processor performance, future work will expand the analysis to include energy efficiency, latency, throughput, and factors such as area efficiency, speed, and robustness under varying conditions, which are vital for embedded systems. Additionally, we plan to move beyond the assumption of ideal conditions by accounting for real-world factors like signal noise, temperature variations, and manufacturing inconsistencies. Lastly, recognizing the reliance on simulation data, we aim to incorporate experimental validation in future research to ensure the practical applicability of our findings in real-world scenarios.

6 Conclusion

In conclusion, this paper analyzed the performance of a 16-bit RISC processor integrated with two multiplier designs: the Vedic multiplier and the shift and add multiplier. Implemented using Verilog on an FPGA board, the study compared key performance metrics such as Worst Negative Slack (WNS) to evaluate timing behavior. The Vedic multiplier showed better setup delay performance with higher WNS but consumed more area and power, while the shift and add multiplier had lower WNS, indicating potential timing challenges, but was more efficient in terms of area and power usage. These results highlight the trade-offs in digital design, where the choice of multiplier depends on the balance between performance, resource utilization, and power efficiency.

References

1. Anju: Performance Comparison of Vedic Multiplier and Booth Multiplier. Int. J. Electron. Communicat. Eng. Technol. **2**(5), 336–339 (2013) BEIESP, ISSN: 2249-054X. [Retrieval Number: E1846062513/2013©BEIESP]
2. Yadav, A., Bendre, V.: Design and Verification of 16-bit RISC Processor Using Vedic Mathematics, 759–764 (2021). https://doi.org/10.1109/ESCI50559.2021.9396965

3. Kumawat, J., Sharma, S.: A 8×8 bit multiplier using Vedic Mathematics. Int. J. Eng. Tech. Res. **2**(3) (March 2014) ISSN: 2321-0869
4. Tirthaji, S.B.K.: Vedic Mathematics. Motilal Banarsidass Publishers, Delhi (1965)
5. Dhillon, H.S., Mitra, A.: A Digital Multiplier Architecture using Urdhava Tiryakbhyam Sutra of Vedic Mathematics. IEEE Conference Proceedings. (2008)
6. Pushpalata Verma, K.K., Mehta: Implementation of an efficient multiplier based on Vedic Mathematics using EDA Tool. Int. J. Eng. Adv. Technol. **1**(5), 2249–8958 (2012)
7. Gaonkar, S., Anitha, M.: Design of 16-bit RISC Processor, Int. J. Eng. Res. Technol. **02**(07) (July 2013)
8. Bandi, V.: Performance analysis for Vedic multiplier using modified full adders. In: Innovations in Power and Advanced Computing Technologies (i-PACT), Vellore, India, pp. 1–5 (2017). https://doi.org/10.1109/IPACT.2017.8245017
9. Shanthi, K.G., Sandhiya, G., Abinaya, K., Sangeetha, A., Aruna, T., Aswini, R.: Performance Analysis of Vedic Multiplier and Modified Vedic Multiplier in Direct Digital Synthesizer, pp. 64–67 (2022). https://doi.org/10.1109/ICESC54411.2022.9885340
10. Agarwal, M.: Performance analysis of Vedic multiplier in signal processing applications. Int. J. Innov. Sci. Res. Technol., 1882–1885 (2024). https://doi.org/10.38124/ijisrt/IJISRT24MAR1194
11. Shaik, S., Kanapala, S., Vijay, V., Pittala, C.S.: Design and performance analysis of low power and energy-efficient vedic multipliers. International Journal of System Assurance Engineering and Management, Springer; The Society for Reliability, Engineering Quality and Operations Management (SREQOM),India, and Division of Operation and Maintenance, Lulea University of Technology, Sweden. **14**(3), 894–902 (2023)
12. Sridevi, A., Sathiya, A.: Design of energy efficient N-bit vedic multiplier for low power hardware architecture. Int. J. Electron. **1–21** (2024). https://doi.org/10.1080/00207217.2024.2382494

All-Optical XOR Gate and Half Adder Implementation Using SPM and SOA

Aakriti Garima, Vasundhara(✉), and Abhilash Mandloi

Sardar Vallabhbhai National Institute of Technology, Surat, Gujarat, India
vasundhara.nit@gmail.com, asm@eced.svnit.ac.in

Abstract. Essential components for all-optical communication and computation are optical gates. They enable data processing in the optical domain by using non-linear optical phenomena to apply logical operations to light signals. Advanced photonic applications find optical gates beneficial due to their fast speed, parallelism and high bandwidth. The design concepts, modelling methods and practical results of an all-optical XOR logic gate that utilises self-phase modulation (SPM), a non-linear optical fiber feature, are presented in detail in this study. Additionally, an all-optical Half Adder is realised at 10 Gb/s, using proposed XOR gate and currently available AND gate simulation. The simulation results precisely align with standard benchmarks, on the basis of time domain analysis.

Keywords: Optical fiber · Self phase modulation (SPM) · Semiconductor Optical Amplifier (SOA) · Gain Saturation · Optisystem

1 Introduction

The core elements of photonic systems are optical gates, which regulate light signals in optical networks to facilitate fast switching, modulation and signal processing. These can be developed using semiconductor optical amplifiers (SOA), electro-optic modulators, non-linear optical effects and photonic crystals. When compared to electronic gates, optical gates provide better energy efficiency, faster switching speeds and larger capacity for data handling. Applications in the fields of signal processing, telecommunications, medical imaging and sensor technologies drive innovation and breakthroughs in a variety of sectors. Optical fibers exhibit non-linear effects when the light intensity is high. The primary non-linear effect is the Kerr effect, where the refractive index n changes with the optical intensity I, as depicted in eq. 1:

$$n = n_0 + n_2 I \tag{1}$$

where n_0 is the linear refractive index and n_2 is the non-linear refractive index coefficient. SPM is a key non-linear effect, where the phase of the optical signal

Supported by Sardar Vallabhbhai National Institute of Technology, Surat.

D. Gupta et al. (Eds.): PCEMS 2024, CCIS 2491, pp. 120–127, 2025.
https://doi.org/10.1007/978-3-031-90577-3_11

is modulated by its own intensity. This leads to a nonlinear phase shift $\Delta\phi$, as mentioned in eq. 2:

$$\Delta\phi_{\mathrm{SPM}} = \gamma P_0 L_{\mathrm{eff}} \tag{2}$$

where:

- γ is the non-linear coefficient of the fiber, given by: $\gamma = \frac{n_2\omega_0}{cA_{\mathrm{eff}}}$
  - n_2 is the non-linear refractive index coefficient,
  - ω_0 is the angular frequency of the optical signal,
  - c is the speed of light in a vacuum,
  - A_{eff} is the effective area of the fiber.
- P_0 is the optical power of the signal.
- L_{eff} is the effective length of the fiber, defined by eq. 3:

$$L_{\mathrm{eff}} = \frac{1 - e^{-\alpha z}}{\alpha} \tag{3}$$

  - α is the attenuation coefficient of the fiber,
  - z is the physical length of the fiber.

2 SOA-Induced Non-linearities

B. R. Ravi et al. in [1] have developed all-optical XOR gate by exploiting the non-linear characteristics XPM and XGM of SOA as a basis for Mach-Zehnder Interferometer (MZI). Data should be in optical domain to attain faster data transfer speeds where MZI can be very helpful. Ajay Kumar et al. in [4] performed the simulation of XOR gate by incorporating SOA-MZI using MATLAB. A. Saharia et al. in [2] designed AND gate using SOA and Band Pass Filter (BPF), operating at high speed of 10 Gb/s. Here, non-linear characteristics of SOA such as XGM and four wave mixing (FWM) is utilised to implement various logic functions. The results are analyzed through optical power meter and optical time domain visualiser. While an all-optical half adder based on non-linear characteristics of SOA, have been designed by R. Mehra et al. in [3], where sum and carry outputs are obtained by realising XOR and AND gate outputs respectively, using BPF and SOA. They further experimented with the power obtained for deciding 0/1 levels and got satisfactory results at 80 Gb/s. To support ultra-fast computing, M. Michael et al. in [5], designed all-optical logic gate with reflective SOA, as they exhibit properties like small power consumption, reduced latency and enlarged switching speed.

3 Operative Mechanism

3.1 SPM Based XOR Gate

In this study, an all-optical XOR gate using OptiSystem, is designed and simulated, as shown in Fig. 1. CW Lasers are used to generate the input signals.

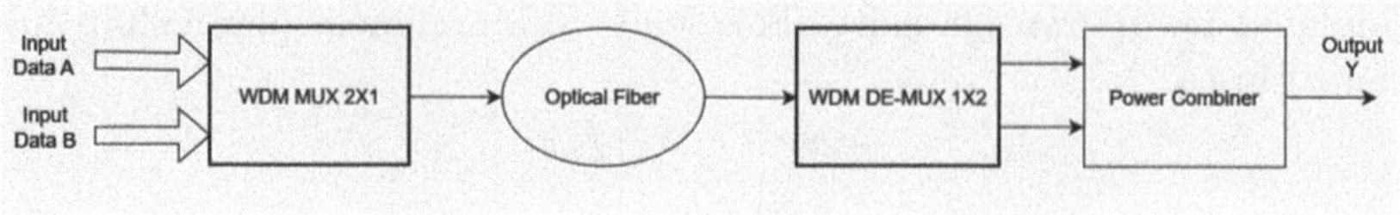

Fig. 1. Block schematic of SPM-based XOR gate.

These are tuned at frequencies 192 THz and 192.1 THz respectively, with power -5 dBm each.

The inputs are combined using a 2x1 multiplexer (MUX) with a bandwidth of 10 GHz and then passed through an optical fiber of length 5 km. The optical fiber induces SPM, introducing a phase shift proportional to the signal intensity. To preserve the phase information, the signal is split back into two paths using a demultiplexer (DEMUX), with bandwidth of 10 GHz.

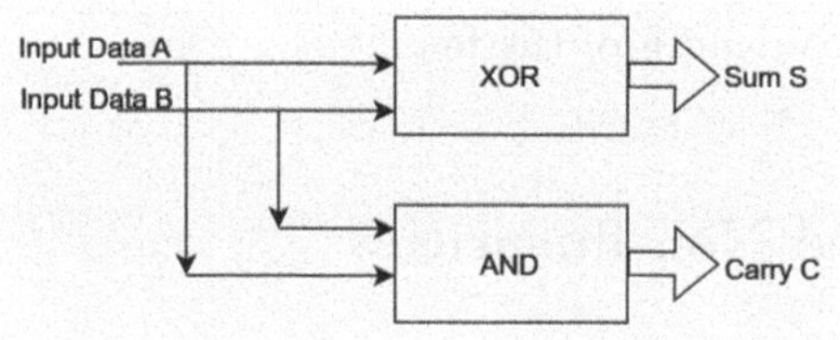

Fig. 2. Block schematic of SPM-based Half Adder

These signals are then recombined using a power combiner to create a modulated signal corresponding to the XOR operation. This combined signal is amplified using a traveling-wave SOA, which exploits gain saturation. Injection current for SOA is 0.19 A. The SOA provides higher amplification for lower input intensities and lower amplification for higher input intensities, crucial for differentiating the input combinations. For visualizing the output, the amplified signal is converted to an electrical signal using a photodetector PIN. Finally, a bandpass gaussian filter, with frequency 12 GHz and bandwidth 15 GHz, cleans the signal, ensuring clear representation of the XOR output. The relationship between bandwidth and bit rate of gaussian BPF is shown in eq. 4.

$$Bandwidth = 1.5 * Bitrate \tag{4}$$

This setup effectively utilizes SPM and SOA gain saturation to achieve the XOR logic function in an optical communication system, using the parameters shown in Table 1.

3.2 SPM Based Half Adder

An all-optical Half Adder is designed and simulated using OptiSystem. It is a combinational arithmetic circuit that adds two numbers and produces two

Table 1. Parameters used in proposed system

Name	Value
Attenuation (dB/km)	0.3
Dispersion $(ps/nm/km)$	10.73
n_2 (m^2/W)	$3.1e^{-040}$
SOA Injection current (A)	0.19
SOA Differential gain (m^2)	$2.78e^{-020}$

outputs, sum bit (S) using XOR gate and carry bit (C) using AND gate, as shown in Fig. 2.

4 Results and Analysis

4.1 Simulation Setup

The simulation setup of XOR gate, as shown in Fig. 3, compares two digital input signals and produces a high output only when the inputs are different. This means for a 2-input XOR gate, exactly one input must be high for the output to be high. The proposed model confirms the same resultant waveform of the output signal. It is clear from the output waveform, that XOR operation has been successfully realized. While with the help of XOR gate and AND gate (as shown in Fig. 4), we have designed and simulated Half Adder at 10 Gb/s, as shown in Fig. 5. The waveforms of input signals data A and data B are shown in Fig. 6a and Fig. 6b respectively. Our proposed system yields outputs as Sum(S) and Carry(C), as shown in Fig. 7a and Fig. 7b respectively.

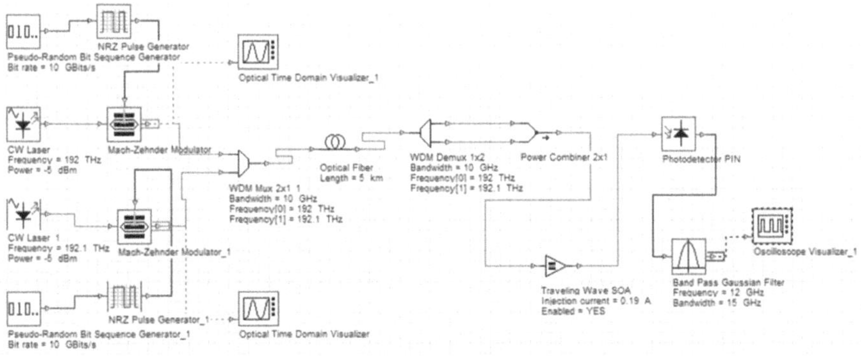

Fig. 3. Simulation setup of an all-optical XOR gate on OptiSystem

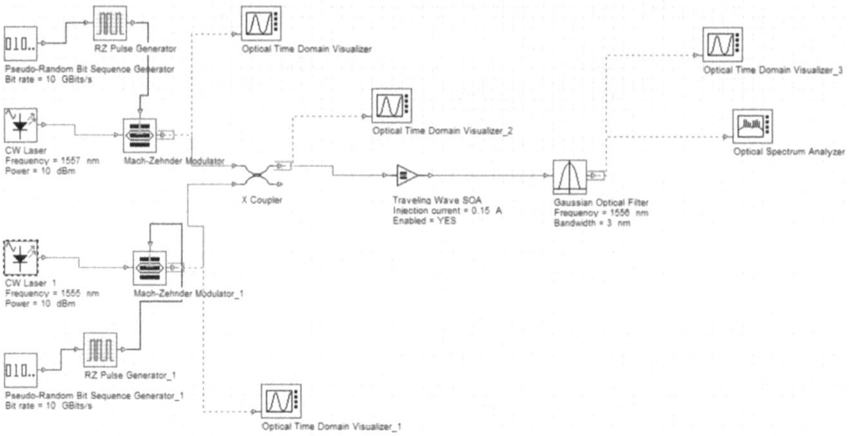

Fig. 4. Simulation setup of an all-optical AND gate on OptiSystem

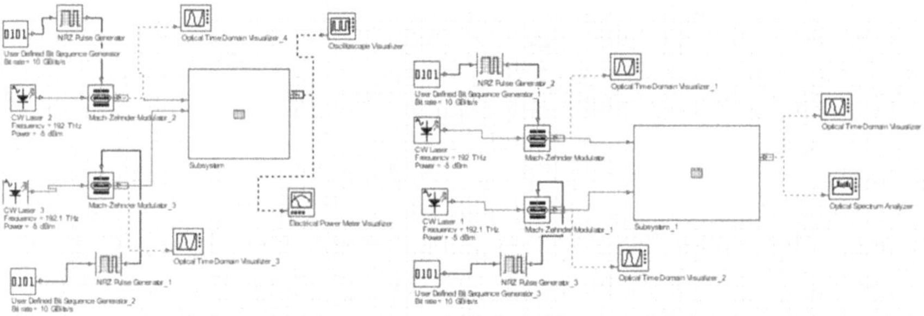

Fig. 5. Simulation setup of an all-optical Half Adder on OptiSystem

4.2 Analysis

Case 1: When both inputs A and B are (0,0), it means that low-intensity optical signals are generated. The MUX combines them into a low-intensity signal, which travels through the fiber with minimal SPM effects. The SOA is operating in its linear region due to low input power, provides slight gain without significant amplification. The PIN photodetector converts the optical signal to a low electrical signal, which is then passed through a gaussian BPF, thus maintaining a low intensity and noise-free signal, resulting in output '0'. **Case 2:** When input A is '0' (low intensity), and B is '1' (high intensity), the MUX combines them into a moderate intensity signal, which undergoes SPM in the fiber, causing a phase shift. Due to the moderate input power, the SOA operates in its linear region, providing linear (proportional) amplification without significant gain saturation effects. It is then converted into a high electrical signal, representing a clear high state '1'. **Case 3:** When input A is '1' and B is '0', similar working mechanism as above is observed for inputs (1,0) as well, generating a high pulse output, represented as '1'. **Case 4:** When both inputs A and B are '1' (high intensity), the MUX combines them into an even higher intensity signal, which undergoes significant non-linear phase shifts (SPM) in the fiber. The power combiner generates a high intensity signal for the SOA, causing significant gain saturation and non-linear changes. **Gain Saturation:** In the presence of high input power, the gain provided by the SOA decreases because the available carriers for amplification get depleted. This results in a lower gain for high-intensity inputs. Thus, generating a low output pulse as '0'. Comparative analysis of proposed method with other existing work in shown in Table 2.

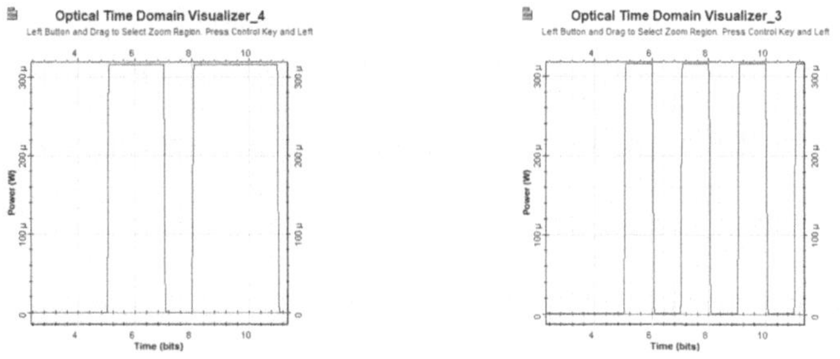

(a) Waveform of Input A(0000110111) (b) Waveform of Input B(0000101010)

Fig. 6. .

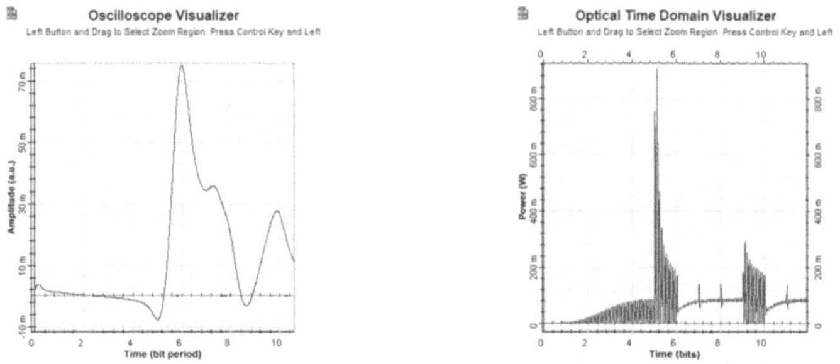

(a) Waveform of XOR gate output, Sum S (0000011101)
(b) Waveform of AND gate output, Carry C (0000100010)

Fig. 7. .

Table 2. Comparison between implementation of a logic gate by different authors

Parameters	Fok et al. [7]	J. M. Oliveira et al. [6]	Proposed design
Data rate (Gbps)	10	10	10
Interactive mediun	HNLF	SOA	SOA+HNLF
Non-linear effect	FWM	XGM	SPM
Quality factor (dB)	10	3.2	19.54
BER	Not mentioned	0.5×10^{-003}	3.67×10^{-103}

5 Conclusion

In this study, we have simulated a high-speed 10 Gbps all-optical XOR gate and Half Adder utilizing SPM. Our design is validated through theoretical research and simulations. By appropriately configuring these parameters, optical fibers can be used to create high-speed all-optical gates that offer good performance, low cost and minimal complexity.

Acknowledgments. I thank Dr. Abhilash S. Mandloi, Ms. Vasundhara and the staff of DoECE of SVNIT, Surat, for their technical assistance, valuable guidance, suggestions and constant motivation for carrying out this noble work.

References

1. Ravi, B.R., Theresal, T., Prince, S.: All-optical basic gates and XOR gate using SOA based mach-zehnder interferometer. In: 2014 International Conference on Communication and Signal Processing, Melmaruvathur, India, pp. 1327–1332 (2014). https://doi.org/10.1109/ICCSP.2014.6950065.
2. Saharia, A., Sharma, R.: An approach for the realisation of NAND, NOR, & AND gate using semiconductor optical amplifier & band pass filter. In: 2014 International Conference on Computational Intelligence and Communication Networks, Bhopal, India, pp. 978–984 (2014). https://doi.org/10.1109/CICN.2014.206.
3. Mehra, R., Jaiswal, S., Dixit, H.K.: All optical half adder design based on semiconductor optical amplifier. In: 2013 Tenth International Conference on Wireless and Optical Communications Networks (WOCN), Bhopal, India , pp. 1–5 (2013). https://doi.org/10.1109/WOCN.2013.6616183.
4. Kumar, A., Kumar, S., Raghuwanshi, S.K.: Implementation of XOR/XNOR and AND logic gates by using mach-zehnder interferometers, Optik, vol. 125, no. 19, pp. 5764–5767 (2014).ISSN 0030-4026, https://doi.org/10.1016/j.ijleo.2014.07.037. (https://www.sciencedirect.com/science/article/pii/S0030402614007785)
5. Michael, M., Britto, E.C., Saranya, Vignesh.: Design and analysis of high-speed all-optical logic gate with various pulse generators. In: 2023 Second International Conference on Electronics and Renewable Systems (ICEARS), Tuticorin, India, pp. 35–40 (2023). https://doi.org/10.1109/ICEARS56392.2023.10085134.

6. Oliveira, J.M., Sousa, F.B.d., Sousa, F.M.d., Oliveira, J.E.d., Costa, M.B.C., Costa, M.B.C.: A new system for all-optical and logic gate on semiconductor optical amplifier based michelson. In: 3rd International Symposium on Instrumentation Systems, Circuits and Transducers (INSCIT). Bento Gonçalves, Brazil 2018, pp. 1–6 (2018). https://doi.org/10.1109/INSCIT.2018.8546703
7. Fok, M.P., Prucnal, P.R.: Polarization effect on optical xor performance based on four-wave mixing. IEEE Photon. Technol. Lett. **22**(15), 1096–1098 (2010). https://doi.org/10.1109/LPT.2010.2050465

Design and Optimization of a 6 GHz Circular Patch Antenna Using Machine Learning Algorithms

Kalyani Bobade[1], Ankita H. Harkare[2(✉)], Ishika Patne[1], Anshika Jain[1], Bhumika Neole[2], and Vaijyanti H. Panse[3]

[1] Shri Ramdeobaba College of Engineering and Management, Nagpur, India
[2] Ramdeobaba University, Nagpur, India
harkareah@rknec.edu
[3] Visvesveraya National Institute of Technology, Nagpur, India
https://www.rknec.edu/

Abstract. Machine learning (ML) has become prime in various industries because it is filled with various and abundant data and has excellent processing. Present-day research underlines its importance in todays technology. The paper focuses on machine learning algorithms in antenna design as well as learning algorithms. Also, it looks for the basic differences between AI and machine learning, and the wide range of their technological approaches. It also compares the outcomes of machine learning in antenna design with those of traditional methods. The ML study shows that the random forest algorithm provides the best prediction results. A circular patch antenna is designed and fabricated using a coaxial feed that operates at 6.15 GHz. The antenna with a 15 percent bandwidth and 8.5 dBi gain is measured.

Keywords: Patch antenna · Machine learning · Narrow band

1 Introduction

Every day we require communication, absolutely effective communication. Therefore, it has become a very important part of our lives and business, education, and other sectors where the exchange of information and ideas is necessary. Communication is mainly about transmitting data from the sender to the receiver and vice-versa. Even though it looks simple, a large amount of data can be transmitted at faster speeds by cables or optical fibers, which are easily installed. Wired communication is not always good in unfavorable connectivity situations. Hence, wireless communication is needed in such a scenario to make sure that the sharing and relay of information are shared. In underground mines, communication is a key feature for safety and productivity. To hold up that market and miners lifestyles, it demands more stable communication systems.

Supported by organization Ramdeobaba University, Nagpur, India.

D. Gupta et al. (Eds.): PCEMS 2024, CCIS 2491, pp. 128–137, 2025.
https://doi.org/10.1007/978-3-031-90577-3_12

Simple, low-profile, cost-efficient, and power-saving antenna, such as a circular patch antenna, can meet this demonstrated need [1].Designing a circular patch antenna, particularly for frequency bands like 6 GHz, requires components such as patch radius, substrate properties, feed position, and others. Classical design methods often count on empirical formulas and repeated simulations, which are time-consuming and may not always lead to optimal performance. To solve this problem, machine learning (ML) algorithms are an effective tool for antenna design and optimization. Machine learning can efficiently inspect the extensive design space, enhance the antenna performance and parameters, and identify complex relationships between design parameters,. By applying these ML techniques, engineers can positively decrease the design cycle time and can achieve above antenna performance compared to regular methods. Without extensive electromagnetic simulations, machine learning can predict resonant frequency of the patch antenna designs, it can also predict the electromagnetic properties of new substrate materials used in patch antennas, aiding in material selection and development. By training on a dataset of previously designed antennas and their performance metrics, ML models can predict the performance of new antenna designs. Machine learning (ML) has enormous potential in the domain of antenna design and antenna behaviour as it allows for significant speeding up while maintaining high accuracy. ML models are used in prediction of the dielectric properties of materials used in DRAs, which is crucial for accurate design and simulation. In this research, we show a complete approach to the design and optimization of a 6.15 GHz circular antenna using machine learning. We used supervised learning and optimization techniques to predict and improve the antenna's performance parameters, return loss, gain, bandwidth, and radiation pattern. This shows that the included machine learning can speed up the antenna design process and help in achieving optimized results. In the latest machine learning is the process of finding significant information from data [2]. These algorithms have capacity for optimization, but how well they work depends on the quality and quantity of the data that is gathered [3].

The following contributions are made by this paper:

- Creating datasets in CST software using parametric analysis.
- Improvement of the simulation outcomes (S_{11}) by utilizing a range of machine-learning algorithms.
- Comparison of machine learning algorithms used in the proposed work based on their results.

An overview of machine learning is provided in this paper, with a particular emphasis on its potential applications in antenna design. Many learning algorithms that are used in machine learning are explored together with the topic itself. The results of each algorithm utilized and the techniques used to design the antennas utilizing machine learning in each are reported [4].

2 Design of Circular Patch Antenna

This section describes the design, fabrication, and testing of a circular patch antenna with a coaxial feed. The material used for the fabrication of the antenna is PEC copper. FR4 is used as a substrate with a dielectric constant of 4.4 and a thickness of 0.16 cm (about 0.06 inch). The design and dimensions of a circular patch antenna using the design equation are illustrated in Fig. 1. Figure 2 shows fabricated antenna.

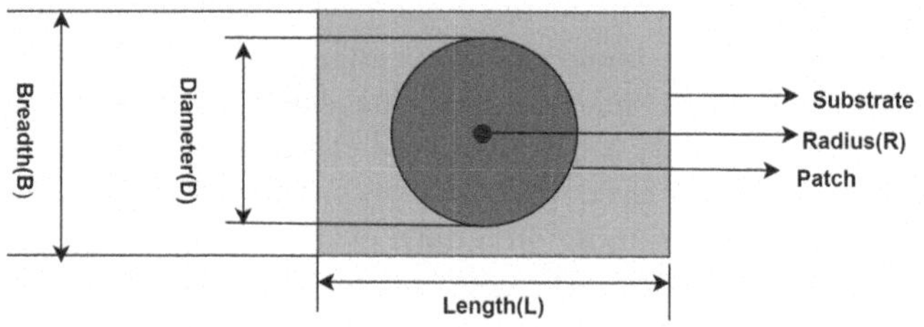

Fig. 1. Geometrical Layout of Proposed Antenna=2.5cm, L=2.7cm (about 1.06 in), D=1.3cm (about 0.51 in), R=0.6cm

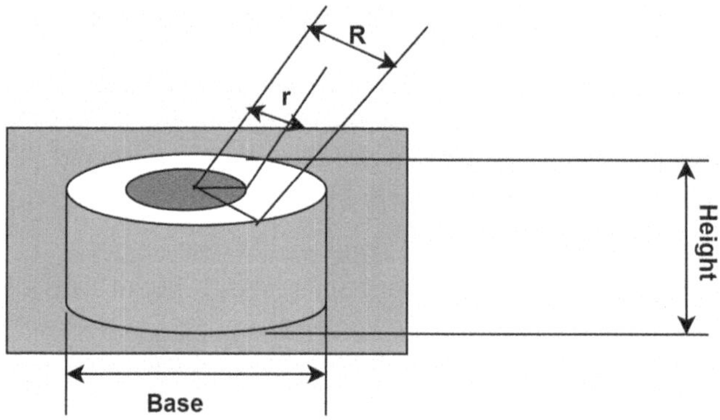

Fig. 2. Geometrical Layout of Proposed Antenna; Height=0.7cm (about 0.28 in), Base=0.5cm (about 0.2 in), R=0.3cm (about 0.12 in).

$$Er(r,\phi,\theta) = E_0 J_1 \frac{2.405r}{a} \cos(\omega t) \tag{1}$$

In a circular patch antenna operating at 6.15 GHz, the electric field distribution optimizes its radiation pattern, impedance characteristics, and polarization. Predominantly concentrated in the dielectric substrate, the E-field forms standing waves with the fundamental TM_{11} mode, exhibiting a sinusoidal radial variation and azimuth symmetry, peaking at the center and diminishing towards the edges, resulting in a broadside radiation pattern. Eq. (1) [5] describes the radial E-field component.

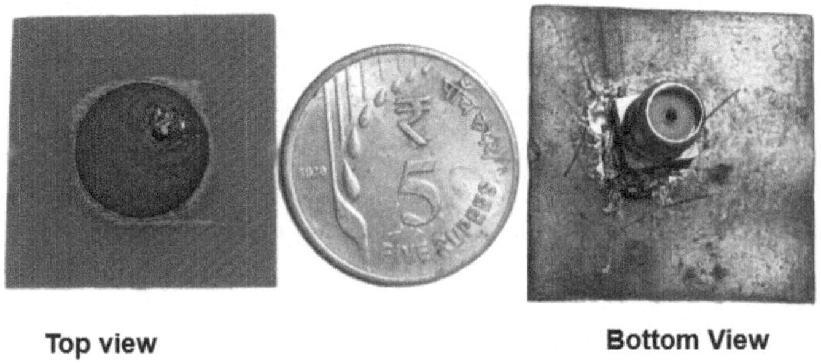

Fig. 3. Circular Patch Antenna

Where the peak field is E_0, the first-order Bessel function is J_1, the radial distance is r, the patch radius, and the angular frequency is ω. Finite Element Method (FEM) or Finite-Difference Time-Domain (FDTD) these methods can be used in numerical simulations, this accurately models this distribution, considering some factors including feed configuration, substrate, and permittivity. Simulations of a 6 GHz antenna show maximum E-field at the center, decreasing radially, with symmetric distribution around the center and field strength dependent on input power and impedance matching. Figure 4 shows the E-fields of the patch antenna. Understanding this distribution, specifically for TM_{11} mode, is essential for improving the performance of antenna, and merging this with the machine learning method allows for effective design clarification to meet particular performance criteria.

The results of the patch antenna are discussed in Sect. 3. The design provides a simulated and measured bandwidth of 15 % and a peak gain of 8.5 dBi at 6.15 GHz.

3 Measured and Simulation Results

As shown in Fig. 3, the proposed MIMO antenna prototype has been fabricated and measured. The measured and simulated impedance bandwidth obtained is 15 percent. This can be observed from the return loss vs frequency plot shown in Fig. 5. It can be observed that the measured and simulated results completely agree. The radiation pattern is important for its directional performance in circular patch antenna. In its fundamental mode (TM_{11}), the antenna produces a

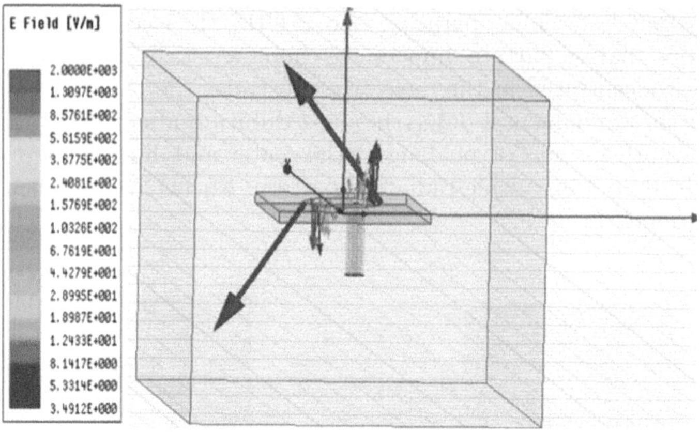

Fig. 4. E-plane of Antenna

broadside pattern, with the main lobe pointing perpendicular to the patch, resulting in uniform radiation in the horizontal plane and minimal radiation along the axis normal to the patch, creating a doughnut-shaped distribution. In the antenna the circular uniformity of the patch ensures unidirectional radiation in the azimuth plane, it also makes it suitable for uniform coverage, including satellite communications and mobile. Additionally, the antenna can achieve circular polarization, which helps reduce signal degradation caused by multi-path interference. This can be observed in Fig. 6.

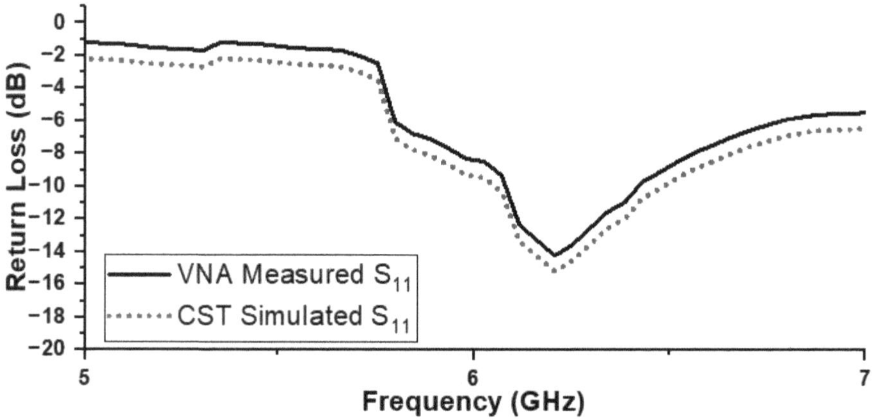

Fig. 5. Return loss of proposed antenna

Figure 7 shows the peak gain obtained at 6.15 GHz as 8.5 dBi. The gain improvement techniques were applied for high gain [6].

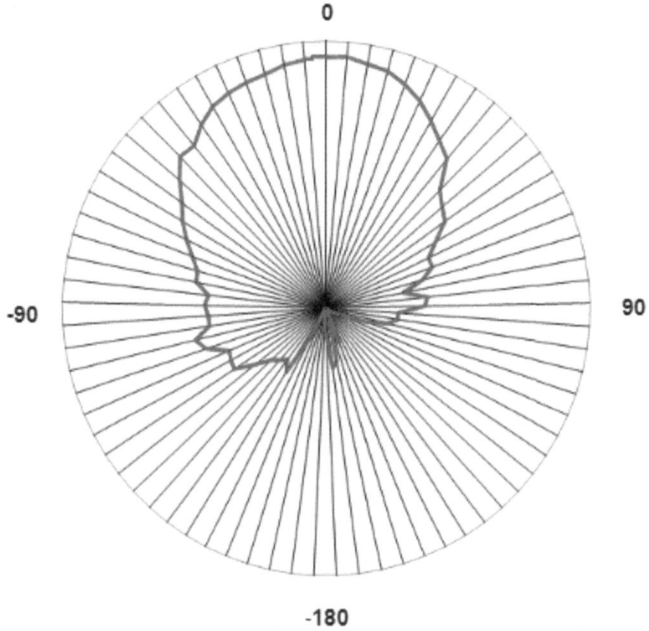

Fig. 6. Radiation pattern of purposed antenna

The measured and simulated results completely agree. Machine learning algorithms can prove beneficial for optimizing the results without human intervention. In order to provide a dataset for the predictions, an electromagnetic wave simulation of a circular patch antenna is carried out in this study using the CST Microwave Studio program. 80,000 datasets were collected for the training from CST Microwave Studio, and 10,00,00 datasets were used as test cases. S_{11} was predicted and various parameters of study were calculated. This is discussed in the following section.

4 Optimization of Design Using Machine Learning Algorithms

Certain machine learning models are trained using the datasets that are generated. These trained models are additionally put to the test for prediction accuracy. The suggested work's machine learning algorithms are K Nearest Neighbor (KNN), Random Forest (RF) and Decision Tree (DT).

4.1 Decision Tree

The supervised learning technique applies to both regression and classification problems. In this tree-structured classifier or regress, internal nodes represent

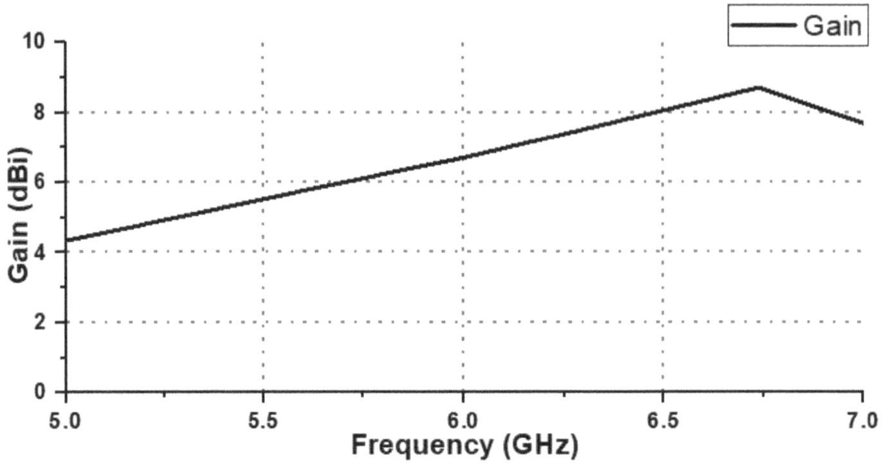

Fig. 7. Gain of purposed antenna

dataset properties, branches indicate decision rules, and each leaf node provides an answer. A decision tree comprises two nodes: the decision node and the leaf node. Leaf nodes represent the results of decisions and do not have any further branches. Decision nodes are used to make decisions and have multiple branches. These branches make use of aspects of the dataset to make decisions or to run tests. A decision tree is a pictorial representation of available options or choices predicted by a set of parameters. It is a tree that branches into sub-trees based on responses to queries. To make informed decisions for each subset, decision tree algorithms recursively divide the training set based on higher feature values [7]. Decision trees use pruning to avoid over fitting for the removal of a few sub-nodes; the "max depth" is important in controlling the complexions of decision tree algorithms, as a deeper tree with more sub-trees can lead to more perfect decision-making [8].

4.2 K-Nearest Neighbor

A simple machine learning technique known as K-nearest neighbour (KNN), an approach that works off the assumption that similar data points can be found near one another. It is widely disposable since it is non-parametric, meaning it does not make any underlying assumptions about distribution of data.This algorithm allows a new data point to the category that is comparable to the categories of its nearest neighbors, drawing the inference that the new data point and the previous data points are comparable. Throughout the training phase ,KNN keeps hold of its knowledge. When new data is experienced, it uses the mean of its k nearest neighbors to predict new values (for regression problems) or classify the data point into a category (for classification problems). KNN calculates the distance between two points using the Euclidean distance formula [9].

4.3 Random Forest

Random Forest is a commonly used machine-learning technique that works for classification as well as regression tasks. Enhancement of model performance and reduced complexities are major features of Random Forest. As the name suggests, Random Forest combines the output of multiple decision trees to reach a single result and averages them to upgrade the dataset's overall accuracy. Random Forest predicts outcomes using the majority votes of many decision trees. Its ease of use and flexibility have fueled its adoption widely [10].

4.4 Measuring Parameters of Optimization

The study uses several metrics to evaluate the performance of the model and for the same following aspects are calculated, mean squared error (MSE), root mean squared error(RMSE), mean absolute error (MAE), mean absolute percentage error (MAPE), and R^2 (R-squared) score. The R^2 score estimates the model's predictive effectiveness and is calculated using a specific formula below [11–13].

$$R^2 = 1 - \frac{Residualsumofsquares}{Totalsumofsquares} \tag{2}$$

When comparing estimated and real data, the MSE calculates the difference. Greater model quality is indicated by a lower MSE given in Eq. (3).

$$MSE = \frac{1}{n}\sum_{i=1}^{n}(values_i - predictedvalue_i)^2 \tag{3}$$

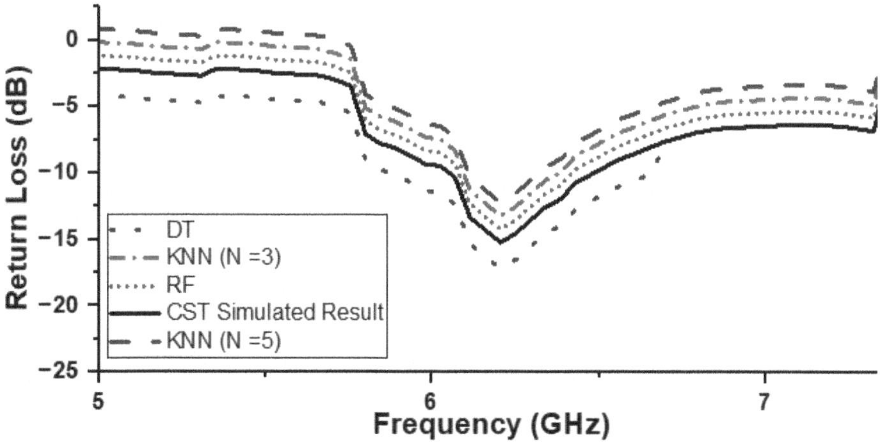

Fig. 8. Return loss of purposed antenna

RMSE is a statistical measure of the distribution of prediction errors, often known as residuals. The mean of the absolute difference between the supplied (actual) and anticipated values is known as the MAE. The formula for the MAE is given in Eq. (4).

$$MAE = \frac{1}{n}\sum_{i=1}^{n} abs(values_i - predictedvalue_i) \quad (4)$$

Table 1. Return loss comparison of various ML algorithms with actual antenna

Algorithm	R^2 Score	MSE	MAE
DT	0.848	0.251	0.134
KNN(N=3)	0.965	0.750	0.338
KNN(N=5)	0.953	1.132	0.418
RF	0.953	1.132	0.418

Table 1 gives a comparative study of the ML algorithm.

The curve fitting for S_{11} for various algorithms is shown in Fig. 8. It can be seen from Fig. 8 that random forest(RF) gives the best S_{11} results which is closest to the actual simulated result. This can also be proven from the values of Table 1.

5 Conclusion

This paper concludes that the entire band efficiency is over 65 percent with the obtained gain being 8.5 dBi for the circular-shaped patch antenna that operates at a frequency of 6.15 GHz with a coaxial feed, thus offering very resourceful solution to the communication engineers working on C band applications. The significance of this study relates with giving Antenna Engineers a dependable, proficient communication engineering solution mainly at C-bands of operation whereas the techniques used show how ML algorithms were used in order to come up with optimization methods for curve fitting of the return loss curves.

References

1. Shakeeb, M.: Circularly polarized microstrip antenna (Doctoral dissertation, Concordia University)
2. El Misilmani, H.M., Naous, T.: Machine learning in antenna design: an overview on machine learning concept and algorithms. In: 2019 International Conference on High Performance Computing and Simulation (HPCS), pp. 600-607. IEEE (2019)
3. Mohri, M., Rostamizadeh, A., Talwalkar, A.: Foundations of machine learning. MIT press (2018)
4. Russell, S.J., Norvig, P.: Artificial intelligence: a modern approach. Pearson (2016)

5. Pandey, A., Singh, A.P., Kumar, V.: Design and optimization of circularly polarized dielectric resonator-based MIMO antenna using machine learning for 5G Sub-6 GHz. AEU-Int. J. Electron. Commun. **162**, 154558 (2023)
6. Harkare, A.H., Kothari, A.G., Bhurane, A.A.: Evolution of gain enhancement techniques in dielectric resonator antenna: applications and challenges. Int. J. Microw. Wirel. Technol. **14**, 1–5 (2023)
7. Mishra, S., et al.: Dual port ring cylindrical dielectric resonator antenna optimization using ML algorithm. Waves in Random and Complex Media. **8**, 1–2 (2022)
8. Yang, L., Moubayed, A., Hamieh, I., Shami, A.: Tree-based intelligent intrusion detection system in internet of vehicles. In: 2019 IEEE Global Communications Conference (GLOBECOM), pp. 1–6. IEEE (2019)
9. Keller, J.M., Gray, M.R., Givens, J.A.: A fuzzy k-nearest neighbor algorithm. IEEE Trans. Syst. Man Cybern. **4**, 580–5 (1985)
10. Mishra, S., et al.: Dual port ring cylindrical dielectric resonator antenna optimization using ML algorithm. Waves Random Complex Media. **8**, 1–2 (2022)
11. Harkare, A.H., Abegaonkar, M.P., Kothari, A.G., Bhurane, A.A.: Compact circularly polarized MIMO dielectric resonator antenna for complete X-band coverage. In: 2023 IEEE Microwaves, Antennas, and Propagation Conference (MAPCON), pp. 1–5. IEEE (2023)
12. Harkare, A.H., Kothari, A.G., Bhurane, A.A., Abegaonkar, M.P., Patel, P.: A compact dielectric resonator antenna with wideband circular polarization characteristics for C and Xband applications. Microw. Opt. Technol. Lett. **66**(7), e34241 (2024)
13. Harkare, A., Kothari, A.G., Bhurane, A.: Novel rectangular multiband dielectric resonator antenna with slot for broadband applications. Int. J. Mech. Eng. **6**, 2868–72 (2021)

Design of Triple Band Ultrathin Polarization Controlled Metamaterial Microwave Absorber with Multiple Resonating Structures

Gaurav Chaitanya[1,2(✉)], Rachit Gupta[1], Paritosh Peshwe[1], and Ashwin Kothari[3]

[1] Indian Institute of Information Technology, Nagpur, India
chaitanya.gaurav@gmail.com
[2] Acropolis Institute of Technology and Research, Indore, India
[3] Visvesvaraya National Institute of Technology, Nagpur, India

Abstract. The design of a microwave absorber resonating in C-band and X-band having ultrathin thickness with polarization controlled behavior, have been reported. The unit cell of reported absorber embodied of multiple resonating structures resulting four absorption peaks at 6.36 GHz (in C band), 9.76 GHz, 11.48 GHz and 11.76 GHz (in X band). The overall absorptivity is obtained above 97%. 173.6 MHz (6.28 GHz–6.45 GHz), 298.6 MHz (9.62 GHz–9.92 GHz) and 710 MHz (11.25 GHz–11.96 GHz) are the FWHM (Full Width at Half Maximum) bandwidths. 0.101 λ0 is ultrathin thickness of the absorber with respect to lowest resonating frequency i.e. at 6.36 GHz. The fourfold symmetry about the axis of the unit cell controls the polarity of the absorber and the absorber behaves is stable over the outspread angle up to 45 degree for TM and TE polarized wave under oblique incident angle. The absorption behavior has been demonstrated by plotting the surface-current distribution at the resonating frequencies.

Keywords: Metamaterial · Absorber · Ultrathin · Polarization-controlled · Triple-band

1 Introduction

In recent times, metamaterial has grappled a lot of surveillance of researchers due to its magnificent properties in sub wavelength scale [1]. Metamaterial has been established as a boon in the discipline of microwave engineering owing to its explicit applications in absorbers, antenna, super lens, filter, cloaking, wireless power transfer, energy harvesting, RFID, etc. [2]. Among several possible applications, metamaterial based absorbers are of significant research interests as they offer low-profile, high absorption, light weight, simple fabrication mechanism, cost effective and other exclusive properties which can be remolded by just altering and scaling the dimensions of unit cell. Microwave absorbers find wide series of utilizations to encounter undesired EM energy, in defense inventories like fighter jet planes, torpedoes and missiles, anechoic chambers, Electromagnetic interference reduction, RADAR, and stealth technology etc. The investigations and amelioration in the design of absorbers based on metamaterial solves many practical constraints of conventional absorbers.

D. Gupta et al. (Eds.): PCEMS 2024, CCIS 2491, pp. 138–146, 2025.
https://doi.org/10.1007/978-3-031-90577-3_13

In 2008, when Landy *et al.* accorded a single-band perfect metamaterial absorber [3], several single, dual, triple and multi-band metamaterial absorbers were proposed and published in the literature [4–8]. With this realization that most of the real world applications are on bend surfaces, researchers started focusing and immersing into ultra and super thin absorbers which could easily bent. In [9], J. Tak *et al.* reported a wearable ultrathin metamaterial absorber. In [10], N. Hakla *et al.* and S. K. Sharma et al. documented conformal metamaterial absorbers for flat and curved surfaces application. Flexible and Polarization-controlled five-band metamaterial absorber is pre-offered by J. Wang *et al.* in [11]. A.K. Singh *et al.* proposed metamaterial absorbers having polarization insensitive triple- and dual-band ultrathin flexible characteristics [12].

In this paper, ultrathin polarization controlled triple band absorber with multiple resonating structures has been proposed. The presented unit cell design is very low profile with substrate thickness of 0.101 $\lambda 0$ at the lowest resonating frequency of 6.36 GHz. The four-fold symmetry in the design of resonating structures of the unit cell enables the polarization controlled behavior. The absorption behavior of the triple band absorber is examined by surface-current distribution.

2 Proposed Unit Cell Design

The unit cell geometry of the absorber with multiple resonating structures with the triple band behavior is showcased in Fig. 1. 0.48 mm thick PET substrate ($\varepsilon r = 3.5$ and $tan\delta = 0.015$) having copper deposition thickness of 0.035 mm have been used for the design. Various resonating structure have been designed on the top of the substrate to achieve the absorption bands at the resonating frequencies. Figure 1 shows the design parameters and the field direction of the designed unit cell.

Floquet port excitations and Master-Slave boundary conditions have been applied on unit cell geometry to observe the characteristics of the absorber and excitation of the electromagnetic wave along the z-axis is given for the computation of the absorptivity. Impinging of the electromagnetic wave on the absorber surface results in its transmitting and reflecting. The reflection coefficient S_{11} and transmission coefficient S_{12} must be as minimal as possible to get the highest absorption. The expression to calculate absorptivity is as follows-

$$A(\omega) = 1 - |S_{11}(\omega)|^2 - |S_{21}(\omega)|^2 = 1 - |S_{11}(\omega)|^2 \quad (1)$$

where $|S_{11}(\omega)|^2$ and $|S_{21}(\omega)|^2$ are reflected and transmitted power respectively. As the geometry is complete copper backed, the transmission coefficient S_{12} will be zero as shown in Eq. 1. The infinite array of the presented unit cell of the metamaterial absorber on the High Frequency Structure Simulator software ver. 14.0 is analyzed using floquet port excitations and Master-Slave boundary conditions. The absorptivity response and reflectivity of the designed triple band absorber is illustrated in Fig. 2.

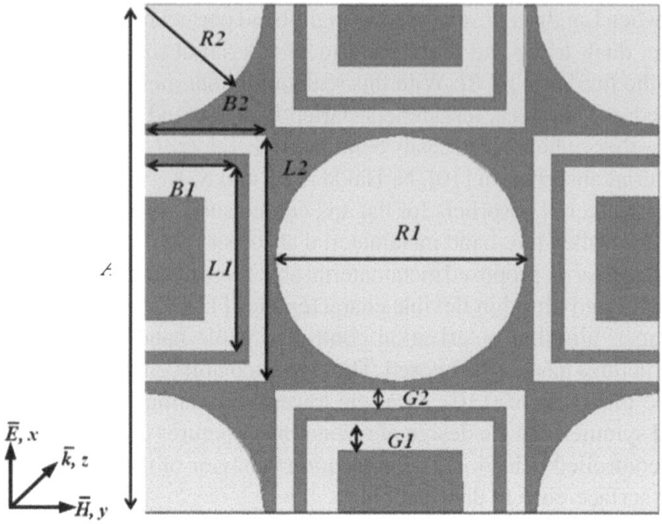

Fig. 1. Proposed unit cell design A = 16.5 mm, $B1$ = 3 mm, B2 = 4 mm, $L1$ = 6 mm, $L2$ = 8 mm, $R1$ = 8 mm, $R2$ = 3.9 mm, $G1$ = 0.6 mm and $G2$ = 1 mm

The individual behavior characteristics of each resonating structure are depicted in Fig. 3. More than 97% of absorptivity is observed in the designed triple band absorber structure which shows four absorption tips at 6.36 GHz (in C-band), 9.76 GHz, 11.48 GHz and 11.76 GHz (in X-band). 173.6 MHz (6.28 GHz–6.45 GHz), 298.6 MHz (9.62 GHz–9.92 GHz) and 710 MHz (11.25 GHz–11.96 GHz) are the FWHM bandwidths.

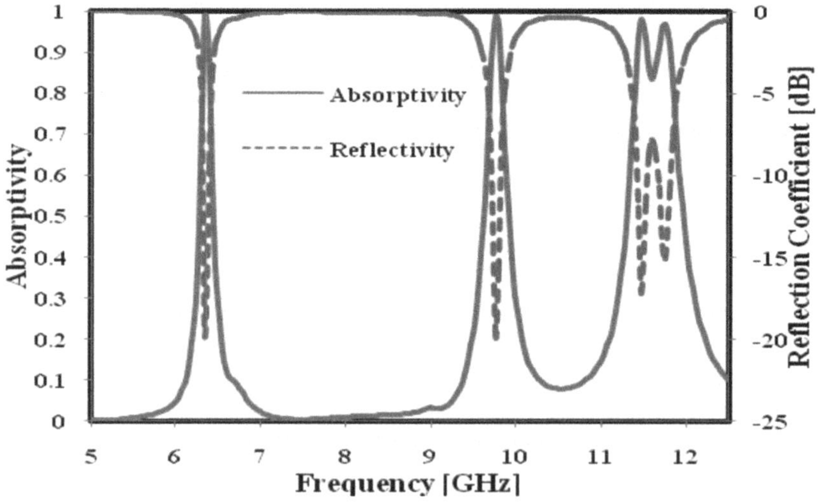

Fig. 2. Absorptivity and Reflectivity (S11) in dB and of the presented absorber

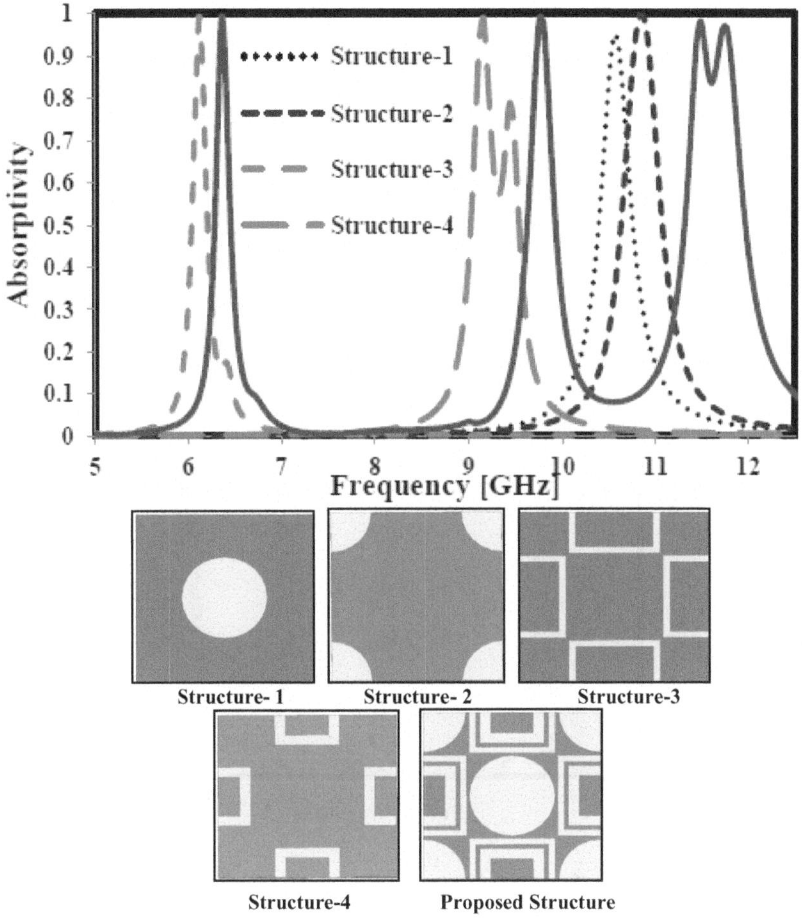

Fig. 3. Absorptivity of individual resonators

3 Critical Analysis

3.1 Different Oblique Incidence and Polarization Angles Responses

The simulated results of the proposed structure which has been examined for varied oblique incidence angles (θ) for both Transverse electric (TE) and Transverse magnetic (TM) polarization are illustrated in Fig. 4 (a) and (b). In transverse electric polarization, incident electric field is directed towards x-axis, where as the incident magnetic field towards y-axis along with wave vector which is directed towards z -axis vary with an angle 'θ'. In transverse magnetic polarization, incident magnetic field is directed along y-axis, where as the incident electric field and wave vector vary with an angle 'θ' and directed towards x and z axes respectively. The design is studied up to 45° due to the four fold symmetry with respect to *xy*-plane and it is observe that in both of the TE and TM polarization triple band absorptivity response is sustained up to indicated incident angle.

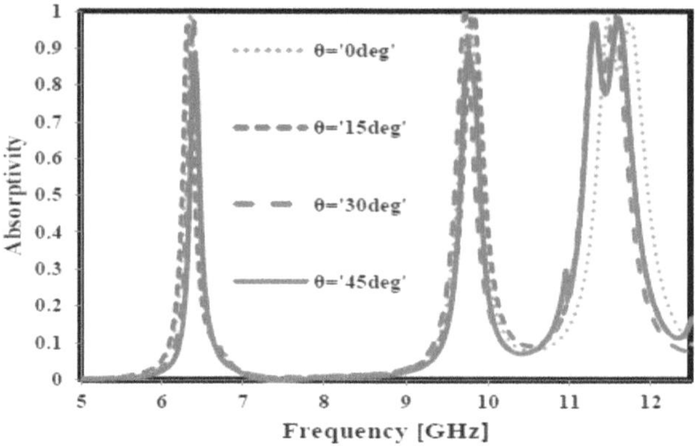

(a)

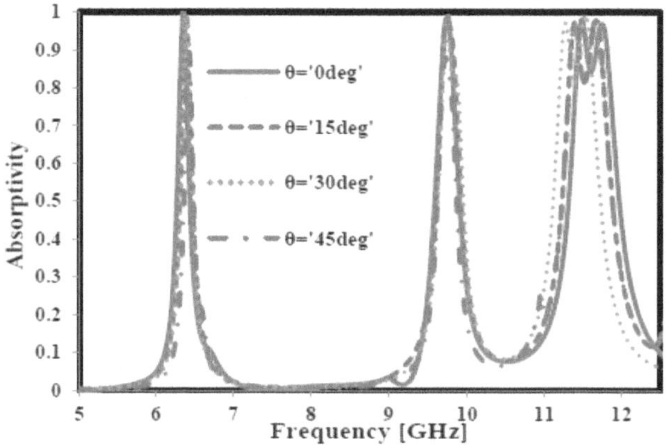

(b)

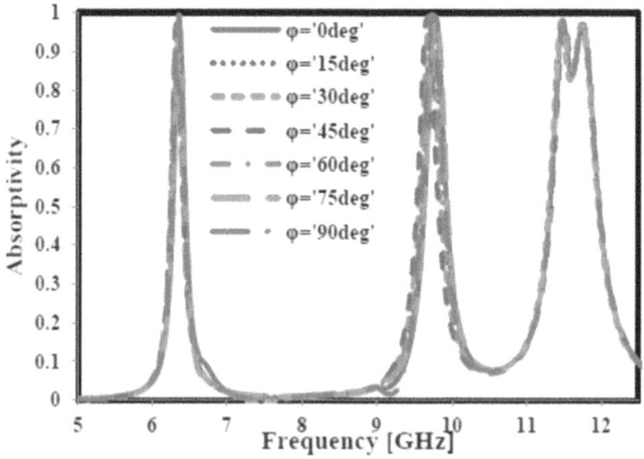

(c)

Fig. 4. Simulated absorption curves at varied oblique incidence angles under (a) Transverse Electric polarization and (b) Transverse Magnetic polarization. (c) at varied polarization angles

Keeping the direction of the electromagnetic wave steady, the directions of H–field and E–field are examined at varied polarization angles to study the polarization performance of the absorber. The polarization controlled behavior of the presented absorber is verified with the similar responses computed at the entire polarization angle ($\phi = 0°$ to $90°$ with step size $15°$) as illustrated from Fig. 4(c). The impedance values at the resonant frequencies are depicted in the Fig. 5 which shows an optimum matching of the structure. The real and imaginary values of impedances at four absorbing frequencies are tabulated in Table 1.

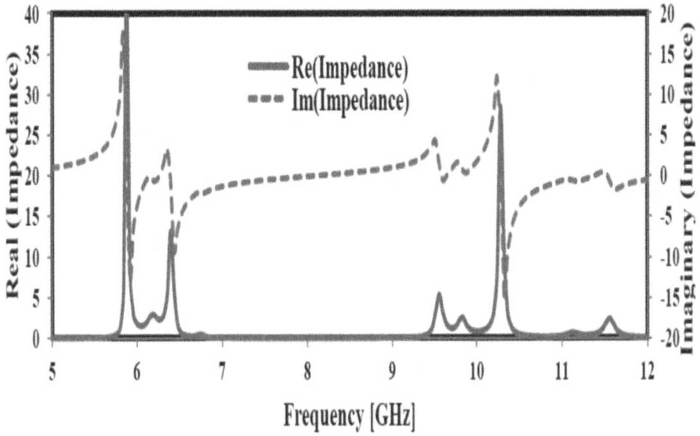

Fig. 5. Impedance of the multiple resonator triple band absorber

Table 1. Real and imaginary values of the impedances for the four absorption frequencies

Frequency [GHz]	Real (Z)	Imaginary (Z)
$f1 = 6.36$	1.19	−0.001
$f2 = 9.76$	0.89	0.12
$f3 = 11.48$	0.95	0.32
$f4 = 11.76$	1.04	−0.27

3.2 Absorption Phenomenon

Figure 6 shows the surface current distributions that have been examined at four absorption peak frequencies to verify the absorption characteristics of the presented absorber. The surface current is dominant on central circular resonating structure-1 and circular-arc structure-2 at frequencies of 11.48 GHz and 11.76 GHz respectively as presented in Fig. 6 (a) and (b). At the same time, at frequencies values of 6.36 GHz and 9.76 GHz, maximum current density is observed at structure- 3 and structure-4 respectively which is illustrated in Fig. 6 (c) and (d). This concentration of fields on the top metal patch is responsible for the excitation of electric field. The directions of surface current at the mentioned structures were anti-parallel with direction of the surface current at the bottom ground. When opposite currents flows on top and bottom metal patches of the proposed unit cell structure, a circulating current is formed. This current is responsible for excitation of magnetic field. The absorptivity is maximal when the excitation of electric and magnetic takes place at the same time.

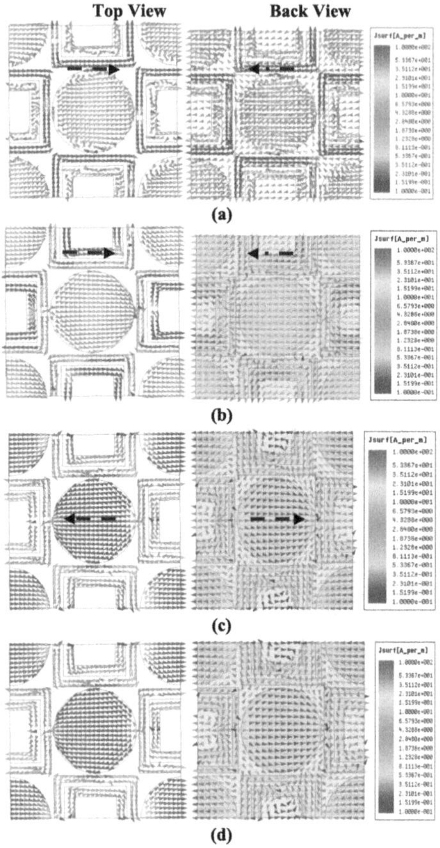

Fig. 6. Distribution of Surface-current of presented absorber at resonating frequencies (a) 6.36 GHz (b) 9.76 GHz, and (c) 11.48 GHz (d) 11.76 GHz.

4 Conclusion

In the proposed model, ultrathin polarization controlled triple band absorber with multiple resonating structures has been presented. The presented absorber is a fitting structure for flattened and bent surfaces because of the ultrathin thickness (0.101λ0) and the low-profile geometry. The absorber unit cell comprises of multiple resonating structures which demonstrates the four absorptions peaks at varied frequencies in C and X band. More than 97% absorptivity is observed in the presented absorber design demonstrate four absorption tips at 6.36 GHz (in C band), 9.76 GHz, 11.48 GHz and 11.76 GHz (in X band). 173.6 MHz (6.28 GHz–6.45 GHz), 298.6 MHz (9.62 GHz–9.92 GHz) and 710 MHz (11.25 GHz–11.96 GHz) are the FWHM (Full Width at Half Maximum) bandwidths. The presented absorber structure is polarization controlled and stable for outspread angle of incidence for both transverse electric and transverse magnetic polarization. The absorption behavior has been demonstrated by plotting the surface-current distribution at the resonating frequencies.

References

1. Kaur, K.P., Upadhyaya, T.K., Palandoken, M.: Dual-band polarization-insensitive metamaterial inspired microwave absorber for LTE-band applications. Progress Electromagn. Res. C **77**, 91–100 (2017)
2. Mishra, N., Choudhary, D., Chowdhury, R., Kumari, K., Chaudhary, R.: An investigation on compact ultra-thin triple band polarization independent metamaterial absorber for microwave frequency applications. IEEE Access **5**, 4370–4376 (2017)
3. Landy, N.I., Sajuyigbe, S.S., Mock, J.J., Smith, D.R., Padilla, W.J.: A perfect metamaterial absorber. Phys. Rev. Lett. **100**, 207–402 (2008)
4. Bhattacharyya, S., Ghosh, S., Chaurasiya, D., Srivastava, K.V.: Wide angle broadband microwave metamaterial absorber with octave bandwidth. IET Microw. Antennas Propag. **9**(11), 1160–1166 (2015)
5. Bhattacharyya, S., Ghosh, S., Srivastava, K.V.: Equivalent circuit modeling of an ultra-thin dual-band microwave metamaterial absorber. In: Proceedings of the Asia-Pacific Microwave Conference, Sendai, Japan, pp. 1244–1246, 2014
6. Wang, N., et al.: Novel quadruple-band microwave metamaterial absorber. IEEE Photon. J. **7**(1), 1–6 (2015)
7. Sood, D., Tripathi, C.C.: Broadband ultrathin low-profile metamaterial microwave absorber. Appl. Phys. A **122**(4), 1–7 (2016). https://doi.org/10.1007/s00339-016-9884-2
8. Hoa, N.T.Q., Lam, P.H., Tung, P.D.: Wide-angle and polarization independent broadband microwave metamaterial absorber. Microw. Opt. Technol. Lett. **59**(5), 1157–1161 (2017)
9. Tak, J., Choi, J.: A wearable metamaterial microwave absorber. IEE Antennas Wirel. Propag. Lett. **16**, 784–787 (2017)
10. Hakla, N., Ghosh, S., Srivastava, K.V., Shukla, A.: A dual-band conformal metamaterial absorber for curved surface. In 2016 URSI International Symposium on Electromagnetic Theory (EMTS), Espoo, pp. 771–774, 2016
11. Wang, J., Yang, R.: Polarization-controlled and flexible single-/penta-band metamaterial absorber. Materials (2018)
12. Singh, A.K., Member, S., Abegaonkar, M.P., Member, S., Koul, S.K.: Dual- and triple-band polarization insensitive ultrathin conformal metamaterial absorbers with wide angular stability. IEEE Trans. Electromagn. Compat. **PP**, 1–9 (2018)
13. Ghosh, S., Srivastava, K.V.: Broadband polarization-insensitive tunable frequency selective surface for wideband shielding. IEEE Trans. Electromagn. Compat. **60**(1), 166–172 (2018)
14. Kalraiya, S., Ameen, M., Chaudhary, R.K., Gangwar, R.K.: Polarization Independent Conformal Metamaterial Absorber using Modified Resonators for Dual Band Applications, IMARC, pp. 3–6

15. Kalraiya, S., Chaudhary, R.K., Abdalla, M.A., Gangwar, R.K.: Polarization independent dual-band conformal metamaterial absorber for X- Band microwave application. Electron. Lett. (2019)
16. Ghosh, S., Nguyen, T.T., Lim, S.: Recent progress in angleinsensitive narrowband and broadband metamaterial absorbers. EPJ Appl. Metamaterials (2019)
17. Chaitanya, G., Suroliya, A., Neema, J., Chandachoriya, A.: Polarization controlled Ultrathin Quad Band Metamaterial Absorber for C and X Band. In: 10th IEEE International Conference on Communication Systems and Network Technologies (CSNT 2021), Bhopal on 18–19 June 2021
18. Chaitanya, G., Peshwe, P., Kothari, A.: Triple Band Polarization Independent Ultrathin Metamaterial Microwave Absorber for C and X Band Applications
19. Wang, B.-X., Chongyang, X., Duan, G., Wei, X., Pi, F.: Review of broadband metamaterial absorbers: from principles, design strategies, and tunable properties to functional applications. Adv. Funct. Mater. **33**(14), 2213818 (2023)
20. Goyal, N., Panwar, R.: Minkowski inspired circular fractal metamaterial microwave absorber for multiband applications. Appl. Phys. A **129**(4), 293 (2023)
21. Barde, C., Gupta, N.K., Ranjan, P., Roy, K., Sinha, R.: Angle-independent wideband metamaterial microwave absorber for C and X band application. Int. J. Microw. Wirel. Technol. 1–9 (2023)

Permittivity Evaluation Based Algorithm for Detecting Optimal Thickness Range of Composite Materials at Microwave Frequencies

Shankar Aenagandula(✉) and Neeraj Rao

Electronics and Communication Engineering Department,
Visvesvaraya National Institute of Technology, 440010 Nagpur, MH, India
shankar.aenagandula@gmail.com
https://vnit.ac.in/engineering/ece/

Abstract. Accuracy in permittivity evaluation plays an important role in composite material characterization. But, this permittivity evaluation gets affected, if there is a variation in the thickness of the sample under test (SUT) at a given frequency band. Hence, finding the suitable thickness range of SUT for evaluating the correct permittivity is a challenging task for low-dielectric low-loss composite materials. This paper presents an algorithm based on the Robert-von Hippel (RVH) method and exhaustive search optimization for finding the suitable thickness range of SUT at a given microwave frequency band ($4-40\,GHz$). The simulation calculates the optimized thickness ranges for the low-dielectric low-loss composite material samples for the C-band, X-band, Ku-band, K-band, and Ka-band. It is also observed that there is no change in the RVH permittivity of the material by varying the loss tangent in the range from 0 to 0.1.

Keywords: Robert-von Hippel method · low-dielectric low-loss composite material · permittivity evaluation

1 Introduction

The measurement of the material properties at microwave frequencies has a vital role in the applications such as composite materials characterization, RF circuit design, materials production, aircraft manufacturing and also agricultural products [1,2]. The material properties at microwave frequencies are divided into electric (ϵ) and magnetic (μ) properties. The electric property is consider as the complex permittivity ($\epsilon_r = \epsilon_r^{'} - j\epsilon_r^{''}$), the real part called as dielectric constant (permittivity) $\epsilon_r^{'}$ of a material, characterises the ability to store the electric field energy and the imaginary part gives the loss factor $\epsilon_r^{''}$. The loss factor $\epsilon_r^{''}$ of any material characterises its ability to dissipate the energy in the form of heat. The loss tangent ($\tan\delta = \frac{\epsilon_r^{''}}{\epsilon_r^{'}}$) is used to quantify the dielectric loss. Most of

D. Gupta et al. (Eds.): PCEMS 2024, CCIS 2491, pp. 147–156, 2025.
https://doi.org/10.1007/978-3-031-90577-3_14

the composite materials are non-magnetic ($\mu_r = 1$). Therefore, the important parameter for the composite material is complex permittivity (ϵ_r).

Composite materials offer significant advantages over metals. They are lightweight, flexible, corrosion-resistant, and possess high fatigue and impact strength [3]. Consequently, these materials are being considered as replacements for traditional materials in aerospace, automotive, and other industries. Composite materials are categorized into low-loss ($\tan\delta < 0.1$) and high-loss ($\tan\delta > 0.1$) types [4], with most composites being low-loss at microwave frequencies [5].

The dielectric properties of composite materials are evaluated using the Robert-von Hippel (RVH) method, also known as the Transmission Line or One-Port method [6–8]. This method determines permittivity based on the reflection coefficient obtained at the input port of a short-circuited waveguide as shown in Fig. 1. However, the reflection coefficient varies with the thickness of the sample under test (SUT) placed inside the waveguide and the operating frequency [9].

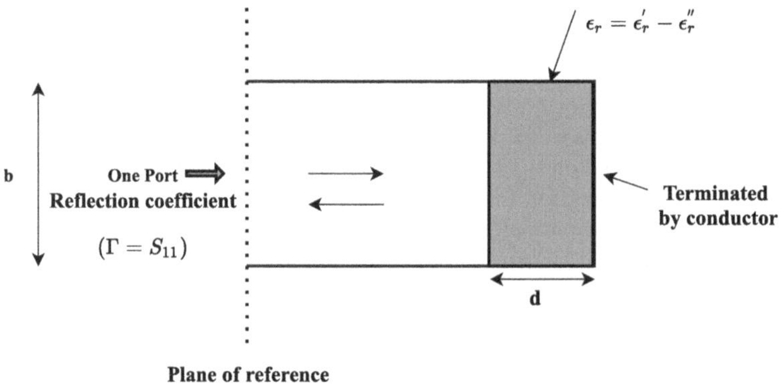

Fig. 1. The short-circuited rectangular waveguide with SUT of the thickness d mm.

Therefore, it is crucial to determine the proper thickness range of the SUT so that the RVH permittivity closely matches the actual permittivity of the SUT within an acceptable tolerance. To address this, we proposed an algorithm that combines the RVH method with exhaustive search optimization to find the suitable thickness range of the SUT for a given microwave frequency band ($4-40\,GHz$). The frequency bands of operation for the waveguide are C-band ($4–8.2\,GHz$), X-band ($8.2-12.4\,GHz$), Ku-band ($12.4-18\,GHz$), K-band ($18–26.5\,GHz$) and Ka-band ($26.5-40\,GHz$).

To validate the algorithm, simulations were performed. These simulations used exhaustive search optimization for permittivity values ranging from 1 to 10 across different frequency bands to identify the optimal thickness range for the SUT.

2 Robert-von Hippel Method

The RVH method or transmission line method is used to evaluate the permittivity of SUT [6]. As per this method, The dielectric sample of a given thickness d (mm) is placed inside the waveguide at the short-circuited end as shown in Fig. 1. When electromagnetic wave impinges the SUT at the short-circuited end, standing waves are generated inside the waveguide. The reflection coefficient is measured at the input port. Then, RVH method uses this reflection coefficient to calculate the permittivity of the SUT.

From this setup, the reflection coefficient (Γ) is measured and with the knowledge of frequency (f), the complex dielectric constant (ϵ_r) is obtained as

$$\epsilon_{r_{RVH}} = \epsilon_r^{'} - j\epsilon_r^{''} = \frac{\left(\frac{x}{kd}\right)^2 + \left(\frac{\lambda_g}{2a}\right)^2}{1 + \left(\frac{\lambda_g}{2a}\right)^2} \tag{1}$$

where, x is a solution of the transcendental equation

$$\frac{\tan x}{x} = \frac{1-\Gamma}{jkd(1+\Gamma)} \tag{2}$$

Equation (2) gives an infinite set of complex roots, whereas the single solution is possible by using von Hippel charts [10] or optimization technique [11].

For the low-loss materials (1) and (2) is approximated by Dakin and Works [4].

$$\frac{\tan(\beta_2 d)}{\beta_2 d} = \frac{\tan(\frac{\theta_\Gamma}{2})}{\beta_1 d} \tag{3}$$

$$\epsilon_{r_{RVH}}^{'} = \left(\frac{\lambda_o}{\lambda_c}\right)^2 + \left(\frac{(\beta_2 d)\lambda_o}{2\pi d}\right)^2 \tag{4}$$

$$\tan\delta = \left(\frac{\lambda_g}{\pi dS}\right)\left(1 - \frac{\lambda_0^2}{\epsilon_{r_{RVH}}^{'}\lambda_c^2}\right)\left(\frac{1+\tan^2\left(\frac{\theta_\Gamma}{2}\right)}{1+\tan^2(\beta_2 d) - \frac{\tan(\beta_2 d)}{(\beta_2 d)}}\right) \tag{5}$$

$$\epsilon_{r_{RVH}}^{''} = \epsilon_{r_{RVH}}^{'} \tan\delta \tag{6}$$

where, λ_g is given as

$$\frac{1}{\lambda_o^2} = \frac{1}{\lambda_c^2} + \frac{1}{\lambda_g^2} \tag{7}$$

Γ = reflection coefficient $\left(\|\Gamma\| e^{j\theta_\Gamma}\right)$
$k = \beta_1$ = wavenumber inside the waveguide $\left(\frac{2\pi}{\lambda_g}\right)$
λ_g = guide wavelength
d = thickness of the SUT.
a = broader dimension of the waveguide.

$\lambda_c = 2a$ (Dominant (TE_{10}) mode operation)
$\lambda_o = \frac{c}{f}$, f = operating Frequency,
$c = 299702547$ m/s (speed of light in air).
S = voltage standing wave ratio (VSWR) $\left(\frac{1+|\Gamma|}{1-|\Gamma|}\right)$
β_2 = wavenumber inside the SUT.

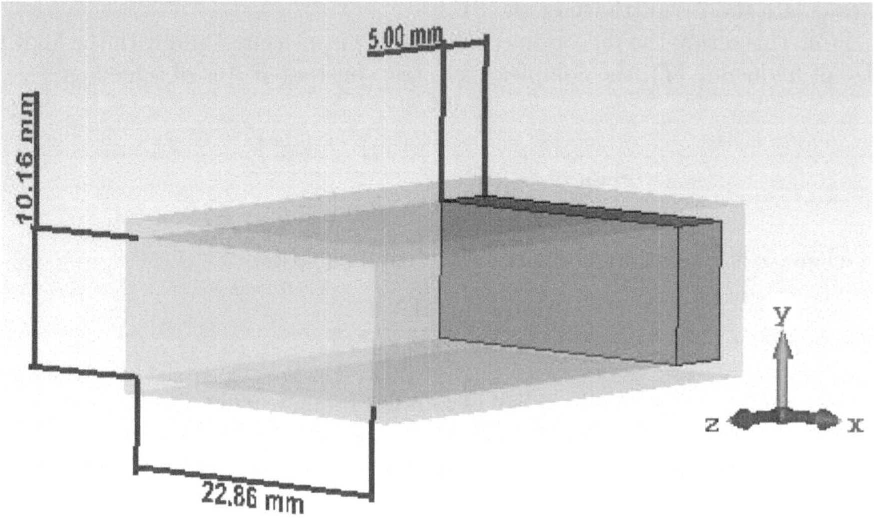

Fig. 2. The CST-MWS design of rectangular waveguide of X-band (WR90) with 5 (mm) thickness SUT to evaluate the dielectric properties.

3 Proposed Algorithm

Algorithm 1 is the proposed algorithm based on RVH method and exhaustive search optimization method for finding the optimal thickness range of SUT at a given microwave frequency band. The step by step explanation for proposed algorithm is given as follows.

Step 1 defines the required inputs as low-loss SUT, maximum permittivity value of SUT ($\epsilon'_{r_{max}}$), maximum thickness (d_{max}) of SUT and microwave frequency band (Δf). Step 2 defines the output as optimum thickness range (Δd^*). In the step 3 one of the operating frequency band (C-band, X-band, Ku-band, K-band, Ka-band) is selected according to the microwave source and avalable waveguide setup. Exhaustive search optimization process is started from step 4 to step 16. A for loop is introduced in step 4 to vary the permittivity of SUT in the range 1 to $\epsilon'_{r_{max}}$. For each value of permittivity a nested for loop is started in step 5 to vary the thickness of SUT in the range 1 mm to d_{max} mm. This nested *for* loop repeats the instructions from step 6 to step 12 until $\epsilon_r^{'} = \epsilon_{r_{max}}^{'}$. Step 6 finds the reflection coefficient by using CST-MWS setup. In step 7 β_2 is calculated from (3). The permittivity ($\epsilon_{r_{RVH}}^{'}$) of SUT is calculated using Robert Von-Hippel method from (4) in step 8. Then, $\epsilon_{r_{RVH}}^{'}$ is compared with actual permittivity $\epsilon_r^{'}$. For the practical applications, the percentage difference between actual permittivity and calculated RVH permittivity must not exceed more than 5% tolerance [1]. This tolerance condition is applied in step 9, *if* tolerance level is less-than or equal to 5% then current thickness value (t) is included in the thickness range for the particular $\epsilon_r^{'}$ else current t value is discarded in step 12. Finally, the common optimized thickness range is calculated in step 16 by taking the intersection of all different thickness ranges obtained for each value of actual permittivity ($\epsilon_r^{'}$). The simulation results for the proposed algorithm are tabulated in the next section.

Algorithm 1. Exhaustive search optimization algorithm for finding the suitable thickness range at microwave frequency bands.

1: **Input: SUT** ($\tan\delta < 0.1$), $\epsilon_{r_{max}}^{'}$, d_{max}, Δf ($4 - 40$ **GHz**)
2: **Output:** Δd^*
3: **Select frequency band:**
   $\Delta f \in$ {**C-band, X-band, Ku-band, K-band, Ka-band**}
4: **for** $\epsilon_r^{'} = 1 : 1 : \epsilon_{r_{max}}^{'}$ **do**
5:   **for** $t = 1 : 1 : d_{max}$ **do**
6:     **Find reflection coefficient** (Γ) **from the setup in the Fig. 2 using CST-MWS.**
7:     **Find** β_2 **from (3).**
8:     **Find** $\epsilon_{r_{rvh}}^{'}$ **from (4).**
9:     **if** $\left|\frac{\epsilon_{r_{rvh}}^{'} - \epsilon_r^{'}}{\epsilon_r^{'}}\right| \leq 5\%$ **then**
10:       $t \in \Delta d_{\epsilon_r^{'}}$
11:     **else**
12:       $t \notin \Delta d_{\epsilon_r^{'}}$
13:     **end if**
14:   **end for**
15: **end for**
16: **Final optimized thickness range for** Δf**:**
   $\Delta d^* = \Delta d_1 \cap \Delta d_2 ... \cap \Delta d_{\epsilon_r^{'}} ... \cap \Delta d_{max}$
17: **End of algorithm**

---

4 Simulation Results and Discussion

We used CST Microwave Studio (CST-MWS) [12], the 3D simulator for RF and microwave applications. The X-band (8.2-12.4 GHz) standard waveguide (WR90) is designed with a short-circuited end and waveguide material is chosen as the perfect electric conductor (PEC). The low loss material of thickness d (mm) is kept at the short-circuited end of the waveguide as shown in Fig. 2. The waveguide is simulated at a dominant (TE_{10}) mode and for testing the material properties, plexiglass ($2.6-j0.0156$) material with thickness 5 mm is chosen.

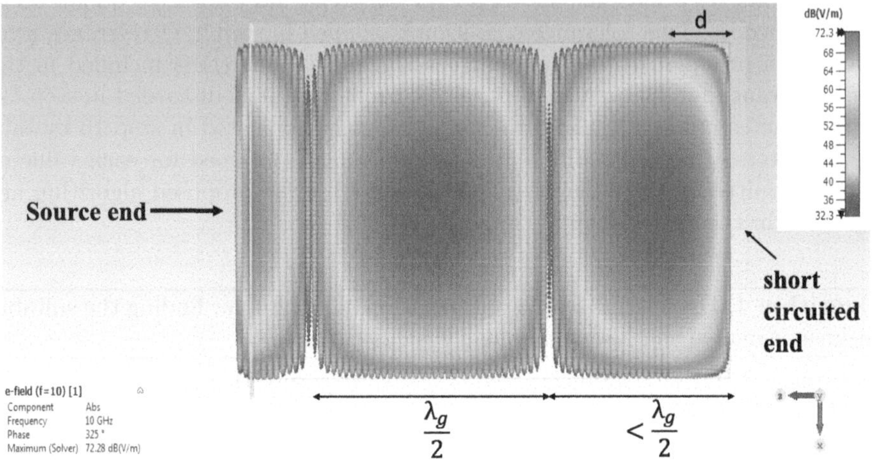

Fig. 3. The E-filed pattern inside the waveguide with thickness d (mm) sample.

The electric field pattern inside the waveguide is shown in Fig. 3. In general, the electric field pattern repeats for every $\frac{\lambda_g}{2}$ distance without the SUT inside the waveguide. However, with the SUT, the first field pattern from the short-circuited end is less than $\frac{\lambda_g}{2}$ distance. This indicates that with the SUT inside the waveguide, field pattern changes which in turn changes the reflection coefficient. This reflection coefficient is used to calculate the material properties.

Table 1. Simulation Parameters

Name and symbol of the parameter	Value
Actual permittivity minimum ($\epsilon'_{r_{min}}$)	1
Actual permittivity maximum ($\epsilon'_{r_{max}}$)	10
Minimum thickness (d_{min})	1 mm
Minimum thickness (d_{max})	20 mm
Loss-factor (ϵ''_r)	0.0156
Frequency band (Δf)	4 to 40 GHz

Table 1 shows the simulation parameters and their values used for simulation of Algorithm 1.

Initially Algorithm 1 is applied for a plexiglass material of different thicknesses from d_{min} to d_{max} over the frequency band (Δf) ($8.2-40$ GHz) using X-band waveguide (WR90) and frequency band (Δf) (4–8.2 GHz) using C-band waveguide (WR187). The RVH permittivity is calculated using (3) and (4). Figures 4 and 5 represents the RVH permittivity of the material over the frequency band ($8.2-40$ GHz) and the frequency band ($4-8.2$ GHz) respectively.

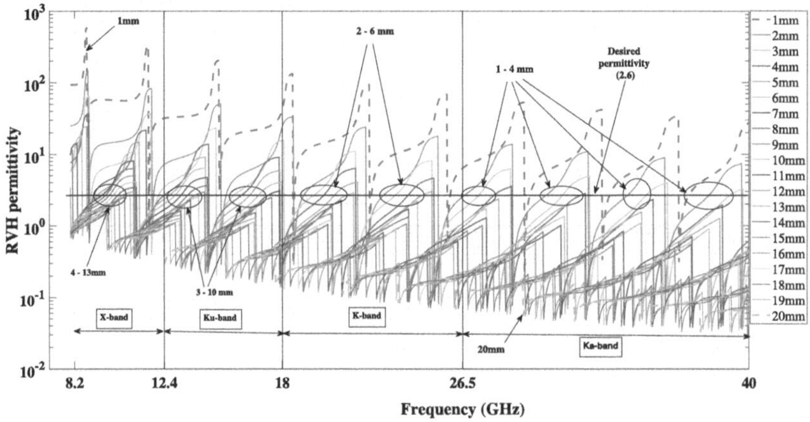

Fig. 4. The RVH permittivity variation over the microwave frequency (8–40 GHz) with different thicknesses ($1\,\text{mm}-20\,\text{mm}$) of the plexiglass ($2.6-j0.0156$) material.

The RVH permittivity values (log-scale) over the frequency bands (X-band, Ku-band, K-band and Ka-band) are shown in Fig. 4 and the frequency band (C-band) is shown in Fig. 5. Each waveform represents a thickness starting from d_{min} (top) to d_{max} (down). From the Figs. 4 and 5 it is clear that the RVH permittivity decreases as the frequency increases for any thickness of the given SUT. Also, it can be observed that there is a range of the thickness of the SUT for each frequency band over which it gives the desired permittivity (2.6). For the permittivity of (2.6), as shown in Fig. 4 X-band is suitable for the thickness of 4 mm–13 mm. Similarly, from the Fig. 5 the desired thicknesses for the C-band is $4\,\text{mm}-20\,\text{mm}$. If the thickness of the SUT is less than 4 mm, neither C-band nor X-band is suitable to detect the desired dielectric permittivity, for such a case, we need to go for higher frequency bands shown in Fig. 4.

Further, the model is simulated according to Algorithm 1 for different materials with the actual permittivity ($\epsilon_r^{'}$) in the range from $\epsilon_{r_{min}}^{'}$ to $\epsilon_{r_{max}}^{'}$ and thickness of the SUT from d_{min} to d_{max}, keeping the same loss factor of 0.0156 over the frequency band (Δf). The optimized thickness ranges for the different frequency bands are listed in Table 2.

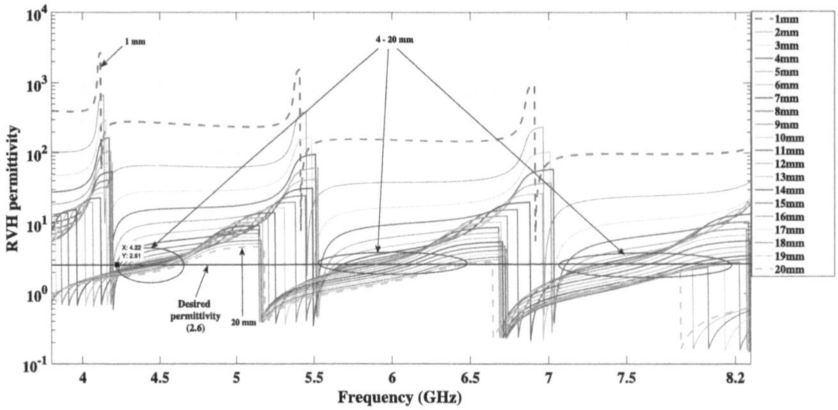

Fig. 5. The RVH permittivity variation over the microwave frequency (4 − 8.2 GHz) with different thicknesses (1mm − 20mm) of the plexiglass (2.6 − j0.0156) material.

Table 2. The Suitable thicknesses for the frequency band

Thickness of SUT (d) (mm) → Actual permittivity of SUT ↓	C-band	X-band	Ku-band	K-band	Ka-band
1-j0.0156	7 to 20	4 to 19	3 to 14	2 to 11	1 to 8
2-j0.0156	7 to 20	4 to 14	3 to 10	2 to 7	1 to 5
3-j0.0156	6 to 18	4 to 13	3 to 10	2 to 6	1 to 4
4-j0.0156	6 to 17	3 to 12	3 to 9	2 to 5	1 to 3
5-j0.0156	5 to 15	2 to 10	2 to 7	1 to 4	1 to 3
6-j0.0156	5 to 14	2 to 10	2 to 7	1 to 4	1 to 3
7-j0.0156	4 to 11	2 to 9	2 to 6	1 to 4	1 to 3
8-j0.0156	3 to 10	2 to 8	2 to 6	1 to 3	1 to 2
9-j0.0156	2 to 10	1 to 7	1 to 5	1 to 3	1 to 2
10-j0.0156	2 to 9	1 to 7	1 to 5	1 to 3	1 to 2
Final thickness range (mm) →	7 to 9	4 to 7	3 to 5	2 to 3	1 to 2

From the Table 2, it is observed that for the low-dielectric low-loss composite materials, the thickness of the sample is 7–9 mm for C-band, 4–7 mm for X-band, 3 − 5 mm for Ku-band, 2 − 3 mm for K-band and 1 − 2 mm for Ka-band suitable to evaluate the permittivity.

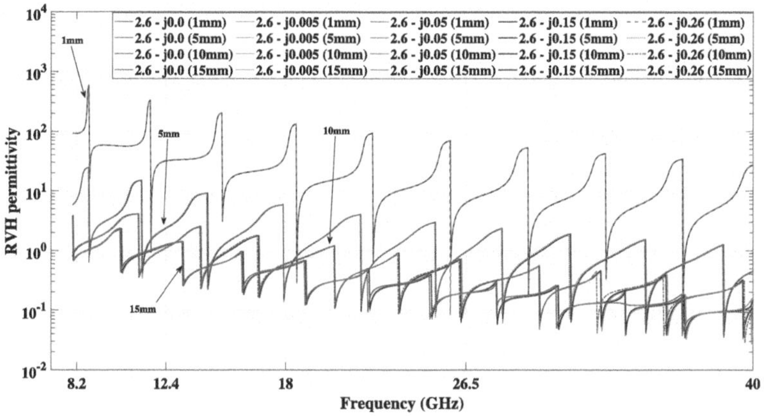

Fig. 6. The permittivity curves over the microwave frequency (8.2–40 GHz) for different loss-factor and thicknesses of the plexiglass material.

In Fig. 6, the RVH permittivity curves are plotted over the microwave frequency band ($8.2 - 40\,GHz$) for varying thickness (1mm, 5mm, 10mm, and 15mm) and loss-tangent (loss-factor) range from 0 to 0.1 (0 to 0.26) for the plexiglass material respectively. Each curve is consisting of overlapping of the five different loss-factor materials of plexiglass. It is also observed that there is no change in the permittivity curve of the material by changing the loss-tangent (loss-factor). Hence, we can conclude that the variation in loss-tangent (loss-factor) of the low-loss materials over the range from 0 to 0.1 (0 to 0.26) is not affecting the permittivity of the material.

5 Conclusion

In this paper, We have proposed an exhaustive search optimization algorithm based on the Robert-von Hippel method to evaluate the optimal thickness range for low-dielectric low-loss composite materials at a given frequency band. From the simulation results we found that variation in loss-factor does not affect the RVH permittivity, hence it is concluded that the common thickness range is independent with respect to variations in loss-factor of the low-loss materials.

References

1. Zoughi, R.: Microwave non-destructive testing and evaluation principles. Springer Science & Business Media, vol. 4 (2000)
2. Nelson, S.: Dielectric properties of agricultural materials and their applications. Academic Press (2015)
3. Kar, K.K.: Composite materials: processing, applications, characterizations. Springer (2016)

4. Dakin, T., Works, C.: Microwave dielectric measurements. J. Appl. Phys. **18**(9), 789–796 (1947)
5. Pereira, G., Schweiger, G.(ed.): Poverty, Inequality and the Critical Theory of Recognition. pp. vol. 3, pp. 83–106. Springer, Cham (2020). https://doi.org/10.1007/978-3-030-45795-2_4
6. Roberts, S., Von Hippel, A.: A new method for measuring dielectric constant and loss in the range of centimeter waves. J. Appl. Phys. **17**(7), 610–616 (1946)
7. Ganchev, S.I., Bhattacharyya, J., Bakhtiari, S., Qaddoumi, N., Brandenburg, D., Zoughi, R.: Microwave diagnosis of rubber compounds. IEEE Trans. Microw. Theory Tech. **42**(1), 18–24 (1994)
8. Venkatesh, M., Raghavan, G.: An overview of dielectric properties measuring techniques. Can. Biosyst. Eng. **47**(7), 15–30 (2005)
9. Chao, S.-H.: An uncertainty analysis for the measurement of microwave conductivity and dielectric constant by the short-circuited line method. IEEE Trans. Instrum. Meas. **1**, 36–41 (1986)
10. Field, R., von Hippel, A.: Dielectric materials and applications, ed. AR von Hippel, John Wiley & Sons. Inc., New York (1954)
11. Iglesias, T., Seoane, A., Rivas, J.: An algorithm for solving roberts-von hippel equation: separation of close solutions. IEEE Trans. Instrum. Meas. **42**(3), 760–763 (1993)
12. Design and engineering simulation: Simulia - dassault systèmes. https://www.3ds.com/products-services/simulia/?utm_source=cst.com&utm_medium=301&utm_campaign=products

Using Generative AI for Rapid Digital Prototyping of Industrial Design Sketches and 3D Renders

Ashish Jha, Vijayasri Kalyan, Ninad Kulkarni(✉), and Ujwal Bhate

Infosys Ltd., Bengaluru, India
{ashish.jha11,vijayasri.kalyan,ninad.kulkarni,
ujwal_bhate01}@infosys.com

Abstract. Ideation phase of the industrial product design process involves drawing multiple forms of sketches, collecting feedback from engineering teams and iterating over the design through sketches and colored renders. These activities involve notable manual effort and time. Recently, Generative AI has demonstrated extensive capabilities towards image generation and has also been adopted to aid in the product design and visualization phases. However, popular general-purpose tools built around such foundational GenAI models typically are unable to cater to the finer domain specific aspects. In this paper, we present an approach to extend Generative AI models to encompass new domain specific visual aspects and components, so that these models can be used for quick iterations over sketches and colored renders by synthesizing specific styles, features and components into an existing sketch, as directed by the designer. The proposed approach allows for rapid visualization and digital prototyping of design ideas as the AI models abstract away typical activities around creating sketch and render feature and design variations, viewpoint variations and drawing/erasing specific components in a sketch. This paper demonstrates the results of the approach on actual data and analyses the efficacy.

Keywords: Generative Design · Product Design · Digital Prototyping

1 Introduction

1.1 Background

The process of industrial design is the practice of designing a product's form and feature prior to its formal production. It is a multi-stage and complex process involving multiple teams and skillsets such as artists, engineers, domain experts and designers [1]. The process kicks off with research to collect information that will drive the subsequent design. The next step is focused on ideation, visualization, and iterative revision of the product form. During this phase, the designer creates several sketches of the product in varying form factors. Through collective inputs from cross functional teams like engineering, business, marketing and other verticals, this design is refined. This phase

D. Gupta et al. (Eds.): PCEMS 2024, CCIS 2491, pp. 157–168, 2025.
https://doi.org/10.1007/978-3-031-90577-3_15

proceeds in iterative cycles where the designer might need to modify the sketch numerous times. Some of the sketches are used to create 3D-colored renders thereby assessing the product in a more realistic appearance. The designer might need to create additional sketches and renders of the same product to visualize variants with subtle changes and different viewpoints, all of which are utilized for the overall design evaluation. An overview of this process has been depicted in Fig. 1.

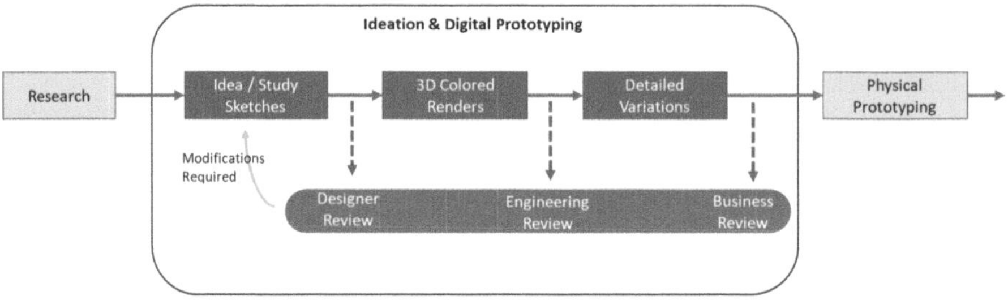

Fig. 1. Overview of the ideation phase of the industrial design process

The iterative nature of this phase involves considerable amount of time and manual effort exerted by the designer to produce numerous sketches. Design software such as AutoCAD [2], SOLIDWORKS [3] and Figma [4] among others have been used extensively by designers to accelerate the process.

1.2 Generative AI for Industrial Design

Of late, Generative AI models have made significant progress in their capabilities around image generation [5]. Not only are these models able to generate realistic, high-fidelity images, but also are able to learn and differentiate different concepts, objects, features and aspects – allowing the image synthesis process to be better controlled and driven towards a desirable outcome, wherein the synthesized image closely adheres to the input instructions [6]. Especially, large diffusion-based models have demonstrated remarkable capability and consistency.

Software and services have been developed to leverage the Generative AI capabilities; popular ones include Midjourney [7], Dall-E [8], DreamStudio [9]. Tools have also been developed with the specific goal of aiding in industrial design, such as Vizcom [10] and Autodesk Generative Design [11].

These tools demonstrate powerful capabilities in design generation by allowing the user to work along with the AI models to create sketches and renders. However, a challenge remains with adapting domain specific features and components. The generative AI models are typically foundational models trained on *extremely* large, generic datasets. They might not have knowledge on specific component types – e.g., types of turbine blades on an engine.

The focus of this paper is to present an approach for using generative AI models in the Ideation and Concept Development phases, enabling rapid digital prototyping. The paper demonstrates how new concepts and features which are domain specific can be introduced to the already powerful foundational models and subsequently applied to generate sketches, modify sketches with specific domain features and convert sketches to 3D color renders. At the time of writing, the prolific use of generative AI models in the creative fine arts has demonstrated their wide-ranging capabilities, however, to the best of our knowledge, the application of these models in the more formal and technical areas of design has not been explored beyond leveraging typical foundational models. The aim of this paper is to present an approach for applying the recent advancements in generative model capabilities, especially those allowing for finer control over the output using the text tokens, towards improving the productivity and efficiency of the iterative industrial design process. This paper does not contrast families of generative models, but instead proposes an approach to leverage generative AI capability.

2 Methodology

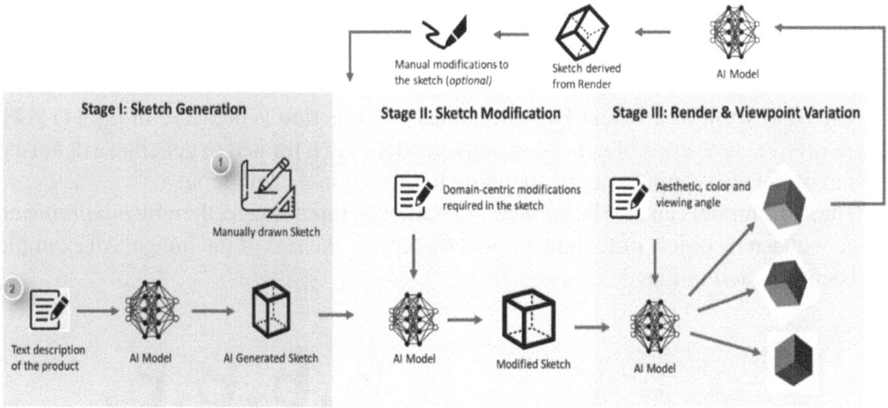

Fig. 2. Overview of the proposed approach for Rapid Digital Prototyping

The proposed approach comprises three distinct stages, as depicted in Fig. 2. The first stage uses AI models to generate sketches based on the description of the product. The designer is able to jump start the entire process by generating a set of sketch ideas and then continuing to refine them in the subsequent stages. The second phase describes methods to modify sketches by synthesizing cross-domain objects, features and aspects into an existing object sketch. This is critical for iterative refinement of the design idea and visualizing the object with different modifications and form factors. The third stage leverages AI models to render the sketch as a realistic 3D image from multiple viewpoints, allowing the designer to make a better assessment of the form factor. Lastly, any of the alternate viewpoint sketches can be converted back into a sketch form for further modifications, either manual or through the AI models.

This section introduces the key deep learning models and elements that are utilized in the presented approach. For the scope of this paper, the base model utilized for image generation is Stable Diffusion [12], which is a type of Latent Diffusion Model [12]. Two model training approaches are used to control the diffusion process and introduce new concepts into the pre-trained diffusion model: ControlNet [13] and LoRA [14]. This section elaborates on these model elements and the remainder of the section explains the proposed approach in detail.

2.1 Diffusion Models for Image Generation

Diffusion Models [15] learn the distribution of features in real images and work by iteratively denoising randomly sampled noise. Latent Diffusion Models include encoders for the image and text modalities which allow the diffusion process to work in the latent space, which is more efficient. Fundamentally, these models are referred to as *text-to-image* models as they ingest textual description as input and synthesize an image that matches the input text. By training the AI model in this regime, it is made to learn the correlations between concepts described in text and their visual representation. Once the model has been trained, it can be used to generate images that combine the learned concepts in novel ways that were not present in the training dataset.

$$L_{DM} = \mathbb{E}_{x,\varepsilon \sim \mathcal{N}(0,1),t}\Big[\|\varepsilon - \varepsilon_\theta(x_t, t)\|_2^2 \Big] \tag{1}$$

The basic form of a Latent Diffusion objective function is depicted in Eq. (1) [12], where $\varepsilon_\theta(x_t, t)$ is a series of denoising autoencoders, each trained to generate a denoised version of the noisy input x_t corresponding to the original clean input x.

The same model can also be applied to a part of an input image, thereby *inpainting* an image with a new object or feature without disrupting the rest of the image. An example has been depicted in Fig. 3.

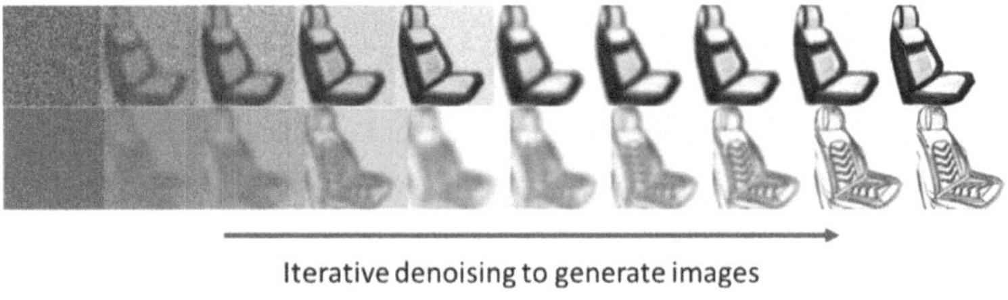

Fig. 3. Example of diffusion denoising

2.2 Model Fine-Tuning with LoRA

Low-Rank Adaptation (LoRA) [14] is a method that drastically reduces the trainable parameters of a model. It was conceived to work with the Transformer [16] layers that are a core part of the model architecture of Large Language Models (LLMs) [17]. It was quickly adopted for use with Diffusion Models. This enables the fine-tuning of LoRA adaptors by training the model with new concepts and visual features. What makes LoRAs even more powerful is their flexibility, ease of training and the ability to leverage multiple LoRA adaptors at the same time, as indicated in Fig. 4. By including or excluding a LoRA adaptor from the generation process, specific domain concepts can be enabled or disabled for the image generation.

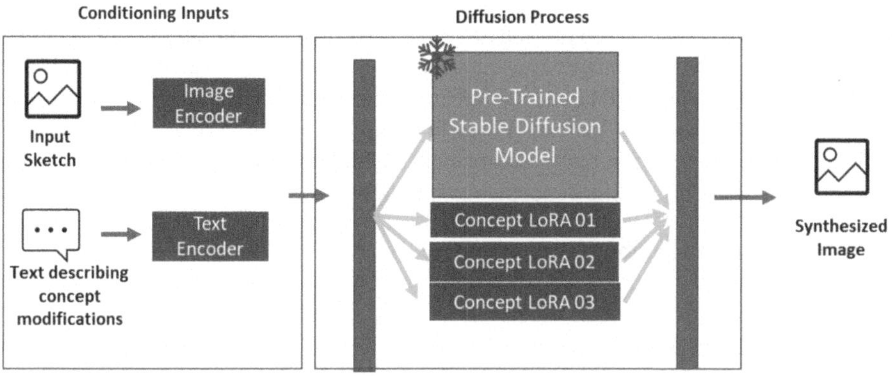

Fig. 4. Overview of the application of trained LoRA adaptors for image synthesis

2.3 Proposed Approach

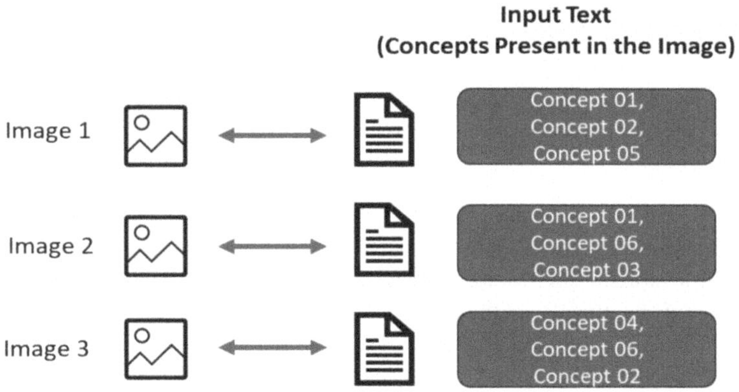

Fig. 5. Simplified form of the LoRA models' training data

The following subsections delve into each of the AI models and how they have been utilized in the context of the proposed approach that has been shown in Fig. 2. Prior to applying the generative AI models, the respective LoRA adaptors need to be trained on a dataset of pairs of images and the corresponding concepts visible in the images. The concepts are maintained in a comma separated text format. The form of a typical training dataset is shown in Fig. 5.

LoRA training is powerful because the trained adaptors can be applied without any significant increase in the hardware or latency required to run the base image generation model. The ability to add new concepts and visual features to a model without compromising the quality of the base model is a key contributor to LoRA's success and adoption. Lastly, the LoRA model layers can be trained with a limited dataset, as long as each new concept is represented by 50–100 images. The number of required images goes up with the number of concepts that are to be trained in a single model. The preparation of the training dataset requires only tagging the respective images with the concepts present in them and does not require granular labels marking specific areas of the image. A combination of these factors makes LoRA training the ideal methodology for generating domain specific concepts in an image.

Stage I: Sketch Generation

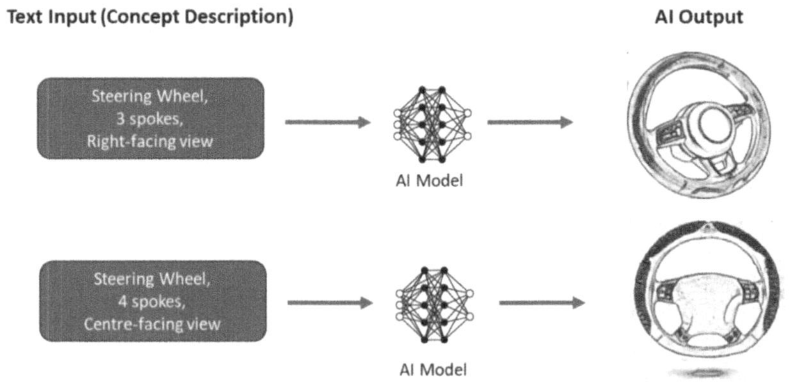

Fig. 6. Example of sketch generation through text description of specific components

The very first step is to generate a sketch of the product with a specific configuration of components or visual features. This is where the generative model comes into play. The designer can leverage a pre-trained Stable Diffusion model in conjunction with LoRA adaptors trained specifically on the domain components and describe their product idea through text, letting the model synthesize multiple images which conform to the input text. Alternately, the designer may prefer to make the initial sketch themselves through other tools. Both approaches are compatible with the subsequent stages. An example is depicted in Fig. 6, wherein car steering wheel is the object to be drawn and the number of spokes is the design concept that defines the structure of the object. Additionally, the viewpoint of the sketch is also given as an input concept. These concepts are made available to the generative model by training a LoRA adaptor to learn these concepts.

Stage II: Sketch Modification

The sketch drawn by the designer or generated by the AI in Stage I is used as input to the model in Stage II. This is the most critical aspect of the proposed approach as it

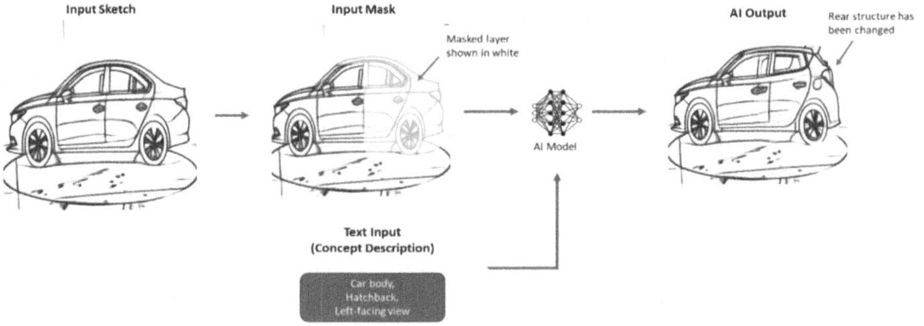

Fig. 7. Example of sketch modification by inpainting part of the input sketch

allows the designer to modify the sketch through text-based instructions. As in Stage I, same LoRA concept adaptors are used for modifying the sketch. However, unlike Stage I, an additional input is provided to the AI model in the form of a sketch and a mask; the mask demarcates the region within the input sketch image which is to be modified. An example is depicted in Fig. 7.

Stage III: Render and Viewpoint Generation

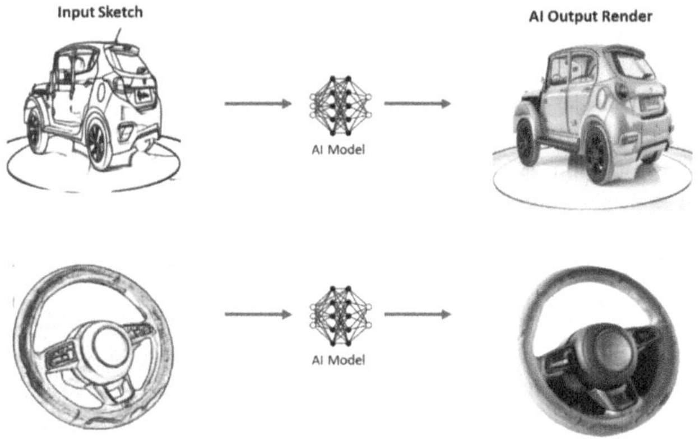

Fig. 8. Examples of AI generated color renders corresponding to input sketches

Once the sketch has been modified (manually and through the AI models) and has the requisite components built into it, a colored render can be generated from the sketch. This is possible through Stable Diffusion model checkpoints notably pre-trained for this purpose or by training a ControlNet model specifically with domain specific sketch-render pairs. An example is shown in Fig. 8.

Further, a novel view synthesis (NVS) model [18] is used to generate colored renders of the object from different viewpoints. By directly synthesizing these renders, a significant amount of manual effort is avoided which would otherwise have been required to draw additional sketches and renders. An example of the process is depicted in Fig. 9.

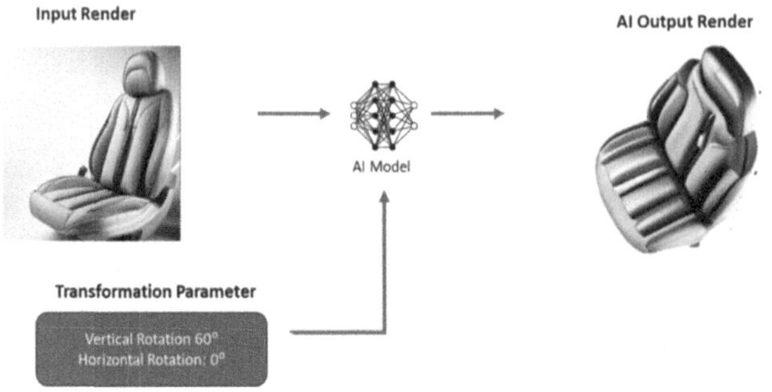

Fig. 9. Example of rotation applied to an input render image.

2.4 Further Sketch Modifications

The models and methods presented above enable generation of sketches with domain specific components, modification of existing sketches and visualizing of color renders in multiple viewpoints. Once this has been achieved, a chosen render at a specific angle of rotation can be converted into a sketch using the AI model. This allows the designer to continue working on the sketch and work in manual edits wherever required. The refined sketch can now once again be applied to Stage II of the proposed approach to iterative refine it and produce new color renders.

2.5 Limitations

This section describes some notable limitations of the proposed method, the foremost of which is the necessity of training data – which comprises a varied set of images and a text description of each image. The concepts which the model is intended to learn must be distinctly represented in the data. In order for the model to successfully isolate the specific concept features, the same visual concept must be present in multiple images with varying conditions described in the text description. The constancy of the specific concept token in the text with the associated visual feature in the images, helps the model in learning this associated. While the number of images required to learn a few novel concepts is not very high, as demonstrated in the experiments in this paper which used ~ 70 images, the required volume goes up quickly with the number and complexity of the concepts. The number of required images can potentially rise to thousands if dozens of concepts are to be trained.

The effectiveness of the model in depicting specific materials and textures is also determined by the quality of images depicting the same materials in the training set.

Inadequate sampling can lead to artifacts or unrealistic synthesis of these aspects in the generate images.

Another limitation is the infrastructure required to utilize these models in real-time, often requiring specialized infrastructure and GPUs. The scale and complexity of the infrastructure depends on the expected latency and number of parallel users of the system.

3 Experiments and Results

To explore the efficacy of the approach, a dataset of car seat sketches was created and used to train multiple LoRA adaptors with new domain specific concepts. The trained LoRA adaptors were used with Stable Diffusion models to produce end-to-end results. The rest of this section introduces the dataset and its constituent concepts, followed by AI generated results.

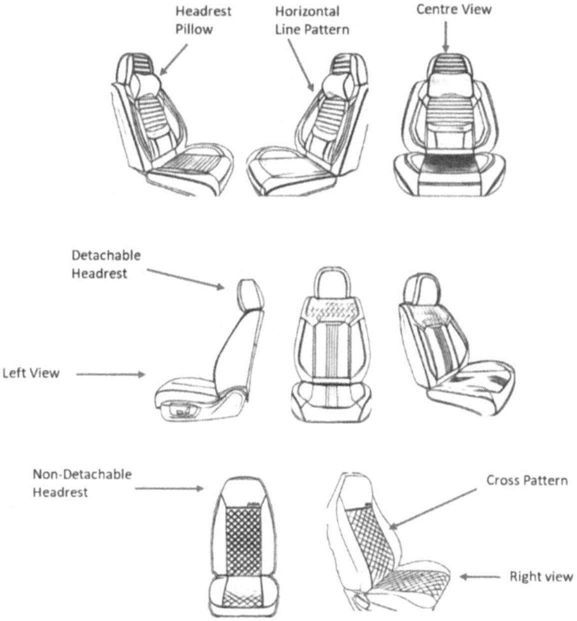

Fig. 10. Sample concepts from the training data. The LoRA adapters learn the correlation between the concept names (text) and their visual features marked in the image above.

Some of the training images have been included in Fig. 10, which also calls out the various concepts introduced to the model – such as headrest variations, seat cover patters and the viewing angle. The training dataset consists of pairs of images and their corresponding concept description in text form. The larger the dataset, the higher is the number of concepts which the model can be trained to learn. For the scope of these experiments, 70 car seat sketches were labelled with their concept properties and used for training.

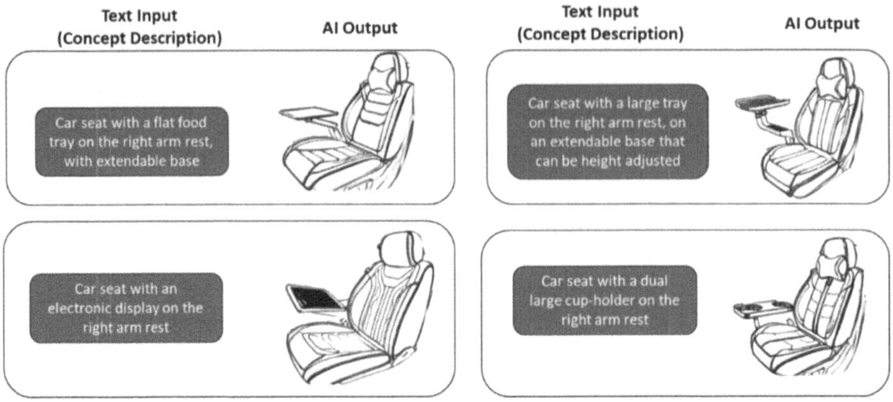

Fig. 11. Stage I example demonstrating the ability to generate sketches from a combination of concepts.

The trained model aids in rapid prototyping by generating the initial set of sketches from the designer's ideas, which could comprise various inter-domain as well as intra-domain concepts and objects, as depicted in Fig. 11. This capability is key to aiding the creative aspect of the design process as the designer can concretely visualize the product design as a sketch right at the offset and incrementally add more concepts to it.

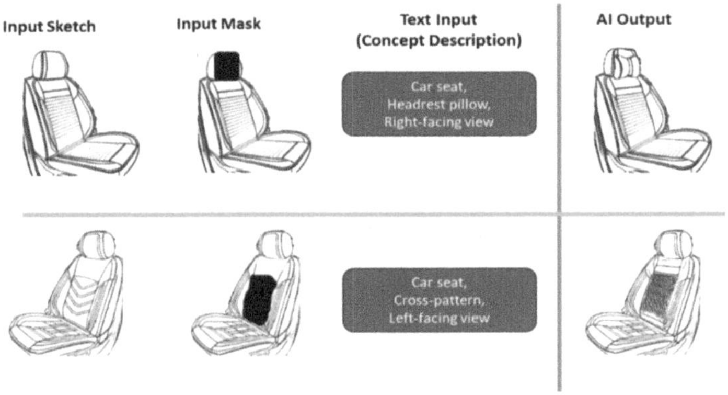

Fig. 12. Stage II examples on car seat sketches. The example above demonstrates inpainting of the seat pattern and headrest.

Using the trained LoRA adaptors, it is possible to modify the sketches as shown in Fig. 12. The inpainting results indicate that structural as well as texture changes can be made to a sketch simply by using the correct text inputs. This capability of the AI models could allow for a high degree of automation in digital prototyping, especially when the designers need to view the product sketch with several permutations and combinations of domain specific changes.

The example in Fig. 13 shows a high quality and realistic AI generated color render corresponding to an input sketch. This view enables a quick assessment of what the

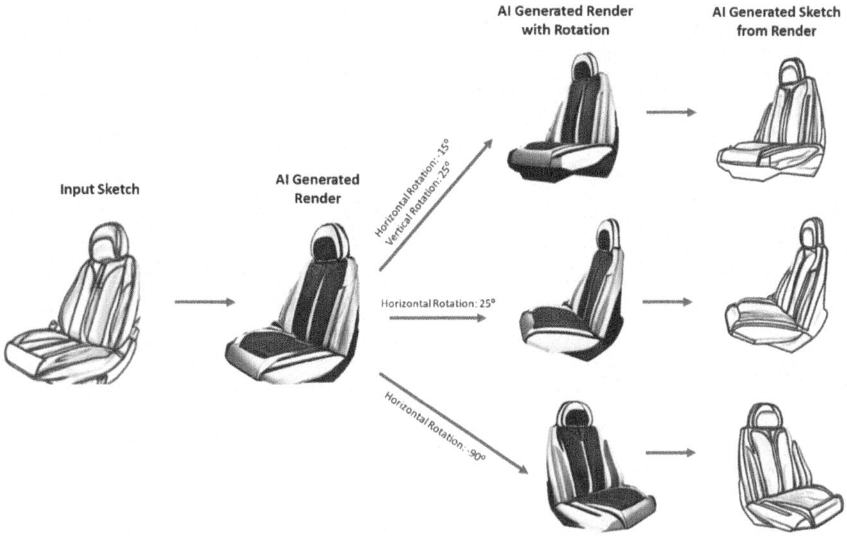

Fig. 13. Stage III examples from the Car Seat Dataset. The example above shows generation of colored renders and their angled variations.

final product may look like. Furthermore, additional views of the same object can be generated by the AI, providing a holistic assessment on the product form. Lastly, sketches are generated from the rotated colored renders which are once again used for further refinement and addition/removal of concepts from the sketch through the AI models.

4 Conclusion

This paper presents an approach to utilizing generative AI models for assisting and accelerating the iterative work around sketches and renders done by industrial designers in the ideation phase, leading to rapid digital prototyping. By leveraging generative AI models through the various methods described in this paper, the designer's efforts are greatly reduced, lowering turn-around time in the crucial design phases. The productivity and ideation output are enhanced through visualization of new concept ideas in concrete form factors, which can be presented to various stakeholders for collection of early feedback and course correction where required.

The presented approach demonstrates the capabilities of such AI models and most crucially, how these can be extended to domain specific features and components which would typically be absent in a general-purpose image generation model. The proposed approach could enhance the turn-around time and productivity across different industries where form factor design is required – from consumer facing product conception and design to industrial and manufacturing parts and components design like in the case automotive, ships, railways and aerospace industries.

The paper elaborates a three-stage approach encompassing the activities of generating sketches and modifying sketches in a controlled manner by describing the concepts and components that are to be drawn in the sketch. The approach also includes details on models which can generate high quality, 3D-colored renders of the sketches from varying viewpoints and also revert the renders back into sketches for the user to continue working on.

Experiments were conducted on a set images and visual concepts covering car seats, car exterior structure as well as car streeting wheels, demonstrating the ability to introduce new concepts and keywords to the models which have a narrow scope and can be utilized to accelerate the digital prototyping process.

While this paper focuses on the visual aspect of the design, the same approach can be extended to jointly consider the engineering aspects of the design, such as placing constraints on the dimensions or apparent weight of the object; we leave the exploration of this area for future work.

References

1. Vizcom Technologies, Inc., "Vizcom". https://www.vizcom.ai/
2. Sohl-Dickstein, J., Weiss, E., Maheswaranathan, N., Ganguli, S.: Deep unsupervised learning using nonequilibrium thermodynamics. In: Proceedings of the 32nd International Conference on Machine Learning, 2015
3. Rombach, R., Blattmann, A., Lorenz, D., Esser, P., Ommer, B.: High-resolution image synthesis with latent diffusion models. In: Proceedings of the IEEE/CVF Conference on Computer Vision and Pattern Recognition (CVPR), 2022
4. Hu, J., et al.: LoRA: Low-Rank Adaptation of Large Language Models, ArXiv, 2021
5. Vaswani, A., et al.: Attention is all you need. In: International Conference on Neural Information Processing Systems, 2017
6. Radford, A., Narasimhan, K., Salimans, T., Sutskever, I.: Improving Language Understanding by Generative Pre-Training, 2018
7. More, A., Chaudhuri, S.: Deep Learning based Novel View Synthesis, ArXiv, 2021
8. Figma, "Figma". https://www.figma.com/
9. Autodesk, "AutoCAD". https://www.autodesk.com/in/products/autocad/
10. Corporation, Dassault Systèmes - SolidWorks, "SolidWorks". https://www.solidworks.com/
11. Midjourney, "MIdjourney". https://www.midjourney.com/home
12. OpenAI, "Dall-E". https://openai.com/dall-e-3/
13. Stability AI, "DreamStudio". https://dreamstudio.ai/
14. Autodesk, "Autodesk Fusion Generative AI". https://www.autodesk.com/solutions/generative-design-ai-software
15. Elasri, M., Elharrouss, O., Al-ma'adeed, S., Tairi, H.: Image generation: a review. Neural Process. Lett. (2022)
16. Cao, P., Zhou, F., Song, Q., Yang, L.: Controllable Generation with Text-to-Image Diffusion Models: A Survey, ArXiv, 2024
17. Zhang, L., Rao, A., Agrawala, M.: Adding conditional control to text-to-image diffusion models. In: Proceedings of the IEEE/CVF International Conference on Computer Vision (ICCV), 2023
18. Pei, E., Campbell, I., Evans, M.: A taxonomic classification of visual design representations used by industrial designers and engineering designers. Des. J. (2011)

PID and LQR Based Control for Quadrotor Landing on a Moving Platform

Prajyot Jadhav, Poojan Gandhi, and Anamika Singh(✉)

Visvesvaraya National Institute of Technology, Nagpur, India
anamikasingh@ece.vnit.ac.in

Abstract. We present a framework enabling unmanned aerial vehicles (UAVs) to autonomously land onto a moving target, utilizing real-time positional data of the target obtained from either a motion-capture system or GPS. Our approach uses onboard sensors, including GPS and IMU, to accurately estimate the state of both the UAV and the moving platform. A robust control architecture utilizing PID and linear quadratic regulator (LQR) techniques is implemented to ensure precise tracking of the moving target during the landing process. We validate the effectiveness and reliability of our methodology by conducting simulations. Through the simulation results, we confirm its capability to achieve autonomous landings on moving platforms.

Keywords: Autonomous Landing · Linear Quadratic Regulator (LQR) · PID · Quad Rotor · Unmanned Aerial Vehicle (UAV)

1 Introduction

Autonomous Unmanned Aerial vehicles (UAVs) are gaining increasing prominence in various industries due to their adaptability and rapid deployment capabilities. They have proven to be highly valuable in a wide range of applications, including but not limited to capturing aerial images for topographical analysis and agricultural purposes, serving as essential first responders during search and rescue missions, and assisting in comprehensive surveying and mapping tasks. UAVs can also work with delivery trucks to shorten package delivery times. However, UAVs have a limited flight time.

The majority of presently employed UAVs typically utilise rechargeable lithium polymer (LiPo) batteries, which offer superior energy density compared to alternative battery technologies. However, it is important to note that these LiPo batteries still exhibit a relatively restricted flight endurance, typically lasting around 10 to 20 min at their maximum capacity. To address these challenges, various solutions are available. These include incorporating a tether for power transmission or enabling autonomous deployment and recovery from a charging station. Tethered flights using a power transmission line, which could potentially offer extended flight endurance but their operational range is inherently constrained by the length of the power line. Enabling autonomous takeoff and landing from a mobile platform equipped with a charging or battery swapping station presents an appealing solution for scenarios requiring sustained flight operations

D. Gupta et al. (Eds.): PCEMS 2024, CCIS 2491, pp. 169–177, 2025.
https://doi.org/10.1007/978-3-031-90577-3_16

over an extended operational range. UAVs and ground vehicles working together would therefore be very beneficial.

Ref. [3] demonstrated quadrotor landing on a moving platform using only onboard sensing and computation. An Extended Kalman Filter (EKF) is used to deal with missing visual detection and to estimate the platform's full state. Trajectories are generated by minimising the jerk and are optimal with respect to a cost function based on the energy necessary to execute it. The quadrotor is directed along the desired trajectory by a nonlinear controller which gives the position and attitude.

Another approach, as explored in Ref. [10] employed Model Predictive Control (MPC) for quadrotor landing on a moving platform, with an EKF tracking the landing target. Utilizing a simplified dynamics model due to slow target rotations, it used a static reference for prediction. Penalizing deviations from both position and estimated target velocity, the approach ensured precise control during landing maneuvers.

Similarly, Ref. [4] utilized MPC, equipping the platform with GPS, IMU, and Wi-Fi for real-time data transmission. Leveraging AprilTag markers for visual tracking and a Kalman Filter for position estimation, it demonstrated another effective approach for quadrotor landing on a moving platform.

Reference [8] adopted a trajectory planning strategy focusing initially on jerk minimization for efficient tag acquisition, later adapting to minimize time spent in turbulent zones as disturbances increased. It incorporated a nonlinear ancillary controller employing sliding control to manage disturbances near the landing platform, alongside an EKF for accurate platform state estimation and visual detection for enhanced UAV-UGV pose estimation

In this paper, we present a system designed to enable the landing of an aerial vehicle on a moving target using real-time positional data of the target, acquired through either a motion capture system or GPS. The control strategies implemented include PID and LQR, considering non-aggressive manoeuvres of the quad rotor system. We validate our approach through comprehensive simulations conducted across various scenarios.

2 System Modelling

Quadrotor dynamics are described using two frames of reference: an inertial frame {O} and a body-fixed frame {B}. The dynamic model of the quadrotor is obtained by using Newton-Euler approach. The quadrotor's design is assumed to be rigid and symmetrical, with the centre of gravity coinciding with the body-fixed frame's origin. Additionally, the propellers are also characterized as rigid, and the thrust and drag forces are proportional to the square of the propeller speed. Roll, pitch, and yaw refer to the rotations about the x, y, and z axes, respectively. The quadrotor configuration is represented by the position of the centre of mass in the inertial frame, the linear velocities in the inertial frame, the orientation in the inertial frame represented by the Euler angles, and the angular velocities in the body-fixed frame. The rate of change of Euler angles can be obtained from the body-fixed angular velocities by,

$$\begin{bmatrix} \dot{\phi} \\ \dot{\theta} \\ \dot{\psi} \end{bmatrix} = \begin{bmatrix} 1 & \sin(\phi)\tan(\theta) & \cos(\phi)\tan(\theta) \\ 0 & \cos(\phi) & -\sin(\phi) \\ 0 & \frac{\sin(\phi)}{\cos(\theta)} & \frac{\cos(\phi)}{\cos(\theta)} \end{bmatrix} \begin{bmatrix} p \\ q \\ r \end{bmatrix} \tag{1}$$

2.1 Quad Rotor Dynamic Model

Although the quad rotor has six degrees of freedom, only four actuators are utilized to generate the control inputs: U =, where U1 represents the overall thrust, while U2, U3, and U4 correspond to the rolling, pitching, and yawing moment, respectively. The relationship between the control inputs and the rotational speeds of the propellers is given by,

$$\left.\begin{array}{l} U_1 = b(\Omega_1 + \Omega_2 + \Omega_3 + \Omega_4) \\ U_2 = bl(-\Omega_1 - \Omega_2 + \Omega_3 + \Omega_4) \\ U_3 = bl(\Omega_1 - \Omega_2 - \Omega_3 + \Omega_4) \\ U_4 = d(\Omega_1 - \Omega_2 + \Omega_3 - \Omega_4) \end{array}\right\} \quad (2)$$

Here, 1 denotes the distance between a rotor and the center of the quadrotor, b and d represent the thrust and drag coefficients respectively. And, Ω1, ···, Ω4 denote the angular speeds of the propellers. The system state vector is defined as X = $[x\ y\ z\ u\ v\ w\ \phi\ \theta\ \psi\ p\ q\ r]^T$. The non-linear dynamics describing the behavior of the quadrotor are represented by the following equations:

$$\left.\begin{array}{c} \dot{\mathrm{x}} = \mathrm{u} \\ \dot{\mathrm{y}} = \mathrm{v} \\ \dot{\mathrm{z}} = \mathrm{w}\dot{u} = -(\cos\phi\sin\theta\cos\psi + \sin\phi\sin\psi)\frac{U_1}{m} \\ \dot{v} = -(\cos\phi\sin\theta\cos\psi - \sin\phi\cos\psi)\frac{U_1}{m}\dot{w} = -(\cos\phi\cos\theta)\frac{U_1}{m} + g \\ \dot{\phi} = p + (\sin\phi\tan\theta)q + (\cos\phi\tan\theta)r\dot{\theta} = q\cos\phi - r\sin\phi \\ \dot{\psi} = \frac{\sin\phi}{\cos\theta}q + \frac{\cos\phi}{\cos\theta}r \\ \dot{p} = \frac{I_y - I_z}{I_x}qr + \frac{U_2}{I_x} \\ \dot{q} = \frac{I_z - I_x}{I_y}pr + \frac{U_3}{I_y} \\ \dot{r} = \frac{I_x - I_y}{I_z}pq + \frac{U_4}{I_z} \end{array}\right\} \quad (3)$$

In these equations, m represents the mass of the quadrotor, g denotes the acceleration due to gravity, and Ix, Iy, and Iz signify the quadrotor's moments of inertia around the three axes.

2.2 State Space Model

The dynamic model of the quadrotor can be linearized around the hovering condition, under the assumption of small attitude angles, with the yaw angle ψ considered as zero. The state vector is modified as

$$\mathrm{X} = [x\ y\ z\ u\ v\ w\ \phi\ \theta\ \psi\ p\ q\ r]^T$$

The state space model is expressed as follows:

$$\dot{\mathrm{x}} = \mathrm{Ax} + \mathrm{Bu}\ \mathrm{y} = \mathrm{Cx} + \mathrm{Du}$$

Where y is the output vector, A is the system matrix, B is the input matrix, C is the output matrix and D is the feed-forward matrix. After linearization of the model described in (3) we get,

$$A = \begin{bmatrix} 0\,0\,0\,1\,0\,0\,0 & 0 & 0\,0\,0\,0 \\ 0\,0\,0\,0\,1\,0\,0 & 0 & 0\,0\,0\,0 \\ 0\,0\,0\,0\,0\,1\,0 & 0 & 0\,0\,0\,0 \\ 0\,0\,0\,0\,0\,0\,0 & -g & 0\,0\,0\,0 \\ 0\,0\,0\,0\,0\,0\,g & 0 & 0\,0\,0\,0 \\ 0\,0\,0\,0\,0\,0\,0 & 0 & 0\,0\,0\,0 \\ 0\,0\,0\,0\,0\,0\,0 & 0 & 0\,1\,0\,0 \\ 0\,0\,0\,0\,0\,0\,0 & 0 & 0\,0\,1\,0 \\ 0\,0\,0\,0\,0\,0\,0 & 0 & 0\,0\,0\,1 \\ 0\,0\,0\,0\,0\,0\,0 & 0 & 0\,0\,0\,0 \\ 0\,0\,0\,0\,0\,0\,0 & 0 & 0\,0\,0\,0 \\ 0\,0\,0\,0\,0\,0\,0 & 0 & 0\,0\,0\,0 \end{bmatrix} \tag{4}$$

$$B = \begin{bmatrix} 0 & 0 & 0 & 0 \\ 0 & 0 & 0 & 0 \\ 0 & 0 & 0 & 0 \\ 0 & 0 & 0 & 0 \\ \frac{1}{m} & 0 & 0 & 0 \\ 0 & 0 & 0 & 0 \\ 0 & 0 & 0 & 0 \\ 0 & 0 & 0 & 0 \\ 0 & 0 & 0 & 0 \\ 0 & \frac{1}{I_x} & 0 & 0 \\ 0 & 0 & \frac{1}{I_y} & 0 \\ 0 & 0 & 0 & \frac{1}{I_z} \end{bmatrix} \tag{5}$$

3 Control Methods

3.1 PID Controller

The classical PID controller is characterized by its simple structure and ease of implementation. In our control architecture, as shown in Fig. 1, we utilize nested loops, with the inner loop dedicated to attitude control and the outer loop responsible for position control. Specifically, the attitude loop is nested within the position loop. From the feedback of the current attitude and angular rates, we derive the control inputs U2, U3 and U4. Particularly in the equations-of-motion outlined later, we compute the value of these control inputs based on the desired attitude. In the outer-loop, we receive the desired position vector and yaw angle. Subsequently, we compare these with the actual position and velocity data. Through this comparison, we determine U1. The thrusts and torques are obtained from the error values as shown in the following equations:

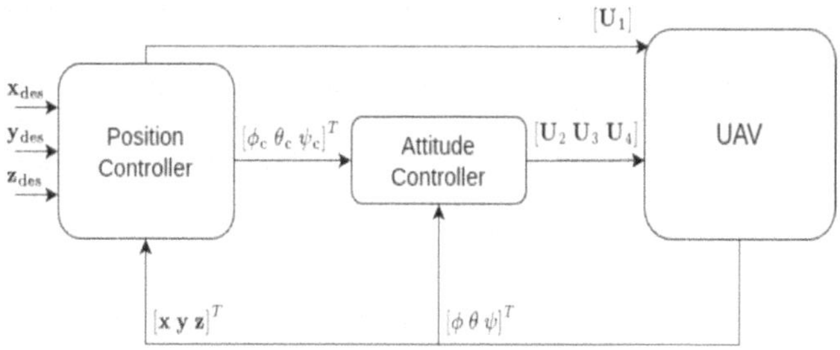

Fig. 1. PID Control Architecture

$$\begin{gathered}
x_c = K_{x,P}(e_x) + K_{x,D}\dot{e}_x + K_{x,I}\int e_x \\
y_c = K_{y,P}(e_y) + K_{y,D}\dot{e}_y + K_{y,I}\int e_y \\
\phi_c = \frac{1}{g}(x_c \sin\psi_{des} - y_c \cos\psi_{des}) \\
\theta_c = \frac{1}{g}(x_c \cos\psi_{des} + y_c \sin\psi_{des}) \\
\psi_c = \psi_{des} \\
\begin{bmatrix} U_1 \\ U_2 \\ U_3 \\ U_4 \end{bmatrix} = \begin{bmatrix} m\left(g + K_{z,P}(e_z) + K_{z,D}\dot{e}_z + K_{z,I}\int e_z\right) \\ K_{\theta,P}(\theta_c - \theta) + K_{\theta,D}\left(\dot{\theta}_c - \dot{\theta}\right) + K_{\theta,I}\int(\theta_c - \theta) \\ K_{\phi,P}(\phi_c - \phi) + K_{\phi,D}\left(\dot{\phi}_c - \dot{\phi}\right) + K_{\phi,I}\int(\phi_c - \phi) \\ K_{\psi,P}(\psi_c - \psi) + K_{\psi,D}\left(\dot{\psi}_c - \dot{\psi}\right) + K_{\psi,I}\int(\psi_c - \psi) \end{bmatrix}
\end{gathered} \tag{6}$$

Where ex = xdes − x is the error in the x-direction. Analogous expressions apply for other dimensions.

3.2 Linear Quadratic Regulator (LQR) Based Controller

The Linear Quadratic Regulator (LQR) provides optimally controlled feedback gains. It enables the design of closedloop systems characterized by stability and high performance. Consider a linear time-invariant system of the form:

$$\dot{x} = Ax + Bu$$

The cost function to be minimized and the control inputs are given by:

$$J = \int_0^\infty \left(\tilde{x}^T \mathbf{Q} \tilde{x} + \tilde{u}^T \mathbf{R} \tilde{u}\right) dt$$

$$u = udes + K(x - xdes)$$

In this formulation, Q signifies the positive semi-definite state weighting matrix, while R denotes the positive-definite cost weighting matrix. The terms $\tilde{x} = x - x_{des}$ and $\tilde{u} = u - u_{des}$ represent the errors in state and input, respectively, where xdes and udes denote the reference state and input. Additionally, $K = -R^{-1}B^TP$, where P is the solution of the Continuous.

Algebraic Riccati equation:

$$0 = \mathbf{PA} + \mathbf{A}^\mathrm{T}\mathbf{P} - \mathbf{PBR}^{-1}\mathbf{B}^\mathrm{T}\mathbf{P} + \mathbf{Q}$$

4 State Machine

The system comprises of three states: takeoff, tracking and landing. Figure 2 illustrates the state machine along with its states and the corresponding transitions triggered by events. We provide a more thorough explanation of each of these stages in the sections that follow.

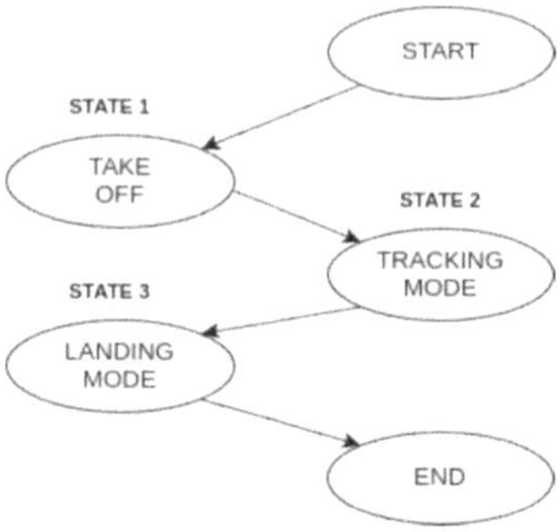

Fig. 2. The flowchart of the state machine

4.1 Takeoff

The quadrotor initially takes off from the ground and stabilizes into a hover position. Upon reaching this hover point, it is commanded to track the moving platform and the state machine is switched to tracking mode.

4.2 Tracking

During this phase, the quadrotor receives the current position of the moving platform as the desired state. Its objective is to track and maintain a position above the moving platform. The desired position is continuously updated at each time instant. Depending on the control method employed (such as PID or LQR), the control inputs are calculated based on the error values. Subsequently, these control inputs are applied to the quadrotor. This phase concludes when the quadrotor successfully flies above the moving platform, and the error in position values falls below a predefined tolerance threshold.

4.3 Landing

Upon entering this phase, the quadrotor switches its operational mode to landing, initiating its descent toward the moving platform. Employing a vertical descent approach, the quadrotor gradually lowers itself onto the platform's surface. Upon achieving a successful landing, the quadrotor disarms itself, bringing the rotation of its motors to a halt, thereby concluding the landing operation.

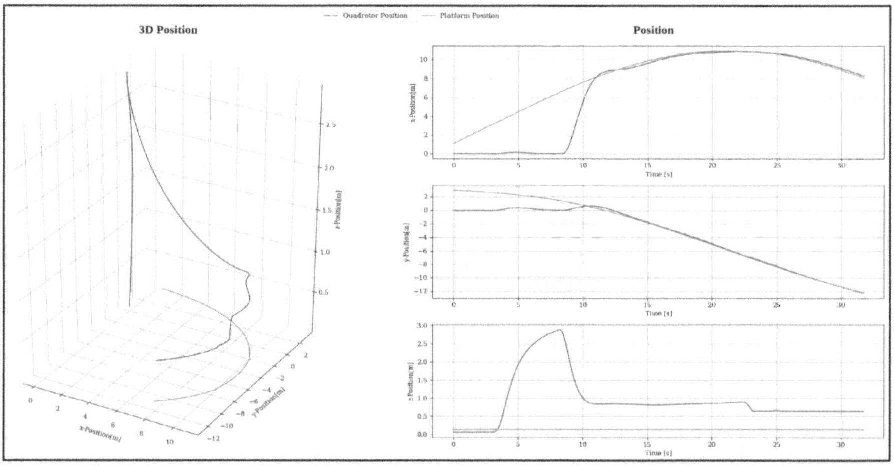

Fig. 3. The results of one of the simulation experiments. We report the data for the position of the quadrotor and the moving platform. The quadrotor takes off at t = 0. The platform moves in a curved path at a speed of 1.55 m/s.

5 Simulation Experiments

5.1 Simulation Environment

To validate the proposed algorithm, we conducted Software-in-the-Loop (SITL) simulations utilizing the PX4 open-source flight control software [7]. The simulations were performed with a 3DR Iris quadrotor model simulated in the Gazebo simulator environment. The simulated quadrotor communicates with the PX4 flight control software via the MAVLink protocol, which defines a standardized set of messages for transmitting sensor data from the simulated environment to PX4, as well as for relaying actuator commands back to the simulated vehicle. To facilitate communication between the simulated quadrotor and the flight controller, we used MAVROS, a ROS package that allows controlling drones via the MAVLink protocol. Within our simulated environment, we utilized a Clearpath Husky as the ground vehicle. On top of the Husky, we mounted a landing platform measuring 100 cm $\times$ 100 cm.

5.2 Experiment Description

We conducted various simulations, to validate our proposed framework. Specifically, we executed simulation experiments involving the movement of the landing platform along different paths, such as straight lines and curved paths. We varied the speed of the landing platform from 0.5 m/s to 3.5 m/s. We conducted experiments in simulation utilizing both the PID and LQR controllers.

Within the simulation environment, we evaluated the performance of our framework across a range of scenarios characterized by steady wind conditions. In Fig. 3 we present the results of one of our simulation experiments. The quadrotor begins the takeoff phase at t = 0. The quadrotor tracks the moving platform until it reaches a point where the position

Table 1. 3DR Iris Quadrotor Parameters

Parameter	Description	Value
m	Mass	1.5 kg
l	Arm length	0.25 m
Ix	Moment of inertia about the x-axis	0.029125 kg·m^2
Iy	Moment of inertia about the y-axis	0.029125 kg·m^2
Iz	Moment of inertia about the z-axis	0.055225 kg·m^2

error falls below a predefined threshold, signaling the appropriate time for landing. The discrepancies in the z-position values between the quadrotor and the moving platform post-landing are due to the landing pad's attachment at an elevation above the Husky.

6 Conclusion and Future Works

This paper presents a system for the landing of an aerial vehicle onto a moving target, utilizing real-time positional data of the target obtained from either a motion-capture system or GPS. PID and LQR based controller was implemented to control the quadrotor for tracking the moving target. The simulation results demonstrate the effectiveness of the proposed approach in enabling the quadrotor to accurately track and land on the moving target.

In our future work, we're aiming to address the following issues with quadrotor control. Firstly, the current methods, like PID and LQR, have trouble stabilizing the quadrotor when it's states are not around the equilibrium state due to the quadrotor's complex non-linear behavior. Secondly, the proposed model does not take into consideration external disturbances like wind velocities. Our goal is to make our controller more robust to handle these issues effectively. To achieve this, we're planning to integrate reinforcement learning techniques with our existing control methods.

References

1. Baca, T., Stepan, P., Saska, M.: Autonomous landing on a moving car with unmanned aerial vehicle. In: 2017 European Conference on Mobile Robots (ECMR), pp. 1–6, Paris, France (2017)
2. Bouabdallah, S., Noth, A., Siegwart, R.: PID vs LQ control techniques applied to an indoor micro quadrotor. In: 2004 IEEE/RSJ International Conference on Intelligent Robots and Systems (IROS), vol. 3, pp. 2451–2456, Sendai, Japan (2004)
3. Falanga, D., Zanchettin, A., Simovic, A., Delmerico, J., Scaramuzza, D.: Vision-based autonomous quadrotor landing on a moving platform. In: 2017 IEEE International Symposium on Safety, Security and Rescue Robotics (SSRR), pp. 200–207, Shanghai, China (2017)
4. Feng, Y., Zhang, C., Baek, S., Rawashdeh, S., Mohammadi, A.: Autonomous landing of a UAV on a moving platform using model predictive control. Drones. **2**, 34 (2018)

5. Foehn, P., Scaramuzza, D.: Onboard state-dependent LQR for agile quadrotors. In: 2018 IEEE International Conference on Robotics and Automation (ICRA), pp. 6566–6572, Brisbane, QLD, Australia (2018)
6. Lee, D., Ryan, T., Kim, H.J.: Autonomous landing of a VTOL UAV on a moving platform using image-based visual servoing. In: 2012 IEEE International Conference on Robotics and Automation, pp. 971–976, Saint Paul, MN, USA (2012)
7. Meier, L., Honegger, D., Pollefeys, M.: PX4: A node-based multithreaded open-source robotics framework for deeply embedded platforms. In: 2015 IEEE International Conference on Robotics and Automation (ICRA), pp. 6235–6240, Seattle, WA, USA (2015)
8. Paris, A., Lopez, B., How, J.: Dynamic landing of an autonomous quadrotor on a moving platform in turbulent wind conditions. J. Field Robot. **36**, 1–15 (2019)
9. Reyes-Valeria, E., Enriquez-Caldera, R., Camacho-Lara, S., Guichard, J.: LQR control for a quadrotor using unit quaternions: Modeling and simulation. In: CONIELECOMP 2013, 23rd International Conference on Electronics, Communications and Computing, pp. 172–178, Cholula, Puebla, Mexico (2013)
10. Tzoumanikas, D., Li, W., Grimm, M., Zhang, K., Kovac, M., Leutenegger, S.: Fully autonomous micro air vehicle flight and landing on a moving target using visual-inertial estimation and model-predictive control. J. Field Robot. **35**, 678–692 (2018)

Overhead Reduction in CSI Estimation for RIS-Enabled MISO Wireless Communications

Anish Kumar Gupta[1(✉)], Vishwas Raval[1,2], and Punitkumar Bhavsar[1]

[1] Visvesvaraya National Institute of Technology Nagpur, Nagpur, India
anishsgsits@gmail.com, punitbhavsar@ece.vnit.ac.in
[2] Central University of Gujrat Vododara, Mosampura, India
vishwas.raval@cug.ac.in

Abstract. Reconfigurable Intelligent Surfaces (RIS) has the capability to modify the wireless communication environment by altering the amplitude and phase of incoming electromagnetic waves, thereby enhancing system capacity and coverage. Precise channel state information (CSI) is crucial for achieving these benefits in RIS-enabled wireless communication systems. However, estimating the channels in RIS-Enabled systems poses significant challenges arises from the large number of RIS elements, which passively reflect signals and unavailability of signal processing capabilities. This article focuses on RIS-assisted multiple-input single-output (MISO) wireless communication systems and proposing a twofold-pattern sparsity orthogonal matching pursuit (TP-OMP) channel estimation (CE) method leveraging channel sparsity in the angular domain to reduce overhead. Monte Carlo (MC) simulations demonstrate a notable reduction in required pilot overhead compared to existing methods.

Keywords: Reconfigurable intelligent surface · Channel state information · Multiple-input single-output · Compressive sensing · channel sparsity

1 Introduction

The 5G mobile communication standard is designed to deliver faster broadband speeds, extensive connectivity, and extremely low latency. This is achieved through the use of advanced technologies like massive MIMO (multiple-input multiple-output) systems and millimeter-wave (mmWave) frequencies, which allow for significantly improved data transmission and network responsiveness. Despite these benefits, these technologies are characterized by high power consumption, hardware complexity, and limited control over the wireless channel. To address these issues, reconfigurable intelligent surfaces (RIS) propose a cost-effective solution with minimal hardware complexity, and can easily be deployed

D. Gupta et al. (Eds.): PCEMS 2024, CCIS 2491, pp. 178–190, 2025.
https://doi.org/10.1007/978-3-031-90577-3_17

into wireless communication systems to create a smart and adaptive radio channel environment. The RIS comprises numerous cost-effective passive elements that can intelligently reconfigure the phases and amplitudes of incoming electromagnetic waves. This capability allows the RIS to direct these waves toward the intended receiver, thereby enhancing the performance of wireless communication [1–4]. Unlike active beamforming, which requires additional active components and higher power, the RIS functions as a passive beamformer.

Most current research on RIS assumes the availability of accurate channel information at the access point (AP) [5]. These studies mainly concentrate on the joint design of active transmit beamforming at the AP and passive reflective beamforming at the RIS to improve both spectral and energy efficiency of the system.. Nonetheless, obtaining perfect channel state information (CSI) in RIS-assisted systems is challenging. There are two key challenges related to CSI estimation in RIS-assisted wireless systems. Firstly, because all RIS elements are passive, they cannot transmit, receive, or process the pilot signals required for precise channel estimation (CE). Second, because an RIS typically comprises a large number of elements often in the hundreds the dimensionality of the channels that need to be estimated is far greater than that in traditional systems. This increase in dimensionality leads to a dramatic rise the pilot signals necessary for CE, complicating the process considerably.The significant passive beamforming gain provided by the RIS comes at the cost of increased overhead for CE. This is because of the additional channels that need to be taken into account between the RIS and its associated AP or users.

Research has presented various techniques for CSI estimation in RIS-enabled wireless communication systems [6]-[10]. For instance, the authors of [6] introduced a least squares (LS) based CE method, assuming an element-by-element ON/OFF channel reflection pattern. [7] proposed a cascaded CE solution that exploits the low-rank nature of MIMO channels and employs A reflection pattern with random ON/OFF states. This method, based on sparse matrix factorization, Significantly reduces the pilot signal requirements. Exploiting the sparse nature of the cascaded channel, the authors in [8] proposed an estimation method based on compressive sensing (CS). This technique effectively estimates the cascaded channel and significantly reduces the training overhead. In [9], a new design is introduced where RIS elements are organized into clusters, with each cluster of adjacent elements sharing the same reflection coefficient. This approach necessitates estimating the composite channel for each cluster sequentially, significantly reduce training overhead and simplify the design of RIS coefficients compared to individual element estimation. Nevertheless, studies [6–9] have estimated the CSI by employing a simple element-by-element ON/OFF reflection pattern, which is costly to implement and inefficient due to not utilizing the full potential of the RIS for CE. To address these issues, the author in [10] fully utilized the reflective properties of the RIS during CE, and employed LS to jointly estimate the base station-users and base station-RIS-users channels in OFDM systems. They adopted a different approach, assuming single-antenna configurations for all users and access points to mitigate prohibitive pilot overhead and complexity associated with higher antenna counts. However, these methods of CSI estimation in RIS-enabled wireless communication still require a large number of pilot signals, which poses a significant challenge in meeting the demands of future wireless communication systems.

Motivated by the drawbacks and challenges associated with the existing methods in estimation the Channel in RIS assisted wireless communication, we propose a twofold-pattern sparsity orthogonal matching pursuit (TP-OMP) method. The different angular cascaded channels show the twofold-pattern sparsity and based on this common non-zero rows and columns can be estimated jointly for all users. Since twofold sparsity is present instead of single sparsity in rows or columns, the proposed method of CE further reduces the required pilot overhead.

2 CSI Estimation Fundamentals in Basic Wireless Communication

Estimating CSI is critical for optimizing data transmission in wireless communication systems. CSI encompasses the characteristics of a radio link, including path loss, scattering, diffraction, fading, and shadowing, which impact signal propagation from transmitter to receiver. This CSI data is used to evaluate the quality of the radio link, determining whether it is in a favorable or unfavorable state [11]. Accurate CSI is crucial as it influences the selection of physical layer parameters and communication schemes. For example, low-order modulation is chosen for poor channel conditions, while high-order modulation is preferred for reliable channels. Incorrect CSI can lead to suboptimal modulation choices and inefficient communication. Additionally, CSI plays a significant role in allocating radio resources and managing interference in wireless systems. Thus, ensuring precise CSI is essential for improving communication performance [12].

2.1 Channel Estimation

The process of CSI estimation in basic wireless communication involves determining the properties of the communication channel, which are crucial for optimizing data transmission and reception. CE allows the communication system to understand the distortions and fading that signals experience as they propagate through the wireless medium. To estimate the CSI, the process typically starts with the transmission of known pilot signals, or training sequences, from the transmitter to the receiver. These pilot signals are carefully designed and known to both the transmitter and receiver. When these signals are transmitted through the wireless channel, they undergo various distortions, such as fading, path loss, and noise. Upon receiving the pilot signals, the receiver compares the received signals with the known transmitted pilots. By analyzing the differences between the transmitted and received pilot signals, the receiver can infer the channel's impulse response or transfer function. This comparison allows the receiver to estimate the impact of the channel on the transmitted signals, which includes the effects of multipath propagation, Doppler shifts, and other impairments. A simple wireless communication system estimates the CSI with following steps:

1. Pilot signal transmission: The transmitter sends out pilot signals at predefined intervals or times. These signals are embedded in the data stream.
2. Reception and extraction: The receiver captures the transmitted signals, including the pilot signals. The receiver then extracts these pilots from the received signal.
3. Comparison and calculation : The received pilot signals are compared with the originally transmitted pilots. The differences between them, caused by the channel, are used to calculate the channel's characteristics. This involves determining the channel's impulse response or transfer function.
4. Channel estimation: Using mathematical and statistical techniques, the receiver estimates the channel's parameters. This estimation provides the CSI, which includes information about the channel's amplitude, phase shifts, and any other distortions affecting the transmitted signal.

Once the CSI is obtained, it is used to adjust the receiver's parameters for optimal signal detection and decoding. This may involve equalization to mitigate the effects of multipath fading and noise, thereby improving the accuracy of data recovery.

2.2 CSI Estimation Methods

Various techniques for CE include the LS method, minimum mean square error (MMSE) method, and maximum likelihood (ML) method, among others. The LS method is one of the simplest and most commonly used techniques. It estimates the channel by minimizing the sum of the squares of the differences between the received pilot signals and the estimated channel responses. Although straightforward, the LS method can be suboptimal in the presence of noise, i.e., it is computationally efficient but can be sensitive to noise and outliers in the received signal. To address this, MMSE method incorporates statistical information about the channel and the noise, providing a more accurate estimation under noisy conditions. By optimizing the mean square error criterion, MMSE estimation provides more robust estimates, particularly in environments with significant noise or interference. ML estimation is another widely used method that determines the channel parameters by maximizing the likelihood of observing the received signal given the transmitted signal. ML estimation assumes a statistical model for both the channel and the noise, providing accurate estimates under ideal conditions but requiring more computational resources compared to LS and MMSE methods.

2.3 Limitations of Conventional CSI Estimation Methods and Solutions in 5G

The conventional methods are useful in basic wireless communication systems where small scale CE is required. However, these methods of CE face several drawbacks when viewed from the prospective of 5G and B5G communication requirements. Typically, these conventional CSI estimation approaches necessitate computationally intensive matrix operations like multiplication, inversion, eigenvalue decomposition, and singular value decomposition. Consequently, traditional methods are only viable for estimating CSI on a small scale in wireless communication systems. On the flip side , in recent years, the surge in mobile devices and traffic has driven the evolution of cellular systems towards 5G wireless communication. To handle the escalating demand and enhance performance, 5G systems incorporate various advanced technologies, including massive MIMO, OFDM, mmWave communication, and RIS. These innovations significantly boost system capacity and quality of service, meeting the growing needs of mobile traffic. However, the adoption of these new techniques introduces considerable challenges for traditional CSI estimation methods in 5G wireless communications. As a result, CSI estimation transforms into a complex, large-scale CE task, requiring intensive matrix operations and leading to high computational complexity. Consequently, obtaining CSI using traditional methods is time-consuming, making it difficult to refresh CSI at the required rate for 5G systems. Additionally, the channel dimensions in 5G is large, the overhead from pilot signals in traditional methods can use up substantial resources, thereby decreasing the bandwidth available for actual data transmission. Therefore, there is a pressing need to develop more efficient CSI estimation techniques for 5G wireless communications.

CE in 5G wireless communication can be significantly enhanced using deep learning techniques. Unlike traditional methods that rely heavily on pilot signals and complex matrix operations, deep learning models can learn the intricate relationships within the channel data, enabling more accurate and efficient estimation. By leveraging large datasets and neural networks, these models can adapt to rapidly changing channel conditions and high mobility scenarios, such as those found in 5G and upcoming 6G systems. This approach not only reduces the computational complexity and resource overhead but also improves the system reliability. The ability of deep learning to generalize from training data and predict channel states with minimal real-time computational load makes it a promising solution for modern and future wireless communication challenges[13,14]. However, CE using deep learning face challenges such as the need for extensive training data and high computational resources, which can limit their practicality in real-time and resource-constrained future wireless communication systems [15]. Additionally, these methods may struggle to generalize across diverse communication environments, impacting their robustness and reliability.

3 CSI Estimation in RIS-Enabled Wireless Communications

3.1 System Model

Figure 1 illustrates the wireless communication system enabled by RIS, where base station (BS) having L antennas is sending information to the K users via RIS. Each user is having single antenna. The RIS is made up of N passive reflecting elements. The channel linking the BS and the RIS is represented by $\mathbf{Q} \in \mathbb{C}^{L\times N}$, while the channel between the RIS and the kth user is denoted as $\mathbf{p_k} \in \mathbb{C}^{N\times 1}$. The direct communication channel from the BS to the kth user is denoted by $\mathbf{r_k} \in \mathbb{C}^{L\times 1}$. All channels are modeled as complex Gaussian random variables with zero mean and unit variance. Let ψ=$[\eta_1 e^{-j\theta_1}, \eta_2 e^{-j\theta_2}, ..., \eta_N e^{-j\theta_N}]'$, represents the RIS reflection coefficient vector. $\phi_n = \eta_n e^{-j\theta_n}$, where $\eta_n \in (0,1)$, $\theta_n \in [\pi, -\pi)$, $n = 1, 2, ..., N$ represent amplitude and phase of reflection coefficient of nth RIS element. The received signal corresponding to all user at the BS is given by

$$\mathbf{y} = \sum_{k=1}^{K} (\mathbf{r}_k + \mathbf{Q}diag(\psi)\mathbf{p_k})x_k + \mathbf{n} \tag{1}$$

, where x_k denotes the signal sent by the kth user to the BS and $\mathbf{n} \in \mathbb{C}^{M\times 1}$ denotes the noise at BS.

The direct channel r_k can be estimated by using traditional methods of CE of wireless communication, so it is assumed to be known. However, channels between BS-RIS, Q and RIS-user, p_k are difficult to measure because passive RIS surface is unable to process the signal, i.e., it can not transmit or/ and receive the signal. Let $\mathbf{A_k} \triangleq \mathbf{Q}diag(\mathbf{p_k}) \in \mathbb{C}^{L\times N}$ be the cascaded channel between BS and k_{th} user. Equation 1 can be written after removing the direct channel as

$$\mathbf{y} = \sum_{k=1}^{K} \mathbf{A_k}\psi x_k + \mathbf{n} \tag{2}$$

Further, for optimizing the power, i.e., $||\mathbf{A_k}\psi||^2$ at receiver, individual channels $\mathbf{Q}$ and $\mathbf{p}$ are not required, instead cascaded channel $\mathbf{A}$ is sufficient to be estimated.

3.2 Basic CE Protocol

The channel r_k is estimated by turning off all the RIS elements in one go thus no incident wave is reflected by RIS surface, the RIS-enabled system can be reduced to a traditional wireless communication system model and r_k is measured by using some conventional methods such as MMSE method , LS method.

Further, to estimate the cascaded channel $\mathbf{A}$, ON/OFF protocol is proposed. It can be observed that each column vector of $\mathbf{A}$ is associated with the single element of RIS and can be expressed as $\mathbf{A} = [\mathbf{a}_1, \mathbf{a}_2, ..., \mathbf{a}_N]$, where $\mathbf{a}_n \in \mathbb{C}^{L\times 1}$ represents the cascaded channel associate with n^{th} element of RIS. Thus in order to estimate the complete channel $\mathbf{A}$, the process of CE is divided into N stages, i.e., equal to number of RIS elements and only one column vector is estimated in each stage. As n^{th} stage is related to the estimation of cascaded channel corresponding to the nth RIS element, the nth RIS element is turned ON in n^{th} CE stage and remaining $N-1$ RIS elements are turned OFF. Then $\mathbf{a}_n$ is estimated using conventional method in each stage and after N stages, we are able to estimate the complete channel matrix $\mathbf{A}$.

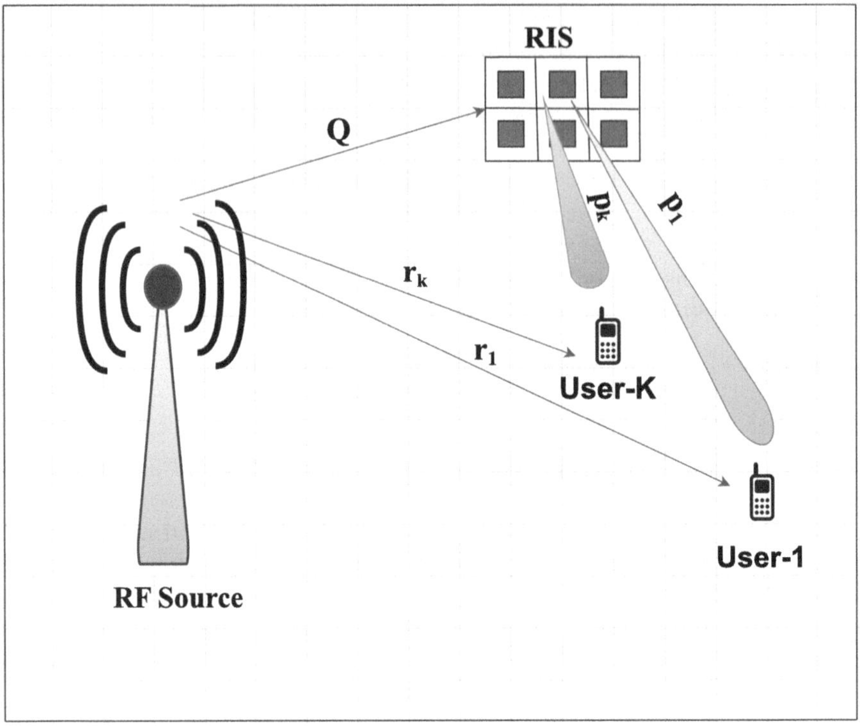

Fig. 1. Illustration of RIS enabled MISO wireless communication.

4 Cascaded CE for MISO Wireless Communications Enabled by RIS

Consider that all users transmit orthogonal pilot signals to the BS through RIS. These pilot signals are transmitted over T time slots. The signal received at the BS in the tth time slot ($t = 1, 2, ..., T$) for the kth user, with the direct channel known and its effect removed, is expressed as

$$\mathbf{y_{k,t}} = \mathbf{Q}diag(\psi_\mathbf{t})\mathbf{p_k}x_{k,t} + \mathbf{n_{k,t}} \tag{3}$$

$$\mathbf{y_{k,t}} = \mathbf{Q}diag(\mathbf{p_k})\psi_\mathbf{t}x_{k,t} + \mathbf{n_{k,t}} \tag{4}$$

where $\psi_\mathbf{t}$ and $x_{k,t}$ are specific instances of the previously defined ψ and x_k for k^{th} user in t^{th} time slot. $\mathbf{n_{k,t}}$ is the $L \times 1$ zero mean Gaussian noise vector with unit variance. Equation 4 can be rewritten as

$$\mathbf{y_{k,t}} = \mathbf{A_k}\psi_\mathbf{t}x_{k,t} + \mathbf{n_{k,t}} \tag{5}$$

Here, $\mathbf{y_{k,t}}$ is a $L \times 1$ received signal vector for k^{th} user in t^{th} time slot. The received signal matrix for k^{th} user after T time slots is given by

$$\mathbf{Y_k} = \mathbf{A_k}\mathbf{\Psi} + \mathbf{N_k} \tag{6}$$

where $\mathbf{\Psi} = [\psi_1, \psi_2, ..., \psi_T]$, $\mathbf{N_k} = [\mathbf{n_{k,1}}, \mathbf{n_{k,2}}, ..., \mathbf{n_{k,T}}]$ and $\mathbf{Y_k} = [\mathbf{y_{k,1}}, \mathbf{y_{k,2}}, ..., \mathbf{y_{k,T}}]$ is a $L \times T$ matrix associated with all the channels for k^{th} user in all time slots T. The cascaded channel $\mathbf{A_k}$ can be represented as

$$\mathbf{A_k} = \mathbf{D_L}\tilde{\mathbf{A_k}}\mathbf{D_N'} \tag{7}$$

where, $\tilde{\mathbf{A_k}}$ is a $L \times N$ angular cascaded channel matrix, $\mathbf{D_L}$ is $L \times L$ matrix and denotes the dictionary unitary matrix at BS, and $\mathbf{D_N}$ is $N \times N$ matrix and denotes the dictionary unitary matrix at RIS. By combining Eq. 6 and Eq. 7, we have

$$\mathbf{Y_k} = \mathbf{D_L}\tilde{\mathbf{A_k}}\mathbf{D_N'}\mathbf{\Psi} + \mathbf{N_k} \tag{8}$$

The cascaded channel can be estimated for each individual user k using the CS CE scheme based on in Eq. 8. However, achieving satisfactory CE efficiency requires a very high pilot overhead.

4.1 TP-OMP Method for Cascaded CE

The angular cascaded channel $\tilde{\mathbf{A_k}}$ can be represented by using Saleh-Valenzuela channel model as

$$\tilde{\mathbf{A_k}} = \sqrt{\frac{LN}{P_Q P_k}} \sum_{p_1=1}^{P_Q} \sum_{p_2=1}^{P_k} \alpha_{p_1}^Q \alpha_{p_2}^k \tilde{\mathbf{b}}(\mu_{p_1}^{Q_r}, \nu_{P_1}^{Q_r}).$$

$$\tilde{\mathbf{a}}(\mu_{p_1}^{Q_t} + \mu_{p_2}^k, \nu_{P_1}^{Q_t} + \nu_{P_2}^k) \tag{9}$$

where $\tilde{\mathbf{b}}(\mu, \nu) = \mathbf{D_L^H}\mathbf{b}(\mu, \nu)$ and $\tilde{\mathbf{a}}(\mu, \nu) = \mathbf{D_N^H}\mathbf{a}(\mu, \nu)$. P_Q, P_k denote the number of paths between BS-RIS and between RIS-k^{th} user respectively. Vectors $\mathbf{a}(\mu, \nu)$, $\mathbf{b}(\mu, \nu)$ represents the steering vectors at RIS and BS respectively. The twofold-pattern sparsity between cascaded channels $\tilde{\mathbf{A_k}}, k = 1, 2, .., K$ for different user can be understood as follows: Each channel $\tilde{\mathbf{A_k}}, k = 1, 2, ..., K$ contains P_Q non-zero rows, with each non-zero row having P_k non-zero columns. Consequently, the total number of non-zero elements in each $\tilde{\mathbf{A_k}}$ is $P_Q P_k < LN$. Since channel $\mathbf{Q}$ is independent of user, the non-zero elements of $\tilde{\mathbf{A_k}}, k = 1, 2, ..., K$ lie on the exactly same rows P_Q, i.e., $U_R^1 = U_R^2 = ... = U_R^K = U_R$, where U_R^k represents the non-zero elements row set of $\tilde{\mathbf{A_k}}$. The matrix $\tilde{\mathbf{A_k}}$ also shows the sparsity in columns pattern, i.e., $U_C^{l,1} \cap U_C^{l,2} \cap ....U_C^{l,K} = U_C^{l,c}, l = 1, 2, 3, ...P_Q$, where, $U_C^{l,k}$ is the non-zero element column set corresponding to the lth non-zero row of $\tilde{\mathbf{A_k}}$.

Further, all cascaded channel $\tilde{\mathbf{A_k}}$, k=1, 2, ..., K show the twofold-pattern sparsity. The channel $\mathbf{Q}$ is independent of user because all users use the common RIS for signal transmission. However, many user share the common channel let $P_c < P_k$ with same angle at the RIS between user and RIS. Thus based on twofold-pattern sparsity, $\mathbf{A_k}$ for many users can be estimated jointly and hence this improve accuracy of CE as well as reduce the pilot overhead.

Furthermore, we integrate the twofold-pattern sparsity with the orthogonal matching pursuit (OMP) algorithm and propose a twofold-pattern sparsity OMP

(TP-OPM) algorithm for cascaded CE. The proposed algorithm is discussed as follows:

Algorithm focuses on estimating channel parameters in RIS-assisted communication systems through a three-stage process that leverages sparsity patterns. In the first stage, the algorithm jointly estimates the totally common row support, U_R, utilizing the row-pattern sparsity. This is achieved by identifying the P_Q row indices corresponding to the non-zero rows. In the second stage, the algorithm targets the imperfect common column support, $U_C^{l,c}$, for each of these non-zero rows. This approach involves estimating the column support for each user and then identifying the indices that appear most frequently across all users. In the third stage, the algorithm improves these estimates by identifying the column supports specific to each user for every non-zero row.Once all sparse matrix supports are detected, the LS method is used to estimate the corresponding matrices. Finally, the angular channel estimates are transformed into spatial channels. This approach reduces the pilot overhead necessary for accurate CE by utilizing the inherent sparsity in the channels. This method adapts dynamically to different scattering environments, making it versatile for various scenarios in RIS-assisted systems.

5 Results and Discussion

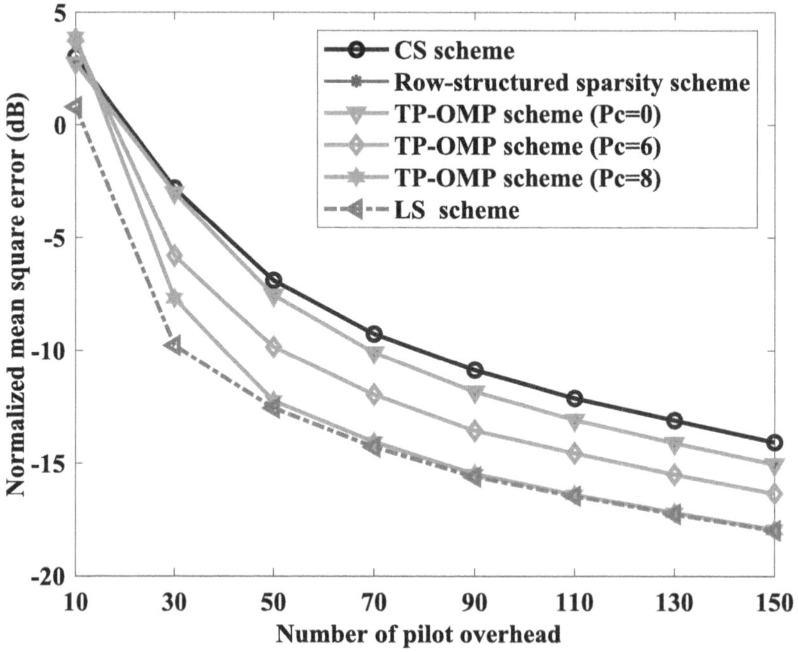

Fig. 2. Normalized MSE performance as a function of number of pilot overhead.

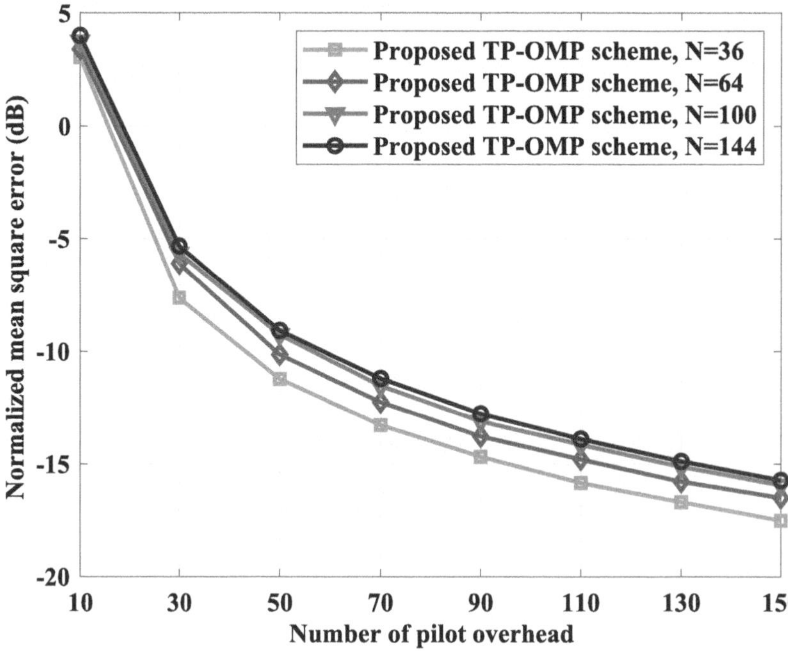

Fig. 3. Normalized MSE with number of pilot overhead with varying RIS elements and $P_c = 6$.

A Monte Carlo (MC) simulation is used in this section to assess the normalized mean square error (NMSE) performance of the proposed RIS-assisted MISO system with varying system parameters and compared the proposed overhead reduction method TP-OMP with the CS method [8] and row structured sparsity based scheme [16]. Unless state otherwise, for simulation, the system parameters are considered as follows: number of antennas at BS, $L = 16$, the number of user, $K = 10$, number of RIS elements $N = 100$, $P_Q = 5$ and $P_k = 8$. The distance between BS-RIS is set as 10 m and that of between RIS-user is set as 100 m .As a point of comparison, we employ the LS scheme as reference, with the assumption that the support for all sparse channels is completely identified.

Figure 2 depicts the NMSE performance of proposed system with varying number of pilot overhead required for cascaded CE and shows the comparison of TP-OMP algorithm with the CS algorithm and row structured sparsity method. As anticipated, increasing the number of pilot signals enhances CE accuracy and reduces NMSE. Figure 2 illustrates how changes in the number of shared paths between the RIS and users affect the NMSE performance of the proposed TP-OMP method. It can be observed from the Fig. 2 that TP-OMP algorithm always consistently outperforms the traditional CS-based approach and the scheme utilizing sparsity along the rows, irrespective of number of shared paths P_c between users and RIS. Compared to CS and row- structured sparsity schemes, the proposed algorithm achieves the same CE accuracy with fewer pilot signals. The

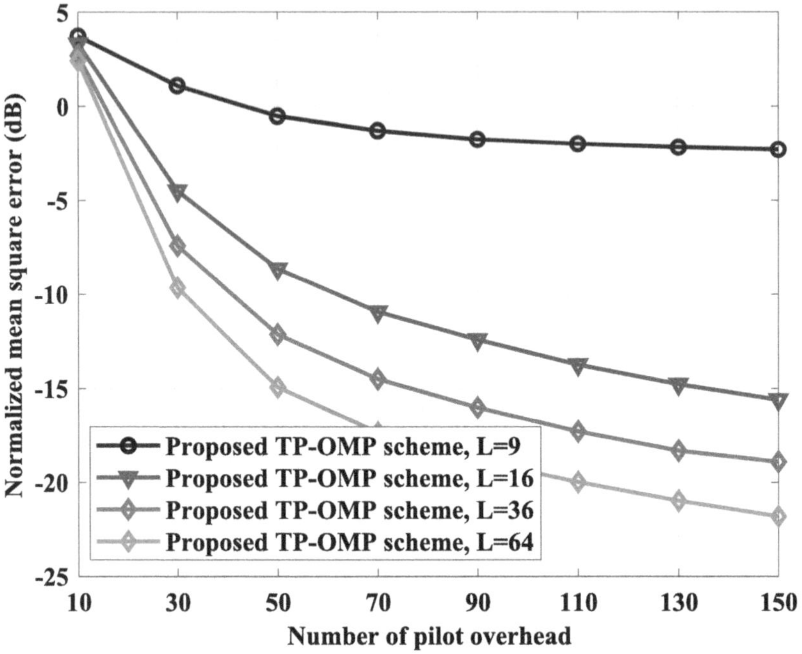

Fig. 4. Normalized MSE with number of pilot overhead with varying BS antennas and $P_c = 4$.

scheme based on row-wise sparsity is a specific instance of the proposed TP-OMP algorithm with $P_c = 0$ as illustrated in Fig. 2. Furthermore, the NMSE performance improves with an increasing number of common paths P_c and approaches the performance of the reference LS scheme.

Figure 3 reveals that increasing the number of pilot signals leads to a decrease in NMSE for all values of N. This indicates an improvement in CE accuracy with more pilot signals.As expected, the proposed TP-OMP scheme exhibits superior performance with lower N values, as evidenced by lower NMSE curves for lower values of N for fix number of transmitted pilot signals. As the number of RIS elements N increases, the number of channels between the BS and RIS, as well as between RIS and users, also increases. Consequently, the size of the cascaded channel matrix expands, necessitating a greater number of pilot signals to achieve the same CE accuracy. This increase in pilot signal requirement is essential to accurately estimate the larger and more complex channel matrix introduced by the additional RIS elements. However, the rate of pilot overhead requirement diminishes with the addition of more RIS elements, making the proposed scheme more suitable for systems with a larger number of RIS elements. This reduction in the rate of increase for pilot signals helps in managing the overall system efficiency, making the proposed TP-OMP scheme advantageous for scenarios involving extensive RIS deployment. Figure 4 illustrates the performance of NMSE as a function of number pilot overhead required for CE with varying number of antennas at BS. The results show a clear advantage of having a larger number of antennas at the BS. The proposed TP-OMP scheme exhibits superior performance with higher L values. This trend underscores the enhanced CE capability provided by a greater number of antennas, which offers improved spatial resolution and better signal representation.

6 Conclusion

In this article, we proposed a TP-OMP algorithm for CE in RIS-enabled MISO wireless communication systems. Our method leverages the sparsity in the rows and columns of the cascaded channels for different users, allowing for a joint estimation of these channels. This approach significantly reduces the pilot overhead required for accurate CE. Further, MC simulation is performed to evaluate the NMSE performance of proposed scheme and compared it with two conventional CSI estimation methods. The simulation results clearly demonstrate the superiority of the proposed method, showing significant improvements in estimation accuracy over the conventional methods. Additionally, the effects of different system parameters on the efficiency of CE were analyzed.

References

1. Basar, E., Di Renzo, M., De Rosny, J., Debbah, M., Alouini, M.-S., Zhang, R.: Wireless communications through reconfigurable intelligent surfaces. IEEE Access **7**, 116753–116773 (2019)
2. Gong, S., Lu, X., Hoang, D.T., Niyato, D., Shu, L., Kim, D.I., Liang, Y.-C.: Toward smart wireless communications via intelligent reflecting surfaces: a contemporary survey. IEEE Commun. Surv. Tutorials **22**(4), 2283–2314 (2020)
3. Liaskos, C., Nie, S., Tsioliaridou, A., Pitsillides, A., Ioannidis, S., Akyildiz, I.: A new wireless communication paradigm through software-controlled metasurfaces. IEEE Commun. Mag. **56**(9), 162–169 (2018)
4. Huang, C., Zappone, A., Alexandropoulos, G.C., Debbah, M., Yuen, C.: Reconfigurable intelligent surfaces for energy efficiency in wireless communication. IEEE Trans. Wireless Commun. **18**(8), 4157–4170 (2019)
5. Gupta, K.A., Bhavsar, P.: Ris assisted ambc communication over spatially correlated channels. IEEE J. Radio Freq. Identif. **8**, 262–269 (2024)
6. Mishra, D., Johansson, H.: Channel estimation and low-complexity beamforming design for passive intelligent surface assisted miso wireless energy transfer. In: ICASSP 2019 - 2019 IEEE International Conference on Acoustics, Speech and Signal Processing (ICASSP), pp. 4659–4663 (2019)
7. He, Z.-Q., Yuan, X.: Cascaded channel estimation for large intelligent metasurface assisted massive mimo. IEEE Wireless Commun. Lett. **9**(2), 210–214 (2020)
8. Wang, P., Fang, J., Duan, H., Li, H.: Compressed channel estimation for intelligent reflecting surface-assisted millimeter wave systems. IEEE Signal Process. Lett. **27**, 905–909 (2020)
9. Yang, Y., Zheng, B., Zhang, S., Zhang, R.: Intelligent reflecting surface meets ofdm: protocol design and rate maximization. IEEE Trans. Commun. **68**(7), 4522–4535 (2020)

10. Zheng, B., Zhang, R.: Intelligent reflecting surface-enhanced ofdm: channel estimation and reflection optimization. IEEE Wireless Commun. Lett. **9**(4), 518–522 (2020)
11. Oyerinde, O.O., Mneney, S.H.: Review of channel estimation for wireless communication systems. IETE Tech. Rev. **29**(4), 282–298 (2012)
12. Yin, X., Cheng, X.: Propagation channel characterization, parameter estimation, and modeling for wireless communications. John Wiley & Sons (2016)
13. Hu, Q., Gao, F., Zhang, H., Jin, S., Li, G.Y.: Deep learning for channel estimation: interpretation, performance, and comparison. IEEE Trans. Wireless Commun. **20**(4), 2398–2412 (2021)
14. Liu, C., Liu, X., Ng, D., Yuan, J.: Deep residual learning for channel estimation in intelligent reflecting surface-assisted multi-user communications. IEEE Trans. Wireless Commun. **21**(2), 898–912 (2022)
15. Huang, H., Guo, S., Gui, G., Yang, Z., Zhang, J., Sari, H., Adachi, F.: Deep learning for physical-layer 5g wireless techniques: opportunities, challenges and solutions. IEEE Wirel. Commun. **27**(1), 214–222 (2020)
16. Chen, J., Liang, Y.-C., Cheng, H.V., Yu, W.: Channel estimation for reconfigurable intelligent surface aided multi-user mmwave mimo systems. IEEE Trans. Wireless Commun. **22**(10), 6853–6869 (2023)

Analyzing Centrality Measures in Network Graphs: Algorithms, Applications, and Heuristics

Pranav Sharma(✉), Prathit Aswar, Suhrid Mulay, Nitin Wartkar, Kaustubh Wankhede, and Ashish Tiwari

Visvesvaraya National Institute of Technology, Nagpur, India
{pranavsharma,paswar,suhridmulay,nwartkar, khwankhede}@students.vnit.ac.in, at@cse.vnit.ac.in

Abstract. Network graph data has multitude of usages. It serves the purpose of identifying and conceptualizing interrelations and dependencies between entities from a large variety of domains including Telecom networks, social network data, GIS, medical data, e-commerce data, collaboration networks, web-links and more. This graph data is later consumed by a variety of end applications for equally varied purposes.

In this paper we present an analysis of various methods of computing values of various kinds of centralities such as network simulation, fluid flow simulation, randomized algorithm, and machine learning-based predictions across a sample of graphs varying in sizes from 34 entities to 40,164 entities, with further emphasis on the application of computed centrality information for the determination of deep communities on the graphs under consideration. We also use a method for computing betweenness centrality for graphs based on a randomized algorithm, and analyze it empirically.

Keywords: Social Networks · Monte Carlo · Graphs · Graph Networks · Randomized Algorithms

1 Introduction

Graph data is consumed by various end applications for equally varied purposes. Analysis of graphs presents an important task for many subdomains across sciences, engineering, and sociology. Network graph data serves the purpose of identifying and conceptualizing interrelations and dependencies between entities. In this paper, we present an analysis and implementation of various methods of computing values of various kinds of centralities across a sample of graphs.

Community detection in a graph forms useful output to reasonably compute groups of related nodes and edges and identify prominent and important ones.

Centrality computation is of immense value in the analysis of networks. Organizations like the FBI regularly use centrality metrics to analyze crime and terrorist networks to direct counter-terrorist activities efficiently, network engineers

D. Gupta et al. (Eds.): PCEMS 2024, CCIS 2491, pp. 191–205, 2025.
https://doi.org/10.1007/978-3-031-90577-3_18

use centrality analysis to determine choke points on a network and reinforce or change topology accordingly. Neurologists use centrality measures to understand brain functions better.[1]

Everett *et al.* [2] look at the betweenness centrality in an ego network, discuss the issue of normalization, and propose an algorithm for calculating the ego centrality. Cohen *et al.* [3] discuss techniques to compute closeness centrality and for directed graphs, and propose efficient algorithms to approximate generalizations of classic closeness centrality to outbound and inbound centralities. The algorithms are designed to perform well on real networks. Z. Sun *et al.* [4] propose a new method of community testing based on the conception of the flow field. The idea is to represent code as a source field and to use flow to form communities. The data goes to connected nodes with the highest field strength. After a few iterations, the nodes with the same data information form a community. U. Brandes [5] propose an algorithm for computing betweenness centrality that provides a significant advantage over existing algorithms. The author proposes a method to speed up computation time by accumulating partial dependencies, bringing the running time from $O(n^3)$ to $O(mn + n^2 \log n)$, where n is the number of vertices in the graph and m is the number of edges in it. We perform our computations on graph data derived by Hamilton *et al.* [6] in their paper on inductive representation learning on large graphs. Khitraya *et al.* [7] propose a game-theory-based approach to assess vertex centrality in directed graphs. Bugedo *et al.*. [8] propose a family of centrality measures based on fine subgraph structures. Wang *et al.*. [9] provides a comprehensive overview of node centrality metrics in biological networks. Ugurlu [10] conducts a comparative analysis of different centrality measures for identifying critical nodes in complex networks.

Computation of centralities for large graph networks is a slow and tedious process[3, 5, 11], it involves a solution of all pair shortest path problems and computation of Eigenvectors of the adjacency matrix of the graph. These problems involve lots of computing power for sufficiently dense or large graphs. Similarly, computation of Shannon's entropy requires calculation of graph centrality for each node of the graph, while this is doable for degree centrality and immediate hopset, it gets computationally expensive for centralities like closeness, betweenness, and LFVC and Eigenvector Centrality.[3, 5, 11]

Our work in this paper is mainly on modeling of centrality scores using burst fluid simulation in graphs, analysis of the algorithm, and its results. We also use random forests of decision trees to predict centralities based on simpler centrality scores and present an analysis of the same for graphs with latent communities. We also use a randomized algorithm for the computation of Betweenness Centrality and empirically analyze it. We also propose mechanisms for reducing the error from the algorithm, along with an analysis of the algorithm for different random samples, and propose an optimal value.

With the use of these heuristic-based techniques we have been able to demonstrate, acceptable results for various applications while expending significantly less computing power than otherwise would be required for a complete process thus facilitating the discovery of neighborhoods and greater structures within information-dense graphs in a timely manner.

2 Graph Centralities

Some of the common centrality scores used in graph networks are mentioned in Table 1, Of these, our work is primarily on heuristics for betweenness centrality.

Table 1. Metrics for various graph data used for analysis

No.	Centralities	Computational complexity
1	Degree Centrality	worst case time: $(O(E))$
2	Betweenness Centrality	worst case time: $(O((V.E) + V^2))$
3	Closeness Centrality	worst case time: $(O((V.E) + V^2))$
4	Ego Centrality	worst case time: $(O(V^3))$
5	Eigenvector Centrality	time: $(O(V^3))$
6	Local Fiedler Vector Centrality	time: $(O(V^3))$
7	Neighborhood Hopset Size Centrality	worst case time: $(O(V^t))$
8	Clustering Coefficient Centrality	worst case time: $(O(V^3))$

1. Degree Centrality: This is the degree of the vertex, that is the number of edges the vertex has.
2. Betweenness Centrality: Betweenness centrality is the ratio of the shortest paths in the graph that pass through a given node to the total number of shortest paths in the graph. The computation for this involves solving all pair shortest path problems.
3. Closeness Centrality: This is the reciprocal of the sum of the length of all shortest paths from a given node. The more central a node is, the lower the sum of lengths of all shortest paths would be. The reciprocal ensures that this value is maximized. Computation for this also involves solving the all-pair shortest path problem.
4. Ego Centrality: Ego centrality is the betweenness centrality for a given node, in its ego network[2, 12].
5. Eigenvector Centrality: Eigenvector centrality is the measure of influence of a vertex in the given graph network, the idea is that a more connected vertex has higher prestige, and thus would have a higher eigenvector centrality.
6. Local Fiedler Vector Centrality: LFVC is a centrality score that uses the Fiedler vector (second smallest eigenvector) of the graph. It measures the sensitivity of the graph's edge or vertex to removal. It also has applications in community detection.[13, 14]

7. Neighborhood Hopset Size: Neighbourhood hopsets are defined as a set of all vertices that can be reached from the source vertex in a set number of hops. A vertex that is closely connected to the graph would have a higher hopset size.
8. Clustering Coefficient: This is the ratio of the number of links present in the neighborhood of the given node to the total number of links that are theoretically possible.

3 Proposed Method of Computing Betweenness Centrality

3.1 Using Monte Carlo Algorithm for Computation of Betweenness Centrality

We propose a method that approximates betweenness value by sampling some source nodes from the graph and computing single source shortest paths from those nodes. This significantly speeds up the computation and gives results near the actual betweenness value. By taking the average of multiple samples, we can minimize the error and achieve a better bound of algorithm failure.

A simple implementation of the algorithm can be described as follows. It relies on a Single Source Shortest Paths function which computes the shortest paths from a given node to all other nodes in the graph (by means of simple searching).

Algorithm 1. Monte Carlo Algorithm for Efficient Computation of Betweenness

Require: $G(V, E)[Graph], k[ControlParameter]$
  $NodeBetweenness \leftarrow \{\}$
  $nodeSet \leftarrow \{\}$
  for $1...k$ **do**
    $rnode \xleftarrow{\$} node \in V$
    $nodeSet.add(rnode)$
  end for
  $shortestPaths \leftarrow nodeSet.map(node \implies singleSourceShortestPath(node))$
  for $path \in shortestPaths$ **do**
    for $node \in path$ **do**

      if $node \notin NodeBetweenness$ **then**
        $NodeBetweenness[node] = 0$
      else
        $NodeBetweenness[node] + = \frac{1}{k}$
      end if
    end for
  end for

3.2 Using Simulations for Scoring Nodes in Graph

Graphs are mathematical abstractions of real-world structures like we encounter every day, from ant colonies to human populations, these real-world networks are often susceptible to outside and inside effects and in many ways act like a living entity themselves (a phenomenon also known as emergence). Like all living entities, these networks too are sensitive to their surroundings and react to changes in the environment. The effect of such modifications on networks is deeply dependent on the structure of the network. A concept captured mathematically by values like centrality and structures like communities. Thus it can be reasoned that a simulation of natural processes on said networks can assist in the process of detecting communities.

In our approach we attempted to model fluid flow through a graph and use it as a preliminary to model multiple phenomena through the network. The simulation is designed to follow some basic axioms:

1. The graph is modeled by a system of tanks and pipes. The nodes/entities serve as tanks and edges serve as pipes.
2. Every node, along with other characteristics, also has a pressure value which represents the pressure exerted by an imaginary fluid present in the node.
3. Every edge of the graph attempts to minimize the pressure differential across it.
4. The flow process through an edge does not introduce nor destroy any amount of the fluid present in the network.

Using these axioms we create a very simplified model of flow through a graph, and we perform very basic simulations using these axioms. The processes simulated were of three kinds:

1. **Basic Source Simulation**: Nodes were selected to be sources that would inject fluid at constant pressure into the network. This approach aids us in understanding how leaders influence and generate communities around them. As illustrated in Fig. 5.
2. **Burst Clear Simulation**: Every iteration node can burst with a certain predetermined probability, or with another predetermined probability be drained of all fluid, or alternatively stay static and be a part of the flow through the network. This models the diffusion of talk through a group, where sources broadcast it to everyone but only sinks who concentrate are in a better position to listen and take in information.
3. **Network Simulation**: Every iteration, with a certain probability, a source-sink pair is generated in the network, the pressure gradient thus generated creates a flow through the graph that closely models the ideal flow of information through a network. Using this process we can model how crucial nodes can be to the flow of information in the network. Thus in essence finding centrality information without computationally expensive exact measures.

The simulation was also created with an extensive visualization system from scratch so that the flow of the fluid could be visualized and metrics noted.

Of particular importance to the system are the quantities of pressure and net flow, pressure giving the immediate influence a node has right now and net flow the long-term accumulated influence. Coloring nodes by pressure also highlights how community formation takes place as a natural part of the simulation in how fluids diffuse through the medium. Some screenshots of the program executing on multiple graphs are presented below.

Note how communities can be segregated on the basis of color, An indicator of the influence leader nodes (dark blue in this case) have on the network. Nodes are colored by their pressure at the current iteration of the simulation. Blue signifies higher and red has lower values.

Figure 6a and 6b correspond to earlier and latter stages of the simulation respectively.

Figure 7a is a network simulation close to the initial state, note source and sinks modeling information flow and Fig. 7b highlights nodes with high net flow values and corresponding high influence on the network The fluid simulations for 75 nodes converge in under a minute. For 100 nodes, the simulations take 1 min to converge, and the simulations don't converge to graphs with 500 nodes even in 10 min. The easy visualization but the high computation time required make fluid simulations a slow but pellucid measure of centrality.

3.3 Predicting Trends of Closeness Centrality

We use random forests[15] to predict trends in closeness centrality for various nodes. The features given to the random forest-based machine learning model are the simpler centralities such as degree centrality, hopset size, and ego network size. We train the model with an 85%-15% train-test split. Closeness is defined in the context of the inclination of a node to belong to the graph and is heavily affected computationally by other nodes surrounding a node. Such a scoring factor amounts to a very sensitive and very high computational score. Due to this sensitivity, a simple predictive model is not viable, as even the smallest form of error can affect the entire scoring.

However, the actual accurate values of such a centrality score are immaterial to rankings so long as order is preserved. The relative trends of the scores are used to define which node is considered more *close*. While the score as an isolated metric provides no greater information it is best viewed in relation to the same score computed from its peers which allows us to determine a ranking that for the most part is accurate and can be improved by the addition of more graph features in the inputs provided to the existing model.

Approach: The idea is to create a model that can replicate the trends of closeness scores so that we can recreate a ranking that is similar to the original. We can use relatively computationally cheaper centrality scores as features for the random forest prediction model. So we created a correlation matrix of centrality scores to find centrality scores that show a correlation between each other. In the above Fig. 3, we can see the correlation of Closeness with Ego and Hopset centrality scores. We use this as a confirmation of our theory that we can create a prediction model. A random forest model was implemented with this centrality set input.

Analysis: For a test train split of 15–85, we were able to retain 85% accuracy in the predictions for closeness centrality, on the medium graph dataset as shown in Fig. 4. The trend is largely carried for centrality scores, with more prominent nodes ranking higher than less prominent nodes. We can use this as confirmation that the random forest can rank the different nodes but is not good at computing actual centrality scores.

Due to a reliance on existing graph-based statistics derived from other measures, our machine-learning model may exhibit sensitivity to incomplete or anomalous data. This can manifest as outliers that are subsequently identified as statistical anomalies. The random forest algorithm can be fine-tuned through the implementation of better initialization constraints for decision tree construction. However, these modifications may lead to increased computational complexity and resource demands. Consequently, a careful balance must be struck between the desired accuracy of the model's output and the computational efficiency of the algorithm

4 Dataset

For testing purposes, graph samples of various sizes have been used to check the functioning of implemented routines and their analysis. Metrics of sample graphs are shown in Table 2 for reference.

5 Benchmark

The computational requirements for various centralities for dense large graphs are shown in Fig. 2. Degree centrality scales slowly, as the in-degrees and out-degrees of all vertices are always known. Neighbourhood hopset scales slowly with the number of nodes, this is because computing hopset is an expected constant time operation. Eigenvector centrality uses a numerical method to compute the eigenvector. Betweenness centrality requires computing solution to the all-pair shortest path problem and hence scales very quickly with the number of vertices in the graph.

Fig. 1b and 1a contain the time required for computation of Degree, Betweenness, and Closeness centrality through classical methods as well as computation

Table 2. Metrics for various graph data used for analysis

Sr no.	Figure	Name of Graph	Nodes	Edges	Connectivity
1	8a	soc-karate.mtx	34	78	True
2	8b	web-edu.mtx	3031	6474	True
3	8c	tech-internet-as.mtx	40164	85123	True
4	8d	grid2_dual.mtx	3136	6112	True
5	8e	mk10-b2.mtx	3150	9446	True

of betweenness centrality using a randomized algorithm. It is clear that the randomized algorithm for betweenness centrality scales significantly better compared to the classical algorithm for betweenness.

6 Analysis of Empirical Results

6.1 Random Sample Method for Computing Betweenness

By using a random sample-based method for betweenness, we observed a significant speedup in execution time. Instead of computing shortest paths for all source nodes, we select a random sample of source nodes and compute the shortest paths from these sources. This speeds up the execution of the algorithm. The values obtained from the random sample betweenness are close to the actual betweenness value, and the accuracy gets better as the number of nodes to be sampled is increased.

This result has been verified for a large number of small and large graphs and holds consistently. This result hints that random sampling is a great estimator of betweenness centrality, but this result needs to be proven mathematically.

7 Conclusion

The process of computing various measures of the influence of individual entities on the structure of graphs, also known as various measures of *centralities* proves extremely useful in the analysis of graphs. Using centralities, the latent structure of a graph can be analyzed and centralities assist in isolating communities in the graph. Computing centralities is an expensive affair and instead of computing accurate centralities, it is beneficial to approximate centralities using faster algorithms and methods. In particular, using a randomized algorithm to compute betweenness centrality yields good results with faster computation time. For random forest-based method, we can observe that a few node clusters show a prediction trend inconsistent with the rest of the graph. Also, it can be observed that the model is not capable of predicting such extreme behavior for anomaly nodes. This inconsistency makes the model prone to anomalies, rendering it incapable of performing predictions of closeness centrality successfully. This is a result of the inherent graph structure in graphs with latent communities. Possible improvements to this approach include adding more features related to the graph and the addition of the latent structure of the graph in the form of a graph vector as input to the random forest model. Another approach that can be explored is the addition of neighborhoods of the nodes as input to the random forest model to better capture graph structure. Fluid and network simulation-based algorithms are beneficial for centrality identification, as the liquid flow model proves extremely useful in predicting trends in centrality and for the identification of prominent nodes. We have analyzed the efficacy of various methods for the computation of betweenness centrality through empirical methods. Of the methods explored in this paper, randomized algorithms are the most promising, and mathematical proof of their soundness and veracity is to be established. For this project, we have implemented and tested multiple algorithms and models inspired by existing literature, as mentioned. All the implementations as well as the dataset used for testing in this paper can be found on our GitHub repository. project-megraphs which is located at https://github.com/appmonster007/project-megraphs.

8 Figures

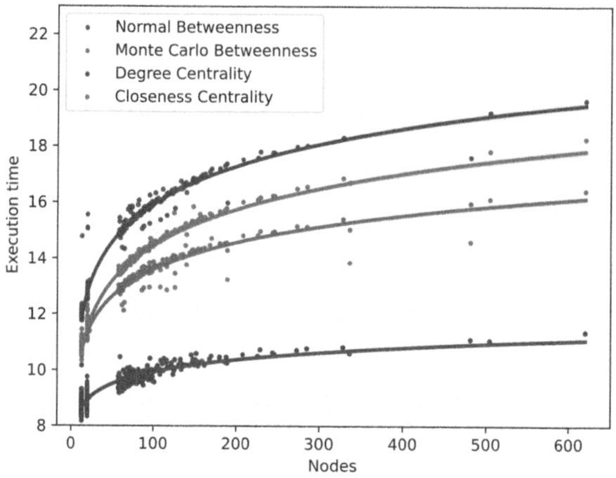

(a) Logarithmic plot of time required for centrality algorithms

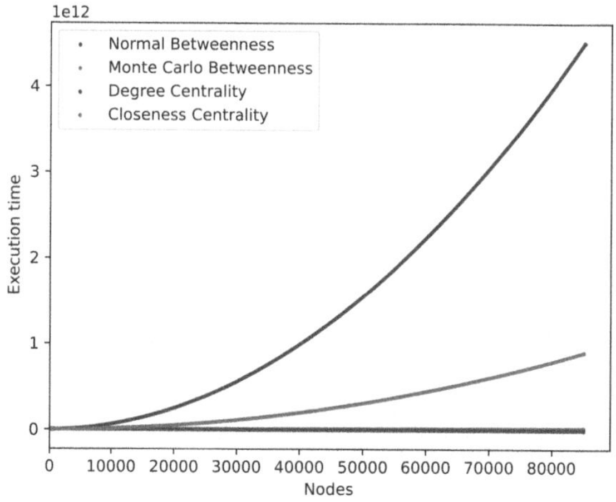

(b) Plot of time required for centrality algorithms

Fig. 1. Comparison between our proposed method and classical algorithms.

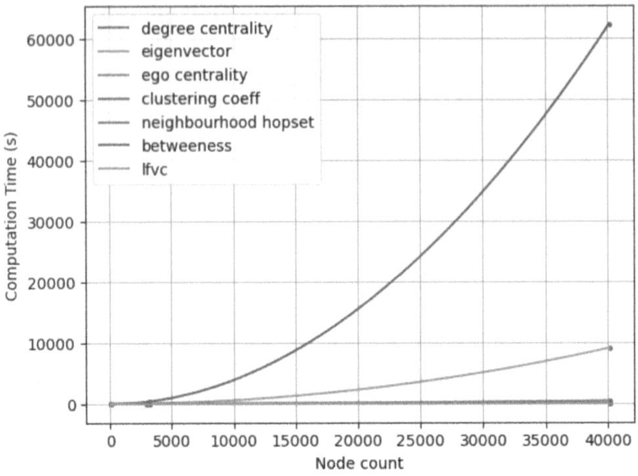

Fig. 2. Computational time complexity of various centrality scores.

	Nodes	Degree	Eigenvector	LFVC	Clustering	Ego	hopset	betweenness	closeness
Nodes	1.000000	0.051127	0.148247	0.012591	0.160107	0.029620	0.038732	0.018631	-0.186180
Degree	0.051127	1.000000	0.371090	0.151924	-0.044454	0.407261	0.518278	0.538369	0.392049
Eigenvector	0.148247	0.371090	1.000000	0.000636	0.104854	-0.028677	0.110031	0.039137	0.084629
LFVC	0.012591	0.151924	0.000636	1.000000	-0.029165	0.068725	0.064899	0.072721	0.061761
Clustering	0.160107	-0.044454	0.104854	-0.029165	1.000000	-0.292621	-0.032594	-0.122388	0.071379
Ego	0.029620	0.407261	-0.028677	0.068725	-0.292621	1.000000	0.306783	0.235596	0.371278
hopset	0.038732	0.518278	0.110031	0.064899	-0.032594	0.306783	1.000000	0.889080	0.478217
betweenness	0.018631	0.538369	0.039137	0.072721	-0.122388	0.235596	0.889080	1.000000	0.282820
closeness	-0.186180	0.392049	0.084629	0.061761	0.071379	0.371278	0.478217	0.282820	1.000000

Fig. 3. A correlation heat matrix for centrality scores of medium-sized graph.

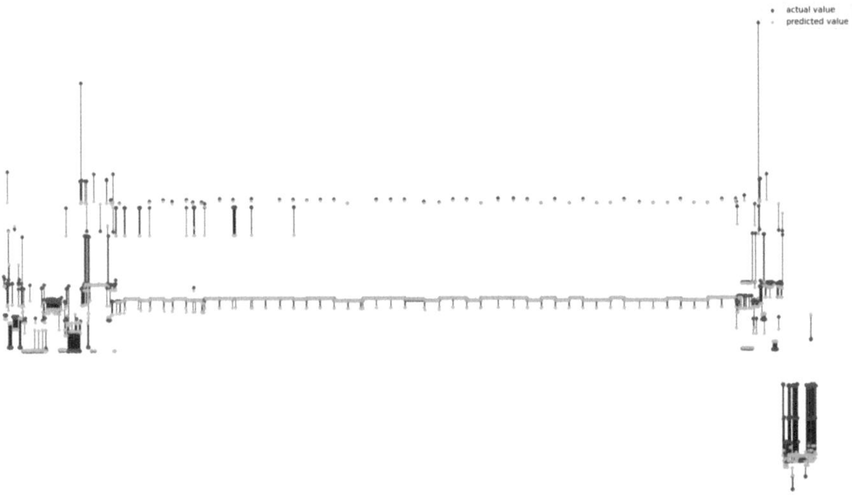

Fig. 4. Comparison of actual and predicted closeness score for a graph.

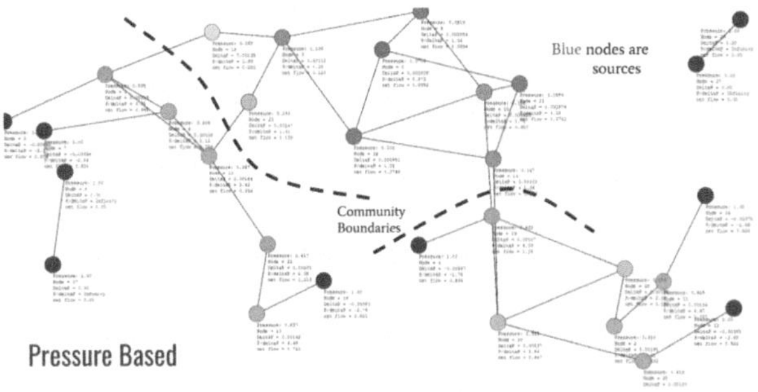

Fig. 5. Pressure-based network simulation.

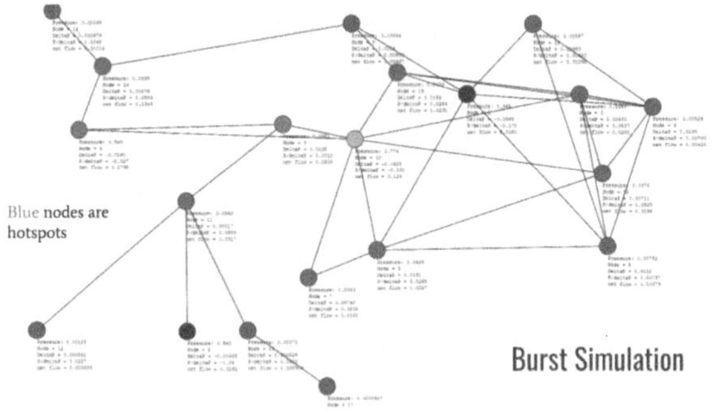

(a)

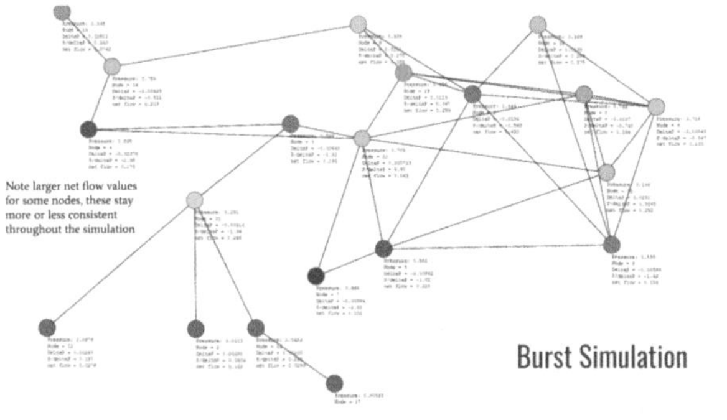

(b)

Fig. 6. Burst Simulation.

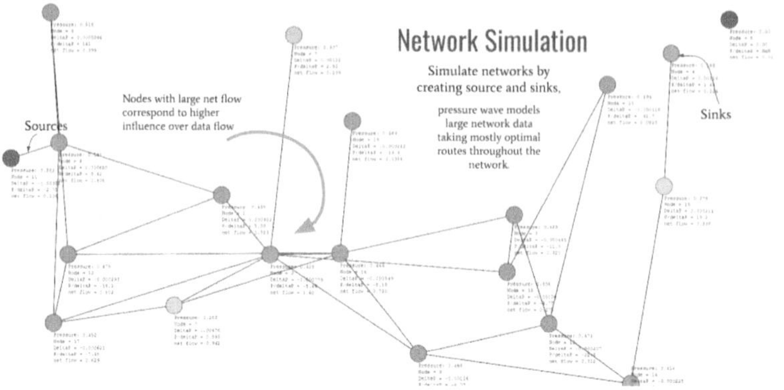

(a)

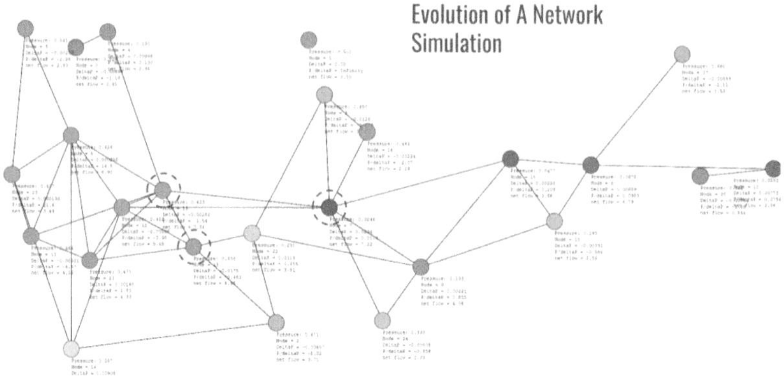

(b)

Fig. 7. Network Simulation.

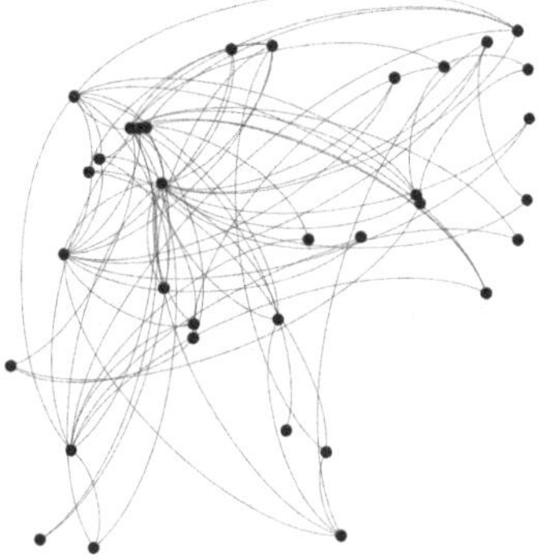

(a) SOC-Karate

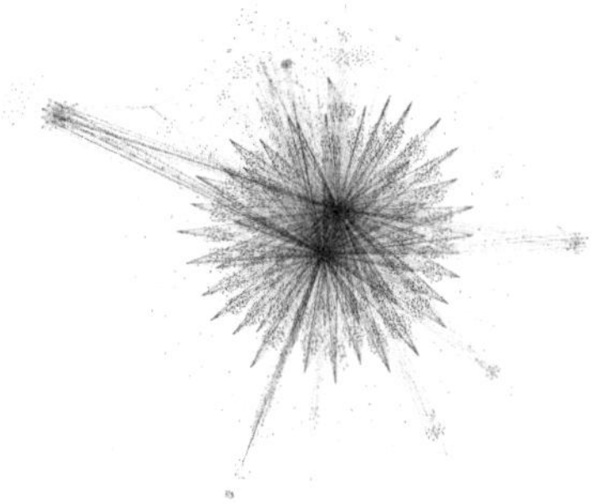

(b) Web-Edu

(c) Tech-Internet-As

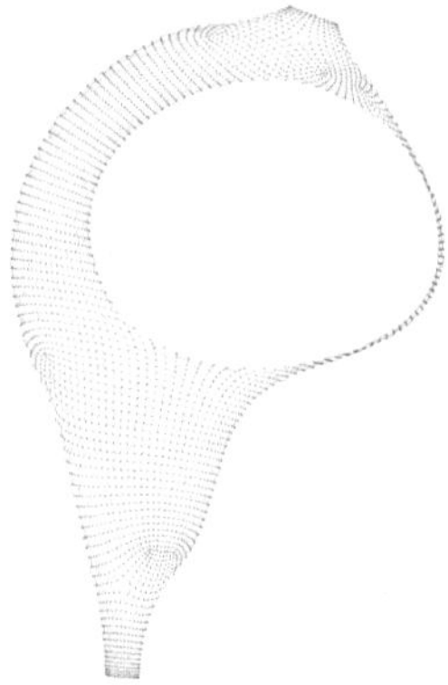

(d) grid2_dual

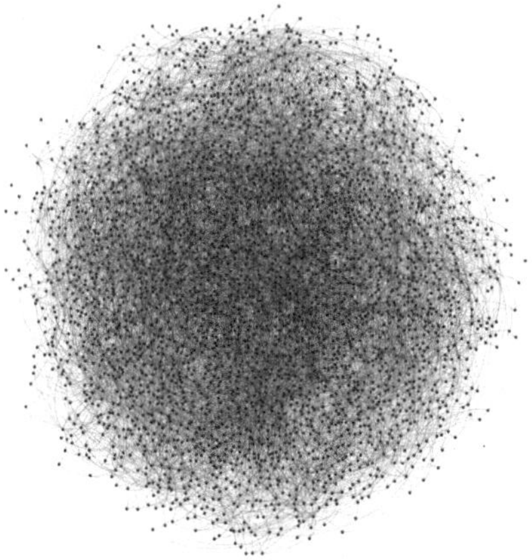

(e) MK10-b2

Fig. 8. Visualisation of various graphs from a dataset.

References

1. Sporns, O. : Graph theory methods: applications in brain networks. Dialogues in Clinical Neuroscience **20**(2), 111–121 (2018). PMID: 30250388. https://doi.org/10.31887/DCNS.2018.20.2/osporns. eprint: https://doi.org/10.31887/DCNS.2018.20.2/osporns. https://doi.org/10.31887/DCNS.2018.20.2/osporns
2. Everett, M., Borgatti, S.P.: Ego network betweenness. Social Networks **27**(1), 31–38 (2005).ISSN: 0378-8733. DOI: https://doi.org/10.1016/j.socnet.2004.11.007. https://www.sciencedirect.com/science/article/pii/S037887330400067X
3. Cohen, E., Delling, D., Pajor, T., Werneck, R.F.: Computing classic closeness centrality, at scale. In: Proceedings of the Second ACM Conference on Online Social Networks, ser. COSN '14, Dublin, pp. 37–50. Association for Computing Machinery (2014). ISBN: 9781450331982. DOIhttps://doi.org/10.1145/2660460.2660465. https://doi.org/10.1145/2660460.2660465
4. Sun, Z., et al.: Community detection in complex networks using flow simulation. In: Proceedings of the 2017 VI International Conference on Network, Communication and Computing, ser. ICNCC 2017, Kunming, pp. 150–155. Association for Computing Machinery (2017). ISBN: 9781450353663. DOI: https://doi.org/10.1145/3171592.3171600. https://doi.org/10.1145/3171592.3171600
5. Brandes, U.: A faster algorithm for betweenness centrality. J. Math. Sociol. **25**(2), 163–177 (2001). https://doi.org/10.1080/0022250X.2001.9990249
6. Hamilton, W.L., Ying, R., Leskovec, J.: Inductive representation learning on large graphs. In: Proceedings of the 31st International Conference on Neural Information Processing Systems, Red Hook, NY, pp. 1025–1035. Curran Associates Inc. (2017) ISBN: 9781510860964
7. Khitraya, V.A., Mazalov, V.V.: Game-theoretic centrality of directed graph vertices. Autom. Remote. Control. **85**(2), 225–237 (2024)
8. Bugedo, S., Riveros, C., Salas, J.: A family of centrality measures for graph data based on subgraphs. ACM Trans. Database Syst. **49**(3), 1–45 (2024)
9. Wang, M., Wang, H., Zheng, H.: A mini review of node centrality metrics in biological networks. Int. J. Netw. Dyn. Intell. **1**(1), 99–110 (2022)
10. Ugurlu, O.: Comparative analysis of centrality measures for identifying critical nodes in complex networks. J. Comput. Sci. **62**, 101738 (2022)
11. Kang, U., Papadimitriou, S., Sun, J., Tong, H.: Centralities in large networks: algorithms and observations. In: Proceedings of the 2011 SIAM International Conference on Data Mining (SDM), Washio, T., Kamath, C., Liu, B., Clifton, C., Liu, H.(eds.) Philadelphia, PA: Society for Industrial and Applied Mathematics (2011). DOI: https://doi.org/10.1137/1.9781611972818. eprint: https://epubs.siam.org/doi/pdf/10.1137/1.9781611972818. https://epubs.siam.org/doi/abs/10.1137/1. 9781611972818
12. Gleich, D.F., Seshadhri, C.: Vertex neighborhoods, low conductance cuts, and good seeds for local community methods. In: Proceedings of the 18th ACM SIGKDD International Conference on Knowledge Discovery and Data Mining, ser. KDD '12, New York, pp. 597–605. Association for Computing Machinery (2012). ISBN: 9781450314626. DOI: https://doi.org/10.1145/2339530.2339628. https://doi.org/10.1145/2339530.2339628
13. Chen, P.-Y., Hero, A.O.: Local fiedler vector centrality for detection of deep and overlapping communities in networks. In: 2014 IEEE International Conference on Acoustics, Speech and Signal Processing (ICASSP), pp. 1120–1124 (2014). DOI: https://doi.org/10.1109/ICASSP.2014.6853771

14. Wang, X., Liu, G., Li, J.: Overlapping community detection based on structural centrality in complex networks. IEEE Access **5**, 25258–25269 (2017). https://doi.org/10.1109/ACCESS.2017.2769484
15. Breiman, L.: Random forests. Mach. Learn. **45**(1), 5–32 (2001). ISSN: 1573-0565. DOI: https://doi.org/10.1023/A:1010933404324. https://doi.org/10.1023/A:1010933404324

Towards Safer Homes: AI-Driven Predictions for Domestic Violence Prevention

Saleha Saudagar(✉), Anuja Jadhav, and Mayura Shelke

Computer Science and Engineering, MIT Art and Design and Technology, University, Pune, India
salehasaudagar@gmail.com

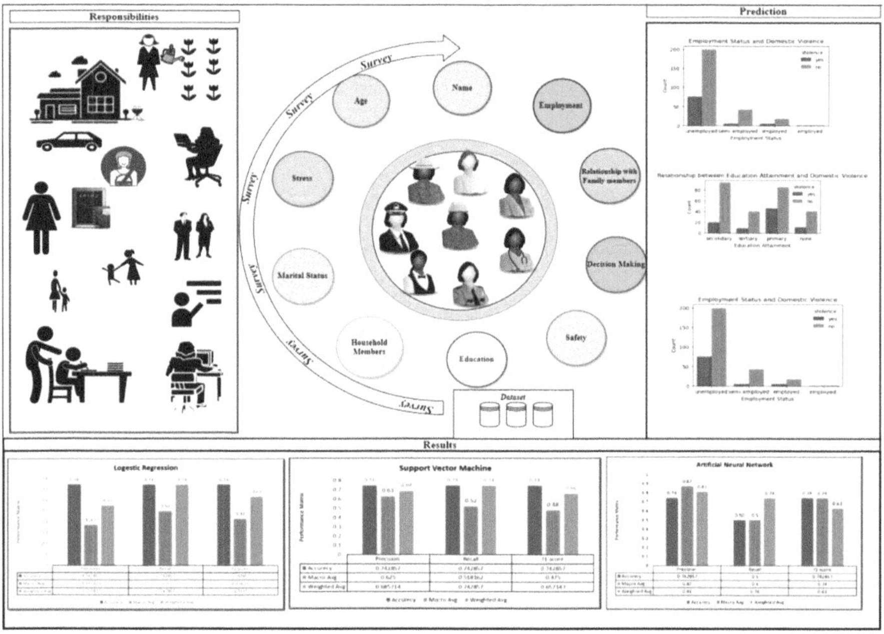

Abstract. Modernization has been the result of the widespread use of digitization in the 21st century, which leads to women equality. Still violence against women remains the major concern, where Artificial Intelligence can help prevent violence against women to some extent, awareness and education will remain the primary solution. It is now mandatory to invest in research and make violence prevention technology-driven and AI-driven. In this paper we are doing surveys of women from different educational and family backgrounds and analyzing the correlation between literacy and domestic violence. The model uses ML and trained here takes the input from the survey and predicts domestic violence against women. We are doing comparative analysis of, The Support Vector Classifier, Logistic regression and Neural Network phenomenon. The Artificial Neural network shows the maximum accuracy of 78%. The violence prediction mode can be helpful for prediction of domestic violence happening against women.

D. Gupta et al. (Eds.): PCEMS 2024, CCIS 2491, pp. 206–220, 2025.
https://doi.org/10.1007/978-3-031-90577-3_19

Keywords: Women safety · Domestic violence · Artificial Intelligence · Artificial Neural Networks · Logistic Regression · Support Vector Machine

1 Introduction

Violence against Women (VAW) is a multifaceted issue which needs to be controlled that demands comprehensive analysis from various perspectives. Studies have utilized web scraping and text mining techniques to collect and analyze VAW-related news, uncovering latent topics and categorizing types of violence through classification algorithms. Additionally, research has delved into the dynamics of VAW in Nigeria, employing multivariate analysis to identify dimensions of physical, social, sexual, and economic violence across different states. Moreover, the importance of education in the fight against domestic and family violence against women has been highlighted, highlighting the importance of issue and pro-grams focused on gender equity and human rights in educational settings [4]. Moreover, the examination of legal instruments in Bangladesh has revealed significant gaps in addressing sexual violence, emphasizing the need for alignment with international standards to effectively prevent violence happening against women in India [5].

By integrating insights from these studies, a more nuanced understanding of VAW can be achieved to inform interventions and policy development. Machine learning techniques have been instrumental in analyzing violence against women across different countries. Studies from Ethiopia [1], Liberia [2], China [3], and India [5] have utilized various machine learning algorithms to predict and understand Significant partner violence (SPV) and domestic violence.

These studies have highlighted the importance of identifying risk factors, predicting vulnerability, and estimating the incidence rates of different forms of violence against women. Algorithms such as CatBoost, Random Forest, LightGBM, and Decision Tree have shown high accuracy rates in predicting IPV and domestic violence, emphasizing the significance of data-driven approaches in developing effective prevention and involvement strategies to fight violence happening against women.

By leveraging machine learning models, researchers can extract valuable insights, identify key features, and enhance policy formulation to address this critical societal issue.

2 Literature Work

In this section analysis of various researches which are focusing on domestic violences, its impact on women's health and relation with literacy are briefly discussed.

A. Violence against women Analysis

The paper analyzes obstetric violence against women, encompassing disrespect, improper body handling, insults, and autonomy neglect during childbirth, aiming to raise awareness and improve public policies in this area [11].

The study analyzed violence against women in Nigeria using multivariate approaches, identifying varying levels of physical, social, sexual, and economic violence across different states [12].

The literacy plays vital role in battling violence happening against women is crucial, focusing on gender equality, human rights, and prevention strategies to transform societal norms and behaviors [13]. The research paper [14] critically analyzes how sexual violence against women is defined and addressed within legal instruments in Bangladesh, highlighting significant gaps compared to international standards.

The study analyzed in [15] violence against women in Pakistani TV drama promos, finding 97% featured violence, mostly in domestic settings, with intimate partners and relatives involved in 40% of cases.

Gender-based violence against women in India is analyzed psychologically and sociologically in the [16], emphasizing the interplay of individual and societal factors to address and prevent such crimes effectively. Psychological violence against women was analyzed in rural and urban areas of Muzaffarabad, showing mild to moderate levels, with a lower likelihood of high severity [17].

B. VAW analysis using Machine Learning

The comparative analysis of violence against women in Islamic and non-Islamic contexts [1], exploring various forms of abuse and seeking effective prevention measures. However, there is a lack of focus on legal frameworks in non-Islamic jurisdictions and limited discussion on intersectionality of violence against women the violence against women defenders in environmental conflicts is analyzed globally [2], revealing recurring themes of displacement, repression, criminalization, violent targeting, and assassinations, especially in industries that extract resources. Here the methods used is Log-linear and binomial regressions for analysis and Data is taken from Environmental Justice Atlas. Author have Identified statistically significant patterns in displacement, repression, criminalization, and targeting. Lack of focus on specific regions or countries and Limited exploration of intersectionality with other social factors.

Domestic violence against women analysis in Semnan using mixed methods, identifying factors like age, education, income, and family dynamics influencing the prevalence of violence. The methods used here are Cross-sectional descriptive quantitative study and Phenomenological qualitative study. Here author lacked in long-term sampling process due to women's resistance and there is an influence of personal experiences on study results [3].

Stakeholder salience analysis reveals that very few stakeholders involved in policy formulation for anti-violence against children and women in Aceh, hindering effective responses to rising violence cases [4]. Here the practice formulation of anti-violence against women and children is conquered by dependent stakeholders.

The author in [5] conducted a critical dissertation analysis of organization like NGO's reports on violence happening against women in Iraq, focusing on linguistic and ideological aspects to reveal societal power dynamics.

Risk factors for domestic violence in India include social norms supportive of violence [6], while women's empowerment and financial independence act as protective factors and facilitators in addressing domestic violence.

Physical violence happening against women by an SO (Significant Other) partner was analyzed using VIVA Survey 2017 data, revealing different profiles of intimate partner violence requiring tailored public policies for prevention and intervention. However, Lack of discussion on intervention strategies and absence of exploration of long-term impact on victims is observed in [7].

A pooled analysis showing a vibrant association between men's deprived mental health and concern of intimate partner violence and known but non-related sexual violence happening against women is indicated [10] (Table 1).

Table 1. Tabular literature study

Sr. No	Reference No	Key Takeaways	Conclusive Remarks
1	[18]	Multimethodological approach with descriptive correlational design. Data collected through online structured questionary validated by experts	Physical violence was most prevalent, followed by psychological and sexual violence against women in Northern Peru. Existing measures showed moderate effectiveness, emphasizing the need for culturally sensitive interventions
2	[19]	Reasonable defense of mistreated women in domestic violence is explored Proposed measures: legal education, psychological intervention, and raising awareness	The paper analyzes women's "violence happening against women" in response to domestic violence, focusing on justifiable defense, prevention measures, legal education, psychological intervention, and raising awareness of domestic violence
3	[20]	22 cases of violence against pregnant women analyzed Most common injuries: abrasions, bruises; locations: head, neck, arms	Features of physical violence happening against pregnant women embrace injuries like scratches and bruises and they are commonly imposed by known men, emphasizing the need for literacy programs and medical involvement
4	[21]	Public service announcements raise awareness and inspire behavioral changes Semiotic analysis. Examination of public service announcements	The paper analyzes the representation of objecting to violence against women through semiotic analysis in public service announcements, emphasizing empowerment, social awareness, and collective responsibility to combat violence

(*continued*)

Table 1. *(continued)*

Sr. No	Reference No	Key Takeaways	Conclusive Remarks
5	[22]	Gender based violence detection using machine Learning algorithms for Evaluation of Decision Trees, Naïve Bayes, SVM, and Neural Networks	Machine learning methodology such as Decision Trees, Ensemble Methods, Naïve Bayes, SVM, and Unsupervised Deep Learning can be utilized to detect patterns of gender violence in digital job offers
6	[23]	YOLOv3 algorithm detects weapons with high accuracy AI predicts crimes to prevent violence against women	Deep learning models like YOLOv3 are utilized for detecting harmful weapons in video recordings to address violence against women, aiding in crime prediction and prevention
7	[24]	ML- based process predicts GBV occurrences for early prevention strategies. Namibia should use ML algorithms to forecast and prevent GBV	ML-based analytical process predicts Gender-Based violence occurrences, aiding in early prevention strategies, with a focus on women and vulnerable groups, as outlined in the paper

3 Methodology

3.1 A Survey Collection

- Survey: Data is collected through surveys from various individuals.
- Survey Analysis: The collected survey data is analyzed to form a raw dataset.

The Survey: The dataset utilized in this study consists of 347 records, obtained through a comprehensive survey conducted using Google Forms. The survey was meticulously designed to gather information on various demographic, socio-economic, and behavioral factors that could potentially influence or indicate domestic violence. The following questions were included in the survey:

- Name
- Age
- Marital Status
- Household Members
- Education
- Employment
- How would you describe the relationships between members of your household?

- Who generally makes decisions about important household matters?
- How safe do you feel in your current living situation?
- How often do you feel stressed in your home environment?
- Are you experiencing Domestic Violence

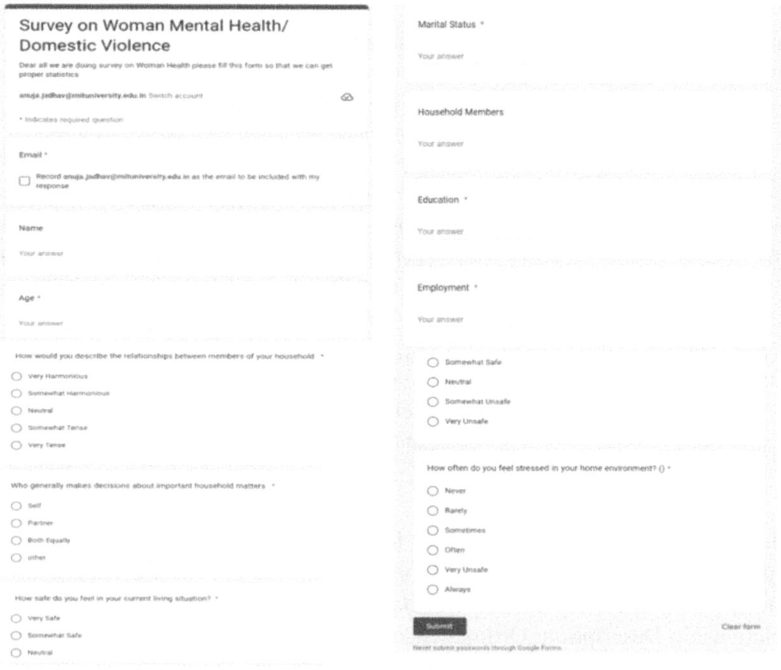

Survey on Woman Mental Health/ Domestic Violence

Dear all we are doing survey on Woman Health please fill this form so that we can get proper statistics

amuja.jadhav@mituniversity.edu.in Switch account

* Indicates required question

Email *

Record amuja.jadhav@mituniversity.edu.in as the email to be included with my response

Name

Your answer

Age *

Your answer

How would you describe the relationships between members of your household *

- Very Harmonious
- Somewhat Harmonious
- Neutral
- Somewhat Tense
- Very Tense

Who generally makes decisions about important household matters *

- Self
- Partner
- Both Equally
- other

How safe do you feel in your current living situation? *

- Very Safe
- Somewhat Safe
- Neutral

Marital Status *

Your answer

Household Members

Your answer

Education *

Your answer

Employment *

Your answer

- Somewhat Safe
- Neutral
- Somewhat Unsafe
- Very Unsafe

How often do you feel stressed in your home environment? () *

- Never
- Rarely
- Sometimes
- Often
- Very Unsafe
- Always

Submit

Clear form

Never submit passwords through Google Forms.

Fig. 1. Survey Form of Domestic Violence Dataset

The Figure no. 1 is showing the form is sent to the women of various fields to generate the dataset for further analysis, some questions were designed to be indirect indicators of domestic violence. These indirect questions were included to capture the context and dynamics within the household without explicitly asking about sensitive issues. Indirect questions can help respondents feel more comfortable and less threatened, leading to more honest and accurate responses. For example, given table questionnaire (Table 2)

- **Survey Analysis:**
  - Name: just for information, it's not useful for prediction.
  - Age: Normalize or standardize the age values if necessary.
  - Marital Status: One-hot encode (e.g., Single, Married, Divorced, Widowed, Separated).
  - Household Members: Encode the relationships (e.g., Number of children, Presence of extended family).

Table 2. Questionnaire for Survey form which shows the purpose of including indirect questions

Question	Purpose
How would you describe the relationships between members of your household?	To assess the overall environment and interpersonal dynamics within the household. Tense or hostile relationships can be indicative of underlying issues such as domestic violence
Who generally makes decisions about important household matters?	This question helps understand the power dynamics within the household. An imbalance of power, where one member consistently dominates decision-making, can be a sign of coercive control, a common aspect of domestic violence
How safe do you feel in your current living situation?	This question directly measures the respondent's perception of safety, which is a critical indicator of domestic violence. Feeling unsafe at home can be a clear sign of abuse
How often do you feel stressed in your home environment?	Frequent stress can be a sign of ongoing conflict or violence in the household. This question helps to identify high-stress environments that may be linked to domestic violence

- Education: Ordinal encode (e.g., No formal education = 0, Primary = 1, Secondary = 2, Higher education = 3). Employment: One-hot encode (e.g., Employed, Unemployed, Student, Retired, Other).
- Relationship Description: Ordinal encode (e.g., Very harmonious = 0, Generally good = 1, Neutral = 2, Some conflicts = 3, Frequent conflicts = 4).
- Decision Making: One-hot encode (e.g., Myself, Spouse/Partner, Jointly, Other household members).
- Safety Feeling: Ordinal encode (e.g., Very safe = 0, Somewhat safe = 1, Neutral = 2, Somewhat unsafe = 3, Very unsafe = 4). Stress Frequency: Ordinal encode (e.g., Never = 0, Rarely = 1, Sometimes = 2, Often = 3, Always = 4).
- Experiencing Domestic Violence: Binary encode (e.g., Yes = 1, No = 0, Prefer not to say = -1).
- Data Preparation:
- Raw Dataset: The initial dataset obtained from the survey analysis. Preprocess Dataset: The raw dataset undergoes preprocessing, which includes:
- Data Cleaning: Removing or correcting any inaccuracies or inconsistencies in the data.
- Next step is Normalization and Feature Engineering

Data Splitting

The preprocessed dataset is split into two parts that is 80% training while 20% testing.

Data splitting ensures that the model can generalize well to new, unseen data by providing a reliable measure of its performance (Fig. 2 and Table 3).

Table 3. Data Splitting

	Purpose	Advantage	Disadvantage
Training Set (80%)	This allows the model to learn from a larger dataset	With more data to train on, the model can develop a deeper understanding of the features and relationships	the evaluation might not fully represent the model's performance
Testing (20%)	This set helps in assessing how well the model generalizes to new data	Testing on unseen data provides a realistic measure of the model's predictive power in real-world scenarios	might lead to less reliable evaluation metrics, potentially influenced by anomalies or outliers

Fig. 2. Detailing of Analysis of Survey

3.2 System Architecture

(Fig. 3).

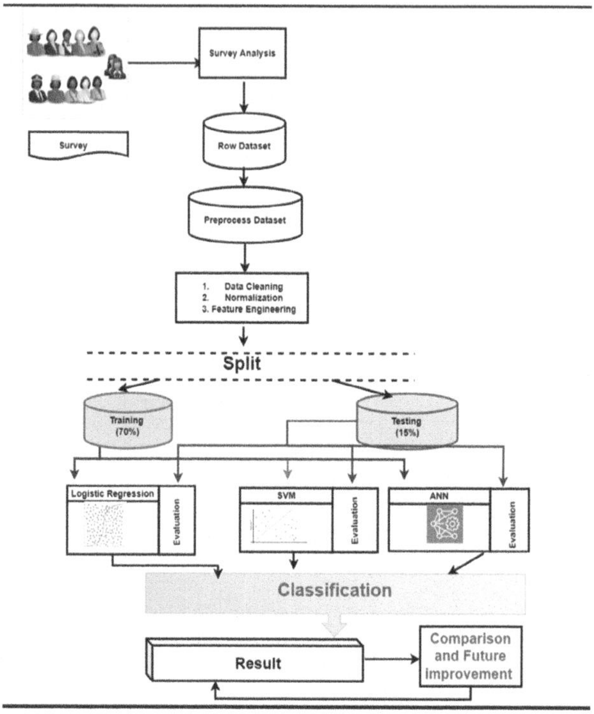

Fig. 3. Shows the diagram outlines a process for using AI-driven predictions to prevent domestic violence, under the theme "Towards Safer Homes." Here's a step-by-step explanation of the diagram.

Model Training and Evaluation:

Logistic Regression: It is trained and then evaluated on the testing set to identify domestic violence happening against women.

$$P(y = 1/x) = \frac{1}{1 + e^{-e(\beta 0+\beta 1X1+\beta 2X2+\cdots+\beta nXn)}\, e^{-e(\beta 0+\beta 1X1+\beta 2X2+\cdots+\beta nXn)}}$$

Here's a breakdown of the equation:

- $P(y = 1/x)$: Experiencing domestic violence given the input x and outcome $y = 1$.
- β_0 are bias terms
- $\beta_1, \beta_2, \ldots., \beta_n$ are coefficients for the input features $x_1, x_2, \ldots, x_n$
- $x_1, x_2, \ldots, x_n$ which represent the impact of each factor on the likelihood of the outcome.
- $x_1, x_2, \ldots, x_n$: The input features or factors that influence the outcome.

Support Vector Machine (SVM): Using the algorithm, training was conducted and then evaluated on the testing set. Support Vector Machines (SVMs) offer a robust alternative for binary classification tasks like domestic violence prediction. Unlike logistic regression, SVMs identify a hyperplane in high-dimensional space which exploits the boundary between the separated classes. This margin represents the decision boundary with the greatest distance to the closest data points from each class (support vectors). SVMs excel in high-dimensional settings and can handle non-linear data using kernel functions, making them adaptable to complex relationships between factors and domestic violence occurrence.

Artificial Neural Network (ANN)

It offers a data-driven approach to domestic violence prediction. Unlike logistic regression's single equation, ANNs employ a layered structure inspired by the brain. Each layer transforms data using activation functions (f), ultimately producing a classification through a complex web of weights (w) and biases (b). This can be represented as:

$$y = f(w^T * h(w^T * x + b) + b) \tag{1}$$

where x is the input data (e.g., income level), w and b represent weights and biases learned during training, h is an activation function in hidden layers, and y is the final predicted class (domestic violence or not). This allows ANNs to capture non-linear relationships and automatically extract important features, potentially providing a more nuanced understanding of domestic violence risk. However, their complex structure makes them less interpretable and computationally expensive compared to simpler models.

4 Results and Discussion

The correlation matrix looks at the connections between socioeconomic and demographic characteristics and domestic violence. The findings indicate that there is a minor association between age (correlation coefficient of 0.17), education (–0.09), income (–0.12), and married status (–0.17) and domestic violence, though not a particularly strong one. There is a little positive association between age and the likelihood of domestic violence, indicating a tiny rise with age. The data suggests a slight negative association between education and income, suggesting that higher income and education levels may (marginally) reduce the risk of domestic violence. A small negative link is also seen for married status, suggesting that some marital statuses may somewhat lower the incidence of domestic violence.

A small negative link is also seen for married status, suggesting that some marital statuses may somewhat lower the incidence of domestic violence. These connections, however, are complex, highlighting the fact that no factor can reliably predict the occurrence of domestic violence because it is influenced by a wide range of interrelated circumstances (Figs. 4 and 5).

The above diagram underscores the significance of educational attainment in influencing the likelihood of experiencing domestic violence, suggesting implications for policies and interventions aimed at addressing and preventing such issues.

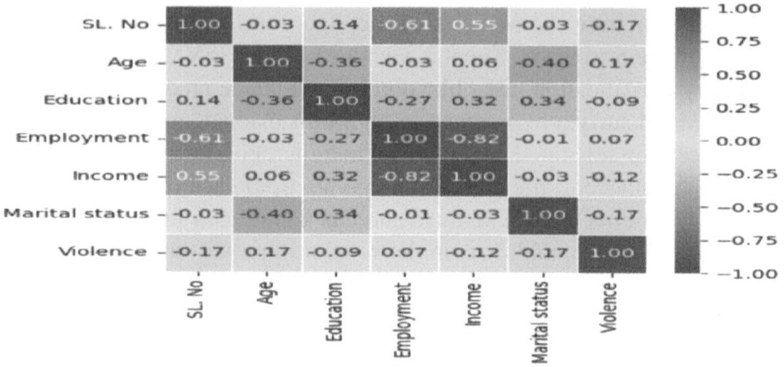

Fig. 4. Correlation Matrix

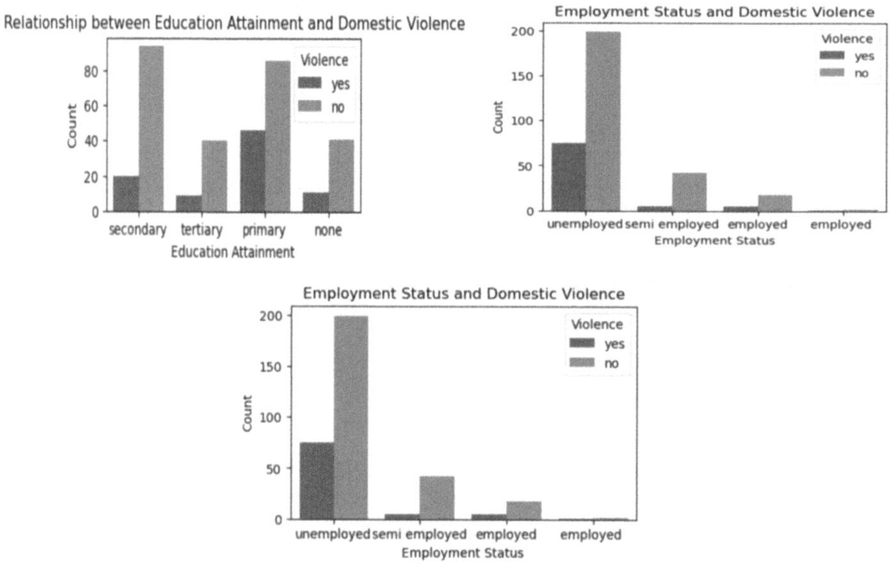

Fig. 5. Relationship between education attainment and Domestic Violence

- Individuals with **secondary education** appear less likely to experience domestic violence compared to those with **primary education** or **no education**.
- **Tertiary education** shows the lowest incidence of domestic violence among the categories.
- There is a notable proportion of individuals with **primary education** who have experienced domestic violence, suggesting an area that may warrant targeted intervention.
- The data hints at a potential protective effect of higher education levels against domestic violence.
- While a significant portion of the unemployed have, most have not been victims of domestic abuse. Those in semi-employment exhibit a lower rate of domestic violence

than do those without jobs. Out of all the groups, those who work have the lowest rate of domestic violence. This figure, which suggests that employment may be related with a lower incidence of domestic violence, can be used to highlight the potential impact of work status on the incidence of domestic violence in articles about domestic violence. It highlights how crucial work and stable finances are to reducing spousal abuse (Tables 4, 5 and 6).

Table 4. Logistic Regression Accuracy matrix

	Precision	Recall	F1-Score
Accuracy	0.742857	0.742857	0.742857
Macro Avg	0.371429	0.5	0.42623
Weighted Avg	0.551837	0.742857	0.633255

Table 5. ANN Accuracy matrix

	Precision	Recall	F1-Score
Accuracy	0.742857	0.5	0.742857
Macro Avg	0.87	0.5	0.74
Weighted Avg	0.81	0.74	0.63

Table 6. Comparative Analysis of Logistic Regression, SVM, and ANN

	Precision			Recall			F1-Score		
Algorithm Used	LR	SVM	ANN	LR	SVM	ANN	LR	SVM	ANN
Accuracy	0.742857	0.742857	0.742857	0.742857	0.742857	0.5	0.742857	0.742857	0.742857
Macro Avg	0.371429	0.625	0.87	0.5	0.518162	0.5	0.42623	0.475	0.74
Weighted Avg	0.551837	0.685714	0.81	0.742857	0.742857	0.74	0.633255	0.657143	0.63

These results highlight the significant association between employment status and domestic violence. Unemployment is linked to a higher risk of domestic violence, whereas employment appears to offer a protective effect. This underscores the importance of economic stability and employment in reducing the incidence of domestic violence, emphasizing the need for policies that promote job security and economic support for vulnerable populations (Fig. 6).

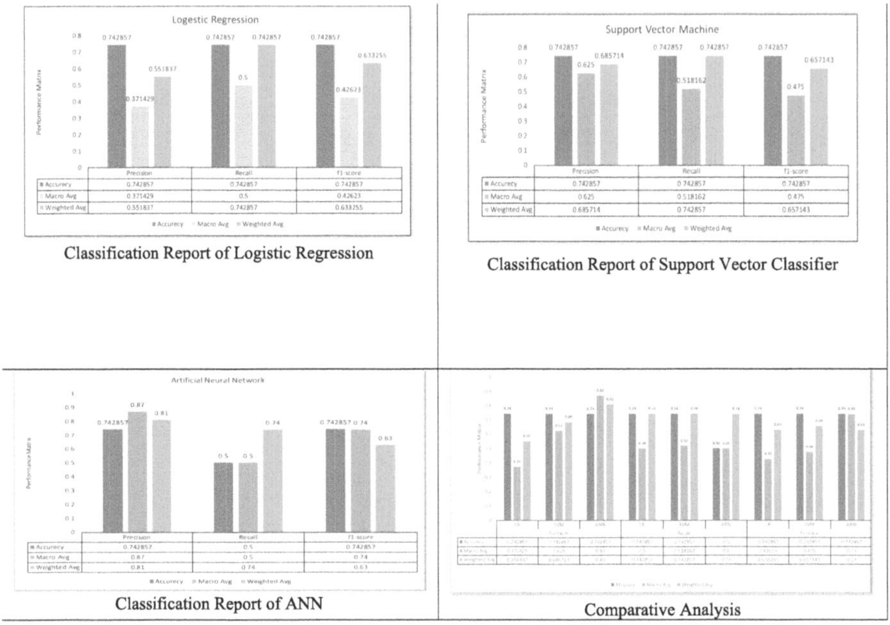

Fig. 6. Classification report LR, SVM, ANN and its comparative analysis.

Conclusion: In the proposed study, we tried to correlate various socio-economic factors and domestic violence happening against women by taking surveys. The Machine Learning model which is trained here takes the input from the survey parameter and predicts domestic violence against women. We did comparative analysis of Support Vector Classifier, Logistic regression and Artificial Neural Networks. The Artificial Neural network shows the maximum accuracy of 78%. The violence prediction mode can be helpful for prediction of domestic violence happening against women. The figures suggest that employment may be related to a lower incidence of domestic violence. Finally, we can conclude that it is crucial to be independent and stable finances can reduce spousal abuse up to some extent.

References

1. Amir, F., Noreen, N., Mushtaq, S.A.: Violence against women: a comparative study of Islamic (with reference to Pakistan) and non-Islamic Jurisdictions. sjesr **6**, 63–70 (2023)

2. Tran, D., Hanaček, K.: A global analysis of violence against women defenders in environmental conflicts. Nat. Sustain. **6**, 1045–1053 (2023). https://doi.org/10.1038/s41893-023-01126-4
3. Shayestefar, M., Saffari, M., Gholamhosseinzadeh, R., et al.: A qualitative quantitative mixed methods study of domestic violence against women. BMC Women's Health **23**, 322 (2023). https://doi.org/10.1186/s12905-023-02483-0
4. Marzaniar, P., Subarsono, A.: Stakeholder salience analysis in the policy formulation of anti-violence against women and children. J. Contemp. Gov. Public Policy **4**, 149–174 (2023). https://doi.org/10.46507/jcgpp.v4i2.158
5. Al-Saaidi, S.K.Z., Abdul-Hussain, I.R.: A Critical Discourse Analysis of a Selected Non-Governmental Organizations' Report on Violence against Womenn Iraq. Arab World Engl. J. **13**(3). 445–466 (2022)
6. Sabri, B., Rai, A., Rameshkumar, A.: Violence against women in India: an analysis of correlates of domestic violence and barriers and facilitators of access to resources for support. J. Evid. Based Soc. Work **19**(6), 700–729 (2022). https://doi.org/10.1080/26408066.2022.2105671
7. Vasconcelos, N., Andrade, F., Gomes, C., Bernal, R., Malta, D.: Physical violence against women by an intimate partner: analysis of VIVA survey 2017. Cien. Saude Colet. **27**, 3993–4002 (2022). https://doi.org/10.1590/1413-812320222710.08162022en
8. Costa, S.W.D.S., et al.: WHOT, a novel tool to assist women victims of violence: a case study in the Brazilian Amazon. IEEE Access **9**, 95046–95060 (2021). https://doi.org/10.1109/ACCESS.2021
9. Saudagar, S., Ranawat, R.: Detecting vehicular networking node misbehaviour using machine learning. In: 2023 International Conference for Advancement in Technology (ICONAT), Goa, India, pp. 1–3, 2023. https://doi.org/10.1109/ICONAT57137.2023
10. Ramsoomar, L., Gibbs, A., Chirwa, E.D., et al.: Pooled analysis of the association between mental health and violence against women: evidence from five settings in the Global South. BMJ Open **13**, e063730 (2023). https://doi.org/10.1136/bmjopen-2022-06373
11. Maria, H., de, Sousa, M., Wirna, M., Alves, D.S.: Obstetric violence: an analysis of abuse against women from different perspectives. Revista Ibero-Americana de Humanidades, Ciências e Educação (2024). https://doi.org/10.51891/rease.v10i5.14087
12. Etim, G.E., Ita, A.N., Ofem, U.J.: Cross cultural dynamics of violence against women in Nigeria a multivariate analysis approach. Int. J. Multidiscip. Res. Anal. (2024). https://doi.org/10.47191/ijmra/v7-i05-06
13. Jéssica, R., Lima, M.: Analysis of the role of education in tackling violence against women. REVISTA INTERDISCIPLINAR E DO MEIO AMBIENTE (RIMA) (2024). https://doi.org/10.52664/rima.v6.n1.2024.e240
14. Subrata, B.: A critical analysis of how sexual violence against women is defined and addressed within existing legal instruments in Bangladesh. J. Gend.-based Violence (2024). https://doi.org/10.1332/23986808y2024d000000028
15. Zaeem, Y., Shazia, I., Toor, S.I.: Content Analysis of Drama Promos: Exploring Instances of Violence Against Women across three Leading Pakistani TV Channels. Global sociological review (2024)
16. Neha, R.Y.: Exploring the nexus of gender-based violence: a psychological and sociological analysis of crimes against women in India. Int. J. Multidiscip. Res. (2023). https://doi.org/10.36948/ijfmr.2023.v05i05.6210
17. Abdullah, F., Usmani, F., Shoaib, M.: Psychological Aspects of Violence against Women: A Quantitative Analysis (2023). https://doi.org/10.55737/qjss.558066961
18. Cabrera Cabrera, X., Puicón Pinday, P.A., Reynosa Navarro, E., Guerra Ayala, M.J.: Violence Against Women in Peru: An Analysis from the Perspective of Workers at Women's Emergency Centers (2023). https://doi.org/10.5281/zenodo.8397778

19. Jie, Y.: Analysis of the Application of Justifiable Defense to the "Violence Against Violence" Behavior of Women in Response to Domestic Violence. Sci. Soc. Res. (2024). https://doi.org/10.26689/ssr.v6i1.6036
20. Szymon, R.., et al.: Characteristics of physical violence against pregnant women - analysis of medico-legal data and literature review. Ginekologia Polska (2024). https://doi.org/10.5603/gpl.95904
21. Aydın, M., Özdengül, A.G.: The Representation of Objecting to Violence Against Women: A Semiotic Analysis. Kadem kadın araştırmaları dergisi (2023). https://doi.org/10.21798/kadem.2023.139
22. Reyes-Delgado, A.T.: Machine learning in the detection of gender violence in digital job offers in the technological sphere (2023). https://doi.org/10.35429/h.2023.13.40.52
23. Shifidi, P.P., Stanley, C., Azeta, A.A.: Machine Learning-Based Analytical Process for Predicting the Occurrence of Gender-Based Violence (2023). https://doi.org/10.1109/etncc59188.2023.10284965
24. Kasiselvanathan, M., Sundari, S., Ramalakshmi, R., Tarunika, R.S.: Foretelling Femicides using Data Mining and Machine Learning (2023). https://doi.org/10.1109/icaiss58487.2023.10250716

Design and Performance Analysis of MIMO VLC for Vehicular Communication Systems

Piyush N. Patel(✉) and Jobin Dann

ECED, SVNIT, Sardar Vallabhbhai National Institute of Technology, Ichchhanath Surat-Dumas, Road, Keval Chowk, Surat 395007, Gujarat, India
{pnp,p22ec003}@eced.svnit.ac.in
https://svnit.ac.in

Abstract. A fresh innovation called vehicular visible light communication makes it possible to wirelessly communicate between cars or between various cars and the infrastructure. This study introduces a mathematical model for Multiple Input Multiple Output (MIMO) visible light communication (VLC) in vehicle-to-vehicle (V2V) communication scenarios. The structure of a vehicular VLC system incorporates a phase shift keying (PSK) modulation scheme, considering diverse weather conditions like haze, snow, and rain. The impact of ambient light sources, leading to inter-symbol interference (ISI), is also taken into account. This study examines the effectiveness of a modulation based on free space optic network using a coupler-based delay line filter. The outcomes across several Free Space Optic (FSO) aspects are evaluated using the quality factor, bit error rate (BER), and eye outline. The research findings demonstrate the achieved results for quality factor and BER under different conditions, highlighting the superior performance of both adjustment techniques and PSK modulation.

Keywords: V2V · MIMO · LED · visible light communication · phase shift keying (PSK) · Quality factor · BER

1 Introduction

Visible light communication (VLC) has received a lot of attention lately from both corporate and academic circles. Its ability to use underutilized resources for wireless transmission in the visible and infrared (IR) spectrums for communication applications is what piques our curiosity. One promising application on the horizon is implementation of the Vehicle-to-vehicle (V2V) communication in fifth-generation networks which is facilitated by the Internet of Things (IoT). [3] The growing demands of Phone internet, IoT, and the constraints of the spectrum of radio frequencies present challenges for 5G wireless communications, necessitating high throughput, extensive connectivity, reliability, and low latency.

D. Gupta et al. (Eds.): PCEMS 2024, CCIS 2491, pp. 221–231, 2025.
https://doi.org/10.1007/978-3-031-90577-3_20

Addressing the limitations of radio frequency (RF) communication systems, it becomes crucial to explore alternative technologies. [3] VLC technology emerges as a viable option, offering several advantages over RF. It employs light as the carrier of the transmission, serving illumination as well as communication concurrently. [10,11] VLC operates in a license-free spectrum, enhances security, is immune to electromagnetic interference, and provides a substantial bandwidth.

Within intelligent transportation systems (ITS), safety applications such as data regarding traffic congestion and accident alerts have gained prominence for their potential to improve road safety as well as address ecological issues, [11] Dedicated short-range communication (DSRC) is recognized as a promising automobile networks within ITS. However, despite ratified standards, commercial application of DSRC has been limited, largely due to early adopters perceiving limited benefits from investing in automobiles with sophisticated electronics and pricey radios installed.

VLC offers benefits over radio frequency (RF) communications in the typical sense, especially in applications for safety. [11] The transmitters and receivers for VLC are less intricate compared to their RF counterparts, leveraging the presence of LEDs already equipped in vehicles, thereby minimizing additional costs. Furthermore, because visible light propagates with strong directivity, VLC enables precise relative location between vehicles. On the other hand, VLC encounters difficulties in V2V situations because to factors including unfavorable weather conditions, sunlight, and ambient light.

2 Literature Review

In [1] In indoor visible light communication (VLC) systems, the efficiency of direct current biased optical orthogonal frequency division multiplexing (DCO-OFDM) systems is investigated in this study.

In [2] This research examines the effects of unfavorable weather on a Free Space Optics (FSO) connection with a range of 500 m, specifically addressing concerns about link capacity and availability.

In [3] the paper presents a visible light communication system installed on vehicles that uses non-return to zero on-off keying modulation, considering various weather conditions. The study includes LED-based and laser diode-based models, analyzing the impact of parameters like transceiver aperture diameters, receiver responsiveness, and beam divergence.

In [4] the paper explores the use of Orthogonal Frequency Division Multiplexing (OFDM) combined with higher order modulation schemes to overcome limitations on achievable data rates.

In [5] In an indoor line-of-sight (LoS) VLC system, the obtained power characteristics of a photodiode are analyzed in this work.

In [6] Designs for Individual Input Configuration with single output (SISO) and multiple input multiple output (MIMO) are simulated, with the MIMO setup showing superior Bit Error Rate (BER) performance.

In [7] This article concentrates on examining the modulation formats employed to generate optical pulses, specifically return to zero (RZ) and non-return to zero (NRZ).

In [8], In order to examine several Multiple-Input Multiple-Output (MIMO) techniques for an indoor Visible Light Communication (VLC) system as a whole the study was written.

In [9] The paper explores the potential of white LEDs for indoor visible-light communication considering the presence of multiple lights in indoor spaces, the paper investigates the impact of interference and reflection.

In [10] In Vehicle-to-Vehicle (V2V) communication settings, the study presents a mathematical model for Multiple-Input Multiple-Output (MIMO) Visible Light Communication (VLC). It considers LED lamps as multiple transmitters and photodiodes (PDs) as multiple receivers, modeling the system in the form of an optical MIMO layout.

In [11] it is a survey that explores the current state of V-VLC technology and highlights unresolved issues and challenges. It provides a comprehensive review of the V-VLC system and delves into the specifics of the VLC communication channel. For those new to the field, this article serves as a useful guide to key literature and trends.

In [12] Free space optics (FSO) offers a new challenge in optical communication as it allows optical signals to be transmitted under haze, mist, and snow, among other weather conditions. To address this, the current research evaluates the system performance by implementing two modulation techniques: Amplitude Shift Keying (ASK) and Phase Shift Keying (PSK).

In [13] Because contemporary light sources have a large bandwidth, this research emphasizes the growing interest in visible light communication (VLC), especially when using modulation techniques like orthogonal frequency division multiplexing (OFDM).

This study focuses on the design and performance evaluation of a Visible Light Communication (VLC) system between vehicles (V2V), accounting for various weather variations and system parameters at different link lengths. An Avalanche Photodiode (APD) at the receiving end of a Phase Shift Keying (PSK) modulated Multiple Input Multiple Output (MIMO) Free Space Optics (FSO) communication system is examined in this research. The main goal is to link the APD to the output for designing and modeling the performance of a MIMO FSO link in order to examine Bit Error Rate (BER) performance under different weather turbulences. A laser diode (LD) acts as an optical transmitter in the suggested system, while an Avalanche Photo Diode (APD) essentially performs the role of an optical receiver. The system takes into consideration various atmospheric conditions such as haze, rain, and snow, with the ambient light source acting as a noise factor. A PSK modulation scheme is employed. The model's evaluation involves the calculation of BER, quality factor (Q-factor), and eye diagram. The evaluation indicates that this Visible Light Communication system for vehicles yields superior results even when subjected to various weather conditions, maintaining consistency across same system configuration and modulation scheme.

The information contained herein has been structured in the following manner: Sect. 3 outlines The model of the system with reference to the Vehicle Visible Light Communication system. Subsequently, the investigation focuses on the first proposed model utilizing LED technology, analyzing the influence of system specifics and evaluating the effect of distance between links across diverse weather circumstances. In Sect. 4, The model for Visible Light Communication (VLC) using Laser Diode (LD) for Vehicle-to-Vehicle (V2V) communications utilizing Multiple-Input Multiple-Output (MIMO) is described, followed by an examination of the relationship between link distance and system characteristics under Different climate circumstances. The findings and comments are presented in Sect. 5, wherein the disparities between the simulation outcomes and the mathematical conclusions reached for various weather circumstances for both suggested models are discussed.

3 System Model

Visible Light Communication (VLC) is an optical wireless approach of communication that uses the electromagnetic spectrum's 380–780 nm wavelengths for medium-range purposes. When implementing VLC in an automobile, photodiodes (PDs) or camera image sensors function as the receivers and LED-based headlights and taillamps as the transmitters. A typical traffic scenario where Vehicle-to-Vehicle (V2V) VLC can be implemented is shown in Fig. 1. As seen, cars are able to connect with other cars nearby, making it easier for the head and tail to engage with each other when they are driving. Furthermore, head-to-head connections are feasible, particularly near junctions. Tail-to-tail linkages are also possible, but less frequently, in specific driving situations.

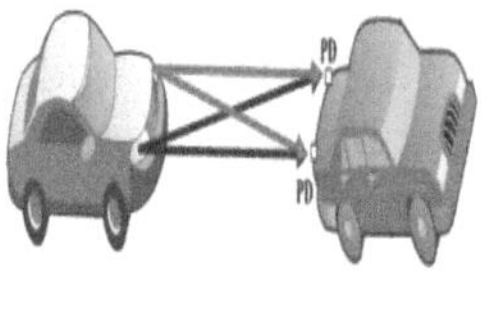

Fig. 1. Simulation design for 4×4 MIMO system

The Free Space Optics (FSO) system comprises three key functional components: the transmitter, receiver, and atmospheric channel. In the transmitter, the data signal undergoes modulation using Phase Shift Keying (PSK), with a transmission rate of two bits per second, particularly under varying turbulence conditions, utilizing a coupler-based delay line filter. Optimal light sources for optical communication must possess suitable wavelength, numerical aperture, high modulation bandwidth, and line-width. The transmission of data through the transmitter onto the recipient can occur through open space, air, or vacuum, depending on the communication medium. The transmission can be affected by a number of meteorological factors, such as haze, snow, rain, and ambient light sources.

An optical detector is used on the receiver side to capture the received data, with options such as PIN photodetectors or Avalanche Photodetectors (APDs). The photodetector, in which electrical signal that is proportional to the square of the optical field is produced by a square-law optoelectronics transducer incident on its surface. Given the typically weak nature of the optical signal after traversing the channel of communication, the photodetectors have to fulfill strict performance standards, including high sensitivity across operational wavelengths, minimal noise, and sufficient bandwidth to accommodate the desired data rate. The optical signal is converted back to an electrical signal at the receiving side and then amplified. Lastly, an optical filter is used to selectively let some wavelengths of the signal to pass through while obstructing others.

4 Model for V2V Communications Using MIMO VLC

As shown in the block diagram in the Fig. 2, here we first use the pseudo random bit sequence generator for the purpose of data which is modulated using the Bpsk modulator and then we use a Mach-Zehnder (MZ) modulator in conjunction with Phase Shift Keying (PSK) modulation to create a Free-Space Optical (FSO) system. This system modulates a Continuous Wave (CW) LASER having an operating frequency of 550 nm THz and an input power of 60 dBm. The PSK pulse generator's encrypted binary digits are first converted into a light wave carrier using this technique. Two bits are sent for each symbol by the PSK pulse generator. The signal which has been modulated undergoes enhancement in system performance as it traverses a delayed filter. The signal that has been optimized is subsequently sent via the FSO channel. The system's robustness is scrutinized, taking into account diverse attenuation values and channel ranges reflective of varied climatic conditions. Photodiodes and a 3R generator are used in the receiver portion to obtain the former electrical signal bit stream. Finally, the system's effectiveness is assessed using a Bit Error Rate analyzer. The implementation of the system model is done in the Optisystem software which is shown in Fig. 3.

4.1 Binary Phase Shift Keying

BPSK modulation employs a coherent detection technique in optical communication, where the laser serves as the carrier. For Binary Phase Shift Keying (BPSK), The Q-function may be used to represent the association among Bit Error Rate (BER) and Signal-to-Noise Ratio (SNR) in an AWGN (Additive White Gaussian Noise) channel. The tail probability of the standard normal distribution is known as the Q-function, and it is frequently written as Q(x). In

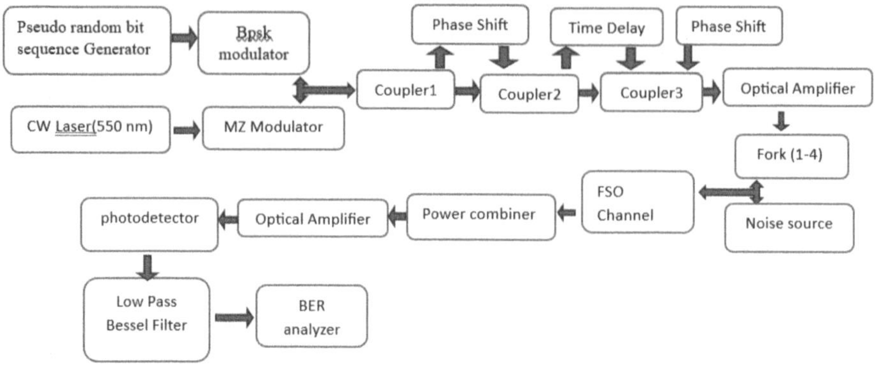

Fig. 2. Block diagram of proposed model.

BPSK modulation, the phase shift assumes values of 0° and 180°. The BER for this technique is determined by a specific equation given below:

$$BER = \frac{1}{2} \cdot erfc\left(\sqrt{\frac{2 \cdot \mathrm{SNR}}{\mathrm{N_0}}}\right) \tag{1}$$

Here, erfc is the complementary error function, SNR is Signal-to-Noise ratio, and N_0 is the one-sided power spectral density of the noise.

The power needs can be easily obtained from the BER expressions. When considering equal BER, and utilizing the normalized average power requirements of Binary Phase Shift Keying (BPSK) compared to Non-Return-to-Zero On-Off Keying (NRZ-OOK), the power demand for BPSK can be formulated as:

$$\frac{P_{\mathrm{BPSK}}}{P_{\mathrm{NRZ}}} = \frac{1}{2\sqrt{2}} \tag{2}$$

In theory, NRZ-OOK necessitates up to $2\sqrt{2}$ times more power than BPSK to achieve a specific BER performance (Fig. 3).

4.2 Channel Model

In our proposed model for MIMO VLC-based V2V communications, optical signals are emitted independently from both sides of the headlamps. On the receiving end, two Photodiodes (PDs) capture these optical signals. The assumption is that the transmission is properly timed, and the calculation for the each PD's received optical power is as follows:

$$P_i = \sum_{i=1}^{4} \frac{A_r I_i cos(\theta)_{ij}}{LER.d_{ij}^2} \tag{3}$$

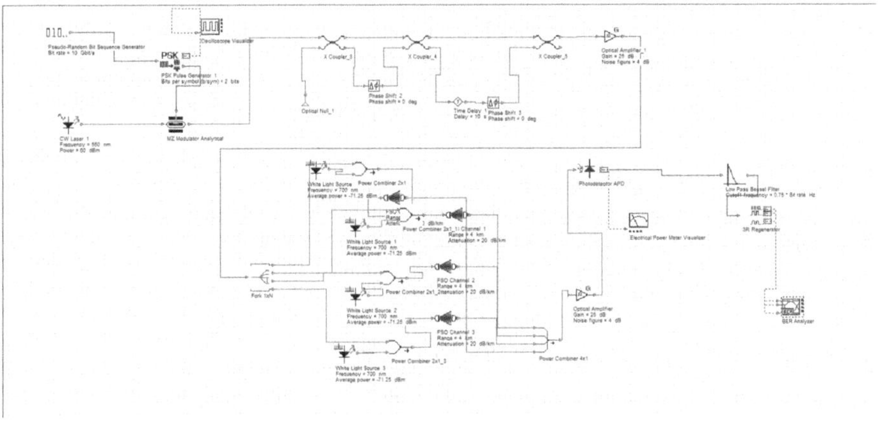

Fig. 3. Design of simulation for 4 × 4 MIMO system

Here, I represents the brightness or in other words luminous intensity of the i-th headlight, θ_{ij} denotes the incident angle of light, and The i-th headlamp's distance from the j-th Photodiode (PD) is denoted by d_{ij}. Multiple Input Multiple Output (MIMO) is one way to conceive this system. In the context of an optical wireless MIMO system, the vector representing The optical power received is represented as:

$$y = Hx + n \tag{4}$$

Here n represents the entire system noise, and H represents the channel matrix with dimensions $N_t \times N_r$. In our Vehicle to Vehicle communications system using VLC and featuring A pair of photodiodes (PDs) with a pair of transmitter, the channel matrix can be articulated as:

$$\begin{pmatrix} h_{11} & h_{12} & h_{13} & h_{14} \\ h_{21} & h_{22} & h_{23} & h_{24} \\ h_{31} & h_{32} & h_{33} & h_{34} \\ h_{41} & h_{42} & h_{43} & h_{44} \end{pmatrix} \tag{5}$$

Taking into account the photoelectric conversion.

$$h_{ij} = \begin{cases} \frac{A_r I_i cos(\theta)_{ij}}{LER.d_{ij}^2} & ,0 < \theta_{ij} < \varphi \\ 0 & , \varphi > \theta_{ij} \end{cases} \tag{6}$$

φ represents half of the photoelectric detector's field of view angle. Given that each channel functions autonomously, the matrix H has complete rank. The carrier frequency cannot be detected by a photoelectric sensor so the impact's variation over time is negligible. As a result, the corresponding values of H stay comparatively constant over time.

4.3 Proposed System Architecture

Table 1. FSO PARAMETERS

FSO CHARACTERISTICS		
DEVICE	SYSTEM PARAMETER	VALUES
PRBS	Bit rate	10 Gbps
CW Laser	Frequency	550 nm
	Power	60 dBm
FSO parameter	Transmitter Aperture Diameter	10 cm
	Receiver Aperture Diameter	20 cm
	Beam Divergence	0.25 mrad
	Attenuation:	
	Haze	10.94–20.68 dB/Km
	Rain	6.97–30 dB/Km
	Snow	40 dB/Km

On the transmitter side, a continuous wave (CW) laser source emitting light at 550 nm is modulated optically and guided by a coupler. The coupler induces changes in the slope of dispersion, effectively lowering turbulence in the atmosphere. The delay line filter's output is then transmitted over the FSO channel, where the environment acts as an interface across the transmitter & the receiver. An additional light source that is combined with the main signal to illustrate the atmospheric interactions. The signal then travels via the receiver part, where it is detected and converted into its electrical form by an optical detector.

Both of the receiver and transmitter parts are connected through the FSO network. An optical amplifier on the receiving end amplifies the optical signal prior to electrical retransmission. In addition, the proposed model uses an Avalanche Photodetector (APD) to detect light signals. The output signal is filtered with a low-pass filter to remove both minor and major interference. The proposed simulation model is assessed using an optical meter and a Bit Error Rate (BER) analyzer. The BER analyzer evaluates the bit error rate and quality factor, and the optical meter measures power. The primary use of the BER analyzer is the eye diagram.

The simulation is conducted utilizing Optisystem 21 software. The simulation architecture of the 4×4 MIMO scheme, comprising four transmitter sections and four receiver sections, is illustrated in Figure. The system's performance is evaluated under varying climatic conditions, including rain, haze, and fog. The simulation is conducted distinct ranges, namely 1 km to 2 km, at a frequency of 550 nm.

5 Results and Discussion

The proposed technique is evaluated using a 4×4 Multiple Input Multiple Output (MIMO) setup, simulating a data transmission rate of 10 Gbps. The simulation parameters and results are summarized in Tables 1, 2, and 3, corresponding to haze, rain, and snow conditions, respectively. Additionally, the corresponding eye diagrams for each weather condition are shown below each table (Figs. 4 and 5).

Table 2. Result analysis (haze condition).

Work done	Bit rate (Gbps)	Attenuation (dB/km)	LOS (m)	Q-factor	BER
PSK DLF ref. [11]	10	20	3900	73.331	0
PSK DLF ref (proposed)	10	20	4000	170.001	0

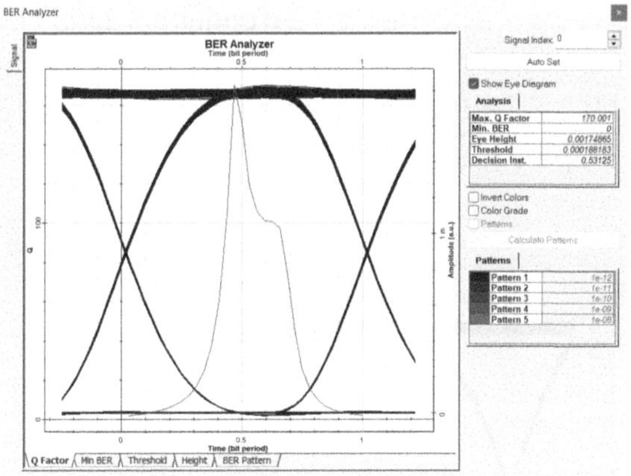

Fig. 4. Eye diagram at 4 km under haze.

Table 3. Result analysis (rain condition).

Work done	Bit rate (Gbps)	Attenuation (dB/km)	LOS (m)	Q-factor	BER
PSK DLF ref. [11]	10	30	2800	48.8145	0
PSK DLF ref (proposed)	10	30	3000	60.8865	0

The model's applicability is extended to distances ranging from 2 to 4 km within challenging free-space environments, with attenuation levels of 20, 30,

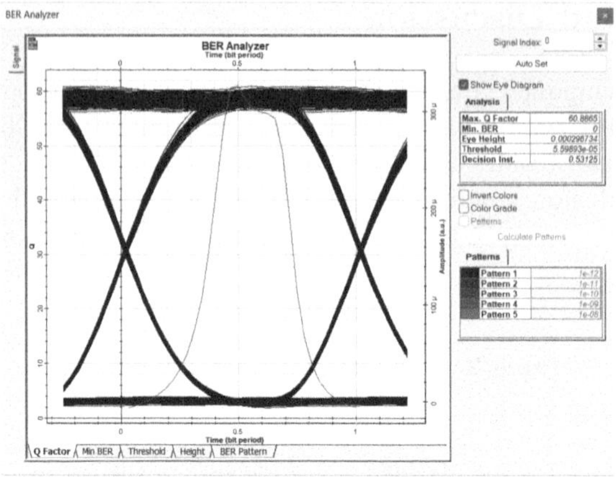

Fig. 5. Eye diagram at 3 km under rain.

Table 4. Result analysis (snow condition).

Work done	Bit rate (Gbps)	Attenuation (dB/km)	LOS (m)	Q-factor	BER
PSK DLF ref [11]	10	40	2100	59.9613	0
PSK DLF ref (proposed)	10	40	2200	113.71	0

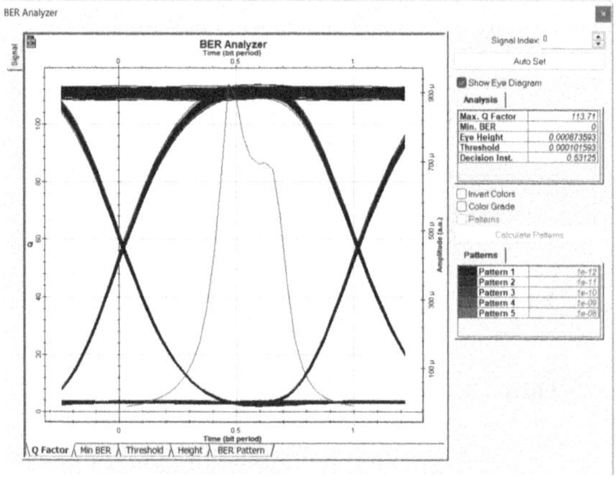

Fig. 6. Eye diagram at 2.2 km under snow.

and 40 dB/km as detailed in the above tables. Performance is measured primarily through Q-factor and BER, demonstrating that the proposed system remains robust across diverse weather conditions (haze, rain, and snow), providing reliable, long-distance communication (Fig. 6 and Table 4).

Further analysis involves both LOS and NLOS components, where reflected light introduces multipath interference. The optical power behavior and its relation to road surface reflection can be represented using a δH matrix, with the luminous intensity of reflected light being a significant factor for future research.

References

1. Abdulkafi, A.A., Alias, M.Y., Hussein, Y.S.: Performance analysis of dco-ofdm in vlc system. In: 2015 IEEE 12th Malaysia International Conference on Communications (MICC), pp. 163–168. IEEE (2015)
2. Arivazhagi, K., Shanthi, T.K.: An effective 8× 7.5 gbps wdm free space optical communication system. Int. J. Eng. Res. Technol. (IJERT). ISSN 2278–0181 (2018)
3. El-Mokadem, E.S., Tawfik, N.I., Aly, M.H., El-Deeb, W.S.: Design and performance evaluation of vehicular visible light communication system under different weather conditions and system parameters. Opto-Electron. Rev. e145580–e145580 (2023)
4. Elgala, H., Mesleh, R., Haas, H., Pricope, B.: Ofdm visible light wireless communication based on white leds. In: 2007 IEEE 65th Vehicular Technology Conference-VTC2007-Spring, pp. 2185–2189. IEEE (2007)
5. Fuada, S., Angga, P.P., Adiono, T.: Analysis of received power characteristics of commercial photodiodes in indoor los channel visible light communication (2017)
6. Gupta, N.: Wireless communication using vlc and mimo technology
7. Ibrahim, Z., Rashidi, C.B.M., Aljunid, S.A., Rahman, A.K., Anuar, M.S.: Nrz and rz analysis for optical cdma based on radio over fiber (rof) technique. In: 2016 3rd international conference on electronic design (ICED), pp. 151–154. IEEE (2016)
8. Khalid, A., Asif, H.M., Mumtaz, S., Al Otaibi, S., Konstantin, K.: Design of mimo-visible light communication transceiver using maximum rank distance codes. IEEE Access **7**, 89128–89140 (2019)
9. Komine, T., Nakagawa, M.: Fundamental analysis for visible-light communication system using led lights. IEEE Trans. Consum. Electron. **50**(1), 100–107 (2004)
10. Liu, W. and He, X.: Performance analysis of mimo visible light based v2v communications. In: 2019 IEEE 89th Vehicular Technology Conference (VTC2019-Spring), pp. 1–4. IEEE (2019)
11. Memedi, A., Dressler, F.: Vehicular visible light communications: a survey. IEEE Commun. Surv. Tutorials **23**(1), 161–181 (2020)
12. Prasad, B., Mallick, B., Patra, K.C., Barpanda, N.K.: Optical wireless systems with ask & psk using coupler-based delay line filter. J. Opt. Commun. **44**(s1), s1217–s1223 (2024)
13. Selvendran, S., Sivanantha Raja, A., Esakki Muthu, K., Lakshmi, A.: Certain investigation on visible light communication with ofdm modulated white led using optisystem simulation. Wirel. Pers. Commun. **109**, 1377–1394 (2019)

Performance and Design Analysis of Visible Light Communication Systems Amidst Optical Background Noise Utilizing Filtering Techniques

Piyush N. Patel(✉) and Jobin Dann(✉)

Sardar Vallabhbhai National Institute of Technology, Ichchhanath Surat-Dumas Road, Keval Chowk, Surat 395007, Gujarat, India
{pnp,p22ec003}@eced.svnit.ac.in
https://www.svnit.ac.in/

Abstract. With the swift progress in Light Emitting Diodes (LEDs), Visible Light Communication (VLC) has attracted considerable research interest in recent years. This study explores indoor VLC utilizing the Optisystem simulation tool to analyze the channel characteristics of an NRZ OOK VLC system. The system, employing a white LED with a wavelength of 450 nm and a photodetector, is designed and simulated. The presence of light noise in the environment, especially by the means of fluorescent lights using traditional ballasts, presents a serious problem for system performance. Ambient noise is reduced by placing a rectangular optical filter following the photodiode at the receiver, leading to improved Bit Error Rate (BER) and the attainment of a clear eye diagram.

Keywords: Non Return to Zero on Off Keying · Bit Error Rate · Visible Light Communication

1 Introduction

Visible light communication, or VLC, has garnered a lot of interest lately from the academic as well as industry sectors. Its ability to use underutilized transmissible capabilities in the visible and infrared (IR) range for communication purposes is what sparks our curiosity. With the increasing demands of IoT, mobile internet, and the limitations of the radio frequency spectrum, the challenges faced by 5G wireless communications include the need for high throughput, extensive connectivity, reliability, and low latency.

Given the constraints of radio frequency (RF) communication systems, it is imperative to explore alternative technologies. VLC technology emerges as a promising option, offering several advantages over RF. It utilizes light as the transmission medium, serving both illumination and communication simultaneously. VLC operates in a spectrum that doesn't require licensing, enhances

D. Gupta et al. (Eds.): PCEMS 2024, CCIS 2491, pp. 232–242, 2025.
https://doi.org/10.1007/978-3-031-90577-3_21

security, is resistant to electromagnetic interference, and offers a significant bandwidth.

The Institute of Electrical and Electronics Engineers (IEEE) has standardized VLC [1]. Early research on visible light communication was done in Japan by the Visible Light Communications Consortium [2]. At the moment, VLC research is being intensively conducted in Asia, Europe, and the Wireless World Research Forum [3]. Therefore, it is anticipated that in the future, lights utilized in every day life will serve both as sources of illumination and as means of communication.

In order to build, simulate, and assess the performance of a VLC system impacted by ambient light noise sources, such a 52 W fluorescent illumination, this article uses the OptiSystem software.

In this paper, we study Visible Light Communication Systems with an emphasis on filtering methods within the framework of Optical Background Noise. The arrangement of the article is as follows: An overview of several studies that have advanced our knowledge of various facets of the Visible Light Communication model is given in Sect. 2. The transmitter, channel, and receiver portions are covered in detail in Sect. 3, which covers the fundamental parts of a VLC system. The simulation approach, including specifics of the applied design, is covered in Sect. 4. Finally, Sect. 5 presents the results and concludes the discussion.

2 Literature Review

In [5], Utilizing the Optisystem simulation application, indoor visible light communication is being investigated while taking into account the useful channel characteristics connected to a white LED. The quality factor and bit error rate (BER) measurements are utilized to analyze the performance of the Visible Light communications network at various bit rates and distances.

In [11], non-return-to-zero on-off keying (NRZ OOK) encoding in a visible light communication system. This device has a photodetector for reception and uses a white LED light source with a wavelength of 450 nm.

In [12] provides a performance evaluation and simulation model of a RZ-OOK VLC network with optical noise as background. An optical rectangular filter is positioned ahead of the photodiode and an electrical low-pass Bessel filter is positioned after the trans-impedance amplifier in order to reduce this noise.

In [2], This article examines in detail the characteristics of the power received at the photodiode end in an indoor line-of-sight (LoS) medium of a VLC system.

In [3], Building a wireless VLC (Visible Light Communication) system with the ability to transfer data via visible light as the medium is the aim of this research project. The system employs an LED array both as the transmitter and receiver, and the transmission occurs through the air medium.

In [8], We do an analysis that includes receiving signal waveform computation, luminance distribution and RMS delay spread. Our method considers the effect of wall reflections as well as the spatial arrangements of transmitters.

In [1], This paper examines the efficiency of inside Direct Current biased optical orthogonal frequency division multiplexing (DCO-OFDM) systems for Visible Light Communication (VLC).

In [9], With the use of the Optisystem V.16 simulation application, the study evaluates the performance of a VLC system utilizing a variety of preequalization strategies. The simulation takes into account realistic characteristics that are obtained from the LED spectrum, requiring equalization.

In [15], This research investigates how wavelength affects an optical line of sight (OLOS) link's performance when it operates through the visible light communication (VLC) spectrum. This study's expansion draws on the discoveries of earlier investigations.

In [6], Noise is included in the design and modeling of a Non Return to Zero On Off Keyed (NRZ-OOK) Visible Light Communication (VLC) network. A PIN photodiode serves as the receiver in this setup, while a 450 nm light-emitting diode serves as the optical source.

In [10] the study explores the performance of indoor VLC using white LEDs, modeled through the OptiSystem tool. The system's effectiveness is evaluated by measuring the quality factor and the logarithm of BER (bit error rate) at different data rates and link distances. The system can transmit data at up to 1 GB/s across a distance of 3 m with a Q factor of 6.508. The white LED is used as the communication source in the investigation, which focuses on NRZ-OOK modulation types. The performance is also assessed under external ambient light conditions, and enhancements are achieved by incorporating a rectangular optical filter.

In [4] In this research, a combination of optical time division multiplexing (OTDM) and wavelength division multiplexing (WDM) is used to study visible light communication (VLC). Additionally, it shows the VLC simulation using a direct sequence optical code division multiple access (DS-OCDMA) system that was created via co-simulation between OptiSystem and Matlab. By employing two degrees of freedom, the method combines the advantages of OTDM and WDM. The bit error rate (BER) is plotted against data rates and user count to illustrate the performance through simulated outcomes. First, the simulation results validate the system's efficacy, and then there is a significant rise in the user base. Furthermore, OCDMA gives each user a unique code, allowing for simultaneous and asynchronous access from numerous users. Long code sequences can be used to reduce the problem of multiple access interference (MAI), but doing so needs extremely high bandwidths. The purpose of this work is to verify the viability of an all-optical conventional correlation receiver for DS-OCDMA access networks via simulation. OptiSystem was used to create the system blocks and run the simulation.

In [7], Using OptiSystem software, a hybrid optical fiber and VLC system is constructed and simulated. A CW laser with a wavelength of 1550 nm is used for optical fiber communication and 450 nm for the VLC system. Using a 40 GHz millimeter wave (mmWave) frequency, 8 Gbps of data is sent over 40 km of optical fiber and 3 m of VLC connection. By altering the propagation distance for both the fiber and VLC connections, the system's performance is examined. The findings show that the BER performance decreases with increasing fiber length. Nonetheless, the suggested hybrid system functions well even at 8 Gbps, a large data rate.

In [14]This study describes the use of OptiSystem software in the design and modeling of a hybrid radio-over-optical fiber and VLC system. A visible light LED at 450 nm is utilized for the VLC system, and a Continuous Wave (CW) laser at 1550 nm is used for optical fiber communication. Using a millimeter wave (mmWave) frequency of 40 GHz, data transmission at 8 Gbps is accomplished via 40 km of optical fiber and 3 m of VLC connection. Furthermore, compared to the Return Zero (RZ) approach, the Non-Return Zero (NRZ) modulation methodology improves performance.

In [13]The FSO communication system implemented in the study is based on WDM and makes use of MIMO technology, which allows the receiver to receive multiple independent copies of the signal. Various MIMO configurations, such as 2×2, 4×4, and 8×8 data streams, are examined in the analysis of the system. Using OptiSystem, key performance variables like BER and Quality Factor are assessed for the WDM-based FSO communication system with MIMO for a variety of MIMO data streams and atmospheric circumstances.

3 Theory of a VLC System

The next a short-range optical wireless communication system known as VLC, or visible light communication, is being hailed as the future of ecologically conscious lighting. It serves a dual purpose by not only providing illumination but also facilitating data transmission. White LEDs are used as the main optical source in VLC, which uses the visible light spectrum between 380 and 780 nm. A free space channel serves as the channel for transmission, and a photodetector is placed at the receiving end. This approach offers several advantages over traditional fiber optic cables, including cost-effectiveness due to the absence of cable requirements, quicker transmission speeds, and immunity to RF interference (Figs. 1 and 2).

3.1 VLC Transmitter

There exist various light sources, with LED, LASER, and sunlight being among the most prevalent. In the realm of Visible Light Communication (VLC), particular interest lies in selecting these sources for transmission. Typical optical sources used in VLC technology are white LEDs, which distinguishes them from more conventional options like fluorescent or incandescent bulbs. White LEDs are anticipated to rule the next generation of lighting products due to their numerous advantages, which include their powerful lifetime of more than 50,000 h, low energy consumption, and advantageous brightness. Some of these features make them suitable not only for illumination but also for data transmission purposes.

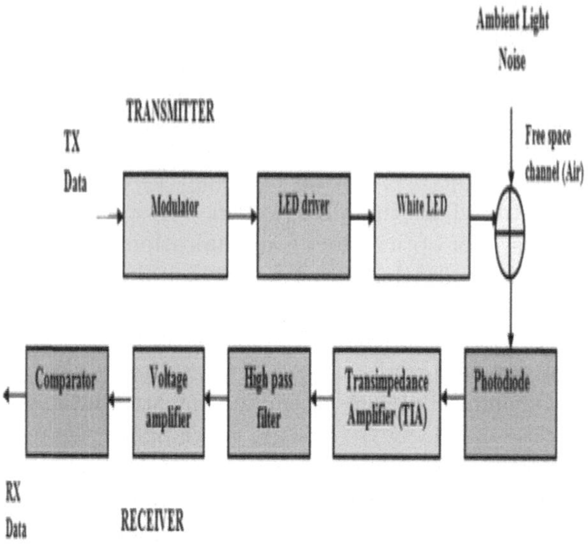

Fig. 1. The proposed Visible Light Communication system utilizing Optisystem is depicted in the block diagram [8].

LED serves as the primary light source in VLC applications. A signal undergoes modulation using an LED, following which the modulated optical signal is transmitted. Additionally, the transmitter side of VLC setups typically incorporates a bit sequence generator and a pulse generator alongside the optical source. NRZ represents the oldest and simplest modulation scheme, achieved by alternately activating or deactivating the LED source. Implementation typically involves utilizing a pseudo-random bit sequence generator and an LED for optical communication. The zero voltage point is directly crossed by the NRZ approach, which is defined by transitions between logic states 0 and 1, and vice versa. On the other hand, transitions in the RZ method briefly stop at the zero voltage level. Originally, binary data was converted from electrical into optical signals using these words, RZ and NRZ, in optical transmission. In this case, an electrical bit "1" denoted an increased optical intensity and a bit "0" indicated a reduced optical intensity.

3.2 Channel

n this study, following the transmitter section, we implement a Free Space Optical (FSO) channel with an appropriate link distance, supplemented by an attenuator. Within this channel, our light source carrying data, alongside any accompanying noise, propagates. In OptiSystem, should we opt for an alternative channel rather than free space, we would utilize an Optical Wireless Communication (OWC) channel, adjusting the attenuation value based on the specific channel type, such as pure water or sea water. FSO channels offer notable advantages, including the elimination of wiring requirements, simplified installation procedures, absence of licensing complications, and minimal associated costs.

In addition to the channel model, the noise model must be included. It is known that a variety of ambient light sources, including sunshine, incandescent lighting, and fluorescent illumination, may cause signal deterioration in the communication system [7]. As a result, a linear optically additive white Gaussian noise (AWGN) medium is commonly used to describe the visible light communication channel. This may be stated as:

$$I(t) = \eta p_i(t) * h(t) + N(t) \tag{1}$$

where I(t) is the light detector current, N(t) is the additive white Gaussian noise, h(t) describes the impulse response, and * stands for convolution. η indicates the photosensitivity of the detector. There are two different circumstances in which VLC may be used: Line of Sight (LOS) and Non-Line of Sight (NLOS). In a Line of Sight (LOS) communication situation, the transmitter creates an unobstructed channel by pointing the light signal in the receiver's direction.

The received power in the Line of Sight (LOS) circumstance is described as follows:

$$P_{r_{LOS}} = H_{LOS}(0)P_t \tag{2}$$

Reflections from walls or other barriers are taken into account in the Non-Line of Sight (NLOS) scenario for The direct current (DC) gain of the channel in the lossy-of-signal (LOS) state usually determines the received signal power, and the contribution from reflected routes, represented by Href(0), is as follows:

$$P_r = H_{LOS}(0)P_t + H_{NLOS}(0)P_t \tag{3}$$

$$P_r = H_{LOS}(0)P_t + H_{NLOS}(0)P_t + \sum ref H_{ref}(0)P_t \tag{4}$$

In order to boost the data transmission rate, we have included ambient light disturbances, such as fluorescent lamps operated by traditional ballasts, and a Line of Sight (LOS) model linking the transmitter and receiver has been implemented.

3.3 VLC Receiver

To transform optical signals from the channel into electrical signals, VLC recipients use a photodetector. Two primary types of photodiodes-Avalanche Photodiodes (APDs) and Positive-Intrinsic-Negative (PIN) photodiodes-are used to perform this. PIN photodiodes are the most widely utilized of these because of their affordability, linear response across a broad range, low power consumption, and resistance to drastic temperature changes. This choice aids in enhancing recipient functionality. In order to recover the original data, the VLC reception setup additionally includes amplification, filtration systems, and a demodulator. Because PIN photodiodes have benefits over APDs, we used them as the photodetector in our VLC receiver design. To accurately recreate the original data, the VLC receiver also has an optical rectangular filter, an electrical low-pass filter, a regenerator, and a transimpedance amplifier.

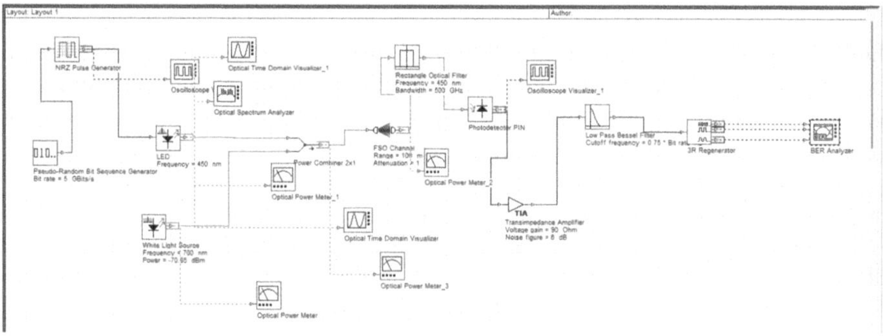

Fig. 2. The simulation setup for a Visible Light Communication (VLC) system utilizes Optisystem, employing a single LED as the primary light source, with additional white light serving as the noise component.

4 Simulation Methodology

The Visible Light Communication system simulation is carried out utilizing Optisystem software developed by Optiwave Corporation. Optisystem serves as a comprehensive simulator at the system level, enabling accurate modeling of optical communication components. This facilitates the analysis of system performance while minimizing expenses, resources, and time investment.

Using the electromagnetic spectrum's 380–780 nm wavelengths, Visible Light Communication (VLC) is an optical wireless communication system intended for medium-range applications. The three main components of the VLC system are the transmitter, the receiver, and the free-space optical channel. An NRZ pulse generator, a PRBS generator running at a rate of 5 Gbit/sec, and an LED emitting light at 450 nm with a modulation bandwidth of 1 THz are all part of the transmitting end of the investigated model. The signal is broadcast over a 100-meter free space medium, taking into consideration artificial light noise from a standard fluorescent bulb with a ballast that emits light at a frequency of 52 W. This light noise affects the functionality of the device.

A Line of Sight (LOS) link is established in order to provide an immediate communication mode between the transmitter and receiver. A rectangular optical filter is positioned in front of the PIN photodetector, which collects light from the fluorescent bulb and the optical signal from the LED, in order to reduce noise from ambient light. Electrical signals are then created from the filtered optical signals. A Line of Sight (LOS) link is established in order to provide an immediate communication mode between the transmitter and receiver. A rectangular optical filter is positioned in front of the PIN photodetector, which collects light from the fluorescent bulb and the optical signal from the LED, in order to reduce noise from ambient light.

The original bit sequence is then recovered using a bit sequence regenerator. The 3R-Regenerator is coupled to a Bit Error Rate (BER) analyzer, which measures the BER values of the received signal. The eye diagram is also created using an eye diagram analyzer. The VLC system's performance is evaluated using the simulated outcomes from this configuration.

Table 1. Parameters and Values

Parameter	Value
Bit rate	5 Gbps
Modulation	NRZ-OOK
Optical source (LED)	Frequency = 450 nm, Bandwidth = 1 THz
Optical background noise	Frequency = 700 nm, Power = 52.572 W
Transmission Channel	Free space optics (FSO)
Distance	100 m
PIN photodetector	Responsivity = 1 A/W
Low pass Bessel filter	Cut off frequency = 0.75 times bit rate
Channel Attenuation	1 dB/km
Rectangle optical filter	Wavelength = 450 nm, Bandwidth = 500 GHz
Silicon PIN photodiode	Responsivity = 0.22 A/W
Transimpedance amplifier	Voltage gain= 90 Ω, Bandwidth = 240 MHz

In Table 1, The bitrate, LED bit sequence wavelength, photodiode responsivity, wavelength and bandwidth of the rectangular optical filter, voltage gain and bandwidth of the transimpedance amplifier, and the low-pass filter's cutoff frequency are among the variables that stay constant throughout the simulation. The visible light spectrum, which spans from 380 nm to 780 nm, is the operating range for the conventional Visible Light Communication (VLC) system. The vendor's datasheet contains standard specifications that the main parameters and values are based on. In particular, the white LED in use has a bandwidth of 1 THz and runs at 450 nm wavelength. The wavelength range in which the photodiode operates is 380–780 nm. The transimpedance amplifier has a 240 MHz

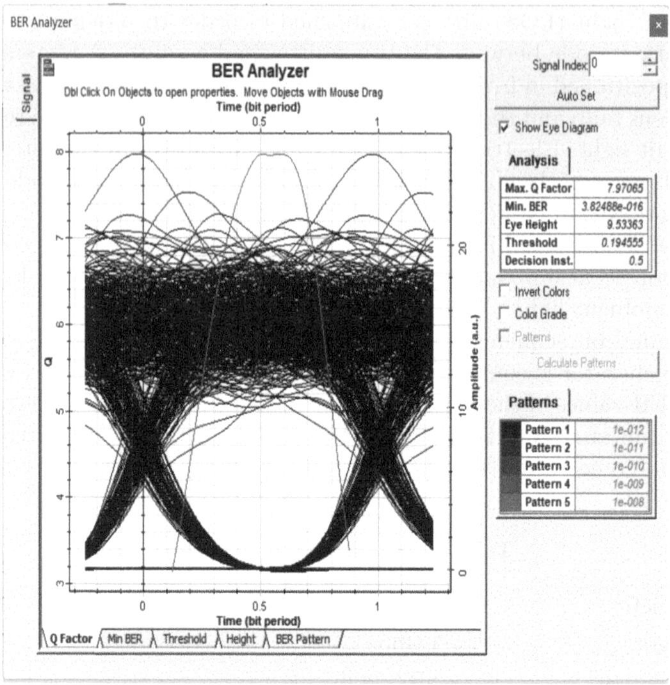

Fig. 3. Eye Diagram Analyzer output or 5 Gbps NRZ-OOK signal.

bandwidth and a voltage gain of 90 Ω. The suggested VLC system simulation is built up using these settings and values, as shown in Fig. 3.

5 Results and Discussion

Table 2. Comparative analysis

Ref	Bit Rate	Modulation	Wavelength	BER	Q-factor
[8], 2015	10 Kbit/s	NRZ-OOK	450 nm	2.4×10^{-33}	11.9
Proposed work	5 Gbit/s	NRZ-OOK	450 nm	3.82×10^{-16}	7.97

We built a visible light communication system with ambient light noise accounting for the characteristics given in Table 1. The Table 2 presents a comparative analysis between a reference study [2] from 2015 and the proposed work, focusing on Bit Rate, Modulation used, Wavelength, Bit Error Rate (BER), and Q-factor. Using a variety of instruments, such as an RF spectrum analyzer, an oscilloscope for visualization, an eye diagram analyzer, and a BER analyzer, we examined the output data. The data signal was conveyed across the free space channel by the white LED's measured optical power of 770.549 mW in conjunction with the optical power of 52.572 W from the ambient light noise. An optical power of 6.516 W was obtained at the receiving end across a 100-meter distance. This corresponds to a received electrical power of 18.539 mW at the photodiode. A Q factor of 7.974 and a BER value of 3.82 × e-15 indicate that the findings showed significant dependability at a distance of 100 m and a data rate of 5 Gbps. These measures demonstrate how well the system can transfer data over long distances.

References

1. Abdulkafi, A.A., Alias, M.Y., Hussein, Y.S.: Performance analysis of dco-ofdm in vlc system. In: 2015 IEEE 12th Malaysia International Conference on Communications (MICC), pp. 163–168. IEEE (2015)
2. Fuada, S., Angga, P.P., Adiono, T.: Analysis of received power characteristics of commercial photodiodes in indoor los channel visible light communication (2017)
3. Gupta, N.: Wireless communication using vlc and mimo technology
4. Kharbouche, A., Madini, Z., Zouine, Y.: Performance improvements of a vlc system, in a v2x context, using a different multiplexing technique. TELKOMNIKA (Telecommun. Comput. Electron. Control) **21**(4), 725–735 (2023)
5. Manivannan, K., Raja, A.S., Selvendran, S.: Performance investigation of visible light communication system using optisystem simulation tool. Int. J. Microwave Opt. Technol. **11**(5) (2016)
6. Maurya, V., Maddila, R.K.: Design and performance analysis of visible light communication system with noise mitigation. In: 2022 International Conference on Engineering and Emerging Technologies (ICEET), pp. 1–4. IEEE (2022)
7. Moorthy, V., Nawawi, N.M., Junita, M.N., Abdullah, S.R., Anuar, M.S., Endut, R.: Simulation analysis of hybrid fiber and radio over vlc system at various propagation distance and data rate. In: AIP Conference Proceedings, vol. 2898. AIP Publishing (2024)
8. Nguyen, H.Q., et al.: A matlab-based simulation program for indoor visible light communication system. In: 2010 7th International Symposium on Communication Systems, Networks & Digital Signal Processing (CSNDSP 2010), pp. 537–541. IEEE (2010)
9. Priyadharshini, R., et al.: Performance analysis of various pre-equalization techniques for visible light communication (2023)
10. Sharma, C., Singh, S., Kumar, R.: Performance analysis of an indoor visible light communication system using optisystem software. In: 2023 Third International Conference on Advances in Electrical, Computing, Communication and Sustainable Technologies (ICAECT), pp. 1–6. IEEE (2023)
11. Sindhubala, K., Vijayalakshmi, B.: Design and performance analysis of visible light communication system through simulation. In: 2015 International Conference on Computing and Communications Technologies (ICCCT), pp. 215–220. IEEE (2015)
12. Sindhubala, K., Vijayalakshmi, B.: Simulation of vlc system under the influence of optical background noise using filtering technique. Mater. Today: Proc. **4**(2), 4239–4250 (2017)

13. Thandapani, K., Gopalswamy, M., Jagarlamudi, S., Sriram, N.B.: Performance analysis of wdm free space optics transmission system using mimo technique under various atmospheric conditions. J. Opt. Commun. **44**(s1), s1653–s1660 (2024)
14. Vaides, M., Nawawi, N.M., Anuar, M.S., Junita, M.N., Yusof, N.R.: Performance analysis of hybrid optical fiber and radio-frequency transmission towards user-end vlc system. In: AIP Conference Proceedings, vol. 3135. AIP Publishing (2024)
15. Vaishali, S.S., et al.: Lambertian luminous intensity radiation pattern analysis in olos indoor propagation for better connectivity. Wirel. Commun. Mob. Comput. **2022**, 1–11 (2022)

UAV Assisted Secrecy Analysis for Future Wireless Communication

Girish A. Koche and Amit Agarwal(✉)

Visvsvaraya National Institute of Techniology Nagpur, Nagpur, India
amit.nitk07@gmail.com

Abstract. We analyze the performance of a wireless communication setup consisting of a source (Alice), a legitimate receiver (Bob), and an eavesdropper (Eve). To enhance secrecy, a UAV-based jammer is employed to generate artificial noise. Two jamming configurations are considered: a single UAV jammer and a multi-UAV jammer configuration that establishes a virtual multiple-input multiple-output (MIMO) system using zero-forcing techniques. For both scenarios, we investigate the average secrecy capacity (ASC) and, based on this, evaluate the secrecy outage probability (SOP) performance. Closed-form expressions for ASC and SOP are derived for both configurations. Simulation and analytical results demonstrate that employing a UAV jammer substantially enhances jamming effectiveness compared to a scenario without jamming.

Keywords: UAV · average secrecy capacity · secrecy outage probability · Zero-forcing · virtual MIMO

1 Introduction

Due to the need for secure transmission channels, physical layer security has attracted a lot of academic interest in wireless communications. Wyner's wiretap channel model [1], which demonstrated that perfect secrecy may be achieved when the main channel performs better than the eavesdropper's wiretap channel, is the source of foundational work in this field. Numerous methods, including two-way communication systems [4], multiple antenna arrays [2,3], and friendly jamming [5–8], have been developed recently to strengthen physical layer security. By adding artificial noise that is intended to compromise the eavesdropper's channel, friendly jamming is the most alluring of these for improving security [5]. In [6], different levels of CSI expertise regarding the primary and wiretapped channels were used to assess the coverage area and efficacy of friendly jammers. Interestingly, most of the research on friendly jamming has focused on using jammers that are placed on the ground.

An emerging area of research focuses on unmanned aerial vehicles (UAVs) due to their capacity to deliver adaptable, high-speed wireless coverage on demand for ground communication networks [9–11]. In [9,10], a single UAV is utilized

D. Gupta et al. (Eds.): PCEMS 2024, CCIS 2491, pp. 243–253, 2025.
https://doi.org/10.1007/978-3-031-90577-3_22

to improve secrecy performance, while [11] explores a zero-forcing approach. However, [11] does not comprehensively compare scenarios involving no jammer, a single UAV jammer, and multiple UAV jammers.

To address this gap, this paper examines and compares a secrecy model that incorporates an air-to-ground jammer alongside a ground communication network composed of a legitimate transmitter-receiver pair and an eavesdropper.

The secrecy performance of a wireless communication system with Alice as the source, Bob as the legitimate recipient, and Eve as the eavesdropper is the main topic of our investigation. We utilize a UAV-based jammer to create artificial noise in order to increase secrecy. Both single-jammer and multi-jammer configurations that comprise a virtual multiple-input multiple-output (MIMO) system are considered. Rayleigh fading channels are used to represent the channels between the source and the receivers, and air-to-ground wireless channels with both line-of-sight (LoS) and non-line-of-sight (NLoS) components are used to model the channels from the UAV jammer to the ground users. Our contributions are as follows:

1. We analyze the secrecy performance for both single and multi-jammer scenarios. We employ zero-forcing precoding in the multi-jammer setup to nullify the jamming interference at the legitimate receiver, enhancing the system's secrecy performance.
2. We provide insights into the effect of UAV jamming on secrecy performance by deriving expressions for the secrecy outage probability (SOP) for both single and multi-jammer settings.
3. We present analytical results and simulations to validate our theoretical findings and demonstrate the effectiveness of the proposed UAV-based jamming strategies by comparing the baseline (i.e., no jammer) scheme.
4. For the specific parameter setting, with the ZF scenario, after 5 dB, we start getting improvement as compared to the no jammer scenario. Specifically, P_{out} significantly decreases from 0.65 at 5 dB to 0.05 at saturation. In contrast, the single Jammer starts showing improvement after 7 dB. Specifically, P_{out} significantly decreases from 0.65 at 7 dB and saturates at 0.2. The no jammer scenario, indicated by the constant P_{out} value of 0.65.

2 System and Channel Model

We examine a situation where an eavesdropper, Eve (E), tries to intercept a communication between a source, Alice (A), and a legal recipient, Bob (B). A UAV-based jammer is used to create artificial noise in order to improve the secrecy performance. A single jammer and a multi-jammer configuration that creates a virtual multiple-input multiple-output (MIMO) arrangement are the two jamming scenarios that we examine. We model the communication channels as independent Rayleigh fading channels from the source to Bob and Eve. Conversely, the air-to-ground wireless channels that carry both line-of-sight (LoS) and non-line-of-sight (NLoS) components are transmitted from the UAV jammer to Bob and Eve. A thorough explanation of this configuration will be given in the parts that follow.

2.1 Ground-To-Ground Channel Model

In the ground-to-ground channel, the fading coefficient is modeled as complex Gaussian, that is, $\mathcal{CN}(0,1)$, and the channel coefficient due to path loss is calculated as $d^{-\alpha_1/2}$, where α_1 is the path loss exponent and d is the distance between the Alice and receiver (Bob or Eve).

2.2 Air-To-Ground Jammer Channel Model

As suggested in [5], we use the air-to-ground channel model. The path loss for LoS and NLoS channel between the UAV jammer and ground users is defined in this model as

$$L(x,y) = \begin{cases} |d_{ix}|^{\alpha_u}\eta_{LoS}, & \text{for LoS link} \\ |d_{jx}|^{\alpha_u}\eta_{NLoS}, & \text{for NLoS link} \end{cases} \tag{1}$$

where $|d_{jx}|$, with $x \in b,e$, indicates the distance between a ground user (either Bob or Eve) and the UAV jammer. η_{LoS} and η_{NLoS} are the attenuation factors for LoS and NLoS links, respectively, whereas α_u is the path loss exponent unique to the air-to-ground channel.

The probability of an LoS connection is given by:

$$P_{\mathrm{LoS}} = \frac{1}{1+\psi\exp(-\omega[\theta-\psi])}, \tag{2}$$

where ψ and ω are constants that depend on the environment, and $\theta = \frac{180}{\pi}\tan^{-1}\left(\frac{h}{r_{jx}}\right)$ represents the elevation angle of the UAV. Here, h denotes the altitude of the UAV, and r_{jx} is the horizontal distance between the ground user and the projection of the jammer onto the ground plane.

For a ground user located at coordinates (x, y) and a UAV located at (x_j, y_j, h), the average path loss can be computed as:

$$\bar{L}(x,y) = P_{\mathrm{LoS}}|d_{jx}|^{\alpha_u}\eta_{\mathrm{LoS}} + P_{\mathrm{NLoS}}|d_{jx}|^{\alpha_u}\eta_{\mathrm{NLoS}}, \tag{3}$$

where $P_{\mathrm{NLoS}} = 1 - P_{\mathrm{LoS}}$ is the probability of an NLoS link.

The resulting average interference from the UAV to the ground user is given by:

$$I_{jx} = \frac{P_j}{\bar{L}(x,y)}, \tag{4}$$

where P_j is the transmit power of the UAV jammer. Fading effects between the jammer and ground users are neglected here, as the analysis centers on the interference caused by the jammer, which primarily depends on the non-fading component of the signal.

3 Secrecy Performance Using a Stationary UAV Jammer

In this section, we evaluate the secrecy performance of the UAV assisted wireless communication setup. Towards this end, in what follows, the signal-to-interference plus noise ratios (SINRs) is calculated for both single jammer and multi-jammer setups.

3.1 Single Jammer

Consider a scenario where communication between a ground-based source (Alice) and recipient (Bob) is threatened by potential eavesdropping from Eve. To counter this threat, a stationary Unmanned Aerial Vehicle (UAV) is deployed to disrupt Eve's surveillance attempts, as depicted in Fig. 1. The received signals at Bob (y_B) and Eve (y_E) can be expressed as follows:

$$y_B = \sqrt{P_A} h_{AB} \cdot S_A \cdot d_{AB}^{-\alpha_1/2} + w_B + \sqrt{P_J} \cdot S_J \cdot h_{JB}, \tag{5}$$

$$y_E = \sqrt{P_A} h_{AE} \cdot S_A \cdot d_{AE}^{-\alpha_1/2} + w_E + \sqrt{P_J} \cdot S_J \cdot h_{JE}, \tag{6}$$

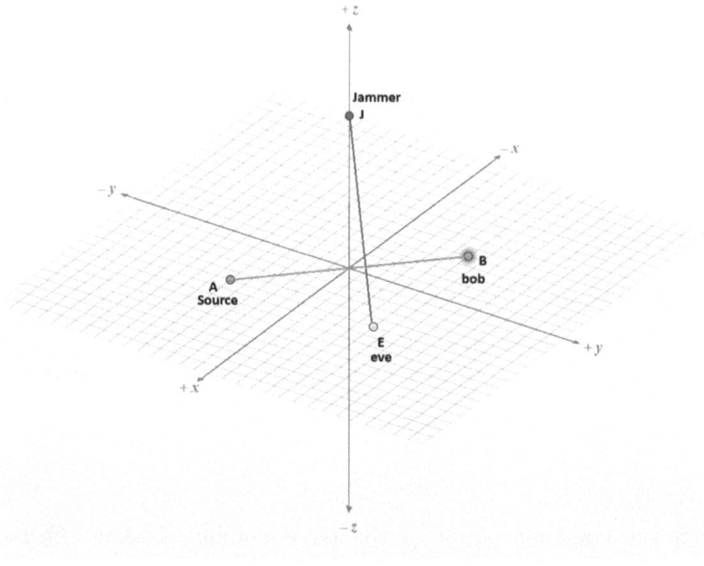

Fig. 1. Single Jammer Setup

where h_{AB} is the fading component channel coefficient between Alice and Bob which follows a complex Gaussian distribution ($\mathcal{CN}(0, 1)$), while $|h_{AB}|$ and $|h_{AE}|$

are Rayleigh distributed, with their squares exponentially distributed with a mean of 1. W_B and W_E represent Additive White Gaussian Noise (AWGN), following a complex Gaussian distribution with variance σ^2. P_J signifies the jamming power deployed by the UAV. S_J represents pseudo-random artificial noise introduced to disrupt Eve's eavesdropping attempts such that $\mathbb{E}|S_J|^2 = 1$. Further, P_A is the source transmit power and S_A denotes the symbol transmitted by the Alice $\mathbb{E}|S_A|^2 = 1$. h_{JB} and h_{JE} represent the channel coefficients between Bob and the UAV, respectively. The corresponding gains $|h_{JB}|^2$ and $|h_{JE}|^2$ is the inverse of path loss and is calculated according to (3).

$$|h_{JB}|^2 = \frac{1}{(d_{JB})^{\alpha}} \frac{1}{(P_{LoS}\,\eta_{LoS} + P_{NLoS}\,\eta_{NLoS})} \tag{7}$$

$$|h_{JE}|^2 = \frac{1}{(d_{JE})^{\alpha}} \frac{1}{(P_{LoS}\,\eta_{LoS} + P_{NLoS}\,\eta_{NLoS})} \tag{8}$$

Thus the interference power at bob and eve is given by

$$I_{JB} = P_J|h_{JB}|^2; \quad I_{JE} = P_J|h_{JE}|^2 \tag{9}$$

Thus, we have the SINRs at bob and Eve

$$\gamma_B = \frac{|h_{AB}|^2 \cdot (d_{AB}^{-\alpha_1}) \cdot P_A}{\sigma^2 + I_{JB}} = \frac{\bar{\gamma}_{AB}|h_{AB}|^2}{1 + \bar{\gamma}_{JB}} \tag{10}$$

$$\gamma_E = \frac{|h_{AE}|^2 \cdot (d_{AE}^{-\alpha_1}) \cdot P_A}{\sigma^2 + I_{JE}} = \frac{\bar{\gamma}_{AE}|h_{AE}|^2}{1 + \bar{\gamma}_{JE}} \tag{11}$$

Here we define the average SNRs at Bob and Eve given by $\bar{\gamma}_{AB} = \frac{P_A(d_{AB}^{-\alpha})}{\sigma^2}$ and $\bar{\gamma}_{AE} = \frac{P_A(d_{AE}^{-\alpha})}{\sigma^2}$, respectively. Similarly, average SNR caused due to jammer is efined as $\bar{\gamma}_{JB} = \frac{I_{JB}}{\sigma^2}$, $\bar{\gamma}_{JE} = \frac{I_{JE}}{\sigma^2}$, respectively.

3.2 Multi Jammer: A Virtual MIMO Setup Employing Zero Forcing

As illustrated in Fig. 2, the two UAV jammers are used which create a 2×3 virtual MIMO setup. Meaning that three transmitter two jammers along with Alice and two users Bob and Eve. Now, we aim to nullify the effect of jmmer interference at Bob by employing zero forcing (ZF) precoding.

Let $g_{AB} = h_{AB} \cdot d_{AB}^{-\frac{\alpha}{2}}$, $g_{AE} = h_{AE} \cdot d_{AE}^{-\frac{\alpha}{2}}$, then the channel matrix is given by:

$$H = \begin{bmatrix} h_{AB} \cdot d_{AB}^{-\alpha/2} \; h_{J_1B} \; h_{J_2B} \\ h_{AE} \cdot d_{AE}^{-\alpha/2} \; h_{J_1E} \; h_{J_2E} \end{bmatrix} = \begin{bmatrix} g_{AB} \; h_{J_1B} \; h_{J_2B} \\ g_{AE} \; h_{J_1E} \; h_{J_2E} \end{bmatrix} \tag{12}$$

Let $S = [S_A, S_{J_1}, S_{J_2}]^T$ denote the message vector, where S_{J_1} and S_{J_2} represent pseudo-random artificial noise, and S_A represents the symbol sent by Alice. Now, let $y = [y_B, y_E]^T$ represent the received vector and $W = [w_B, w_E]^T$ denotes

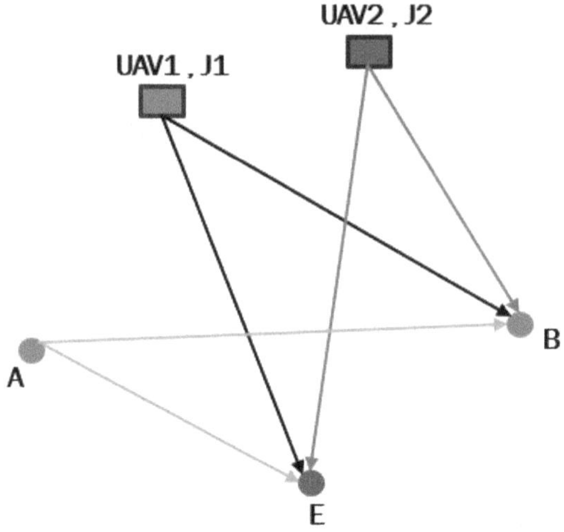

Fig. 2. Multi jammer setup

the noise vector consisting noise at Bob and Eve. Let x be the transmitted pre-coded vector, then y can be written as:

$$y = Hx + W \tag{13}$$

$$y = HQ \cdot \text{diag}(\sqrt{P_{A_1}}, \sqrt{P_{J_1}}, \sqrt{P_{J_2}})S + W \tag{14}$$

here Q is a precoding matrix designed to eliminate the jamming effect on Bob.

Zero Forcing. Zero-Forcing (ZF) is a spatial signal processing technique used for precoding in multi-user MIMO wireless communication systems. It enables a multi-antenna transmitter to completely cancel out multiuser interference. This requires perfect channel state information (CSI) at the transmitter. In the ZF approach, the precoding matrix Q is defined as the pseudo-inverse of the channel matrix, effectively eliminating interference across users in the system. Formally, the ZF precoding matrix Q is given by:

$$Q = \text{diag}(1, h_{J_2B}, -h_{J_1B}) \tag{15}$$

By replacing Q in (14) with $P_{J_1} = P_{J_2} = \frac{P_J}{2}$, we get:

$$y_m = \begin{bmatrix} y_B \\ y_E \end{bmatrix} = \begin{bmatrix} g_{AB}\sqrt{P_A}S_A + w_B \\ g_{AE}\sqrt{P_A}S_A + (h_{J_1E}h_{J_2B} - h_{J_2E}h_{J_1B})\sqrt{\frac{P_J}{2}}S_J + w_E \end{bmatrix} \tag{16}$$

Here, $g_{AB}\sqrt{P_A}S_A$ is the signal part, and w_B is noise part at Bob. Clearly, the interference at Bob due to the jammer is zero.

SNR at Bob and SINR at Eve. At Bob the received SNR is given by:

$$\gamma_B = |h_{AB}|^2 \cdot \frac{P_A}{\sigma^2} \cdot d_{AB}^{-\alpha_1} \tag{17}$$

Note that since interference ay the Bob is zero, thus we call it SNR and not SINR. Next, the received SINR at Eve is given by:

$$\gamma_E = |h_{AE}|^2 \cdot \frac{\frac{P_A}{\sigma^2} \cdot d_{AE}^{-\alpha_1}}{1 + \underbrace{0.5\,|h_{J_1E}h_{J_2B} - h_{J_2E}h_{J_1B}|^2 \cdot \left(\frac{P_J}{\sigma^2}\right)}_{\bar{\gamma}_{JE}}} = \frac{\bar{\gamma}_{AE}|h_{AE}|^2}{1 + 0.5\bar{\gamma}_{JE}} \tag{18}$$

We calculated $|h_{J_1B}|^2$, $|h_{J_2B}|^2$, $|h_{J_1E}|^2$, and $|h_{J_2E}|^2$ as $\frac{1}{L(x,y)}$, which is discussed in (3).

By defining $\bar{\gamma}_{J_1B} = P_J|h_{J_1B}|^2/\sigma^2, \bar{\gamma}_{J_2B} = P_J|h_{J_2B}|^2/\sigma^2, \bar{\gamma}_{J_1E} = P_J|h_{J_1E}|^2/\sigma^2$ and $\bar{\gamma}_{J_2E} = P_J|h_{J_2E}|^2/\sigma^2$, we can rewrite $\bar{\gamma}_{JE}$ as

$$\bar{\gamma}_{JE} = \left(\left|\sqrt{\bar{\gamma}_{J1B}} \cdot \sqrt{\bar{\gamma}_{J2E}} - \sqrt{\bar{\gamma}_{J1E}} \cdot \sqrt{\bar{\gamma}_{J2B}}\right|^2\right) \tag{19}$$

3.3 Secrecy Capacity and Secrecy Outage Probability

Secrecy Capacity. Secrecy capacity is calculated by,

$$Cs = \max\left[0, \log_2\left(\frac{1+\gamma_B}{1+\gamma_E}\right)\right] \quad [\text{b/sec/Hz}]. \tag{20}$$

SOP Derivation for a Single Jammer. The secrecy outage probability (SOP) is expressed as:

$$P_{\text{out}}(R_{\text{th}}) = P(C_S < R_{\text{th}}) \tag{21}$$

The SOP for the UAV-assisted jamming network is derived and is defined as the likelihood that the secrecy capacity is less than a given threshold $R_{\text{th}} > 0$. By substituting (10), (11), and (20) into (21), the SOP becomes:

$$P_{\text{out}}(R_{\text{th}}) = P(C_S < R_{\text{th}}) = P\left(\frac{1+\gamma_B}{1+\gamma_E} < 2^{R_{\text{th}}}\right) \tag{22}$$

$$= P\left(h_{AB} < \frac{(2^{R_{\text{th}}}-1)(1+\bar{\gamma}_{JB})}{\bar{\gamma}_{AB}} + \frac{2^{R_{\text{th}}}\bar{\gamma}_{AE}h_{AE}(1+\bar{\gamma}_{JB})}{(1+\bar{\gamma}_{JE})\bar{\gamma}_{AB}}\right) \tag{23}$$

$$= 1 - e^{-\frac{(2^{R_{\text{th}}}-1)(1+\bar{\gamma}_{JB})}{\bar{\gamma}_{AB}}}\mathbb{E}_{h_{AE}}\left[e^{-\frac{2^{R_{\text{th}}}\bar{\gamma}_{AE}h_{AE}(1+\bar{\gamma}_{JB})}{(1+\bar{\gamma}_{JE})\bar{\gamma}_{AB}}}\right] \tag{24}$$

$$= 1 - e^{\frac{(1+\bar{\gamma}_{JB})(1-2^{R_{\text{th}}})}{\bar{\gamma}_{AB}}} \times \int_0^\infty e^{-\frac{2^{R_{\text{th}}}\bar{\gamma}_{AE}h_{AE}(1+\bar{\gamma}_{JB})}{(1+\bar{\gamma}_{JE})\bar{\gamma}_{AB}}} e^{-h_{AE}}\, d(h_{AE}) \tag{25}$$

$$= 1 - \frac{1}{1 + \frac{2^{R_{\text{th}}}(1+\bar{\gamma}_{JB})\bar{\gamma}_{AE}}{(1+\bar{\gamma}_{JE})\bar{\gamma}_{AB}}} e^{-\frac{(1+\bar{\gamma}_{JB})(2^{R_{\text{th}}}-1)}{\bar{\gamma}_{AB}}} \tag{26}$$

For a clearer understanding, we analyze the scenario where $\bar{\gamma}_{AB} \approx \bar{\gamma}_{AE}$, $\bar{\gamma}_{JE} \gg \bar{\gamma}_{JB}$, and $\bar{\gamma}_{AB} \gg \bar{\gamma}_{JB}$. This leads to an asymptotic expression for the SOP:

$$P_{\text{out}}^{\infty}(R_{\text{th}}) \approx 1 - e^{-\frac{(1+\bar{\gamma}_{JB})(2^{R_{\text{th}}}-1)}{\bar{\gamma}_{AB}}} \tag{27}$$

$$\approx \frac{(2^{R_{\text{th}}} - 1)(1 + \bar{\gamma}_{JB})}{\bar{\gamma}_{AB}} \tag{28}$$

SOP Derivation for Multi Jammer Setup: A Virtual MIMO Setup Employing Zero Forcing. Similarly, we derive the Secrecy Outage SOP for a multi-jammer setup in a virtual MIMO employing zero forcing. The SOP is given as

$$P_{\text{out}}(R_{\text{th}}) = P(C_S < R_{\text{th}}) = P\left(\frac{1+\gamma_B}{1+\gamma_E} < 2^{R_{\text{th}}}\right) \tag{29}$$

$$= P\left(h_{AB} < \frac{(2^{R_{\text{th}}}-1)}{\bar{\gamma}_{AB}} + \frac{2^{R_{\text{th}}}\bar{\gamma}_{AE}h_{AE}}{(1+\bar{\gamma}_{JE})\bar{\gamma}_{AB}}\right) \tag{30}$$

$$= 1 - e^{-\frac{(2^{R_{\text{th}}}-1)}{\bar{\gamma}_{AB}}} E_{AE}\, e^{-\frac{2^{R_{\text{th}}}\bar{\gamma}_{AE}h_{AE}}{(1+\bar{\gamma}_{JE})\bar{\gamma}_{AB}}} \tag{31}$$

$$= 1 - e^{(1-2^{R_{\text{th}}})/\bar{\gamma}_{AB}} \int_0^{\infty} e^{-\frac{2^{R_{\text{th}}}\bar{\gamma}_{AE}h_{AE}}{(1+\bar{\gamma}_{JE})\bar{\gamma}_{AB}}} e^{-h_{AE}}\, d(h_{AE}) \tag{32}$$

$$= 1 - \frac{1}{1 + \frac{2^{R_{\text{th}}}\bar{\gamma}_{AE}}{(1+\bar{\gamma}_{JE})\bar{\gamma}_{AB}}} e^{-\frac{(2^{R_{\text{th}}}-1)}{\bar{\gamma}_{AB}}}. \tag{33}$$

4 Results and Discussions

In this section, we validate the findings discussed in the previous sections through computer simulations. Specifically, we discuss three scenarios a) there is no UAV jammer (which is also baseline) b) Only a single UAV jammer J1 is present c) we introduce one more jammer J2 and create a virtual MIMO employing ZF.

In Fig. 3, the analytical outage probability (P_{out}) is shown as a function of SNR ($\bar{\gamma}_{j_1E}$) in dB. We set the other parameters as $\bar{\gamma}_{\text{AE}} = 10$, $\bar{\gamma}_{\text{AB}} = 30$ dB, $\bar{\gamma}_{\text{J2B}} = 5$ dB, $\bar{\gamma}_{\text{J2E}} = 5$ dB, and $\bar{\gamma}_{\text{J1B}} = 5$ dB. The simulation perfecttly matches with the analytical results in (26) and (33). The results show the performance of three different scenarios: Single Jammer (jammer J1 only), No Jammer (baseline curve), and virtual MIMO with ZF (Jammer J1 and J2 both with equally divided power). For the ZF scenario, after 5 dB, we start getting improvement as compared to the no jammer scenario. Specifically, P_{out} significantly decreases from 0.65 at 5 dB to 0.05 at saturation. In contrast, the single Jammer starts showing improvement after 7 dB. Specifically, P_{out} significantly decreases from 0.65 at 5 dB and saturates at 0.2. The no jammer scenario, indicated by the constant P_{out} value of 0.65, remains unchanged across all jammer power levels.

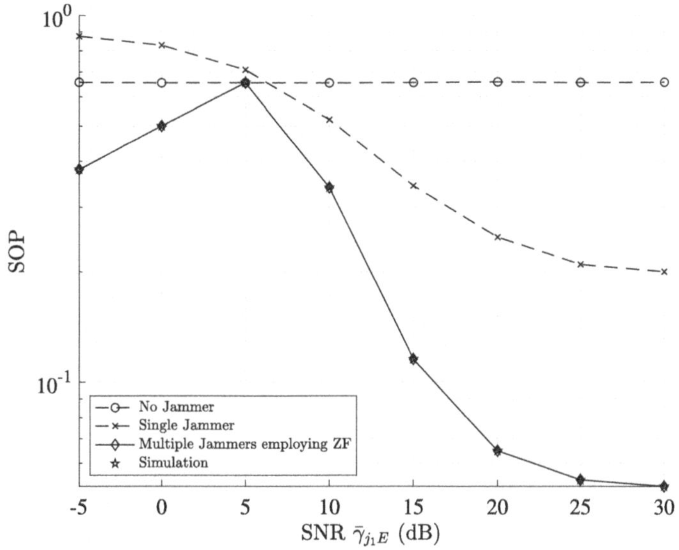

Fig. 3. SOP for different jamming strategies as a function of average Jammer SNR $\bar{\gamma}_{J_1E}$ (dB) when the $R_{th} = 0.5$ b/sec/Hz.

This comparison highlights the effectiveness of the ZF approach, especially at higher jammer power levels, where it provides a notable reduction in outage probability. Specifically, at 20 dB, the ZF scenario achieves an P_{out} of 0.065, significantly lower than the 0.25 observed in the single Jammer scenario and the 0.66 in the no Jammer scenario.

Figure 4 demonstrates the ASC for different jamming strategies as a function of the SNR $\bar{\gamma}_{j_1E}$. The ASC is evaluated for three scenarios: no jammer, single jammer, and multiple jammers employing ZF. We set the same parameters, i.e., $\bar{\gamma}_{\mathrm{AE}} = 10$, $\bar{\gamma}_{\mathrm{AB}} = 30$ dB, $\bar{\gamma}_{\mathrm{J2B}} = 5$ dB, $\bar{\gamma}_{\mathrm{J2E}} = 5$ dB, and $\bar{\gamma}_{\mathrm{J1B}} = 5$ dB. The results reveal significant improvements in the ASC with the introduction of jammers after a certain threshold. For the no jammer scenario, the ASC values remain low at approximately 0.61 b/s/Hz. Introducing a single jammer (J1) significantly enhances the ASC, reaching up to 1.28 b/s/Hz (when saturated). The most substantial improvement is observed with multiple jammers (J1 along with J2) employing the ZF technique, where the ASC climbs to as high as 2.6 b/s/Hz. Specifically, at $\bar{\gamma}_{j_1E} = 20$ dB, a detailed comparison provides significant insights into the effectiveness of each strategy. The ASC for the no jammer scenario is approximately 0.6 b/s/Hz. In contrast, the ASC for the single jammer scenario shows a substantial improvement, reaching around 1.17 b/s/Hz. The most notable improvement is observed with the multiple jammers employing ZF, where the ASC reaches around 2.5 b/s/Hz. These results clearly demonstrate that using multiple jammers with ZF not only enhances the secrecy capacity but also provides a more robust solution against potential eavesdropping, significantly outperforming the other scenarios.

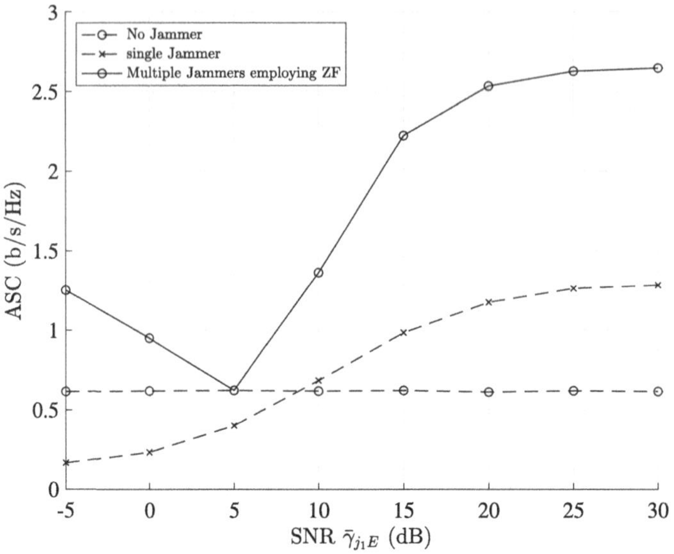

Fig. 4. Average secrecy capacity (ASC) for different jamming strategies as a function of SNR $\bar{\gamma}_{j_1E}$ (dB).

4.1 Conclusion

In this research, we explored a secrecy model involving a UAV-based jammer, which includes a transmitter, a legitimate receiver and an eavesdropper operating within a ground network. The use of UAVs as jammers provides considerable benefits over conventional ground-based jammers, owing to their ability to be deployed flexibly and reposition dynamically. Our findings showed that the implementation of a virtual MIMO based Zero-Forcing (ZF) UAV jammer substantially improves secrecy performance when compared to scenarios without a jammer and single jammer scenario.

References

1. Wyner, A.D.: The wire-tap channel. Bell Syst. Tech. J. **54**, 1355–1387 (1975)
2. Cumanan, K., Ding, Z., Sharif, B., Tian, G., Leung, K.K.: Secrecy rate optimizations for a MIMO secrecy channel with a multiple-antenna eavesdropper. IEEE Trans. Veh. Technol. **63**(4), 1678–1690 (2014)
3. Liu, R., Liu, T., Poor, H.V., Shamai, S.: Multiple-input multiple output gaussian broadcast channels with common and confidential messages. IEEE Trans. Inf. Theory **56**(9), 4215–4227 (2010)
4. Wang, H., Luo, M., Yin, Q., Xia, X.: Hybrid cooperative beamforming and jamming for physical-layer security of two-way relay networks. IEEE Trans. Inf. Forensics Secur. **8**(12), (2013)
5. Goel, S., Negi, R.: Guaranteeing secrecy using artificial noise. IEEE Trans. Wireless Commun. **7**(6), 2180–2189 (2008)

6. Vilela, J.P., Bloch, M., Barros, J., McLaughlin, S.W.: Friendly jamming for wireless secrecy. In: Proceedings of IEEE International Conference Communications, Cape Town, South Africa, May2010
7. Hui, H., Swindlehurst, A.L., Li, G., Liang, J.: Secure relay and jammer selection for physical layer security. IEEE Signal Process. Lett. **22**(8), 1147–1151 (2015)
8. Liu, J., Liu, Z., Zeng, Y., Ma, J.: cooperative jammer placement for physical layer security enhancement. In: IEEE Network, vol. 30, no. 6, pp. 56–61, November–December 2016
9. Al-Hourani, A., Kandeepan, S., Lardner, S.: Optimal LAP altitude for maximum coverage. IEEE Wirel. Commun. Lett. **3**(6), 569–572 (2014)
10. Zhou, Y., Yeoh, P.L., Chen, H., Li, Y., Hardjawana, W., Vucetic, B.: Secrecy outage probability and jamming coverage of UAV-enabled friendly jammer. In: 11th International Conference on Signal Processing and Communication Systems (ICSPCS)
11. Flores Cabezas, X.A., Moya Osorio, D.P., Latva-aho, M.: Distributed UAV-enabled zero-forcing cooperative jamming scheme for safeguarding future wireless networks. Centre for Wireless Communications, University of Oulu, Finland

Real Time Early Curve Accident Warning System

Adithya Mekkattu Ajith, Anit Peter, Anitta P. Markose, Tania Kalloor Hendry(✉), and Jency Thomas

Computer Science and Engineering Department, Muthoot Institute of Technology and Science, Kochi, India
taniahendry02@gmail.com, jencythomas@mgits.ac.in
https://mgmits.ac.in

Abstract. The system presents a Real-Time Early Curve Accident Warning System that is specifically designed to address the safety issues associated with hairpin bends on roads. Additionally, the system labels vehicles as HMV or LMV for better warning of opposite direction vehicles. It is done through the use of cameras set at strategic positions and specially calibrated to record live footages of approaching cars round these critical corners. Furthermore, using modern deep learning algorithms, this recorded footage is continuously processed in real-time to obtain vehicle detection with immediate speed computation. If they are detected passing through these bends at speeds that could be dangerous, a real-time alert will be triggered soon enough by the system hence informing drivers instantly and reducing chances of accidents by a great margin. By fusing high-tech solutions together and introducing some powerful real-time monitoring capabilities, it seeks to solve basic road transport safety concerns within contemporary settings. With an aim towards creating safer roads and communities, the project anticipates playing its part in improving public safety standards across society as a whole.

Keywords: Real-Time Early Curve Accident Warning System · Hairpin bends · HMV · LMV · deep learning algorithms · vehicle detection · speed computation · road transport safety · public safety

1 Introduction

Road crashes, especially at acute curves and tight bends are a major global road safety challenge. Such dangerous parts of the road however become highly treacherous to both drivers and pedestrians in instances such as bad weather or areas that may be prone to landslides. However, despite various initiatives aimed at reducing these risks, there is still an ongoing issue with curve accidents due to poor visibility, speeding past them and other dangerous driving behaviors.

Recent data shows that this problem has worsened over the years as evident by increased number of curve accidents. The World Health Organization (WHO)

D. Gupta et al. (Eds.): PCEMS 2024, CCIS 2491, pp. 254–267, 2025.
https://doi.org/10.1007/978-3-031-90577-3_23

claims that globally, road traffic accidents are among the leading causes of deaths, many of which happen on curvaceous roads. There is a need for urgent interventions to improve safety on dangerous corners while reducing the incidence and severity of accidents occurring.

This paper, therefore, introduces a Comprehensive Real-Time Early Curve Accident Warning System. The innovative mechanism at the heart of this system exploits advanced technology to improve road safety at risky corners. It seeks to track and analyze possible hazards in sharp cornering and hairpin bends with specific emphasis on real-time detection and warning systems.

In this concept, advanced deep learning algorithms are employed in the YOLOv8 framework which is famous for its speed in object detection under real time conditions. Thus by leveraging on YOLOv8 capabilities, the goal of the system will be to achieve high levels of accuracy when it comes to both identifying vehicles on roads and precisely measuring their speeds.

Moreover, the system is also being designed in a manner that will facilitate classification between Light Motor Vehicles (LMV) and Heavy Motor Vehicles (HMV). This classification is meant to enable focused warnings so that all vehicles approaching dangerous turns can receive the necessary alerts thereby ensuring overall road safety improvement.

The processing module of the system has two main aims: one is to identify vehicles in terms of accuracy and the other is to measure their velocities as they move. These functionalities are very important in generating timely and accurate alerts, providing motorists with necessary safety options and mitigating potential dangers related to dangerous corners.

As development moves forward, the aim of this system will be to help integrate into existing infrastructure seamlessly hence facilitating its wide deployment and accessibility. Also, there are efforts towards formulating protocols for working together with relevant authorities like police stations and hospitals so as to ensure immediate response during emergency cases. This innovative system is expected to greatly improve road safety by reducing traffic accidents on turns saving lives and minimizing property loss.

2 Literature Survey

The research paper [1], addresses the imperative need for advanced car crash detection systems amidst the rising frequency of motor vehicle accidents. It introduces a groundbreaking paradigm shift by advocating an innovative strategy that amalgamates video and audio data collected from dashboard cameras, departing from the limitations of conventional single-modal approaches. This revolutionary methodology is underpinned by the belief that leveraging multimodal data sources, each offering unique perspectives, can substantially enhance the accuracy and efficacy of crash detection algorithms. To operationalize this novel approach, the researchers employ cuttingedge deep learning methodologies, notably harnessing gated recurrent units (GRU) and convolutional neural networks (CNN), adept at extracting intricate features from both visual and auditory data streams. A key distinguishing aspect of this research lies in its employment of ensemble learning–a technique amalgamating outputs from multiple classifiers trained on diverse data modalities. This ensemble technique, facilitated through a weighted average ensemble approach, aims to leverage the strengths of individual data modalities, culminating in a more robust and accurate crash detection system. Rigorous experimentation, conducted using real-world car accident YouTube clips, unequivocally demonstrates the superiority of the proposed ensemble-based system over conventional singlemodal classifiers, showcasing remarkable advancements in accurately detecting and identifying car crashes.

Groundbreaking methodologies for automatic car accident detection by leveraging Cooperative Vehicle Infrastructure Systems (CVIS) alongside machine vision technologies are introduced in [2]. With a focus on enhancing response times and overall traffic safety, these methods prioritize swift and accurate accident detection through specialized image datasets like CAD-CVIS and advanced deep neural network models such as YOLO-CA. Experimental evaluations highlight YOLO-CA's exceptional performance in rapidly and precisely identifying accidents, demonstrating its superiority over alternative models. The implications of these advancements are far-reaching, promising to reduce rescue operation durations, enhance safety by alerting nearby drivers, integrate with intelligent transportation systems for real-time monitoring, and aid infrastructure planning through continuous accident detection. Additionally, the establishment of benchmark datasets like CAD-CVIS propels further research, fostering ongoing advancements in accident detection algorithms and models. In essence, these papers underscore the transformative potential of innovative methodologies in automatic car accident detection, emphasizing their role in improving response efficiency, enhancing safety, and driving continual progress in research and technology within the field.

Creating a sophisticated road recognition and warning system tailor-made for vehicles is focused by [3]. By utilizing cameras mounted on vehicles, the system captures and analyzes crucial road data, such as lane markings and obstacles, using computer vision techniques. Its main goal is to interpret this data quickly and take proactive measures, like issuing warnings or implementing preventive strategies, to prevent accidents and promote safer driving practices. While the paper doesn't dive deep into the technical details of the binocular vision system it relies on, it highlights its complexity and functionality. Additionally, the paper discusses essential mathematical models, like safety distance and synchronous motor operations, which form the backbone of the system, ensuring precise responses and enhancing overall effectiveness. Simulation experiments using Matlab/Simulink validate the system's design and functionality across various driving scenarios, assessing its adaptability and responsiveness. However, the paper could benefit from clearer explanation of the evaluation metrics used to measure the system's performance. Nonetheless, the primary focus remains on advancing road safety through an integrated road recognition system, powered by binocular vision technology, capable of addressing risks in diverse real-world driving situations.

Unlike conventional systems that react to immediate dangers, the method discussed in [4] aims to predict risks, issuing early warnings to drivers for proactive measures. Central to this system is the Fusion-Residual Predictive Network (FRPN), a sophisticated framework amalgamating multi-scale features and advanced learning methodologies. Operating in stages, FRPN encodes and predicts images, refining them with a specialized Generative Adversarial Network (GAN) model for clearer hazard identification. The system employs meticulous object detection and tracking, ensuring accurate monitoring within the vehicle's field of view. Crucially, it incorporates a detailed risk assessment mechanism, evaluating hazards based on size changes, center of gravity movements, and angle variations to generate a risk index for timely warnings. Evaluation across diverse datasets confirms the system's superior predictive accuracy and early warning capability compared to existing methods. In essence, this paper introduces a groundbreaking proactive warning system that aims to significantly enhance traffic safety by empowering drivers with early alerts and preventive measures against potential accidents.

A novel object detection algorithm aimed at improving accuracy in complex scenarios with dense target distribution and size variations is introduced in [5]. By replacing YOLOv3's residual structure with FCN-DenseNet, FF-YOLO enhances feature extraction for better object recognition. It adds a fourth detection scale to learn shallower location information and integrates an SPP module to fuse local features, improving detection effectiveness. Termed YOLOv4-5D, the algorithm surpasses YOLOv4 in accuracy, particularly for small objects like traffic signs, while maintaining real-time detection speed, making it ideal for applications like autonomous driving. With enhanced performance across diverse object sizes, FF-YOLO/YOLOv4-5D offers promise for real-world scenarios requiring accurate and rapid object detection, demonstrating significant advancements in object detection technology.

YOLOv4-5D is a cutting-edge object detection framework tailored for autonomous driving. It focuses on enhancing detection accuracy while maintaining real-time performance. By optimizing the backbone network CSPDarknet53 and introducing deformable convolution, YOLOv4-5D improves feature extraction for objects with geometric variations. The PAN++ feature fusion module efficiently combines low-level and highlevel features, enhancing detection across different scales. Notably, the "5D" redesign incorporates a time dimension, improving accuracy in video scenarios. An optimized network pruning algorithm reduces computational resources while maintaining performance. YOLOv4-5D outperforms YOLOv4, achieving a 3.1% increase in mean average precision (mAP) on the BDD test set, particularly for small objects like traffic signs. Maintaining real-time speed, the framework introduced in [6] excels in detecting both large and small objects, making it a promising tool for enhancing autonomous vehicle safety and efficiency.

YOLOv4_AF, an advanced model tailored for vehicle detection and classification in intelligent transportation systems is introduced in [7]. Unlike traditional models, YOLOv4_AF utilizes a novel attention mechanism called Convolutional Block Attention Module (CBAM), allowing it to focus on important image parts while filtering out distractions. This enhances resource allocation for more effective processing. The model's Feature Pyramid Network (FPN) is also improved, leading to better object representation and detection accuracy. Empirical results show YOLOv4_AF outperforms existing models like Faster R-CNN and EfficientDet in accuracy metrics, showcasing its potential for real-world applications. Additionally, its versatility extends beyond vehicle detection, making it adaptable for other object detection tasks. YOLOv4_AF's advancements make it a promising solution for enhancing visionbased object detection systems, especially in transportation and beyond.

The research [8], introduces T-YOLO, a modified version of YOLO-v5 aimed at detecting small and tiny objects, particularly for parking occupancy monitoring. T-YOLO focuses on achieving realtime detection with high precision, crucial for timely responses in parking space management. It incorporates a Multi-Scale Module (MSM) that processes input images at multiple scales, enhancing detection of tiny objects. Additionally, attention modules like Spatial Attention Module (SAM) and Channel Attention Module (CAM) improve precision by focusing on informative parts of the image. T-YOLO maintains a real-time speed of 30 frames per second (fps) and is lightweight, suitable for embedded systems with limited resources. Its multi-scale mechanism enables it to detect objects of various sizes effectively, making it promising for real-world applications requiring accurate and efficient detection of small objects in parking scenarios.

Researchers developed a semi-supervised fuzzy CMean (SSFCM) clustering method for vehicle classification in [9], aiming to enhance accuracy and address imbalanced datasets. It combines supervised and unsupervised data to extract hidden information during clustering, reducing sensitivity to initial centroids. Compared to traditional Fuzzy C-Mean (FCM), SSFCM correlates better with expert classification, achieving approximately 86accuracy using only 10 outperforms other semi-supervised algorithms like k-nearest neighbors and random forest. SSFCM's advantages include accurate vehicle classification, enhanced by semi-supervised learning, and reduced reliance on random centroids. It improves with more labeled data. However, the paper suggests better methods for generating consistent supervised labels and further research on resampling techniques for multi-class problems. Overall, SSFCM offers promising results for automated vehicle classification and analysis in large databases.

A hybrid convolutional neural network (CNN) for vehicle classification in IoT systems, combining data from CCTV and fiber Bragg grating (FBG) sensors is proposed in [10]. This hybrid CNN achieves high detection levels and in-class accuracy of 90%–97%, surpassing conventional CNNs like AlexNet and ResNet. By integrating image and sensor data, it offers a comprehensive representation of vehicles and improves system robustness with intrinsic backup verification. The approach enhances traffic monitoring, management, and resource optimization in IoT applications. Challenges include integrating diverse data sources and addressing variations in image quality and dataset size, which can affect categorization results. Overall, the hybrid CNN shows promise for enhancing traffic control efficiency and enabling automatic vehicle counting and separation in IoT-based traffic systems.

In [11], a super-learner ensemble method for vehicle-type classification in intelligent transportation systems, aiming to improve accuracy and discrimination among vehicle classes is suggested. It combines three deep neural networks–ResNet50, Xception, and DenseNet- using adaptive fusion weights learned during training. Evaluated on challenging datasets, the method achieves high accuracy and comparable performance to state-of-the-art methods, with data augmentation aiding generalization. The approach is praised for its simplicity, robustness, and effectiveness in handling complex datasets without relying on handcrafted features. Challenges include distinguishing similar vehicles and handling unbalanced datasets and false detections. The study calls for future research to address these challenges and enhance performance further. Overall, the paper offers valuable insights into ensemble learning's application in vehicle-type classification, showcasing its potential to advance intelligent transportation systems.

A semisupervised vehicle classification scheme using an ensemble broad learning system (BLS), aiming to enhance accuracy and efficiency is introduced in the article [12]. By combining semisupervised learning and BLS ensemble, it trains base classifiers iteratively, improving discrimination. A dynamic ensemble structure selects high-learning base classifiers, yielding faster training times and high accuracy. Experiments on real datasets show superior performance over single classifiers and mainstream methods like CNN and SVM, particularly in handling imbalanced data. Advantages include handling large-scale datasets with limited labeled samples and faster training. However, parameter tuning is required, and applicability may vary. Overall, the scheme offers a promising solution for efficient and accurate vehicle classification.

The article [13], introduces a video-based system for measuring vehicle speed on urban roads, comprising a motion detector, license plate detector, and feature tracking module. It accurately identifies moving vehicles, locates license plates, and tracks distinctive features to determine vehicle speed. Evaluation on various weather conditions yielded an average speed measurement error of - 0.5 km/h, meeting regulatory standards in over 96% of cases. The system excels in challenging scenarios, outperforming existing methods. It offers promise for traffic monitoring and enforcement applications, providing a nonintrusive solution for urban road speed measurement.

A traffic monitoring system that combines passive infrared sensors (PIRs) and ultrasonic rangefinders for real-time vehicle detection, classification, and speed estimation utilizes dynamic Bayesian networks to fuse sensor data and achieve high accuracy. Equipped with low-power computational devices, traffic sensor nodes transmit data to a central server for processing, reducing energy consumption. The chosen sensors offer advantages such as cost-effectiveness, accuracy, and weather resistance. By leveraging wireless sensor networks, the system in [14] addresses deployment challenges. Overall, it presents a promising solution for improving traffic management and safety.

A novel method for estimating vehicle speed using WiFi signals, aiming to address traffic congestion and accidents is addressed in [15]. By analyzing how moving vehicles affect surrounding WiFi signals, the authors develop a model to estimate speed based on signal characteristics. They conduct experiments with WiFi devices, demonstrating accuracy within 2.6 km/h, meeting speedometer regulations. Compared to traditional methods like radars and cameras, WiFi-based sensing offers seamless monitoring for large-scale scenarios. This approach could enable cost-effective and efficient vehicle speed estimation, potentially revolutionizing traffic management with widespread deployment.

A practical solution for traffic management is offered by an algorithm for estimating mean traffic speed using uncalibrated cameras. By analyzing sequential images and tracking moving vehicles, the algorithm in [16] estimates speed based on geometric relationships and common-sense assumptions. Tested with images from the Washington State Department of Transportation, the algorithm shows promising accuracy, despite challenges like lighting conditions and shadows. Compared to traditional methods like inductance loops, it offers similar speed estimates with minor discrepancies. This algorithm presents a valuable tool for traffic monitoring and traveler information, leveraging existing camera infrastructure for efficient speed estimation.

3 Proposed System

To make hairpin bends safer, we propose an Early Curve Accident Warning System (ECAWS) that works in real-time and includes a speed warning feature. This innovative technology uses cameras placed strategically at hairpin corners to continuously record live road videos. These videos are then analyzed by a powerful module supported by deep learning algorithms to detect incoming vehicles and accurately measure their speed. If there's a potential risk or accident, the system alerts drivers to increase their awareness. Moreover, if an accident occurs, the system automatically alerts nearby police stations and hospitals. This not only provides drivers with crucial safety information but also ensures a coordinated response among authorities, making navigation through tricky corners safer for drivers.

The system architecture is designed to process and analyze data in real-time. Cameras placed at hairpin bends capture video footage, which is then sent to the processing module for analysis. Using deep learning algorithms, the module detects vehicles, estimates their speed, and identifies potential accidents. If a vehicle is speeding or an accident is detected, the system triggers alerts for both drivers and authorities. Additionally, the system stores video footage and data in a database to improve the algorithms over time, identifying patterns and trends in accidents.

The Object Detection Module is essential for identifying vehicles and detecting accidents using advanced computer vision methods. This module interprets traffic flow at sharp bends, providing a basis for safety evaluations. With real-time accuracy in locating and measuring objects, this module allows for proactive approaches to road safety.

The Speed Calculation Module determines vehicles' approach speeds to hairpin bends. By studying recognized cars' travel patterns, this module provides accurate estimates of their speeds. Identifying vehicles approaching curves at unsafe speeds in real-time helps prevent accidents.

The Warning System Module responds quickly to potential safety hazards identified by the Object Detection and Speed Calculation Modules. It triggers warnings when vehicles approach curves at unsafe speeds or when there are vehicles from both sides on the curve. Timely warnings provide drivers with crucial information, enabling them to adjust their driving and navigate through hairpin turns safely.

By integrating these modules into a cohesive system, the Early Curve Accident Warning System aims to significantly improve safety on hairpin bends, reducing the risk of accidents and enhancing overall road safety.

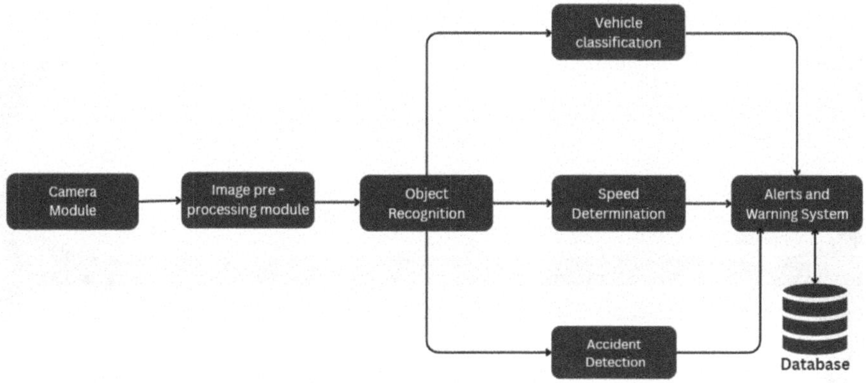

Fig. 1. System Architecture

3.1 YOLOv8

YOLOv8, developed by Ultralytics, stands as a cutting-edge computer vision model, pushing the boundaries of object detection and venturing into other vision AI tasks. It builds upon the success of previous YOLO versions, boasting advancements in speed, accuracy, and versatility. This makes YOLOv8 a powerful tool for various applications, from real-time object identification in videos to image analysis in diverse fields. While the specifics of its inner workings haven't been formally published, YOLOv8 likely leverages recent advancements in deep learning and computer vision. It probably incorporates techniques like convolutional neural networks (CNNs) for feature extraction and attention mechanisms to focus on crucial image regions. This efficient architecture allows YOLOv8 to excel in both speed and accuracy.

YOLOv8 achieves impressive results, balancing speed with high detection accuracy. This makes it suitable for real-time applications where fast processing is crucial. But YOLOv8 goes beyond just object detection. It offers functionalities like instance segmentation, which separates distinct objects within an image, and pose estimation for tasks like human posture recognition. This broadens its applicability across various computer vision domains.

Ultralytics provides a user-friendly Python package and command-line interface for YOLOv8. This makes it accessible to developers of varying expertise, allowing them to integrate YOLOv8's capabilities into their projects with relative ease. YOLOv8 is also designed to be adaptable to different hardware platforms. It can run efficiently on powerful GPUs for demanding tasks or be optimized for deployment on resource-constrained edge devices. YOLOv8's ability to deliver high performance with a user-friendly approach makes it a compelling option for developers working on computer vision tasks. Its versatility and adaptability further solidify its position as a state-of-the-art model with the potential to revolutionize various fields.

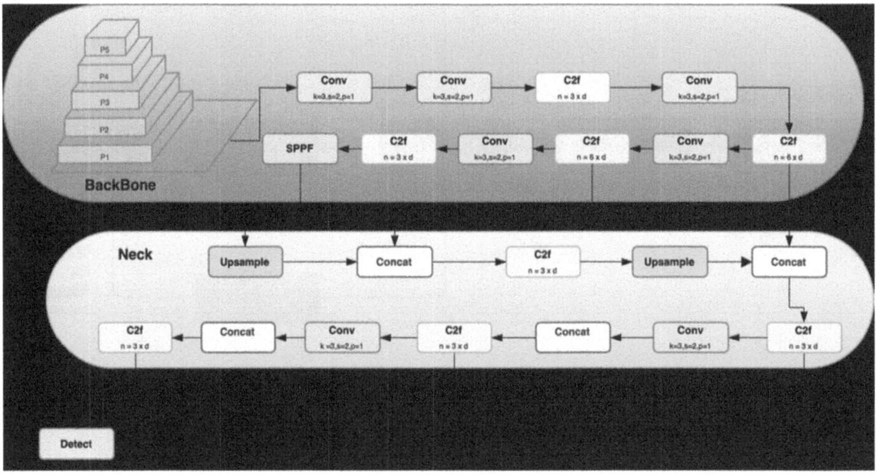

Fig. 2. YOLOv8 Architecture

3.2 Training Strategy

The initial YOLO model was trained to classify two classes: Light Motor Vehicle (LMV) and Heavy Motor Vehicle (HMV). The training dataset comprised 20,321 images, collected from various sources such as Kaggle, Roboflow, and others. These images depicted different types of vehicles and were labeled as either

LMV or HMV. The 8 s phase of training was done to detect accident class. The training dataset contains 22,916 images collected from Kaggle and Roboflow. Before training, the images underwent preprocessing to ensure optimal performance. The preprocessing techniques applied included AutoOrient to correct any orientation inconsistencies and resize to standardize all images to a resolution of 640×640 pixels. Additionally, augmentation techniques were employed to enhance the diversity of the dataset and improve the model's robustness. Augmentation techniques included rotating the images by $90°$ in clockwise, counter-clockwise, and upside-down orientations. Brightness adjustment was applied to simulate varying lighting conditions. Furthermore, a blur effect was applied to mimic real-world scenarios where images may not be perfectly clear.

By utilizing these preprocessing and augmentation techniques, the training dataset was enriched with a wide variety of vehicle images, ensuring that the YOLO model could effectively learn to classify LMVs and HMVs under different conditions and orientations.

4 Results and Discussions

4.1 Vehicle Classification and Accident Detection

After the initial phase of training the YOLOv8 model for classifying Light Motor Vehicles (LMV) and Heavy Motor Vehicles (HMV). the mAP (mean Average Precision at IoU of 0.5) metric shown on the graph is around 0.85 for both training and validation sets. This suggests that the model has an average precision of 86% for detecting and classifying vehicles with an IoU of 0.5. IoU refers to Intersection over Union, which is a metric used to measure how well a detection box overlaps with the ground truth box. The training loss curves for box loss, classification loss, and DFL loss all appear to converge over the course of training. This suggests that the model is learning effectively. The validation loss curves are generally higher than the training loss curves, but they also appear to converge. The mAP curves show that the model has higher precision for objects with a higher IoU. This means that the model is more accurate at detecting vehicles when there is a significant overlap between the predicted bounding box and the actual bounding box. Additionally, the model has an F1 score of 0.83.

The YOLOv8 model successfully learned to detect accidents with an accuracy of 90% and has F1 score of 0.87 in the next phase of training. This model was then combined with the first YOLOv8 model trained to classify Light Motor Vehicles (LMV) and Heavy Motor Vehicles (HMV). The combined system leverages the strengths of both models. The first model efficiently identifies LMVs and HMVs, while the second focuses its processing power on analyzing only those regions to detect accidents. This two-stage approach improves overall efficiency while maintaining good accuracy for accident detection.

4.2 Speed Estimation

The methodology employed by ECAWS for speed estimation involves the strategic placement of two lines at a predetermined distance along the road section.

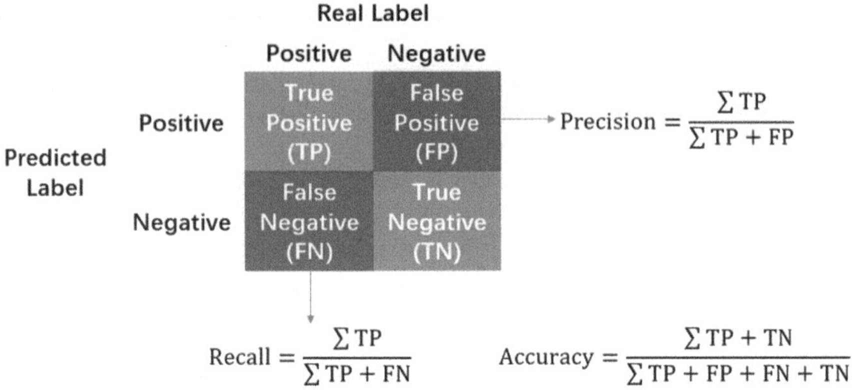

Fig. 3. Formulas for precision, recall and accuracy

$$F1 = \frac{2 \times Precision \times Recall}{Precision + Recall}$$

Fig. 4. Formula for F1 score

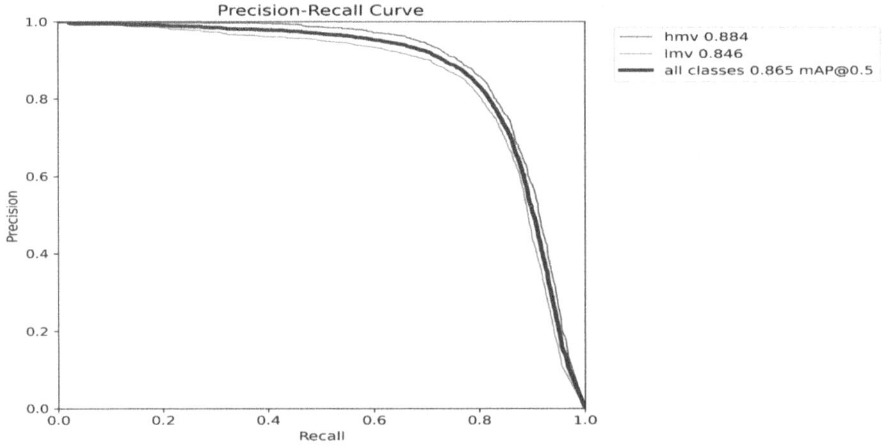

Fig. 5. Precision Recall Curve – Phase 1

Upon a vehicle's entry and exit past these lines, the system records the respective times. Utilizing the recorded time data and the known distance between the lines, the speed of the vehicle is calculated using the distance/time formula. This calculated speed serves as a basis for identifying vehicles exceeding safe speed thresholds.

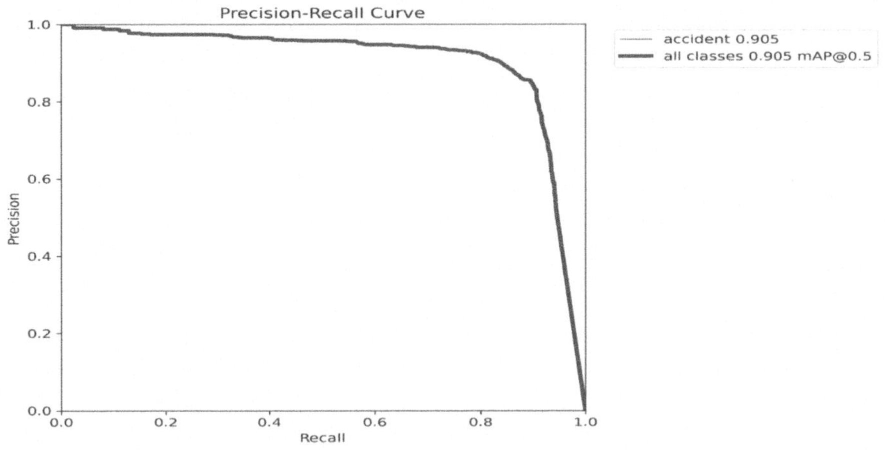

Fig. 6. Precision Recall Curve – Phase 2

Fig. 7. Speed estimation

4.3 Alert System

The Early Curve Accident Warning System (ECAWS) operates seamlessly to notify nearby police stations and ambulance drivers via WhatsApp, ensuring swift emergency response. Upon detection of a potential accident, ECAWS promptly sends alerts containing pertinent details to these designated recipients. The system also facilitates the transmission of real-time accident frames, enabling authorities to accurately assess the situation and allocate appropriate resources.

In addition to alerting emergency responders, the system proactively warns drivers about potential hazards, such as the presence of Heavy Motor Vehicles (HMVs), overspeeding vehicles, and detected accidents. This comprehensive alert system enhances coordination between emergency responders and optimizes their ability to deliver timely assistance, ultimately reducing response times and mitigating the impact of accidents on affected individuals.

5 Conclusion

In conclusion, the Real-Time Early Curve Accident Warning System, augmented with a speed warning component, represents a pioneering solution geared towards significantly enhancing safety on hairpin bends. Through the strategic deployment of cameras capturing real-time road footage and a processing module powered by deep learning, this comprehensive system delivers a multi-faceted approach to road safety. It not only detects vehicles and calculates their speeds but also triggers timely warnings for drivers, fostering a proactive and data-driven safety environment. The system's contributions extend beyond mere hazard detection, encompassing the enhancement of road safety, timely identification of potential risks, efficient traffic management, and the identification of speeding hazards. Its swift response mechanism ensures that drivers and authorities receive critical alerts promptly, further reinforcing its commitment to proactive safety measures. In essence, this innovative system embodies a holistic approach to road safety on hairpin bends, ushering in a new era of efficient and responsive traffic management.

Acknowledgements. We are grateful to Almighty who has blessed us with good health, committed and continuous in- terest throughout the project work.

We express our sincere thanks to our project guide, Ms. Jency Thomas, Assistant Professor, Department of Computer Science And Engineering, Muthoot Institute of Technology and Science for her guidance and support which were instrumental in all the stages of the project work and without whom the project could not have been accomplished

Besides this, several people have knowingly and unknowingly helped us in the successful completion of this project. We express our sincere gratitude to all of them.

References

1. Choi, J.G., Kong, C.W., Kim G., Lim S.: Car crash detection using ensemble deep learning and multimodal data from dashboard cameras. Elsevier (2021)
2. Tian, D., Zhang, C., Duan, X., Wang X.: An automatic car accident detection method based on cooperative vehicle infrastructure systems. IEEE (2019)
3. Han, Z., Liang, J., Li, J.: Design of intelligent road recognition and warning system for vehicles based on binocular vision. IEEE (2018)
4. Zhou, Y.F., Xie, K., Zhang, X.Y., Wen, C., He, J.B.: Efficient traffic accident warning based on unsupervised prediction framework. IEEE (2021)
5. Baoyuan, C., Yitong, L., Sun, K.: Research on object detection method based on FF-yolo for complex scenes. IEEE (2021)

6. Cai, Y., et al.: YOLOv4-5D: an effective and efficient object detector for autonomous driving. IEEE (2021)
7. Zhao, J., et al.: Improved visionbased vehicle detection and classification by optimised YOLOv4. IEEE (2022)
8. Carrasco, D.P., Rashwan, H.A., Garcıa, M. A., Puig, D.: T-YOLO: tiny vehicle detection based on YOLO and multi-scale convolutional neural networks. IEEE (2021)
9. Niroomand, N., Bach, C., Elser, M.: Robust vehicle classification based on deep features learning. IEEE (2021)
10. Kamencay, P., Markovic, M., Dubovan, J., Dado, M., Benedikovic, D.: Neural network-optimized vehicle classification using clustered image and fiber-sensor datasets. IEEE (2023)
11. Hedeya, M.A., Eid, A.H., Abdel-Kader, R.F.: A super-learner ensemble of deep networks for vehicle-type classification. IEEE (2020)
12. Guo, L., Li, R., Jiang B.: An ensemble broad learning scheme for semisupervised vehicle type classification. IEEE (2021)
13. Luvizon, D.C., Nassu, B.T., Minetto, R.: A video-based system for vehicle speed measurement in urban roadways. IEEE (2017)
14. Odat, E., Shamma, J.S., Claudel, C.: Vehicle classification and speed estimation using combined passive infrared/Ultrasonic sensors. IEEE (2018)
15. Wang, J., Tong, J., Gao, Q., Wu, Z., Bi, S., Wang, H.: Device-free vehicle speed estimation with WiFi. IEEE (2018)
16. Dailey, D.J., Cathey, F.W., Pumrin, S. An algorithm to estimate mean traffic speed using uncalibrated cameras. IEEE (2000)

Activity Detection Based Performance Analysis in Cell-Free Massive MIMO System

Mitesh Solanki[1](✉), Vinutha Bhat[2], and Shilpi Gupta[2]

[1] Marwadi University, Rajkot 360003, India
miteshbhai.solanki@marwadieducation.edu.in
[2] Sardar Vallabhbhai National Institute of Technology, Surat 395007, India
sgupta@eced.svnit.ac.in

Abstract. Massive Multiple Input Multiple Output (mMIMO) technology is essential for massive machine type communication (mMTC), where identifying active devices from a large pool via multiple access points (APs) is crucial. In emerging cell-free mMIMO (CF-mMIMO) systems, device activity detection is a critical task. This paper introduces a proximal gradient algorithm for detecting device activity in grant-free random-access scenarios within CF-mMIMO networks, specifically targeting two dominant APs. Each active device sends non-orthogonal pilot sequences to the APs, which then relay the received signals to a central processing unit (CPU) for joint activity detection. The proposed clustering-based proximal gradient algorithm significantly improves performance and efficiency in activity detection, offering reduced execution time compared to existing methods.

Keywords: Cell-Free massive MIMO · Activity Detection · grant free random access · Central Processing Unit

1 Introduction

One of the primary applications of fifth-generation (5G) and beyond 5G wireless networks is massive access, also known as massive connectivity. In cell-free massive Multiple Input Multiple Output MIMO (CF-mMIMO) networks, all access points (APs) are connected to a central processing unit (CPU) via fronthaul for cooperative signal processing. This architecture has recently been recognized as an effective means of consistently delivering high data rates. CF-mMIMO outperforms both small cells and traditional cellular technology due to the elimination of "cell boundaries," which eradicates inter-cell interference [1]. Consequently, CF-mMIMO is considered a viable solution for mitigating inter-cell interference caused by network densification [2, 3].

Device activity detection is a critical task in machine-type communications (mMTC) within emerging cell-free mMIMO (CF-mMIMO) systems. In mMTC, only some devices are in active stage for a given time. Identification of such active devices from the pool is fundamental. In this context, at the receiver side, active devices are identified with coordinating efforts of multiple APs. A random-access scheme based on grant free

D. Gupta et al. (Eds.): PCEMS 2024, CCIS 2491, pp. 268–279, 2025.
https://doi.org/10.1007/978-3-031-90577-3_24

concept was proposed in [4], in which, out of all orthogonal sequences a pilot sequence is selected by every active device randomly. This task is performed to notify APs about their intent of its data transmission. The AP then allocates resources based on its ability to resolve collisions. However, in such scenarios, there may be the high probability that same sequence is selected by more than one active device which may lead to access failures due to collisions. This results in increased average latency, making grant-based protocols inefficient for mMTC applications.

Cell-free massive MIMO systems have gained significant attention for their ability to dramatically improve spectral efficiency and provide seamless coverage by utilizing a large network of geographically distributed antennas [5]. Unlike conventional mMIMO systems, which operate within fixed cell boundaries, cell-free MIMO eliminates these boundaries, enabling cooperative processing among multiple APs to jointly serve users. One of the key challenges in these systems is detecting active user equipment (UEs) in the uplink, a task complicated by the vast scale of antenna arrays and the sporadic nature of user activity. To address this, recent research has focused on advanced signal processing and optimization techniques, aiming to enhance the accuracy and efficiency of activity detection in these complex environments [4, 5].

To work with low-latency requirements of mMTC, a random-access strategy based on grant-free access was proposed in [5, 6]. The limitations of grant-based random-access strategies have prompted the development of several grant-free protocols [7], allowing active devices to connect to wireless networks without the need for contention. In these grant-free protocols, data and pilots are sent to the AP without requiring prior scheduling. In large-scale scenarios, it is crucial for UEs to communicate with APs without handshaking [8]. As coherence time of channel is limited in a wireless system, the number of present pilot sequences which are orthogonal is also limited. Therefore, non-orthogonal signature sequences are assigned to each UE, complicating device activity identification due to interference and the need for non-orthogonal sequences [9]. Estimation of channel and activity detection of UE is analyzed by considering the cost associated with these processes to characterize the achievable rates [10]. The access without grant in a given system is examined under the propagation conditions which are optimally favorable in mMIMO scenario [11–13].

Despite extensive studies on activity detection from various perspectives, most existing approaches for CF-mMIMO assume that accurate collection of received signals is to be done at the CPU [14–17]. However, in practical scenarios, the CPU and APs are connected via fronthaul with limited capacity, necessitating the compression or quantization of signals at the APs before forwarding to the CPU. This compression/quantization process introduces errors that degrade detection performance at the CPU. To tackle this issue, [18] proposed a design where each AP uses a uniform quantizer independently, and the CPU employs a detection algorithm. Unfortunately, this separate design does not account for the interaction between the quantizers and the detection algorithm, making it challenging while achieving end-to-end optimal performance.

A promising solution to the activity detection challenge in CF-mMIMO systems lies in coordinate descent methods, which decompose the high-dimensional detection task into smaller, more manageable subproblems. By iteratively updating one variable while holding the others constant, this technique gradually improves the accuracy of detecting active UEs' signals. However, its iterative nature can slow convergence, particularly in large-scale systems with numerous antennas and UEs [5]. To address this, there is a pressing need for more robust, efficient algorithms that reduce execution time without compromising accuracy. Optimizations such as parallel processing can significantly speed up the method, enabling it to scale effectively while maintaining low computational complexity, making it well-suited for the demands of CF-mMIMO systems.

This paper introduces a comprehensive framework aimed at achieving efficient communication and accurate activity detection over fronthaul links with limited capacity. The framework incorporates compression and quantization modules at the APs, with corresponding decompression and detection modules at the CPU. By applying quantization to a more efficient representation of the high-dimensional signal, rather than its raw form, the compression and decompression processes are significantly optimized [5]. To ensure overall system efficiency, the framework is designed with joint optimization of all modules, preventing any adverse impact from one module on the others. This approach ensures a balanced and effective integration of compression, quantization, and detection for enhanced performance.

Motivated by the limitations of existing methods, this paper presents a robust framework for activity detection in CF-mMIMO systems, designed to reduce computational load and improve preprocessing efficiency. The proposed solution integrates compression and quantization modules at the APs, along with decompression and detection modules at the CPU. By applying quantization to an optimized representation of the high-dimensional signal, rather than its original form, and jointly optimizing all modules, the framework ensures that each component performs efficiently without degrading overall accuracy. A key novelty of this paper is the implementation of a proximal gradient approach for activity detection, which markedly improves detection accuracy compared to conventional techniques, setting it apart as a more effective solution. This approach not only boosts detection activity performance but also offers greater computational robustness, leading to improved outcomes in both accuracy and efficiency, particularly in capacity-limited, large-scale scenarios.

The rest part of this paper is structured as follows: System model is presented in Sect. 2. Section 3 discusses a centralized algorithm and a clustering-based activity detection method under perfect fronthaul links. Section 4 focusses on simulation results, while Sect. 5 presents a comparison with existing work. At the end, Section 6 has concluded with highlighting the findings of the work.

2 System Model

Consider a cell-free network with K user equipment's (UEs) and M access points (APs), where each AP has N antennas and each UE has a single antenna. The APs are connected to a central processing unit (CPU) via fronthaul. In the massive access scenario of mMTC, traffic is sporadic, with only a few UEs active at any given time. Geographically, UEs and APs are distributed such that only a small number of APs are in close proximity to each UE, resulting in strong links with just a few APs.

To address this, the CF-mMIMO architecture was introduced, eliminating traditional cell boundaries and enabling all devices to communicate with every AP. In this cell-free

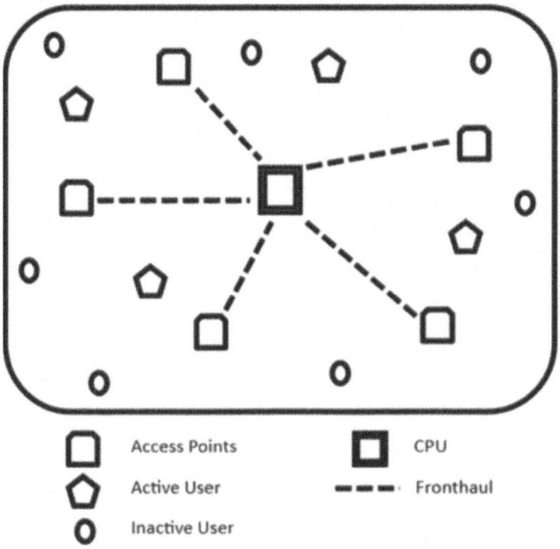

Fig. 1. Cell-Free Massive MIMO architecture for mMTC.

framework, all APs are interconnected with a CPU via a fronthaul network, as illustrated in Fig. 1. Our proposed algorithm focuses on a cluster of two APs to enhance performance in this setup. Flowchart for proximal gradient algorithm is presented in Fig. 2, where Q_m is noise covariance matrix and Q_y is covariance matrix of received signal. The channel gain of nth antenna of mth AP for k^{th} device [8] is determined by

$$g_{m,n,k} = \sqrt{\beta_{m,k}} h_{m,n,k} \tag{1}$$

where the large-scale fading coefficient between the kth user and m^{th} APs is denoted as $\beta_{m,k}$, small scale fading co-efficient is represented by $h_{m,n,k} \sim CN(0,1)$.

At the CPU, large scale fading coefficients are known. The number of channels uses within each coherence interval is τ_c. Given the large number of UEs, assigning orthogonal pilot sequences to every single UE is impossible, typically $\tau_c \leq K$. Hence, non-orthogonal unique signature sequence is assigned. $s_k \epsilon \mathbb{C}^{L\times 1}$ belonging to complex numbers to each UE K. Signature length is denoted by L. At the CPU, all the signature sequence are known.

At the nth antenna of the mth AP, the signal $y_{m,n} \epsilon C^{L\times 1}$ received [8] is given as

$$y_{m,n} = \sum_{k=1}^{K} a_k \rho_k^{\frac{1}{2}} g_{m,n,k} s_k + \mathcal{W}_{m,n} = S\mathcal{D}_a \mathcal{D}_\rho^{\frac{1}{2}} g_{m,n} + \mathcal{W}_{m,n}, \tag{2}$$

where $S = [s_1 s_2 \cdots s_K] \in C^{L\times K}$ represents the signature sequence set. ρ_k denotes transmitted power by user k, $\mathcal{D}_a = diag(a)$, $\mathcal{D}_\rho = diag(\rho_1 \rho_2 \cdots \rho_K)$, $g_{m,n} = \left[g_{m,n,1}, g_{m,n,2} \cdots g_{m,n,K}\right]^T \in C^{K\times 1}$ is the channel vector to the nth antenna of the mth AP from all K users and the independent additive white Gaussian noise vector is $\mathcal{W}_{m,n} \sim CN\left(0, \sigma^2 I_L\right)$. Thus, at the mth AP, the signal $Y_m \epsilon C^{K\times N}$ received [8] is given

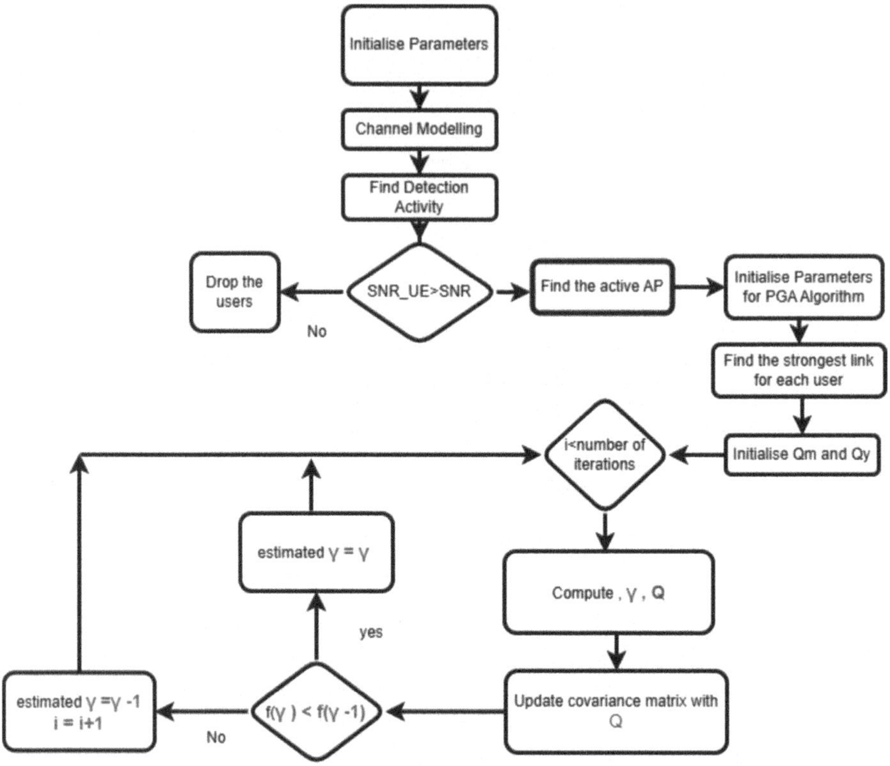

Fig. 2. Flowchart for Proximal Gradient Algorithm Activity Detection

as

$$Y_m = S\mathcal{D}_a\mathcal{D}_\rho^{\frac{1}{2}}G_m + \mathcal{W}_m, \tag{3}$$

where $G_m = [g_{m,1}g_{m,2}\cdots g_{m,N}]\epsilon C^{K\times N}$ is the channel matrix between the K users and the mth AP. The noise matrix is $\mathcal{W}_m = [\mathcal{W}_{m,1}\mathcal{W}_{m,2}\cdots\mathcal{W}_{m,N}]\epsilon C^{L\times N}$

$$Y = \begin{bmatrix} S & 0 & 0\,0 \\ 0 & S & 0\cdots \\ \cdots & \cdots & \cdots\cdots \\ 0 & 0 & \cdots S \end{bmatrix} \begin{bmatrix} \mathcal{D}_a\mathcal{D}_\rho^{\frac{1}{2}} & 0 & \cdots 0 \\ 0 & \mathcal{D}_a\mathcal{D}_\rho^{\frac{1}{2}} & \cdots 0 \\ \vdots & \vdots & \ddots & \vdots \\ 0 & 0 & \cdots & \mathcal{D}_a\mathcal{D}_\rho^{\frac{1}{2}} \end{bmatrix} + \mathcal{W} \tag{4}$$

where $\mathcal{W} = \left[\mathcal{W}_1^T \mathcal{W}_2^T \cdots \mathcal{W}_M^T\right]^T$. It is seen from Y that every column is distributed and columns are independent. Covariance matrix Q is

$$Q = \begin{bmatrix} S\mathcal{D}_\gamma \mathcal{D}_{\beta_1} S^H & 0_L & \cdots 0_L \\ 0_L & S\mathcal{D}_\gamma \mathcal{D}_{\beta_2} S^H & \cdots 0_L \\ \vdots & \vdots & \ddots & \vdots \\ 0_L & 0_L & \cdots & S\mathcal{D}_\gamma \mathcal{D}_{\beta_M} S^H \end{bmatrix} + \sigma^2 I_{L,M}, \tag{5}$$

where $D_{\beta M}$ is the diagonal matrix $\mathcal{D}_\gamma = diag(\gamma), \mathcal{D}_{\beta_m} = diag(\beta_m), \beta_m = (\beta_{m,1}, \beta_{m,2}, \cdots, \beta_{m,K})$ where $\gamma = (a_1\rho_1, a_2\rho_2, \cdots, a_k\rho_K)$. The process of maximizing $p(Y|\gamma)$ or equivalently minimizing $-log(p(Y|\gamma))$ yields the highest probability estimate of γ which [5, 8] is given as

$$p(Y|\gamma) = \prod_{m=1}^{M} \frac{1}{|\pi Q_m|^N} \exp\left(-Tr\left(Q_m^{-1} Y_m Y_m^H\right)\right) \tag{6}$$

where $Q_m = SD_\gamma D_{\beta_m} S^H + \sigma^2 I_L$.

$$\gamma^* = \arg\min_{\gamma} \sum_{m=1}^{M} log|Q_m| + Tr\left(Q_m^{-1} \frac{Y_m Y_m^H}{N}\right) \tag{7}$$

3 Clustering Based Activity Detection

For the cluster with two dominant APs, it is required to find two strongest links for each user and compute the gradient of cost function. Considering a cluster of T < M dominant APs for every UE [8], a_m, b_m are defined as

$$\begin{aligned} a_m &= \beta_{m,k} s_k^H Q_m^{-1} s_k \\ b_m &= \beta_{m,k} s_k^H Q_m^{-1} Q_{Y_m} Q_m^{-1} s_k. \end{aligned} \tag{8}$$

The co-variance matrix Q_Y is updated using proximal operator given by

$$Q_Y = sign(Q_{Y_m}) max(0, abs(Q_{Y_m}) - \lambda) \tag{9}$$

where λ is the regularization parameter. Penalty factor is considered as 0.2 and the cost function is reduced by taking penalty factor into account.

$$\mathcal{F}_\gamma = log([Q_{Y_m}]) + \frac{1}{N} Tr\left(Q_{Y_m}^{-1} Q_{Y_m}\right) + Penalty \tag{10}$$

4 Simulation Results

In order to simulate a network and prevent boundary effects, square area that is encircled by the edges is taken into consideration. The M APs are distributed independently and arbitrarily. In the network, the K users are evenly distributed. For simulation, an area with $K = 400$ devices, the probability of activation, $\epsilon = 0.1$ and the signature sequence with length $L = 40$ is considered. The maximum transmit power is 200 mW for a device. Noise variance $\sigma^2 = -109$ dB. Every user device's signature sequence is considered to be derived from the Gaussian distribution, CN(0, I_L).

Missed detection probability versus false alarm probability is displayed via ROC curves. Comparison of Co-ordinate descent and proposed Proximal gradient algorithms is shown in Fig. 3. For activity detection, data from several dominating APs (T = 2) are taken into consideration. Since there are closed form equations available for solving a 3rd degree polynomial equation, this approach is computationally efficient for T = 2.

To provide a fair comparison, three cases are taken into account: M = 20, 25, 30 where the AP density is varying. This is shown in Fig. 4.

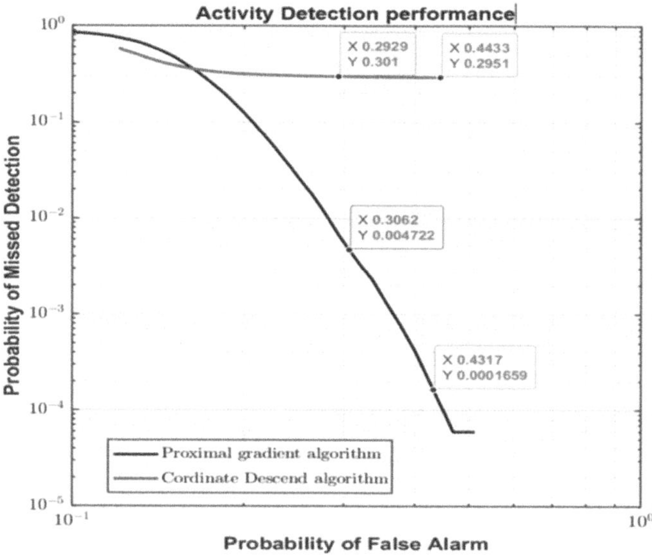

Fig. 3. Activity detection performance comparison of co-ordinate descent and Proximal gradient algorithms with cell of area $2 \times 2\,\text{km}^2$, $M = 25$, $N = 4$, $L = 40$, $K = 400$, $\epsilon = 0.1$.

Because of the enhanced spatial resolution of the devices, activity detection performs better with increasing no of antennas as illustrated in Fig. 5.

Impact of various activation probabilities onto algorithm's ability to detect activity is investigated in Fig. 6. Different five cases are taken in into account with $\epsilon = 0.1, 0.2, 0.3, 0.4, 0.5$.

The performance gain with varying number of antennas at APs is analyzed for N = 4,6, 8,10, and 12 and presented in Fig. 7. It can be observed that as number of antennas

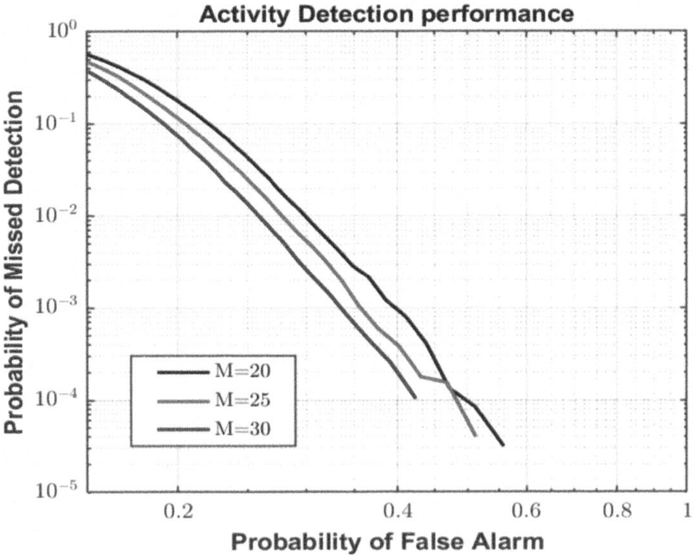

Fig. 4. Activity detection performance of Proximal gradient algorithms with $2 \times 2\ \text{km}^2$, cell area, $N = 4$, $L = 40$, $K = 400$, $\epsilon = 0.1$.

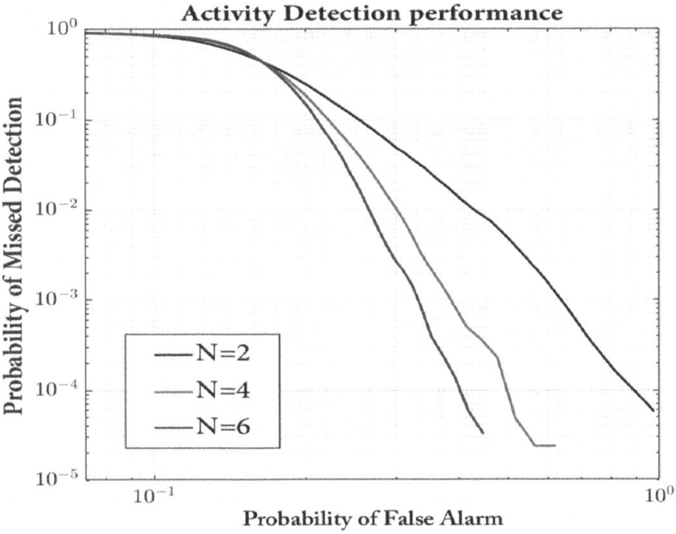

Fig. 5. Activity detection performance of Proximal gradient algorithms with $2 \times 2\ \text{km}^2$, cell area, $M = 20$, $L = 40$, $K = 400$, $\epsilon = 0.1$.

per AP increases, probability of missed detection decreases. However, performance is not increasing by significant amount when N is 12.. Also increasing the antennas at APs increases complexity and channel hardening or favorable propagation does not improve significantly beyond a limit. Thus, using Proximal gradient algorithm which uses the line

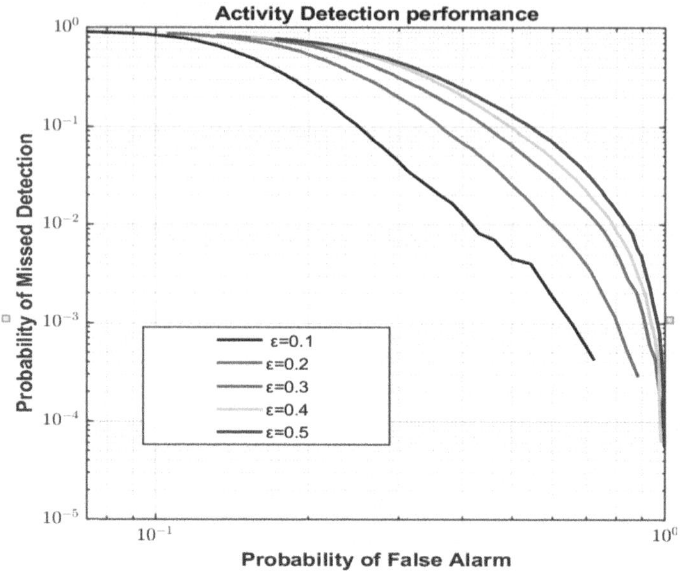

Fig. 6. Activity detection performance of Proximal gradient algorithms with $2 \times 2\,\text{km}^2$, cell area, $M = 20$, $L = 40$, $K = 400$, $N = 4$, for different activation probabilities

searching for activity detection, better performance is obtained and also the execution time is reduced.

In summary, the probability of missed detection increases when the device activation probability increases because higher activation probabilities can lead to higher rates of false alarms or incorrect triggers, reducing the overall reliability of detecting genuine targets or conditions. Also, the interference increases and the difficulty of distinguishing between active and inactive users results in missed detection. Therefore, even when the total number of active devices exceeds the length of the signature sequence, the proposed algorithm supports activity detection.

5 Comparison

Comparative finding regarding execution time for Proximal Gradient and Co-ordinate Descent algorithms is presented in Table 1. While comparing, it is observed that proposed algorithm takes less execution time. Less execution time is reported due to the less signal processing involved in the work.

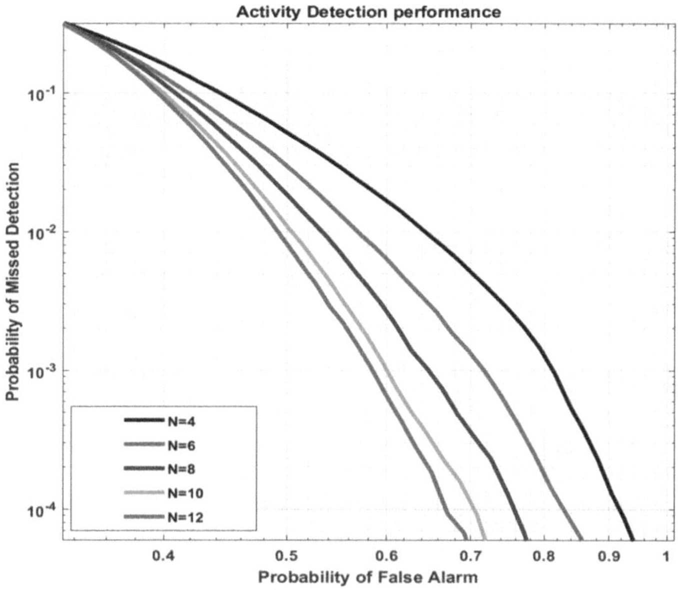

Fig. 7. Activity detection performance of Proximal gradient algorithms with $2 \times 2\,\text{km}^2$, cell area, $M = 20, L = 40, K = 400$, for different number of AP's antenna

Table 1. Execution time comparison between Proposed Algorithm and Co-Ordinate Descent algorithm.

Parameters	Proposed algorithm (Time in sec)	Co-ordinate Descent Algorithm (Time in sec)
$M = 4, N = 2, K = 400, Samples = 10000$	1797	6812
$M = 25, N = 4, K = 400, Samples = 1000$	729	1099
$M = 20, N = 2, K = 600, Samples = 1000$	2081	3704
$M = 30, N = 2, K = 600, Samples = 1000$	3464	8498
$M = 40, N = 2, K = 600, Samples = 1000$	4737	6715

6 Conclusion

The grant-free random-access scenario is examined within cell-free massive MIMO networks. A method for activity detection was proposed, and activity detection problem was tackled. With its ability to withstand shadow fading effects and increased macro diversity, the CF-mMIMO network can offer superior coverage for mMTC applications. Simulation outcomes demonstrate that the macro diversity gain provided by cell-free networks enhances activity detection performance. Proximal gradient algorithm outperforms the co-ordinate descent algorithm in terms of performance also in terms of execution time.

References

1. Zhang, J., Chen, S., Lin, Y., Zheng, J., Ai, B., Hanzo, L.: Cell-free massive MIMO: a new next-generation paradigm. IEEE Access **7**, 99878–99888 (2019)
2. Ngo, H.Q., Ashikhmin, A., Yang, H., Larsson, E.G., Marzetta, T.L.: Cell-free massive MIMO versus small cells. IEEE Trans. Wireless Commun. **16**(3), 1834–1850 (2017)
3. Interdonato, G., Björnson, E., Quoc Ngo, H., Frenger, P., Larsson, E.G.: Ubiquitous cell-free massive MIMO communications. EURASIP J. Wirel. Commun. Netw. **197**, (2019)
4. Henriksson, M., Gustafsson, O., Ganesan, U.K., Larsson, E.G.: An architecture for grant-free random access massive machine type communication using coordinate descent. In: 2020 54th Asilomar Conference on Signals, Systems, and Computers, Pacific Grove, CA, USA, pp. 1112–1116 (2020)
5. Ganesan, U.K., Björnson, E., Larsson, E.G.: Clustering-based activity detection algorithms for grant-free random access in cell-free massive MIMO. IEEE Trans. Commun. **69**(11), 7520–7530 (2021)
6. Bana, A.-S., et al.: Massive MIMO for internet of things (IoT) connectivity. Phys. Commun. **37**, 1–17 (2019)
7. Shahab, M.B., Abbas, R., Shirvanimoghaddam, M., Johnson, S.J.: Grant-free non-orthogonal multiple access for IoT: a survey. IEEE Commun. Surv. Tutorials **22**(3), 1805–1838 (2020)
8. Dai, L., Wang, B., Yuan, Y., Han, S., Chih-lin, I., Wang, Z.: Non-orthogonal multiple access for 5G: solutions, challenges, opportunities, and future research trends. IEEE Commun. Mag. **53**(9), 74–81 (2015)
9. Ganesan, U.K., Bj¨ ornson, E., Larsson, E.G.: An algorithm for grant-free random access in cell-free massive MIMO. In: IEEE 21st International Workshop on Signal Processing Advances in Wireless Communications (SPAWC), pp. 1–5 (2020)
10. Jovović, I., Forenbacher, I., Periša, M.: Massive machine-type communications: an overview and perspectives towards 5G (2015)
11. Wang, H., Wang, J., Fang, J.: Grant-free massive connectivity in massive MIMO systems: collocated versus cell-free. IEEE Wirel. Commun. Lett. **10**(3), 634–638 (2021)
12. Marzetta, T.L., Larsson, E.G., Yang, H., Ngo, H.Q.: Fundamentals of Massive MIMO. Cambridge University Press (2016)
13. Chen, Z., Bjornson, E.: Channel hardening and favorable propagation in cell-free massive MIMO with stochastic geometry. IEEE Trans. Commun. **66**(11), 5205–5219 (2018)
14. Shao, X., Chen, X., Ng, D.W.K., Zhong, C., Zhang, Z.: Cooperative activity detection: sourced and unsourced massive random-access paradigms. IEEE Trans. Signal Process. **68**, 6578–6593 (2020)
15. Ding, T., Yuan, X., Liew, S.C.: Sparsity learning-based multiuser detection in grant-free massive-device multiple access. IEEE Trans. Wireless Commun. **18**(7), 3569–3582 (2019)

16. Wang, Z., Liu, Y.-F., Wang, Z., Yu, W.: Scaling law analysis for covariance-based activity detection in cooperative multi-cell massive MIMO. In: Proceedings of the IEEE International Conference on Acoustics, Speech and Signal Processing (ICASSP) (2023)
17. Chen, Z., Sohrabi, F., Yu, W.: Sparse activity detection in multicell massive MIMO exploiting channel large-scale fading. IEEE Trans. Signal Process. **69**, 3768–3781 (2021)
18. Haghighatshoar, S., Jung, P., Caire, G.: Improved scaling law for activity detection in massive MIMO systems. In: IEEE International Symposium on Information Theory (ISIT), pp. 381–385 (2018)

Application of OTFS Modulation in Autonomous Vehicle Driving Systems for 6G Wireless Communication

Balakrishna Gugulothu[1], Amit Agarwal[1(✉)], and Sourabh Tiwari[2]

[1] Department of ECE, VNIT NAGPUR, Nagpur, Maharashtra, India
amit.agarwal@ece.vnit.ac.in

[2] Department of ECE, University Institute of Engineering, Chandigarh University, Mohali, Punjab, India

Abstract. Autonomous vehicle (AV) technology relies on robust, high-speed wireless communication systems to ensure safety and efficiency, particularly as we move towards 6G networks. Orthogonal Time Frequency Space (OTFS) modulation has emerged as a key solution to address the challenges posed by high-mobility environments, such as rapid Doppler shifts and time-varying channels. This paper investigates the application of OTFS modulation in AV driving systems, proposing a comprehensive system architecture for its integration within a 6G framework. Through performance evaluations and a case study, the paper demonstrates OTFS's ability to provide reliable, low-latency communication between AVs and infrastructure, positioning it as a transformative technology for future autonomous transportation systems.

Keywords: OTFS · Autonomous Vehicles · B5G · Wireless Communication · Bit Error Rate · Sum Rate

1 Introduction

Autonomous vehicles (AVs) represent a paradigm shift in transportation, offering the promise of safer, more efficient, and convenient mobility solutions. With the potential to reshape urban landscapes and revolutionize various industries, AVs rely heavily on robust and high-speed wireless communication systems to operate effectively. As the world transitions from 5G to 6G networks, there arises a pressing need to develop communication technologies capable of meeting the stringent requirements of AV scenarios. Traditional modulation schemes, while effective in conventional wireless communication, face significant challenges in high-mobility environments characterized by rapid Doppler shifts and time-varying channels. In response to these challenges, Orthogonal Time Frequency Space (OTFS) modulation has emerged as a promising waveform, offering unique advantages ideally suited for AV communication within 6G networks. By leveraging the inherent properties of OTFS modulation, such as resilience to Doppler effects and multipath fading, researchers aim to address the communication challenges inherent

D. Gupta et al. (Eds.): PCEMS 2024, CCIS 2491, pp. 280–293, 2025.
https://doi.org/10.1007/978-3-031-90577-3_25

in AV driving systems. This paper aims to explore the application of OTFS modulation in AV driving systems within the context of 6G wireless communication. Through a comprehensive investigation, the paper seeks to evaluate the efficacy of OTFS modulation in overcoming the communication hurdles faced by AVs, ranging from high mobility to dynamic channel conditions. Additionally, the paper endeavours to propose a novel system architecture that integrates OTFS modulation seamlessly into AV driving systems, thereby laying the foundation for next-generation autonomous transportation networks.

1.1 Literature Survey

Orthogonal Time Frequency Space (OTFS) modulation has garnered significant attention in recent years as a promising waveform for future wireless communication systems, particularly in high-mobility scenarios such as autonomous vehicle (AV) communication within 6G networks. The development of OTFS technology has been motivated by the need to overcome the limitations of traditional modulation schemes in environments characterized by high Doppler shifts and time-varying channels. One of the key challenges in utilizing OTFS modulation for AV communication lies in efficient channel estimation and signal detection techniques. Sabapathy et al. [1] provide a comprehensive survey of various channel estimation and signal detection methods tailored for OTFS modulation in 6G networks. These techniques play a crucial role in mitigating the effects of channel impairments and ensuring robust communication links between AVs and base stations. Moreover, the practical feasibility of deploying OTFS modulation in AV communication systems has been explored through hardware implementation studies. Isik et al. [2] demonstrated the FPGA implementation of OTFS modulation, highlighting its potential for real-world deployment in 6G communication systems, including AV networks. This research underscores the importance of developing hardware-efficient solutions to enable the widespread adoption of OTFS technology in AV communication. Furthermore, the integration of OTFS modulation with emerging technologies such as reconfigurable intelligent surfaces (RIS) has shown promise in enhancing communication performance in high-Doppler channels. Harshavardhan et al. [3] explored the use of RIS-aided OTFS modulation to improve reliability in challenging propagation environments. This approach opens up new possibilities for improving AV communication systems by leveraging advanced technologies for signal enhancement and beamforming. Assessing the efficiency of multi-carrier waveforms, including OTFS, in 6G wireless communication systems has been a topic of interest in recent research. Raslan et al. [4] evaluated the benefits of OTFS for AV communication and highlighted its potential to address the challenges of high mobility. In conclusion, the reviewed literature underscores the growing interest and investment in leveraging OTFS modulation for autonomous vehicle communication within 6G wireless networks. Research efforts focus on advancing channel estimation, signal detection, hardware implementation, and integration with emerging technologies to realize the full potential of OTFS in AV networks. These developments pave the way for transformative advancements in AV technology and contribute to the realization of safe and efficient autonomous transportation systems. Literature Survey: Orthogonal Time Frequency Space (OTFS) Modulation Orthogonal Time Frequency Space (OTFS) modulation has emerged as a promising waveform for next-generation wireless communication systems, offering advantages in high-mobility scenarios such as autonomous vehicle (AV) communication within 6G networks. The following survey explores recent research efforts and advancements in OTFS modulation, focusing on key studies from the literature. One area of significant interest in OTFS research is the development of efficient modulation schemes tailored for high-mobility environments. Studies such as Chu et al. [5] delve into the concepts, applications, benefits, and challenges of OTFS modulation for non-terrestrial networks, highlighting its potential to enable reliable communication in dynamic scenarios. The integration of OTFS modulation with emerging technologies, such as hardware-in-the-loop (HIL) experiments using software-defined radios (SDR), has also been investigated. Nauman et al. [6] conducted HIL experiments to evaluate OTFS modulation performance, shedding light on its suitability for 6G communication systems, including AV networks. Efforts to enhance the efficiency and performance of OTFS modulation techniques have led to the exploration of novel receiver designs and algorithms. Zhang et al. [7] proposed a joint channel estimation and symbol detection receiver for OTFS in vehicular networks, demonstrating improved performance in challenging communication scenarios. Furthermore, the application of OTFS modulation in specific use cases, such as integrated sensing and communication for moving vehicle detection, has been investigated. Zielinski et al. [8] explored wireless OTFS-based integrated sensing and communication, showcasing its potential for applications in AV systems and beyond. Advancements in signal processing techniques for OTFS modulation have also been a focus of research. Zhu et al. [9] proposed a phase-coded time-domain interleaved OTFS waveform with an improved ambiguity function, aiming to enhance the robustness and reliability of OTFS communication systems. Moreover, studies have addressed practical challenges related to OTFS modulation, such as peak-to-average power ratio (PAPR) estimation and analysis. Kumar et al. [10] investigated PAPR estimation techniques in OTFS 6G waveforms, providing insights into managing power efficiency in OTFS-based communication systems. The integration of OTFS modulation with other communication technologies, such as massive MIMO (multiple-input multiple-output), has also been explored. Uthayakumar et al. [11] investigated hybrid multi-dimensional modulation techniques in non-orthogonal spatial-delay-doppler domains, aiming to enhance spectral efficiency and reliability in future communication systems. Additionally, research has focused on addressing hardware impairments and optimizing OTFS modulation performance in practical scenarios. Doosti-Aref et al. [12] conducted performance analysis of OTFS under hardware impairments in millimeter-wave vehicular communication networks, highlighting the importance of robustness and resilience in OTFS-based systems. In conclusion, the surveyed literature highlights the diverse research efforts and advancements in OTFS modulation, spanning modulation schemes, receiver designs, integration with emerging technologies, signal processing techniques, and practical implementations. These developments pave the way for realizing the full potential of OTFS modulation in future wireless communication systems, including AV networks and beyond.

Orthogonal Time Frequency Space (OTFS) Modulation Orthogonal Time Frequency Space (OTFS) modulation has garnered significant attention as a promising waveform for next-generation wireless communication systems, offering unique advantages in challenging scenarios such as high-mobility environments. This survey explores recent advancements and research efforts in OTFS modulation, focusing on key studies from the literature. One area of interest in OTFS research is the development of efficient modulation techniques to address the specific requirements of high-mobility communication systems. Zhu et al. [9] proposed a phase-coded time-domain interleaved OTFS waveform with an improved ambiguity function, aiming to enhance the robustness and reliability of OTFS communication in dynamic environments. Efforts to optimize OTFS modulation performance in practical scenarios have led to investigations into peak-to-average power ratio (PAPR) estimation and analysis. Kumar et al. [10] conducted research on PAPR estimation techniques in OTFS 6G waveforms, providing insights into managing power efficiency and spectral regrowth in OTFS-based communication systems. Moreover, the integration of OTFS modulation with other communication technologies, such as reconfigurable intelligent surfaces (RIS), has been explored. Basharat et al. [13] investigated the potential of RIS-assisted backscatter communication to enhance OTFS performance, particularly in challenging propagation environments. Studies have also focused on addressing hardware impairments and optimizing OTFS modulation performance in practical scenarios. Doosti-Aref et al. [12] conducted performance analysis of OTFS under hardware impairments in millimeter-wave vehicular communication networks, emphasizing the importance of robustness and resilience in OTFS-based systems. Furthermore, advancements in channel estimation and signal detection techniques tailored for OTFS modulation have been a focus of research. Zheng et al. [14] proposed a low-complexity sparse Bayesian learning channel estimation method based on generalized approximate message passing (GAMP) for OTFS systems in vehicle-to-everything (V2X) scenarios. Research efforts have also explored the application of OTFS modulation in integrated sensing and communication systems for IoT networks. Kaushik et al. [15] investigated synergies between OTFS technology and key 6G technology enablers, highlighting its potential for enabling efficient and reliable communication in IoT applications. In addition, studies have addressed the efficiency and performance of OTFS modulation in specific scenarios, such as satellite communication systems. Shi et al. [16] explored the use of OTFS-enabled low Earth orbit (LEO) satellite communications to mitigate severe Doppler effects, demonstrating its effectiveness in overcoming channel impairments. Overall, the surveyed literature underscores the diverse research efforts and advancements in OTFS modulation, spanning modulation schemes, integration with emerging technologies, channel estimation techniques, and practical implementations. These developments contribute to the growing body of knowledge on OTFS modulation and pave the way for its widespread adoption in future wireless communication systems.

2 System Architecture

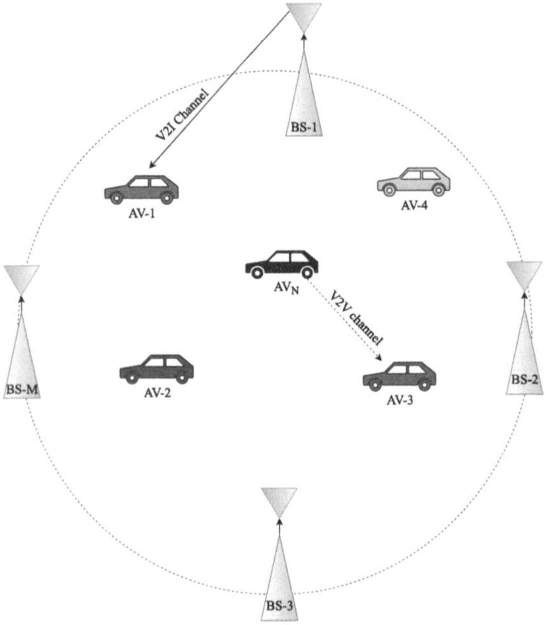

Fig. 1. Application of OTFS Modulation in Autonomous Vehicle Driving Systems

The proposed system architecture integrates ***Orthogonal Time Frequency Space (OTFS)*** modulation into the ***6G wireless communication*** framework to support ***autonomous vehicle (AV)*** operations. This architecture is designed to address the challenges of ***high mobility***, ***Doppler shifts***, and ***multipath fading*** in AV systems. It is composed of three key components: ***Base Stations***, ***Vehicle-to-Infrastructure (V2I) Communication***, and ***Vehicle-to-Vehicle (V2V) Communication*** (Fig. 1).

The ***Base Station*** serves as the central node of the system, responsible for managing communication between AVs and the 6G core network. It performs critical tasks like OTFS signal processing, resource allocation, and mobility management. Additionally, it employs advanced algorithms to minimize interference, ensuring stable, high-quality communication even in dynamic environments, mitigating Doppler effects, and maintaining reliable connections with high-speed vehicles.

The ***V2I Communication*** facilitates communication between AVs and roadside infrastructure such as traffic signals, road signs, and traffic management systems. This interaction is enabled by ***Roadside Units (RSUs)***, which are equipped with OTFS modulation capabilities. RSUs transmit real-time traffic updates and safety instructions to AVs, enhancing road safety and efficiency. Moreover, they collect and analyze data to optimize traffic flow and increase situational awareness.

The ***V2V Communication*** allows AVs to communicate directly with each other, sharing information on vehicle positions, speeds, and driving intentions. This direct communication is critical for ***collision avoidance***, ***cooperative driving***, and ***platooning***. Each AV is equipped with an onboard communication unit that supports OTFS modulation, reducing latency by enabling peer-to-peer communication without relying on the base station.

The system supports seamless operation through ***dynamic spectrum allocation***, adapting to changing network conditions and AV requirements. ***Quality of Service (QoS)*** management prioritizes important communications, such as safety alerts, to ensure timely and reliable transmission. Furthermore, robust ***security protocols*** are employed to safeguard communication links from cyber threats, maintaining data integrity and confidentiality. Overall, the architecture facilitates continuous communication between AVs, RSUs, and base stations, enhancing reliability and reducing latency, which are critical for safe and efficient AV operations in 6G networks.

2.1 System Model

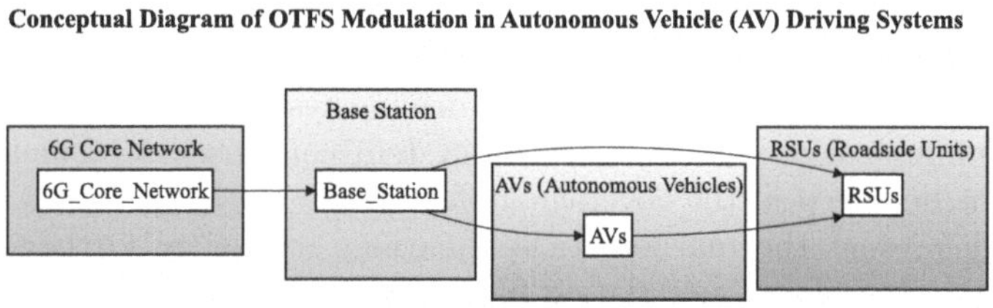

Fig. 2. Concept Diagram of OTFS Application in Vehicular Technology

The system model, depicted in Fig. 2, illustrates the application of OTFS modulation in AV driving systems within a 6G wireless communication framework. The model includes the following primary components: the 6G Core Network, Base Station, Roadside Units (RSUs), and Autonomous Vehicles (AVs). The ***6G Core Network*** acts as the backbone of the communication system, connecting base stations and providing high-speed, low-latency communication across the network. It is essential for managing data traffic and ensuring network stability.

The ***Base Station*** serves as the central hub that coordinates communication between AVs and the core network. It manages signal processing, resource management, and mobility tracking, ensuring AVs maintain strong and reliable connections even in high-speed conditions. The base station also plays a critical role in minimizing interference and optimizing signal quality for both V2I and V2V communications.

The ***RSUs (Roadside Units)*** are strategically deployed along roadways to facilitate ***V2I communication***. These units are equipped with OTFS modulation capabilities, acting as fixed nodes that transmit traffic updates, safety information, and manage communication between AVs and infrastructure. They also collect and process data to improve road safety and traffic management.

The ***AVs (Autonomous Vehicles)*** are equipped with onboard communication units supporting both ***V2I*** and ***V2V*** communication using OTFS modulation. AVs communicate directly with each other, allowing for real-time sharing of information such as positions and driving maneuvers. This peer-to-peer communication reduces latency, improving communication efficiency and reducing reliance on base stations. The system model ensures that AVs can maintain robust communication links even in environments with high mobility, rapid Doppler shifts, and varying channel conditions.

3 Problem Formulation

The objective of this paper is to investigate the application of Orthogonal Time Frequency Space (OTFS) modulation in autonomous vehicle (AV) driving systems within the context of 6G wireless communication. Specifically, the study aims to evaluate the efficacy of OTFS modulation in addressing the communication challenges faced by AVs, including high mobility, Doppler shifts, and multipath fading. The paper also proposes a system architecture for integrating OTFS into AV driving systems and conducts performance evaluation to assess the reliability, latency, and throughput of the proposed system.

Consider an autonomous vehicle (AV) driving system operating within a 6G wireless communication framework. The system comprises N vehicles and M base stations deployed in the coverage area.

Here, we present the mathematical equations related to Orthogonal Time Frequency Space (OTFS) modulation for autonomous vehicle (AV) driving systems within a 6G wireless communication framework.

3.1 Signal Representation

In OTFS, the transmitted signal $x(t)$ is represented in the delay-Doppler domain as a function of time t, delay τ, and Doppler frequency f_d. Mathematically, the signal representation can be expressed as:

$$x(t, \tau, f_d) = \sum_{n=0}^{N-1} \sum_{m=0}^{M-1} X[n, m] \cdot \phi_{\mathrm{DD}}(t - nT, \tau - mT) \cdot e^{j2\pi f_d (nT - \tau)} \tag{1}$$

where:

- N is the total number of vehicles in the AV network.
- M is the total number of base stations deployed in the coverage area.
- $X[n, m]$ represents the transmitted symbol at time n and delay m.
- T is the symbol duration.
- $\phi_{\text{DD}}(t, \tau)$ is the delay-Doppler spreading function.

3.2 Channel Model

The channel between the transmitter and receiver experiences time and frequency dispersion effects due to vehicle mobility and environmental conditions. The channel impulse response h_{ij} represents the gain between users i and j.

3.3 Received Signal

The received signal $y(t)$ is affected by additive white Gaussian noise (AWGN) $w(t)$ and channel distortion:

$$y(t) = h_{ij} * x(t) + w(t) \tag{2}$$

where:

- $w(t)$ is the AWGN with power spectral density σ^2.

3.4 Sum Rate of Cluster l (R_l)

The sum rate of cluster l in OTFS can be calculated as:

$$R_l = \sum_{k=1}^{K} \log_2 \left(1 + \frac{P_{\text{BS}} \cdot \text{GR}_{ij} \cdot |h_{ij}|^2}{\sigma^2} \right) \tag{3}$$

where:

- K is the total number of subcarriers available for transmission.
- P_{BS} is the transmit power allocated to each base station.
- GR_{ij} is the gain ratio between users i and j.
- h_{ij} is the channel gain between users i and j.
- σ^2 is the noise power.

3.5 BER Calculation

The BER is calculated as the probability of incorrect bit detection at the receiver. For OTFS modulation, the BER can be expressed as:

$$\text{BER} = \frac{1}{NM} \sum_{n=0}^{N-1} \sum_{m=0}^{M-1} \mathbb{P}(\hat{X}[n, m] \neq X[n, m]) \tag{4}$$

where:

- $\hat{X}[n, m]$ is the estimated symbol at the receiver.
- $\mathbb{P}(\hat{X}[n, m] \neq X[n, m])$ is the probability of error for each symbol.

3.6 Variables

To formulate the problem, the following variables are defined:

- N: Total number of vehicles in the AV network.
- M: Total number of base stations (BSs) deployed in the coverage area.
- K: Total number of subcarriers available for transmission.
- P_{total}: Total available transmit power.
- P_{BS}: Transmit power allocated to each base station.
- P_{V2I}: Transmit power allocated to vehicle-to-infrastructure (V2I) communication.
- P_{V2V}: Transmit power allocated to vehicle-to-vehicle (V2V) communication.
- R_l: Sum rate of cluster l.
- GR_{ij}: Gain ratio between users i and j.
- h_{ij}: Channel gain between users i and j.
- σ^2: Noise power.

3.7 Error Detection and Correction

Various error detection and correction techniques, such as maximum likelihood detection, minimum mean square error (MMSE) estimation, and zero-forcing (ZF) detection, can be employed to improve BER performance in OTFS systems. By formulating the BER problem in this manner, we can analyze the performance of OTFS modulation in autonomous vehicle driving systems and evaluate its suitability for 6G wireless communication applications.

For OTFS in autonomous vehicle driving systems, the delay-Doppler domain transformation allows for efficient representation and processing of signals in high-mobility scenarios. The transformation involves mapping the time-domain signal to delay and Doppler domains, facilitating robust communication in dynamic environments. Additionally, power allocation strategies are employed to optimize spectral efficiency and minimize interference, ensuring reliable V2I and V2V communication. The mathematical models and algorithms used in OTFS enable effective handling of channel impairments and mobility-induced variations, enhancing the overall performance of autonomous vehicle networks in 6G wireless communication systems.

4 Simulation and Result Analysis

4.1 Simulation Setup

The simulation setup involves several system parameters crucial for evaluating OTFS modulation in AV driving systems. The modulation scheme used is OTFS, with a specific number of subcarriers denoted by M. The number of delay taps is represented as n_{Taps}, with delay tap values set as $\text{DelayTaps} = [0, 1, 2, 3, 4]$. Similarly, the Doppler tap values are $\text{DopplerTaps} = [0, 1, 2, 3, 4]$, and the noise power is $N_0 = 1$. For the simulation parameters, the Signal-to-Noise Ratio (SNR) ranges from $-10\,\text{dB}$ to $10\,\text{dB}$, and the number of Monte Carlo iterations is set to 100 (Table 1).

Table 1. Simulation Parameters

Parameter	Description	Value
EbdB	Array of Eb/No values in dB	–10:2:10
Eb	Energy per bit	Calculated from EbdB
No	Noise power	1
SNR (dB)	Signal-to-Noise Ratio in dB	Calculated from Eb and No
M	Number of transmit antennas	32
N	Number of receive antennas	16
nTaps	Number of channel taps	5
DelayTaps	Array of delay taps for channel model	[0 1 2 3 4]
DopplerTaps	Array of Doppler taps for channel model	[0 1 2 3 4]
Ncp	Length of cyclic prefix	Maximum of DelayTaps
ITER	Number of iterations for Monte Carlo simulation	100

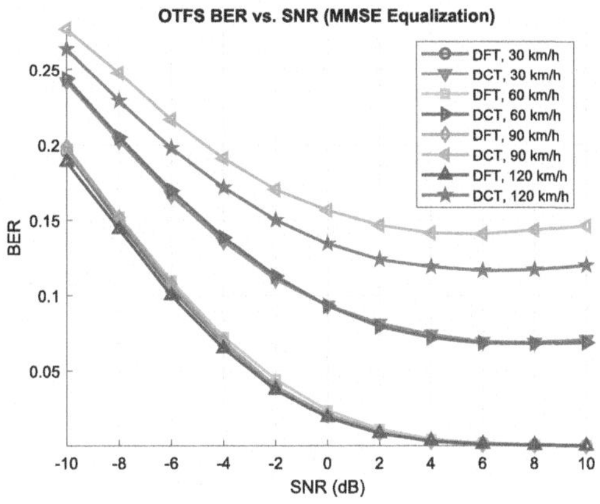

Fig. 3. Bit error rate Vs SNR performance of the OTFS based AV system

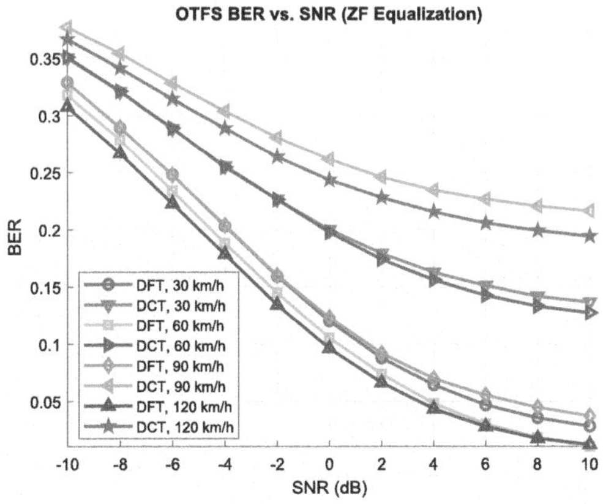

Fig. 4. Bit error rate Vs SNR performance of the OTFS based AV system

4.2 Channel Modeling

The channel modeling includes the generation of the Channel Impulse Response (CIR) using a complex Gaussian random vector h with n_{Taps} taps. The effective channel matrix (Heff) is then calculated, representing the channel response in the delay-Doppler domain. This calculation involves using transmit and receive precoding matrices, along with the specified delay and Doppler taps.

4.3 Signal Transmission and Reception

In signal transmission and reception, the process starts with the generation of binary data bits ($XddBits$) followed by square root energy normalization. These bits are then transformed to the time-frequency (TF) domain using Discrete Fourier Transform (DFT) matrices. During transmission, the precoded signal is sent over the channel, where it encounters Gaussian noise, which is added to the received signal.

4.4 Equalization Techniques

Two primary equalization techniques are employed: Minimum Mean Square Error (MMSE) and Zero Forcing (ZF) equalization. MMSE equalization estimates the transmitted symbols from the received signal by minimizing the mean square error. Similarly, ZF equalization estimates the transmitted symbols by nullifying the interference, aiming to maintain the integrity of the received signal.

4.5 Bit Error Rate (BER) Calculation

For BER calculation, the process involves comparing the decoded symbols with the transmitted symbols to determine the error rate. This calculation is performed for each SNR point for both MMSE and ZF equalization techniques, allowing for a comprehensive evaluation of their performance.

4.6 Results and Discussion

The BER performance analysis includes plotting BER versus SNR curves for both MMSE and ZF equalization techniques, comparing their performance under varying SNR conditions in Fig. 3 and Fig. 4. This analysis highlights the effectiveness of the equalization techniques in OTFS systems. The observations provide insights into the impact of channel characteristics, such as delay and Doppler spreads, on BER performance. Additionally, the discussion addresses the trade-offs between complexity and performance of the equalization techniques and explores the suitability of OTFS modulation for various communication scenarios.

The comparative analysis highlights the advantages of OTFS over OFDM in AV communication systems. OTFS modulation not only enhances BER performance, reducing error rates, but also improves throughput, ensuring more efficient data transmission. These improvements are essential for the reliable operation of autonomous vehicles, where low latency and high data rates are critical for safety and efficiency. The findings underscore the potential of OTFS to address the unique challenges of high-mobility communication, paving the way for its integration into next-generation AV systems within 6G wireless networks.

5 Conclusion

This paper has presented a comprehensive overview of the integration of Orthogonal Time Frequency Space (OTFS) modulation into autonomous vehicle (AV) driving systems within a 6G wireless communication framework. The paper has highlighted the advantages of OTFS modulation in addressing the unique challenges of high-mobility environments, Doppler shifts, and multipath fading encountered by AVs. Through performance evaluations and case studies, it has been demonstrated that OTFS offers higher throughput, and greater reliability compared to traditional communication methods such as Orthogonal Frequency Division Multiplexing (OFDM).

The paper has illustrated practical scenarios where AVs are controlled by a centralized base station using OTFS modulation. The potential impact of OTFS extends beyond individual AVs, transforming entire transportation ecosystems and paving the way for smarter cities and a more connected world. The adoption of OTFS modulation in AV systems represents a significant advancement in transportation technology, promising a future of safer, more efficient, and more intelligent transportation systems.

6 Challenges and Future Directions

6.1 Future Research Directions

Addressing the above challenges requires focused research and development efforts in several key areas.

- **Advanced Synchronization Techniques:** Future research should explore advanced synchronization techniques tailored for OTFS in high-mobility scenarios. Methods such as adaptive synchronization algorithms that can dynamically adjust to changing conditions could improve the robustness and reliability of the system.
- **Enhanced Channel Estimation Algorithms:** Developing more accurate and efficient channel estimation algorithms is crucial. Machine learning and artificial intelligence (AI) techniques could be leveraged to predict and adapt to channel variations in real-time, enhancing the performance of OTFS in rapidly changing environments.

- **Cost-Effective Hardware Solutions:** Research into cost-effective and robust hardware solutions for OTFS implementation is essential. Innovations in semiconductor technology and the development of specialized processors for real-time signal processing could reduce costs and improve performance. Additionally, efforts to miniaturize and ruggedize the hardware would facilitate its deployment in AV systems.
- **Interference Mitigation Strategies:** Future work should focus on advanced interference mitigation strategies. Techniques such as cooperative communication, where AVs share information to collectively manage interference, and the use of AI to predict and avoid interference scenarios, could enhance the reliability of the communication system.
- **Scalability Solutions:** Scalability can be addressed by exploring decentralized network architectures and edge computing solutions. By offloading some processing tasks to edge devices or implementing a hierarchical network structure, the system can better handle a large number of AVs. Additionally, dynamic resource allocation and efficient spectrum management techniques will be crucial in maintaining network performance as the number of connected vehicles grows.
- **Security and Privacy:** Ensuring the security and privacy of data transmitted in AV systems is another critical area for future research. Developing robust encryption methods and secure communication protocols tailored for OTFS can protect against cyber threats and ensure the integrity of the communication network.

References

1. Sabapathy, S., Prabhu, J.S., Maruthu, S., Jayakody, D.K.: Profuse channel estimation and signal detection techniques for orthogonal time frequency space in 6g epoch: a survey. IEEE Access **11**, 129963–129993 (2023)
2. Isik, M., Nkomo, M., Das, A., Dandekar, K.R.: Fpga implementation of otfs modulation for 6g communication systems. In: 2023 IEEE Future Networks World Forum (FNWF), pp. 1–7 (2023)
3. Harshavardhan, G., Bhat, V.S., Chockalingam, A.: Ris-aided otfs modulation in high-doppler channels. In: 2022 IEEE 33rd Annual International Symposium on Personal, Indoor and Mobile Radio Communications (PIMRC), pp. 409–415 (2022)
4. Raslan, W.A., Abdel-Atty, H.M., El-den, B.M.: Assessing the efficiency of multicarrier waveforms in next-generation 6g wireless communication systems. In: 2023 International Telecommunications Conference (ITC-Egypt), pp. 260–265 (2023)
5. Chu, T.M.C., Zepernick, H.-J., Hüük, A., Westerhagen, A., Granbom, B.: Otfs modulation for non-terrestrial networks: concepts, applications, benefits, and challenges. In: 2023 16th International Conference on Signal Processing and Communication System (ICSPCS), pp. 1–10 (2023)
6. Nauman, M., Lopacinski, L., Maletic, N., Scheide, M., Krstic, M., Grass, E.: 6g and beyond: hardware-in-the-loop experiments with otfs modulation using SDR. In: 2023 Workshop on Microwave Theory and Technology in Wireless Communications (MTTW), pp. 72–77 (2023)

7. Zhang, X., Wen, H., Yan, Z., Yuan, W., Wu, J., Li, Z.: A novel joint channel estimation and symbol detection receiver for orthogonal time frequency space in vehicular networks. Entropy **25** (2023)
8. Zielinski, T.P., et al.: Wireless otfs-based integrated sensing and communication for moving vehicle detection. IEEE Sens. J. **24**, 6573–6583 (2024)
9. Zhu, J., et al.: A phase-coded time-domain interleaved otfs waveform with improved ambiguity function. In: 2023 IEEE Globecom Workshops (GC Workkshpps), pp. 395–400 (2023)
10. Kumar, A., Chakravarty, S., Sao, C., Syed, T.M., Nanthaamornphong, A.: Peak to average power ratio est. and analysis in otfs 6g waveform by using selective mapping algorithm. SoutheastCon **2024**, 1070–1074 (2024)
11. Uthayakumar, T., Mei, J., Wang, X.: Hybrid multi-dimensional modulation in non-orthogonal spatial-delay-doppler domains for beyond 5g, and 6g communications. In: 2022 IEEE 95th Vehicular Technology Conference: (VTC2022-Spring), pp. 1–6 (2022)
12. Doosti-Aref, A., et al.: Performance analysis of otsm under hardware impairments in millimeter-wave vehicular communication networks. ArXiv arxiv:2309.04161 (2023)
13. Basharat, S., Hassan, S.A., Mahmood, A., Ding, Z., Gidlund, M.: Reconfigurable intelligent surface-assisted backscatter communication: a new frontier for enabling 6g iot networks. IEEE Wirel. Commun. **29**, 96–103 (2021)
14. Zheng, Y., et al.: Gamp-based low-complexity sparse bayesian learning channel estimation for otfs systems in v2x scenarios. Electronics (2023)
15. Kaushik, A., et al.: Integrated sensing and communications for iot: synergies with key 6g technology enablers. ArXiv arxiv:2309.13542 (2023)
16. Shi, J., et al.: Otfs enabled leo satellite communications: a promising solution to severe doppler effects. IEEE Netw. **38**, 203–209 (2024)

Geophysical Monitoring System for Landslide-Prone Areas Using OhmPi and ESP32 Integration

M. Aswathy, Aryadevi Remanidevi Devidas(✉), and H. Balaji(✉)

Center for Wireless Networks and Applications (WNA), Amrita Vishwa Vidyapeetham, Amritapuri, Vallikavu, India
{aswathym,aryadevird,balajih}@am.amrita.edu

Abstract. Landslides are the most frequent natural hazards in mountainous regions, posing significant threats to property and lives in the Himalayas and the Western Ghats of India. The Western Ghats rank as India's second most landslide-prone area, following the Himalayas. Effective monitoring of landslide-prone areas, particularly in regions susceptible to extreme weather events like the Western Ghats in India, necessitates advanced geophysical and environmental monitoring systems. Conventional techniques often lack the temporal and geographical resolution required to thoroughly analyze subsurface dynamics. We introduce an integrated geophysical monitoring system built on the OhmPi platform, enhanced with the capabilities of the ESP32 microcontroller. This device enables precise three-dimensional imaging of subsurface conditions through measurements of Electrical Resistivity Tomography (ERT), Spontaneous Potential (SP), and Induced Polarization (IP). Additionally, it includes external environmental sensors for temperature, pore water pressure, rainfall, and soil moisture, providing a comprehensive monitoring strategy. This technology enhances the effectiveness of early warning systems by facilitating real-time data collection, processing, and transmission over Wi-Fi or Bluetooth. Tested and prototyped in landslide-prone areas, the integrated system shows promise for improved risk management and community safety in high-risk regions.

1 Introduction

In recent years, India has experienced a notable increase in large landslides, driven by more frequent and intense monsoon rainfall [1] events. This surge underscores the urgent need to deepen our understanding of moisture-induced landslide dynamics, which exhibit significant spatial and temporal variability. Landslides in tropical mountainous regions present a formidable challenge due to the complexity of slope stability, compounded by seasonal monsoons and diverse geological formations. Traditional geophysical monitoring methods often struggle to provide accurate and timely assessments in such environments. However, geophysical techniques, particularly Electrical Resistivity Tomography (ERT), have

D. Gupta et al. (Eds.): PCEMS 2024, CCIS 2491, pp. 294–308, 2025.
https://doi.org/10.1007/978-3-031-90577-3_26

emerged as comprehensive tools for monitoring subsurface hydrological processes within unstable slopes, offering exceptional spatial resolution. Recent advancements in ERT present promising avenues for enhanced monitoring. This study explores the dynamic nature of resistivity thresholds to refine the monitoring of slope stability, especially in tropical mountainous regions. With the help of ERT and the sophisticated OhmMapper Imaging System (OhmPi), this study aims to record changes in subsurface resistivity over time. Integrating OhmPi technology, renowned for its high-resolution resistivity imaging, with ERT offers a means to surpass the limitations of traditional methods and attain deeper insights into subsurface conditions.

The enhanced OhmPi system supports 3D subsurface imaging and integrates various environmental sensors, including Spontaneous Potential (SP) and Induced Polarization (IP) measurements [2]. Utilizing the ESP32 microcontroller, the system benefits from built-in Wi-Fi and Bluetooth capabilities, lower power consumption, and cost-effectiveness, making it a versatile tool for geophysical applications [3]. This integrated approach demonstrates the OhmPi system's capability to provide near-real-time, high-resolution subsurface imaging and environmental data, making it suitable for landslide prediction, groundwater monitoring, and other geophysical applications. By developing dynamic resistivity thresholds sensitive to environmental variations, this research aims to enhance the accuracy and reliability of ERT-based slope stability monitoring, laying the groundwork for a more adaptive and responsive geophysical monitoring approach [4]. This research is poised to play a pivotal role in effective landslide hazard mitigation in dynamic and challenging terrains.

The rest of the paper is organized as follows: Sect. 2 examines the Literature review in this area. Section 3 details the Overview of the present Landslide Early Warning System deployed in Western Ghats. Section 4 narrates the Methodology, Sect. 5 is System design, Sect. 6 talks about the System design, and Sect. 7 is the conclusion and future work.

2 Literature Review

This section covers advanced techniques and technologies for landslip monitoring and prediction, including the basics of Electrical Resistivity Tomography (ERT), its evolution, and integration with other geophysical methods. It also discusses the use of the Internet of Things (IoT) and remote sensing technologies, alongside the development of open-source, accessible hardware solutions. Singh Rawat and Barthwal discussed a real-time monitoring system for landslides that employs an IoT-based architecture with various sensors (such as temperature, humidity, soil moisture, acceleration, and GPS) to gather critical meteorological, hydrological, geological, and geophysical data. However, they have not addressed the long-term sustainability and maintenance requirements of the system [5]. Amarasinghe et al., discussed a comprehensive review of 200 articles on rain-induced landslides in tropical regions, analyzing the relationship between rainfall patterns and landslide occurrences. It identifies key factors influencing landslides in these areas and discusses a structured risk assessment and management framework and work is not integrating with advanced modeling techniques to enhance the accuracy and applicability of landslide risk assessments in tropical climates. [6].

Nicola Casagli et al., discussed an application of remote sensing techniques such as interferometric SAR (InSAR) and multitemporal interferometric SAR (MTInSAR) for the detection, monitoring, and prediction of landslides. It emphasizes the role of these technologies in analyzing radar imagery to monitor and predict landslide movements by assessing displacement time series and employing ground-based and aerial systems for real-time monitoring, but they have not studied the system's cost-effectiveness [7]. Kyrillos et al., Evaluate surface-monitoring methods for large-scale landslides and delve into local-scale subsurface monitoring strategies that track movement, stress, water, temperature, and potential warning signs. The analysis also covers data-gathering methods and developing physical models and prototype field systems. The findings emphasize that the choice of monitoring method depends on specific case conditions, with considerations given to accuracy, spatial and temporal resolution, cost, durability, and field applicability. The author does not consider the interpolation of nonlinear relationships in Subsurface Data [8].

Arnaud Watlet et al., highlight how this data, alongside weather and pore water pressure readings, helps in understanding subsurface hydrological processes that could lead to slope failures, thus contributing to more effective local landslide early warning systems and it does not mention the Long-term Monitoring and Sustainability of the Monitoring Systems and the cost-effectiveness of the system [9]. Wu-Nan Tsai et al., Electrical resistivity tomography (ERT) to provide early warnings for landslides by monitoring the physical changes in a slope caused by rainfall infiltration. The technique employs a hybrid array of electrodes both in boreholes and on the surface, which measures changes in water saturation to predict slope stability. demonstrates ERT's effectiveness in correlating rainfall events with changes in subsurface resistivity, offering valuable insights for landslide prediction and early warning systems. The authors did not consider the integration of ERT Data with other Geotechnical Measurements like soil tension, and pore water pressure [10]. Denis Jongmans et al., explained the progress in geophysical monitoring of landslides, emphasizing the use of electrical resistivity imaging and continuous seismic recordings. It situates these developments within the broader context of geophysical methods applied to landslide analysis, highlighting how specific geophysical parameters can serve as precursors for the activation or reactivation of landslides. However, the cost analysis is not explained [11].

Aurore Carrier et al., studied the use of Induced Polarization (IP) alongside traditional Direct Current (DC) resistivity measurements to enhance landslide characterization and monitoring. This study does not establish a site-specific model that relates the measured geophysical properties to the actual soil and water characteristics [12]. Hojat et al., mentioned that ERT is effective in tracking hydrogeological changes and identifying fracture zones before collapse. Results indicated that landslide instability occurred where water saturation exceeded 45% particularly at boundaries of differing water saturations. However, they do not mention the cost-effectiveness of the system [13]. Vincenzo Lapenna et al., explained the use of time-lapse electrical resistivity tomography (TL-ERT) in monitoring landslides, emphasizing recent advancements in data processing and inversion methodologies. TL-ERT is highlighted as a robust and cost-effective method for identifying water-saturated zones and groundwater pathways within landslide bodies. It also proves valuable in tracking changes in top-soil moisture and the dynamics of wetting fronts during heavy rainfall events. This study reveals a complex system and needs to understand the complex relationship between resistivity measurements and specific hydrogeological parameters [14].

A. O. Cruz Júnior et al., developing an affordable, customizable, and programmable multichannel geo resistivity meter using open hardware and free software, offering a cost-effective alternative to expensive commercial systems. The prototype featured electrode multiplexing to streamline the switching process and reduce cable complexity, along with synchronous demodulation to enhance noise rejection, particularly useful in urban settings, the author failed to provide the field testing and validation of the system [15]. Watlet et al., proposed an OhmPi project that aims to develop an open-source, low-cost resistivity meter tailored for use by humanitarian organizations and the research community. Highlighting its flexibility, the OhmPi allows for novel acquisition strategies and complex monitoring experiments, distinguishing it from less adaptable commercial systems. It has not considered the Long-term performance and environmental testing of the system [16]. Clement et al., The development and testing of OhmPi, a low-cost, open hardware resistivity meter designed for environmental research, particularly suited for small-scale experiments. Priced under 500 dollars, OhmPi can handle automatic measurements using up to 32 electrodes. Initial tests using a soil-analog electrical circuit demonstrated the meter's capability to provide reliable and robust measurements across a broad range of resistance values. Research into OhmPi's long-term stability and performance under repeated use and different environmental conditions would be valuable [17].

Perrone et al., review the application of Electrical Resistivity Tomography (ERT) in the study of landslides, a technique that has gained traction due to technological advancements in data acquisition and tomographic inversion algorithms. It discusses the effectiveness of both 2D and 3D ERT in exploring a variety of landslide phenomena across different geological and geographical contexts, highlighting the method's strengths and limitations The research review does not cover the accuracy of ERT interpretations in such cases and potentially establish material-specific calibration methods [18]. Amarasinghe et al., present a comprehensive review of 200 articles on rain-induced landslides in tropical regions, analyzing the relationship between rainfall patterns and landslide occurrences. It identifies key factors influencing landslides in these areas and discusses a structured risk assessment and management framework and it is not in the site-specific conditions and monitoring data [19]. Most research has focused solely on real-time monitoring systems using ERT signals, neglecting both cost-effectiveness and the integration of ERT data with other geotechnical measurements. This research aims to integrate advanced geophysical monitoring techniques, such as Electrical Resistivity Tomography (ERT), with innovative platforms like OhmPi, which hold significant potential for enhancing landslide prediction and environmental monitoring. Future efforts should concentrate on long-term field validation and performance testing of these systems. Additionally, it is crucial to address their cost-effectiveness and scalability to ensure widespread adoption, particularly in resource-limited settings.

3 Overview: Landslide Early Warning System in Western Ghats

Every year, landslides pose a risk to human life and cause substantial financial losses in India. To tackle this critical issue, a real-time landslide monitoring and early warning system is in operation in Munnar, one of the location in the Western Ghats of India for the past 13 years [20]. Figure 1(a) shows the deployed IoT based geophysical sensors system which gather signals from the field via a Data Acquisition (DAQ) system that forwards the information to an IoT cloud for real-time analysis. Figure 1(b) shows the data from an Electrical Resistivity Tomography (ERT) system deployed near to the Landslide Early Warning System at Munnar to gather subsurface signals.

The IoT system monitors changes in soil moisture, temperature, and pressure with the aid of geophysical sensors, and the ERT system helps to map the underlying geological structures and identify potential risks. The ERT technology adds value to the IoT system based landslide early warning by mapping subterranean structures and locating possible weak spots. That will help to improve our comprehension of the mechanics of landslides, and forecast possible failures. In the present deployed system, these data are captured and analysed separately.

The goal of the proposed OhmPi system is to deliver robust, accurate, and reliable environmental assessments and landslide risk forecasting by combining these two functions into a single integrated system.

4 Methodology

Extreme weather events, like continuous rain, often trigger landslides, making early warning systems and effective monitoring essential to mitigate their impact on vulnerable areas. Traditional ground monitoring techniques, though effective, are costly, complex, and have limited accessibility. To address these challenges, we propose a versatile and cost-effective geophysical monitoring system based on the OhmPi platform with ESP32 microcontrollers. This system integrates various sensors, including Spontaneous Potential (SP) and Induced Polarization (IP) measurements, to comprehensively understand subsurface materials and processes. It features an Electrical Resistivity Tomography (ERT) system, external sensors for soil moisture, temperature, and pressure, a backup power system, and signal conditioning circuits. The software includes firmware for data acquisition and processing, algorithms for calculating resistivity, and wireless protocols for data transmission to a remote server or cloud platform. Designed for easy deployment and scalability, this system is suitable for applications like seismic forecasting, groundwater analysis, and environmental monitoring. By enhancing accuracy and reliability in ecological assessment and landslide risk prediction, this system significantly improves risk management and protects vulnerable communities, representing a major advancement in geospatial technology.

Data Collection and Sensor Integration: The system incorporates temperature, pressure, and soil moisture sensors to monitor areas susceptible to landslides. These sensors offer real-time environmental data that is essential for risk assessment and landslip prediction. One of the main factors affecting slope stability is variations in water content, which is detected by soil moisture sensors. This information is enhanced with pressure and temperature sensors, providing a thorough picture of the surroundings.

Data Acquisition Using OhmPi and ESP32: The OhmPi platform combined with the ESP32 microcontroller powers the data collection system. Sensor data is gathered, processed, and ready for system transmission. Because of its low cost and high-resolution 3D imaging capabilities, the OhmPi platform was chosen. Real-time data transfer is made possible by the integrated Wi-Fi and Bluetooth capabilities of the ESP32 microcontroller.

(a) (b)

Fig. 1. (a) DAQ system with geophysical sensors in the Field.(b) ERT testing in the field.

Electrical Resistivity Tomography (ERT) Implementation: Electrodes are positioned along the survey line to perform Electrical Resistivity Tomography (ERT). An electrical current is supplied to assess changes in subsurface resistivity, and the resulting potential difference is recorded. Slope stability analysis requires the precise spatial resolution of subsurface conditions, which ERT offers (Fig. 2).

- Develop a versatile and cost-effective geophysical monitoring system: Most commercial devices are closed-source, limiting their ability to interface with additional sensors. OhmPi, however, can be easily controlled by any sensor and produce values that can initiate measurements, making data collection more intelligent and adaptable.
- Including Spontaneous Potential (SP) and Induced Polarization (IP) measurements: The system provides a comprehensive understanding of subsurface processes and properties.
- Enhance good system connectivity and data management: By using the ESP32 microcontroller, has built-in Wi-Fi, Bluetooth, and lower power consumption.
- Providing scalability and ease of deployment: It is more compatible and easy to install and used to verify various geophysical monitoring needs, such as landslide prediction, groundwater assessment, and environmental monitoring (Fig. 3).

We are developing an OhmPi platform-based integrated geophysical monitoring system. High-resolution 3D imaging of subsurface conditions is provided by the upgraded OhmPi system, which uses the ESP32 microcontroller to perform Electrical Resistivity Tomography (ERT), Spontaneous Potential (SP), and Induced Polarisation (IP) measurements. The system provides a comprehensive monitoring strategy by integrating external environmental sensors for temperature, pore water pressure, rainfall, and soil moisture. Integrating sensors makes real-time data collection,[16] processing, and wireless transmission to a distant server for additional analysis and visualization possible. Our method combines sophisticated data processing and communication software with accurate hardware design, including signal conditioning, multiplexer control, and reliable power supply management. The system's capacity to detect crucial subsurface alterations brought on by monsoon rains was shown through field deployment in the Western Ghats, boosting the resilience of early warning systems. This study opens the door to scalable and affordable landslip monitoring methods, which could revolutionize risk management strategies in susceptible areas.

4.1 System Flow

The data acquisition process begins with initializing the power supply to ensure all components are powered. The system checks sensor connectivity to verify proper function, then collects data from external sensors monitoring soil moisture, temperature, pressure, and rainfall. Analog signals from these sensors are

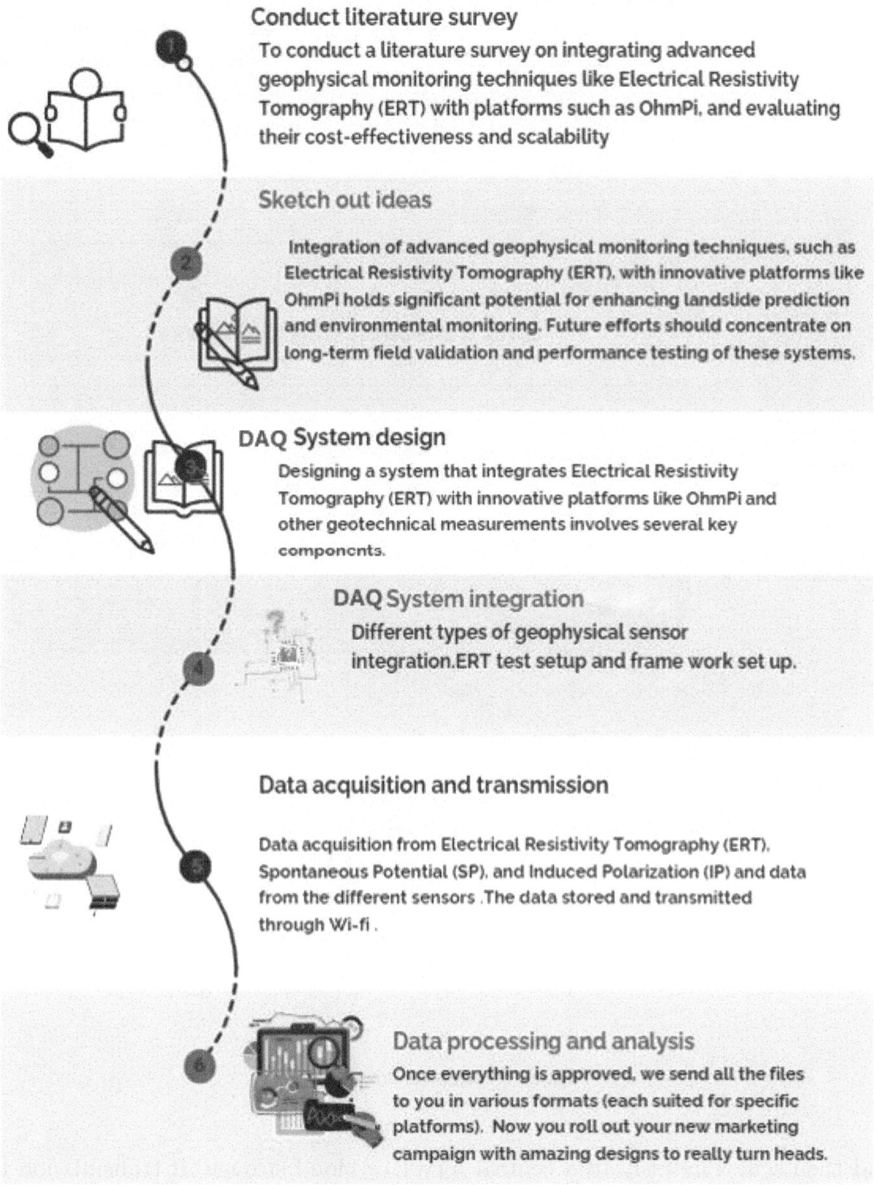

Fig. 2. System Processing.

converted to digital, processed through a multiplexer, and conditioned using voltage dividers and filters for accuracy (Fig. 4).

The data transmission process starts by initializing the ESP32 module for Wi-Fi connectivity. Data is pre-processed on the ESP32 for efficient transmission

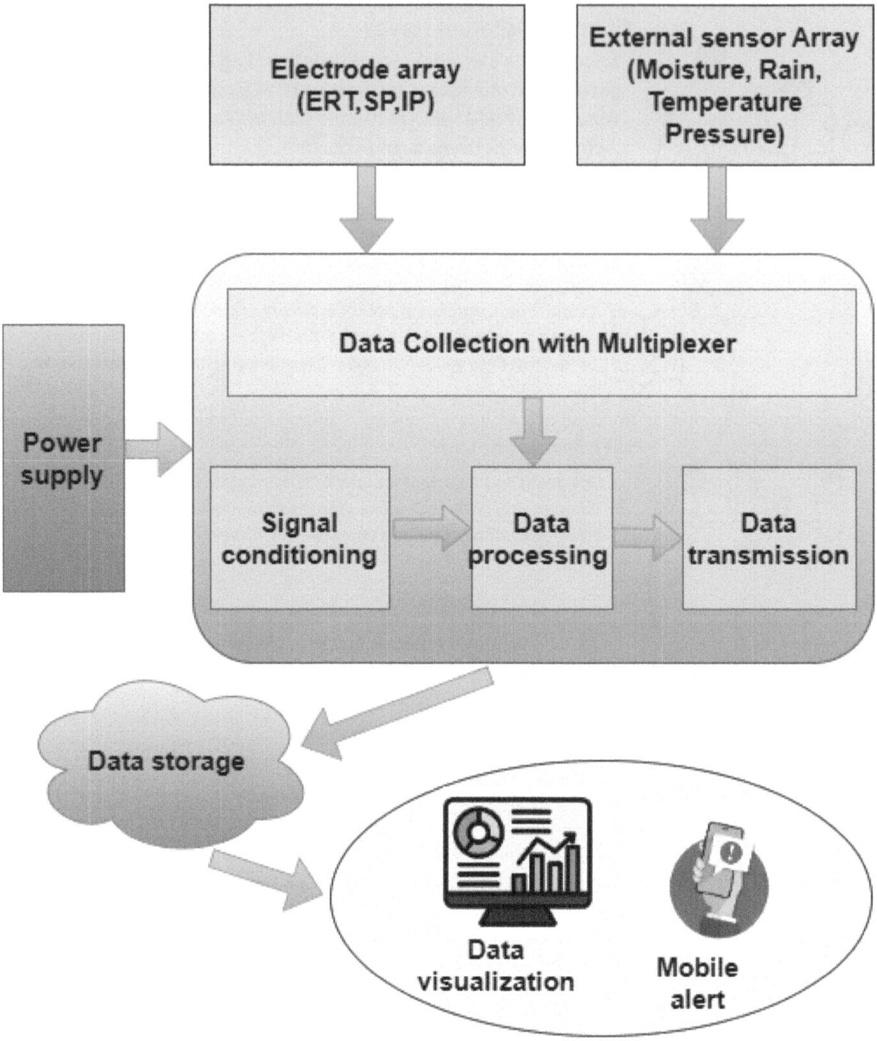

Fig. 3. System architecture.

and then sent wirelessly to a central server or cloud storage. If transmission is successful, data is stored and visualized on a web or mobile app, with alerts and notifications set for critical changes, providing real-time monitoring and analysis.

4.2 Sensors

Integrating different sensors with the monitoring system allows for comprehensive environmental monitoring, providing valuable data for landslide prediction and risk assessment.

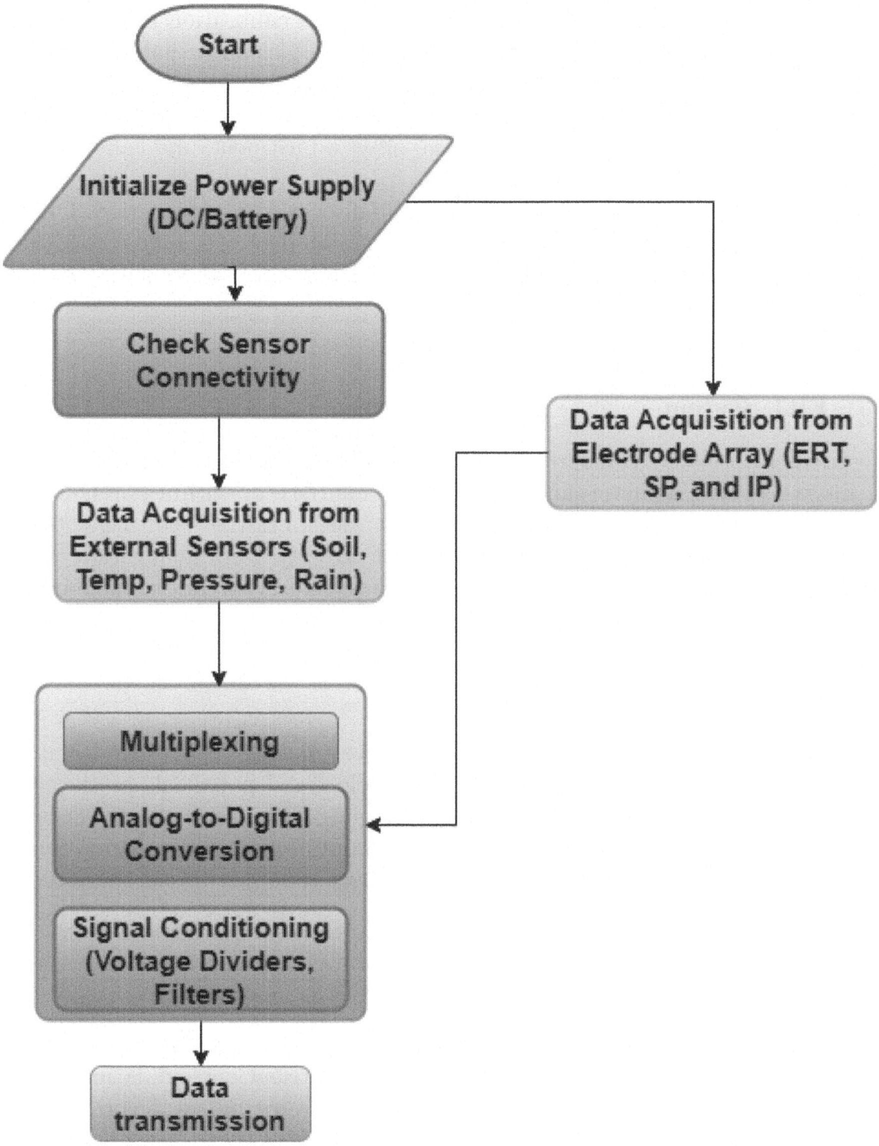

Fig. 4. Data Acquisition Flow diagram.

- Soil Moisture Sensor: Measures the volumetric water content of soil. Provides an analog voltage or digital signal proportional to soil moisture. Signal conditioning may involve amplification, filtering, and calibration to ensure accurate measurement.
- Temperature Sensor: Measures the temperature of the surrounding environment. Output is a voltage or digital signal corresponding to temperature.

Signal conditioning may involve linearization and calibration to convert raw sensor data into temperature readings in degrees Celsius or Fahrenheit.
- Pressure Sensor: Measures atmospheric pressure or pressure from soil or water. Provides an analog or digital signal proportional to pressure. Signal conditioning may include amplification, offset adjustment, and calibration to ensure accurate pressure measurement.
- Rain Gauge: Measures the amount of rainfall over a specific period. Output is a digital signal or pulse for each unit of rainfall detected. Signal conditioning may involve debouncing circuits to filter out noise and ensure accurate counting of rain events (Fig. 5).

4.3 ERT System

Electrical Resistivity Tomography (ERT) involves placing electrodes at regular intervals along a survey line on the ground surface or within boreholes. An electrical current is applied through two electrodes, and the resulting potential difference is measured using the remaining electrodes. Data acquisition is performed using multiple electrode configurations, such as Wenner or dipole-dipole, to capture variations in resistivity with depth.

4.4 Spontaneous Potential

The Spontaneous Potential (SP) method involves placing two electrodes at the surface or within boreholes to detect natural electrical potentials produced by electrochemical processes. Unlike other techniques, SP measurement tracks the potential difference between the electrodes without applying an external electrical current. Analyzing anomalies in SP data over time can reveal the presence of mineral deposits, cracks, or clay layers.

4.5 Induced Polarization

Induced Polarization (IP) utilizes a setup in the ERT with electrodes positioned along a survey line. During measurement, an electrical current is introduced into the ground, and subsequent voltage responses over time are recorded. Data acquisition entails capturing the amplitude and phase of these voltage responses at various time intervals. The resulting IP data offer insights into the polarizability of subsurface materials, indicating the presence of clay minerals, sulfide minerals, or other polarizable materials.

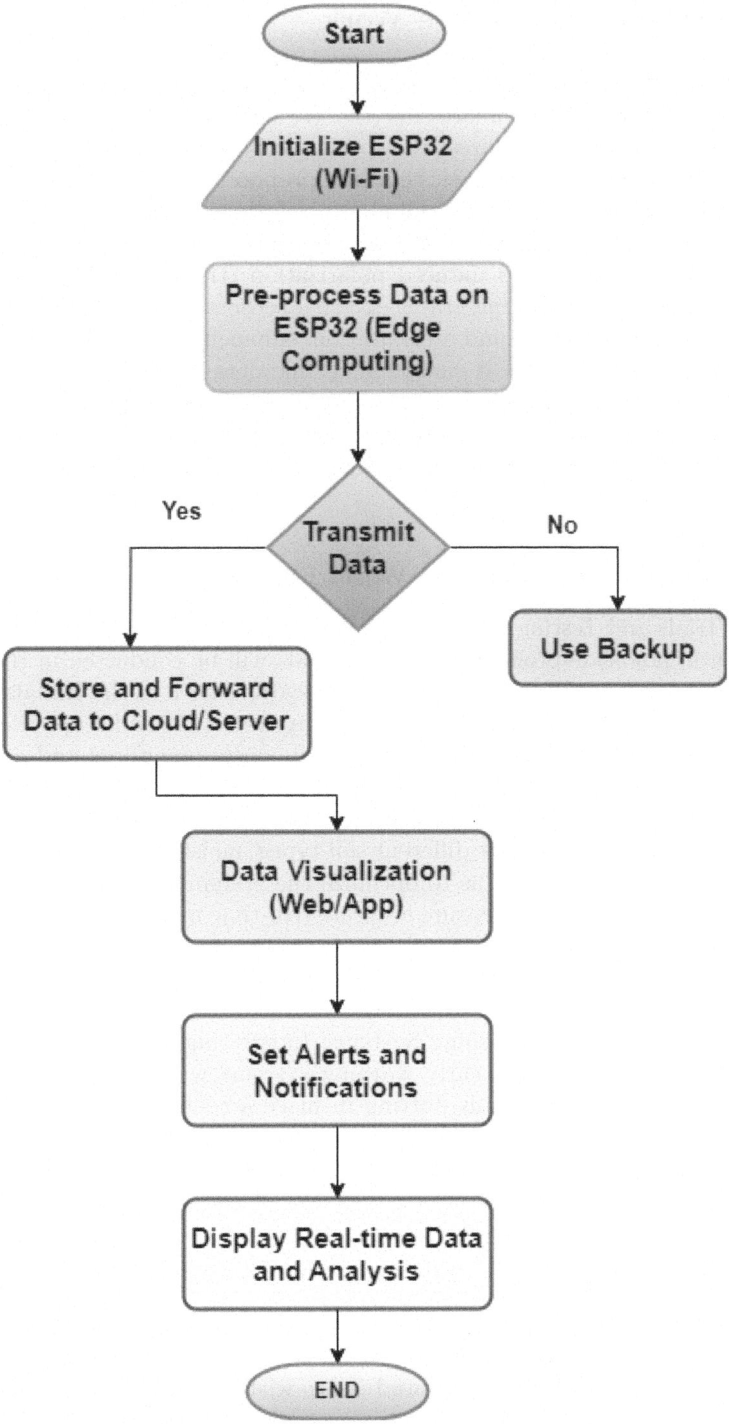

Fig. 5. Data Transmission Flow Diagram.

5 Conclusion and Future Work

Improved geo-environmental models are crucial for managing earthquake-prone areas, especially in severe climates like the Western Ghats in India. Traditional methods often struggle to address the spatial and temporal challenges of underground development. To address this, we propose an integrated geophysical monitoring system based on the OhmPi platform and the ESP32 microcontroller. This system combines electrical resistivity tomography (ERT), spontaneous potential (SP), and induced polarization (IP) to provide detailed 3D subsurface imaging. It also integrates external sensors to monitor soil moisture, temperature, pore pressure, and precipitation, enhancing earthquake forecasting. Using Wi-Fi and Bluetooth, it enables real-time data acquisition and transmission for more efficient warning systems. While still in the testing phase, this system shows promising potential for improving disaster management and community safety by advancing seismic mitigation strategies with innovative technologies.

5.1 Implementation and Real-World Experiments

- Field Trials and Testing: The next phase involves rigorous field testing of the system in landslide-prone areas. Initial trials will be conducted in the Western Ghats, followed by testing in diverse locations such as the Himalayas and coastal slopes. During these trials, data from the sensors and geophysical measurements (ERT, SP, and IP) will be continuously monitored and compared with historical landslide data to assess the system's accuracy and reliability.
- System Calibration and Optimization: The collected data will be used to refine sensor calibration for different soil types, moisture levels, and environmental conditions, as well as to optimize the system's algorithms. This calibration phase is crucial to ensure accurate real-time monitoring across various terrains. Additionally, the system's power management and data transmission methods will be optimized to extend battery life and enhance communication range in remote areas.
- Integration into Early Warning Systems: Integrating the monitoring system with the current landslip early warning systems will be the last phase of implementation. This entails putting in place a centralized system for data gathering and analysis, allowing automated notifications for communities that are at risk, and permitting real-time risk assessments.

5.2 Technical Challenges and Integration Solutions

- Data Synchronization and Interference: Combining observations from multiple sensors and geophysical techniques in real-time poses challenges with data synchronization, as well as potential issues with electrical noise and interference. To address this, a data acquisition framework using the ESP32 microcontroller was developed, employing a synchronized sampling system that triggers each sensor's data collection in sequence. Signal conditioning circuits with filters were also added to reduce electrical noise before processing.

- Environmental Factors and Sensor Reliability: Environmental factors that can affect a sensor's performance and dependability include moisture intrusion and temperature swings. Ruggedized shells with waterproof and temperature-resistant features were employed to increase robustness. To better account for these environmental differences, automatic self-calibration methods will be the main focus of future advancements.
- Data Processing and Storage: Efficient processing and storage are necessary due to the substantial amount of high-resolution data from ERT, SP, and IP measurements as well as real-time sensor data. To solve this, on-site data preparation using edge computing and the ESP32 module was employed. This included filtering, noise reduction, and preliminary analysis to find changes in subsurface conditions. After processing, the data is sent to a cloud-based server for additional analysis and long-term storage.

Acknowledgment. We express our honest gratitude to our beloved Chancellor Sri. Mata Amritanandamayi Devi (Amma), for the inspiration and motivation in all our endeavors. We would like to acknowledge our teammates and associates in the Amrita Center for Wireless Networks applications

References

1. Ramesh, M.V., Ushakumari, P.: Threshold-based data aggreg0.23ation algorithm to detect rainfall-induced landslides. In: Proceedings of the 2008 International Conference on Wireless Networks (ICWN'08), vol. 1, pp. 255–261 (2008)
2. Aswathy, M., Praveen, K., Balaji, H.: A novel data acquisition board for heterogeneous sensor network. In: 2023 8th International Conference on Communication and Electronics Systems (ICCES), pp. 487–491. IEEE (2023)
3. Liya, M.L., Aswathy, M.: Impact of basic communication modules and techniques in humanitarian crisis-a brief survey. In: 2024 International Conference on Computer, Electrical & Communication Engineering (ICCECE), pp. 1–8. IEEE (2024)
4. Slater, L., Binley, A., Versteeg, R., Cassiani, G., Birken, R., Sandberg, S.: A 3D ERT study of solute transport in a large experimental tank. J. Appl. Geophys. **49**(4), 211–229 (2002)
5. Rawat, P.S., Barthwal, A.: LANDSLIDE MONITOR: a real-time landslide monitoring system. Environ. Earth Sci. **83**(8), 1–18 (2024)
6. Amarasinghe, M.P., Kulathilaka, S.A.S., Robert, D.J., Zhou, A., Jayathissa, H.A.G.: Risk assessment and management of rainfall-induced landslides in tropical regions: a review. Nat. Hazards, 1–53 (2023)
7. Casagli, N., Intrieri, E., Tofani, V., Gigli, G., Raspini, F.: Landslide detection, monitoring, and prediction with remote-sensing techniques. Nat. Rev. Earth Environ. **4**(1), 51–64 (2023)
8. Ebrahim, K.M., Gomaa, S.M., Zayed, T., Alfalah, G.: Recent phenomenal and investigational subsurface landslide monitoring techniques: a mixed review. Remote Sens. **16**(2), 385 (2024)

9. Watlet, A., et al.: 4D electrical resistivity to monitor unstable slopes in mountainous tropical regions: an example from Munnar, India. Landslides **20**(5), 1031–1044 (2023)
10. Tsai, W.N., et al.: Electrical resistivity tomography (ERT) monitoring for landslides: case study in the lantai area, yilan taiping mountain, Northeast Taiwan. Front. Earth Sci. **9**, 737271 (2021)
11. Jongmans, D., Fiolleau, S., Biévre, G.: Geophysical monitoring of landslides: State-of-the-art and recent advances. In: Understanding and Reducing Landslide Disaster Risk, vol. 3 Monitoring and Early Warning 5th, pp. 75–84 (2021)
12. Carrier, A., Meric, O., Bottelin, P.: Characterizing landslide dynamics from time-lapse time domain induced polarization and ground-based imaging: a case study of the MontGombert landslide (French, Alps). Landslides **21**(2), 353–369 (2024)
13. Hojat, A., et al.: Geoelectrical characterization and monitoring of slopes on a rainfall-triggered landslide simulator. J. Appl. Geophys. **170**, 103844 (2019)
14. Lapenna, V., Perrone, A.: Time-lapse electrical resistivity tomography (TL-ERT) for landslide monitoring: recent advances and future directions. Appl. Sci. **12**(3), 1425 (2022)
15. Cruz Júnior, A.O., da Ponte-Neto, C.F., Wiermann, A.: Design and construction of an automated and programmable resistivity meter for shallow subsurface investigation. Geosci. Instrument. Methods Data Syst. **12**(1), 15–23 (2023)
16. Watlet, A., et al.: Latest developments of OhmPi: towards small-scale ERT surveys and monitoring experiments in the field. In: NSG2023 29th European Meeting of Environmental and Engineering Geophysics, vol. 2023, no. 1, pp. 1–5. European Association of Geoscientists & Engineers (2023)
17. Clement, R., Fargier, Y., Dubois, V., Gance, J., Gros, E., Forquet, N.: OhmPi: an open source data logger for dedicated applications of electrical resistivity imaging at the small and laboratory scale. HardwareX **8**, e00122 (2020)
18. Perrone, A., Lapenna, V., Piscitelli, S.: Electrical resistivity tomography technique for landslide investigation: a review. Earth Sci. Rev. **135**, 65–82 (2014)
19. Daily, W., Ramirez, A., Binley, A., LeBrecque, D.: Electrical resistance tomography. Lead. Edge **23**(5), 438–442 (2004)
20. Ramesh, M. V., Thirugnanam, H., Singh, B., Nitin Kumar, M., Pullarkatt, D.: Landslide early warning systems: requirements and solutions for disaster risk reduction-India. In: Progress in Landslide Research and Technology, vol. 1, no. 2, pp. 259–286. Springer, Cham (2023)

A 2.4 GHz PLL Design Using a Source Switched Charge Pump and An Active Loop Filter

Sayali Pralhad Joshi(✉), Vibha Vyas, Vaishali Ingale, and Vanita Agarwal

COEP Technological University, Pune, India
{sayalipj22.extc,hod.extc,vvi.extc,vsa.extc}@coeptech.ac.in

Abstract. In our proposed work a 2.4 GHz PLL was designed in 180 nm technology. PLLs were created specifically for transceivers and processors to generate frequencies. The proposed PLL has acquired lock in 600 ns. The reference frequency was taken as 75 MHz. This design made use of a LC type VCO. This paper is unique from the point of view that usually a PLL design is realized using a passive loop filter. But due to the limited range of this filter, this design has employed the use of an active loop filter. An active loop filter has been designed using a Telescopic Opamp with 40 dB gain at lower frequencies. With a higher control range we can lock the VCO at a higher frequency band. The range of the control voltage was 1.6–1.8 V. The phase noise measured was –156 dBc/Hz at 100 MHz offset. The total power dissipation is quite low at 36.84 mW for the entire design.

Keywords: PLL - Phase Locked Loop · ALF - Active Loop Filter · VCO - Voltage Controlled Oscillator · CP - Charge Pump · PFD - Phase Frequency Detector · ICMR - Input Common Mode Range · CMRR-Common Mode Rejection Ratio

1 Introduction

Phase Locked Loop (PLL) is a closed-loop negative feedback synthesizer that is used to synchronize the local oscillator with the reference clock. PLLs, are frequently employed in high-performance digital systems to produce precisely timed on-chip clocks. Modern wireless communication systems such as Bluetooth and Zigbee use Phase Locked Loop (PLL) largely for jitter reduction, clock synthesis, skew, and synchronization. The local oscillator is voltage dependant and is most susceptible to noise. The noise is mostly present in the form of phase noise or jitter. The output signal from the oscillator can either have a phase lag or a phase lead. This could cause variations in the clock signal, which would cause receivers to lose information. The PLL designed in this paper is for the frequency of 2.4 GHz. [3–5, 7, 13] are being referred in this paper to understand the general working of PLL and all its comprising blocks. In order to minimize

D. Gupta et al. (Eds.): PCEMS 2024, CCIS 2491, pp. 309–321, 2025.
https://doi.org/10.1007/978-3-031-90577-3_27

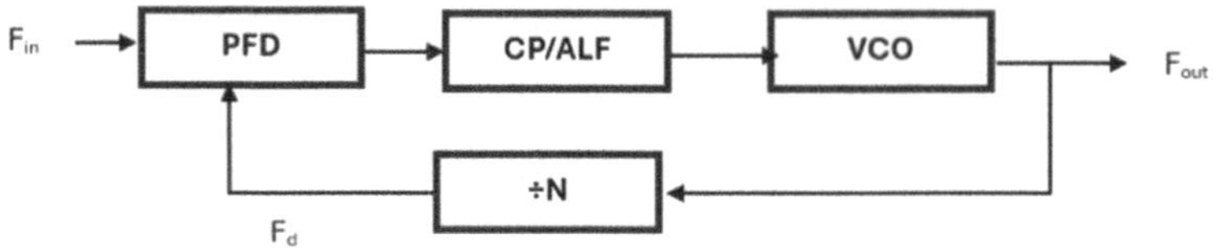

Fig. 1. PLL Block Diagram

phase error, the PLL closed loop compares the oscillator output's phase to the reference signal. The Type II PLL block majorly consists of the PFD, CP, Loop Filter, VCO and a N-Divider. In Fig. 1 we can observe the block diagram closely. Major Contributions of the work presented are:

- The use of Gate switched Charge Pump
- The loop filter is realized using Active Loop Filter to increase the range of VCO control voltage range. 2 stage Op Amp has been used to serve this purpose.
- LC type Voltage Controlled Oscillator has been used.

This is how the paper is structured. Section 2 covers the literature survey and circuit architecture where all the blocks are explained with respect to the resources available. A brief working of these blocks is also mentioned. Section 3 explains the methodology and design particulars used in this paper. Section 4 describes the simulation, results, and analysis. Section 5 gives the conclusion and a comparison table to highlight all of the current research done on this topic.

2 Literature Survey and Circuit Architecture

This section gives us a basic understanding of the PLL structure based on the available literature. The generic Type I PLL uses a phase detector. This is basically a mixer that gives a proportional voltage from the phase differences between two signals. Analog and digital phase detectors are the two varieties available [7]. The reference and VCO signals are multiplied by the analog phase detector. A double frequency term and a *cos* function of the phase difference are present in the output. Usually, a EX-OR gate is used to create a digital phase detector. But the process of phase detection is a non linear process. If the lock frequency is very far from the current oscillator frequency, the amount of time required to acquire lock is indefinite. In order to solve this issue, newer PLLs incorporate a frequency detector along with the phase detector [7]. The Frequency detector brings win and $wout$ close to each other. Subsequently, the Phase detector takes the control of the acquire range when the difference becomes small. If the signals are periodic, we can combine these two loops and use a circuit called as "Phase/Frequency Detector". When the phases of the reference clock and the feedback clock deviate by a certain amount, the PFD generates an error output signal that reflects this phase difference [12]. The input reference frequency and the VCO's feedback frequency are the respective inputs for this sequential circuit. The outputs are namely UP and DOWN. [11] uses a crystal of 75 Mhz. This shows that the reference frequency from the crystal oscillator should be of the order of a few MHz.

The Voltage Controlled Oscillator is the main part of PLL. An ideal VCO has its output frequency equal to a linear function of the control voltage. Most transceivers require oscillators with tunable frequencies. The loop filter's control voltage aids in the VCO's tuning. Ring oscillators and LC oscillators are the two forms of VCOs. We have utilized an LC voltage-controlled oscillator in this paper. The VCO is most susceptible to process variation. The oscillation frequency will therefore vary greatly as a result [18]. The LC topology's oscillation frequency is

$$f_{osc} = \frac{1}{2\pi\sqrt{LC}} \tag{1}$$

Varactors or Voltage-dependent capacitors are used to tune the frequency. Generally, reverse biased pn junction diodes were used as varactors [2,4]. But there were many drawbacks associated with it, like high resistivity of the n-well material and the constant substrate capacitance that reduced the Quality factor and limits the control voltage range. Hence, MOS varactors are introduced. MOS varactors are employed in oscillators nowadays. The structure of the MOS varactors is such that the NMOS transistor is inserted inside a n-well, forming an "accumulation-mode varactor", as shown in Fig. 2.

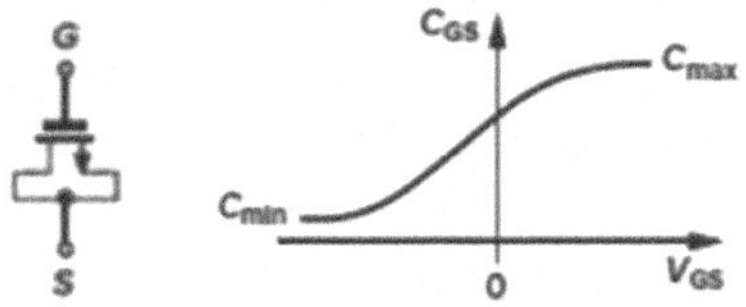

Fig. 2. Accumulation mode Varactor modelled as NMOSCAP and its Cgs vs Vgs graph [13]

The capacitor's first terminal is made by ohmically shorting the substrate, drain terminal, and the source terminal, while the second terminal is formed by the gate. A simple LC tank oscillator demonstrates a decaying effect as a result of energy being lost through the inductor in the form of heat. Hence a cross linked negative-Gm oscillator circuit architecture has been used.

Next block is the charge pump. Through the use of a charge pump circuit, the PFD's sequential logic state is transformed into an analog signal that may be utilized to drive a VCO. It is made up of two symmetrical current sources, "UP" and "DOWN," that are controlled by the PFD output [6]. The charge pump is used to reduce the finite phase error that is inherently present in The Type I PLL. Our aim is to make the loop gain infinite, so that any error can be removed. This can be achieved by using Charge Pump and loop Filter combination. Additionally, the clocked output from the phase/ frequency detector needs to be converted to a proportional DC output. It consists of two current sources that are used interchangeably according to the outputs from PFD. [6] discusses the various techniques that are currently being used to design Charge Pump. The CP 's transfer function is given as below.

$$\frac{V_{out}}{\Delta\phi}(s) = \frac{I_p}{2\pi(R_p + \frac{1}{C_p s})} \tag{2}$$

The loop filter is connected to the output of Charge Pump [13]. It converts the current from the charge pump to equivalent DC voltage and filters out any higher frequency components. The loop filter decides the loop bandwidth of the circuit. It is the crucial part of the PLL as it decides the speed of lock acquisition. We use either a passive 2nd order RC filter or an active loop filter. The loop filter decides the bandwidth of the PLL and the hop-time in case of a Fractional-N PLL. To achieve high gain Common gate and common source are cascaded to create a Telescopic Opamp architecture that provides a high gain [15,16]. Telescopic amplifiers are commonly utilized because of their efficiency, high voltage gain, high swing, and small design [17].

In order to match the output oscillator's frequency to the reference frequency, a frequency divider is therefore necessary. The integer divider is in the order of $\div$ 2,$\div$4,$\div$8 and$\div$16. If N increases by 1, $fout$ increases by $fref$ [13].

3 Methodology

This section discusses the suggested circuit topology that is applied in this work. When choosing the circuit topologies to be used, certain application parameters are taken into consideration for PLL design. Bluetooth and Zigbee both use 2.4 GHz frequencies in their operations. The four primary building blocks were individually designed, tested, and then integrated into a closed loop. Figure 3 portrays the structural circuit designed for this paper.

3.1 Proposed Phase Frequency Divider

Comparing the feedback frequency and the reference frequency is done first in the phase frequency divider block. Conventional sequential logic typically uses D flip flops to do this. The dead band zone at zero, however, is one of its drawbacks. This originate from D flip flop's finite rise and fall times, which leave a tiny, uncaptureable region of phase difference. Therefore, it cannot be identified if the phase difference is less than $|\ \phi\ |$. Thus a random phase error accumulates in the VCO in the form of 'jitters'. Additionally there is no corrective feedback present. The design used in this paper, cited from [12] minimizes the delay caused by the reset path and makes use of pass transistor logic. The Reference clock is compared to the feedback clock' inverted value and then passed through transmission gate logic. An analogous logic of this is followed by the 'UP' signal. In all, the circuit made use of ten transistors. The reference clock frequency is

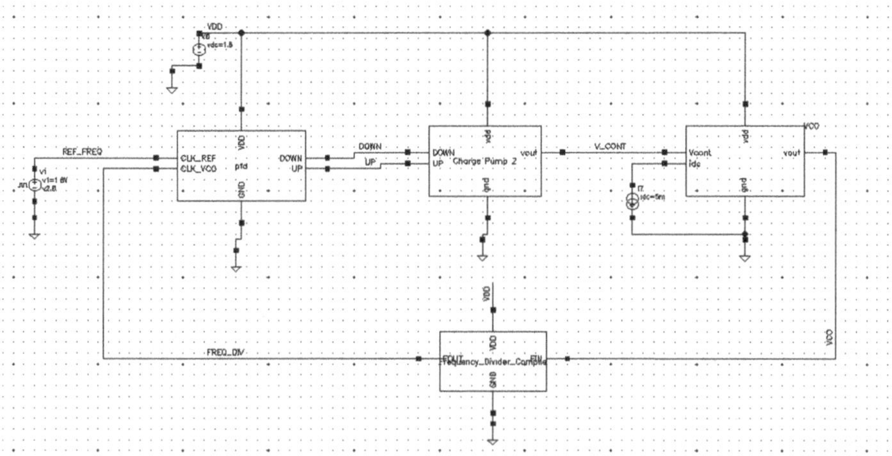

Fig. 3. The structural model of the PLL circuit

75 MHz. The circuit of Phase Frequency detector is show in Fig. 4. The PFD's gain is:

$$K_{PFD} = \frac{1}{2\pi} \tag{3}$$

3.2 Proposed Voltage Controlled Oscillator

The voltage-controlled oscillator is the first block to be designed. LC oscillators are recommended because of their stable response since the intended frequency is fairly high. Monolithic inductors have made it possible to use passive components to construct oscillators. During the design phase, an inductor value of about 1.5 nH was assumed. It was determined that the capacitance value was around 34 pF using the calculation above. An NMOS varactor has been utilized in order to have a variable capacitance. It's symbol is showcased in Fig. 2 shows the graph between Cgs and Vgs and the symbol of a varactor. Thus with changing capacitance, the desired frequency of oscillation can be varied. Two NMOS varactors were used back to back, each of value 20 pf. The practical value of the free running frequency (when control voltage is '0') is equal to 2.3 GHz. The cross coupled NMOS and PMOS transistors are required to have a high gain to create sustainable oscillations.

[13] explains that an ideal VCO is such whose output frequency is a linear function of it's control voltage.

$$w_{out} = w_0 + K_{VCO} \times V_{cont} \tag{4}$$

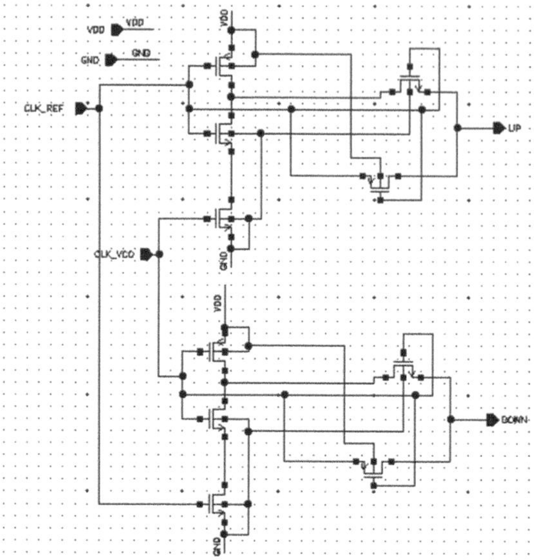

Fig. 4. Phase Frequency Detector design

Voltage (V)	Frequency (Hz)
0	2236886254
0.1	2342798238
0.3	2493765586
0.5	2881927433
0.8	2852497361
0.9	2821988937
1	2699055330
1.3	2634351949
1.6	2482560015
1.8	2320400965

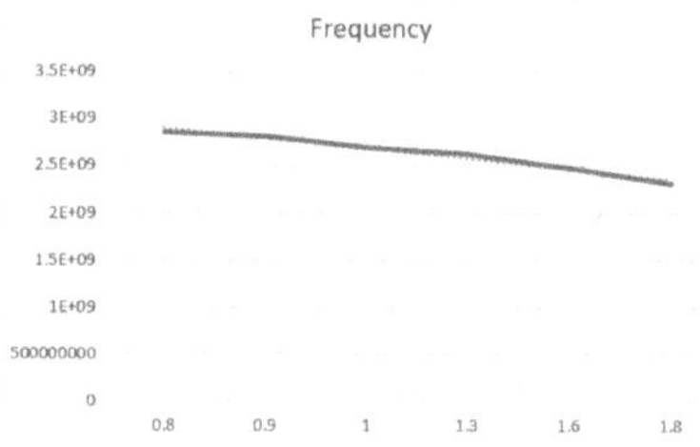

Fig. 5. Graph between measured frequency and applied VCO Control Voltage

The gain of the VCO was measured to be 400 MHz/V. The values were plotted on a graph as shown in Fig. 5. The bias current for this VCO was chosen to be 5 mA. A PMOS current source was employed in the design. The inductor is never ideal. There always exists some series resistance to it. This series resistance is modelled in the form of an *Rp*. The equivalent parallel resistance is measured to be 178.8 Ω.

$$Rp = Q^2 \times Rs \tag{5}$$

Here Q represents the quality factor. We know that every LC tank always exhibits a decaying oscillatory behavior because of the presence of the series resistance, therefore this modeling is helpful. If a resistance $-Rp$ is placed in parallel to Rp,then equivalent resistance $Rp \mid\mid -Rp$ across the oscillator is equal to ∞. The tank can oscillate infinitely. If there is a significant positive feedback or the loop gain is sufficiently negative, this negative resistance can be reached. The cross coupled NMOS transistors create a sufficient positive feedback. The equation of Voltage vs Current is as follows:

$$\frac{Vx}{Ix} = \frac{-2}{gm} \tag{6}$$

In order to complement the NMOS pair, an equivalent PMOS pair is attached. A graph between Vgs and resulting frequency was plotted to draw the range of Vcontrol. For the tank's oscillations to build up the necessary condition is $Rp - 2/gm \geqslant 0$. Here gm represents the transconductance of the PMOS and NMOS transistors. After optimization, the 'W' was chosen to be 40 equivalent parallel resistance u and 'L' to be 0.2 equivalent parallel resistance u. The circuit topology created is shown in Fig. 6.

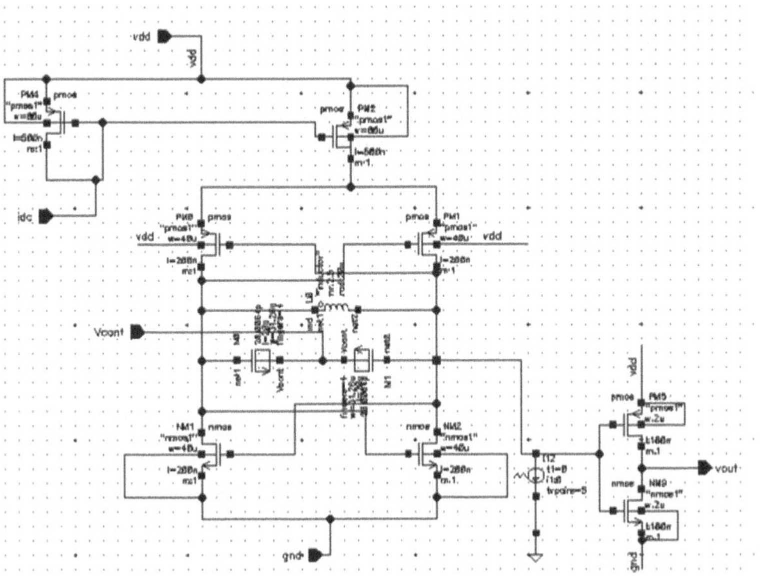

Fig. 6. LC Voltage Controlled Oscillator design

3.3 Proposed Source Switched Charge Pump

Charge pumps are used to lower the constant finite phase error. Charge pumps and loop filters together function as integrator. A variety of topologies are used to implement the charge pump. [14] discusses these topologies. The circuit architecture used here is Source switched Charge Pump due to its higher speed Control voltage range and better control over the operating region of MOSFET. The current sources of the charge pump are connected to the source terminal, hence the name Source Switched Charge Pump. The advantage of this architecture is the use of only one current source. The current source has a value of 4equivalent parallel resistance mA. The circuit is shown in Fig. 7. The operation is explained as follows. A wide swing current mirror is used to copy the current from the PMOS branch having PM0, PM1 transistors to the second branch containing NM0 and NM1 transistors. Thus an equivalent bias voltage Vbp and Vbn is generated which is then applied to the last branch. The 'UP' and 'DOWN' signal control the transistors PM4, NM2 which perform the switching action. The control voltage in closed loop ranges between 1.6–1.8.

3.4 Proposed Active Loop Filter

A passive loop filter of the second order with capacitor values of 10nF was initially used to build the PLL. The lock frequency was found to still be 200 MHz below what was needed. The lock frequency was measured to be 2.2 GHz. This highlighted one of the passive loop filter's drawbacks: its limited control voltage range. Instead of it, active filters are utilized to overcome this. [2] discusses the advantage of ALF along with CP. [1,5] explains the transfer equations of passive and active loop filters and its affect on the output phase noise. ALF requires the use of Op Amp. Telescopic Opamp design has been implemented to make the active loop filter. The gain equation of the Opamp is given in (7). The Op Amp isolates the VCO control voltage from Charge Pump voltage. OP-AMP design takes gain, ICMR, slew rate, and CMRR into account. At lower frequencies, the OP-AMP's gain was measured to be approximately 40 dB. There is about a –20 dB/decade roll-off rate at cut off frequency 3 MHz. This OP-AMP that is utilized in the design of active loop filter is portrayed in Fig. 8. Figure 7 is the charge pump design. The output voltage from the CP and ALF ranges between 1.6–1.8. The Vbias voltage for the NM3, NM2, NM4, PM0, PM1 was calculated to be around 650 mV, 1 V and 1.2 V.

$$Av = gm1[gm3ro3ro1)||(gm5ro5ro7)] \tag{7}$$

The loop bandwidth has been calculated to be around 5000 Hz.

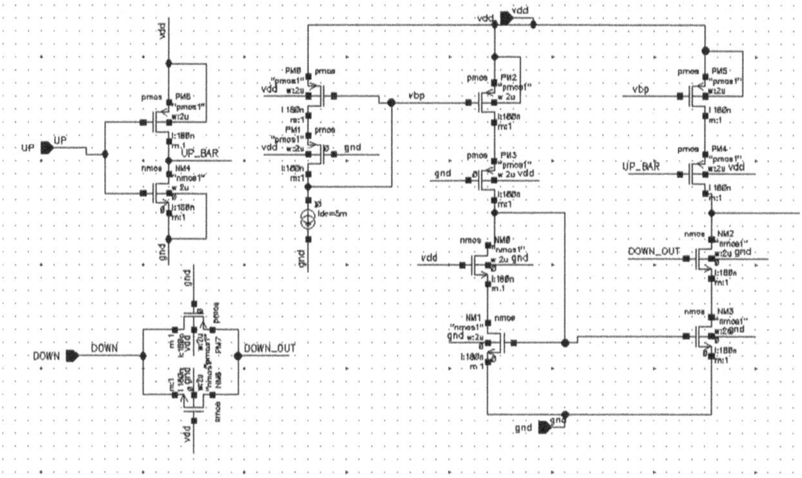

Fig. 7. Charge Pump Circuit for PLL

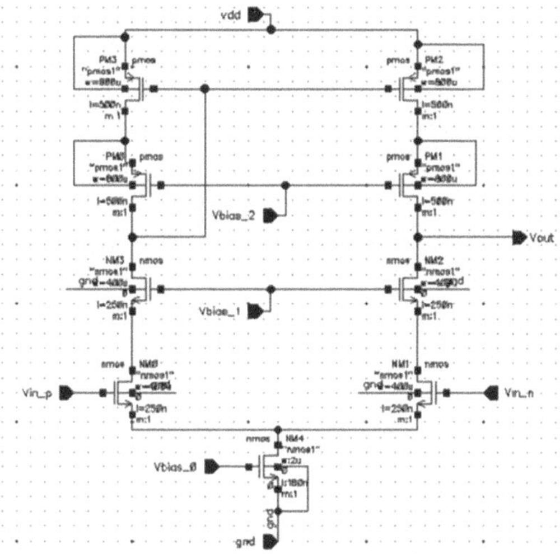

Fig. 8. Telescopic Op-Amp design for active Loop Filter

3.5 Proposed Frequency Divider

The frequency divider was designed using the conventional D flip-flop and inverter. The frequency of the VCO is divided by 32. Therefore, five such divide-by-two blocks were added to the feedback in order to match the reference frequency and the feedback frequency.

In a closed loop setup, the PLL passive loop filter's transfer function is as follows. Here the variable is phase, which is the integration of frequency (Fig. 9 and 10).

The transfer function is as follows:

$$H(s) = \frac{\frac{I_p K_{VCO}}{2\pi C_p}(R_p C_p s + 1)}{s^2 + \frac{I_p}{2\pi}\frac{K_{VCO}}{M} R_p s + \frac{I_p}{2\pi C_p}\frac{K_{VCO}}{M}} \tag{8}$$

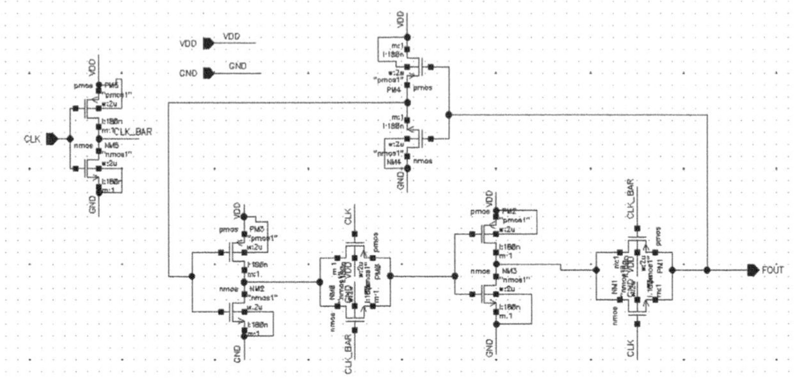

Fig. 9. Frequency Divider Circuit for PLL (5 such blocks connected in series for ÷ 32)

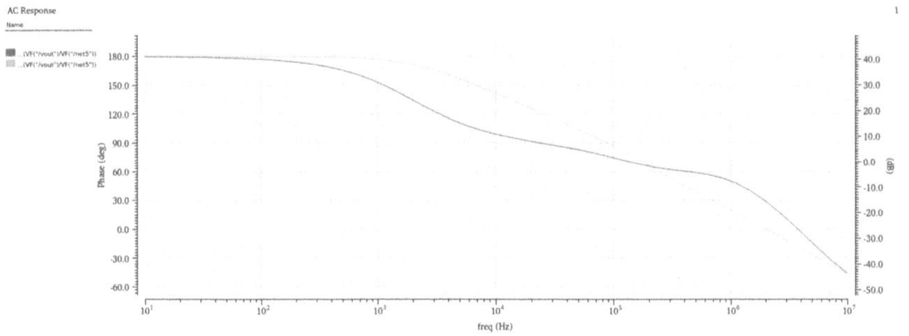

Fig. 10. Op Amp's gain and phase plot

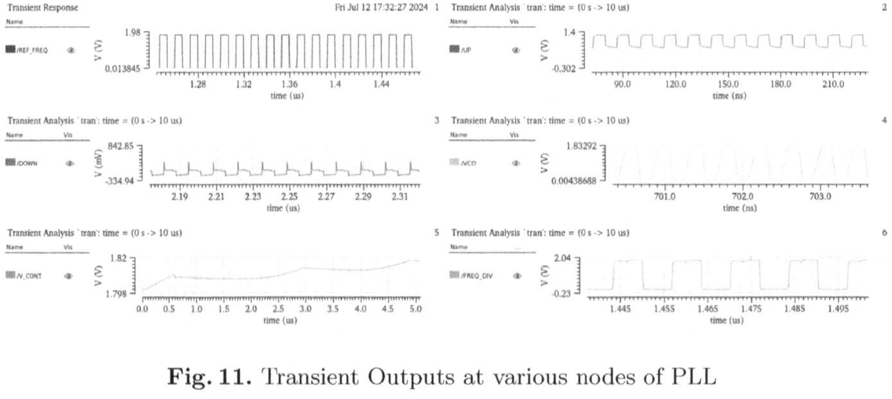

Fig. 11. Transient Outputs at various nodes of PLL

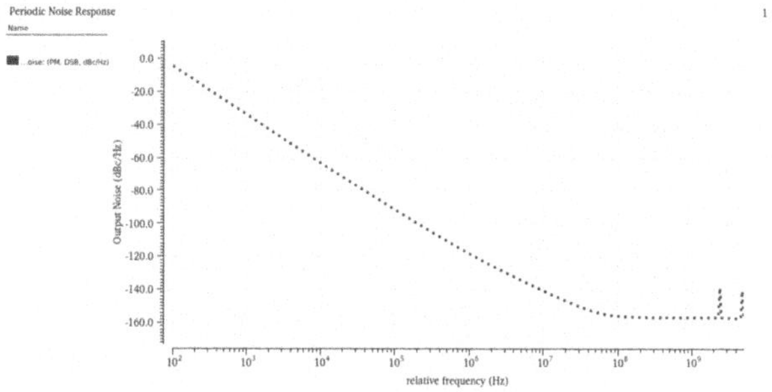

Fig. 12. Phase Noise graph of VCO

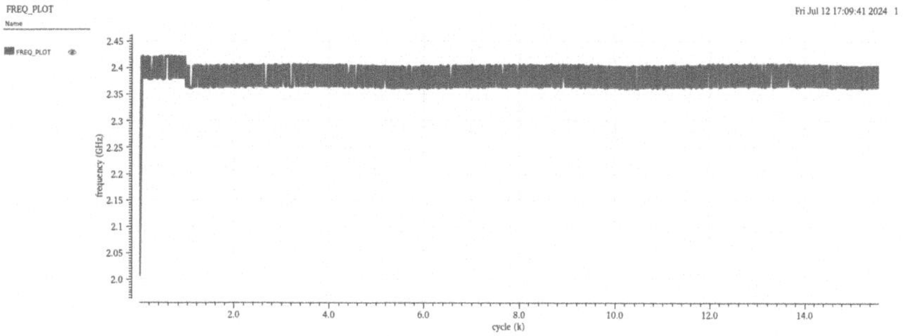

Fig. 13. Frequency of VCO with respect to time

4 Simulation Results and Analysis

This section presents the results of the simulation for the proposed design that was implemented in 180 nm technology in Cadence Virtuoso. The simulation was performed on Spectre tool. The specifications for this design were to design a PLL for the frequency of 2.467 GHz a reference frequency of 75 MHz. The use of Telescopic Opamp helped us to achieve a higher gain. This in turn gave a higher lock frequency.

The results obtained were 2.467 GHz with a reference frequency of 75 MHz. The feedback frequency is measured to be 73.4 MHz.

Table 1. Process variation

Process Corner	VCO Frequency
NN	2.39 GHz
SS	2.38 Ghz
SF	2.375 GHz
FS	2.405 GHz
FF	2.46G Hz

Table 2. Comparison of previous work with the proposed PLL design

Parameter	Ref [2]	Ref [8]	Ref [9]	Ref [10]	This Work
Process (nm)	65 nm	65 nm	350	90 nm	180 nm
VCO Type	Ring	LC	LC	Ring	LC
Supply (V)	0.4	0.68	2.6	1.8	1.8
Lock Frequency (Hz)	350 MHz	2.4 GHz	2.4 GHz	1 GHz	2.47 GHz
Phase Noise (dBc/Hz)	-90 dBc	–110 dBc/Hz	–115 dBc/Hz @10 MHz	–	–150 dBc/Hz
Power (mW)	0.109 mW	0.68 mW	88.4 mW	4.2 mW	33.894 mW
Lock Time	–	–	–	100 ns	100 ns

The lock time is measured to be 630 ns. The phase noise that is generated is always due to the L, C components. The phase noise was –156 dBc/Hz @ 100 MHz offset. The beat frequency was set at 2.4 GHz while measuring the pnoise of the oscillator. Additionally, tests like DC operating point, transient noise, pss, and pnoise were also performed. The power dissipation was 36.84 mW. For the active loop the gain was measured to be 40 dB, with a gain bandwidth of 3 MHz. The Table 2 compares the results of this work with that of [2, 8–10]. We can see an improvement in phase noise of the oscillator. The output at each node of the PLL is observed in Fig. 11, namely UP, DOWN, Active Loop filter output, VCO output and the feedback frequency. Figure 12 represents the phase noise plot of the VCO. Figure 13 shows the frequency plot with respect to time. The process variation was simulated with respect to FF, SF, FS, SS corners. The Table 1 shows the results of Process Variation.

5 Conclusion and Future Scope

This paper presents a design for a PLL with a frequency of 2.4 GHz. The design was done in Cadence Virtuoso in 180 nm technology. This PLL design is used in Zigbee and Bluetooth modules. Frequency synthesis has been achieved in this design from 75 MHz to 2.4 GHz. Lower phase noise has been measured in this design. It can further optimized by replacing the integer-n divider with fractional-n divider using sigma delta modulator to create a PLL that switches frequencies at smaller intervals. A delta sigma modulated fractional divider can be used for the same.

References

1. Han, Y., Zhang, Q., Li, C.-K., Li, X.-D.: Analysis of the influence of the loop filter in the phase locked loop on the output phase noise. In: 2018 15th International Computer Conference on Wavelet Active Media Technology and Information Processing (ICCWAMTIP), Chengdu, China, pp. 185–189 (2018). https://doi.org/10.1109/ICCWAMTIP.2018.8632591.

2. Moon, J.-W., Choi, K.-C., Choi, W.-Y.: A 0.4-V, 90–350-MHz PLL with an active loop-filter charge pump. IEEE Trans. Circuits Syst. II Express Briefs **61**(5), 319–323 (2014). https://doi.org/10.1109/TCSII.2014.2312800
3. Analog Dialogue 52-07 , July 2018 1 Phase-Locked Loop (PLL) Fundamentals. I. Collins. Published 2018, Engineering, Physics (2018)
4. Barrett, C.: Fractional/Integer-N PLL Basics, Technical Breief, Texas Instruments (1999)
5. Banerjee, D.: PLL Performance, Simulation and Design, 5th edn. (2017)
6. Ahmad, N., et al.: Charge pump and loop filter for low power PLL using 130 nm CMOS technology. J. Phys.: Conf. Ser. **1049**, 012060 (2018)
7. Razavi, B.: Design of Analog CMOS Integrated Circuits, 2nd edin,. McGraw Publication (2017)
8. Paidimarri, A., Ickes, N., Chandrakasan, A.P.: A 0.68 V 0.68 mW 2.4 GHz PLL for ultra-low power RF systems. In: IEEE Radio Frequency Integrated Circuits Symposium (RFIC). Phoenix, AZ, USA, pp. 397–400 (2015). https://doi.org/10.1109/RFIC.2015.7337789
9. Fialho, V., Vaz, J.: A 2.4 GHz CMOS integer-N phase-locked loop. Instituto Superior de Engenharia de Lisboa, Rua Conselheiro Emídio Navarro 1, 1959-0071 Lisboa, Portugal (1959)
10. Patil, P.T., Ingale, V.: Design of a low power PLL in 90 nm CMOS technology. In: IEEE 5th International Conference for Convergence in Technology (I2CT). Bombay, India, pp. 1–3 (2019). https://doi.org/10.1109/I2CT45611.2019.9033824
11. Pu, X., Kumar, A., Nagaraj, K.: Area-efficient low-noise low-spur architecture for an analog PLL working from a low frequency reference. IEEE Trans. Circuits Syst. II Express Briefs **59**(6), 331–335 (2012). https://doi.org/10.1109/TCSII.2012.2195064
12. Abdul Majeed, K.K., Kailatha, B.J.: A novel phase frequency detector for a high frequency PLL design. In: International Conference on Design and Manufacturing, IConDM 2013 (2013)
13. Razavi, B.: RF Microelectronics, 2nd edn. McGraw-Hill Publication (2011)
14. Liu, X., Willson, A.N.: A pA-leakage CMOS charge pump for low-supply PLLs. In: 53rd IEEE International Midwest Symposium on Circuits and Systems. Seattle, WA, USA, pp. 1037–1040 (2010). https://doi.org/10.1109/MWSCAS.2010.5548821
15. da Silva, M.D.C., Severo, L.C., Girardi, A.G.: "Design of a telescopic operational amplifier using a semi-automatic synthesis (2014)
16. Kouhalvandi, L., Aygün, S., Güneş, E.O., Kırcı, M.: Design of a high gain telescopic-cascode operational amplifier based on the ZTC operation condition. In: 2017 24th IEEE International Conference on Electronics, Circuits and Systems (ICECS), Batumi, Georgia, pp. 538–541 (2017). https://doi.org/10.1109/ICECS.2017.8292125.
17. Shedge, D.K., Wani, P.W., Itole, D.A., Pokharkar, S.B., Sutaone, M.S.: CMOS telescopic cascode operational amplifier. In: 2013 International Conference on Green Computing, Communication and Conservation of Energy (ICGCE), Chennai, pp. 87–89 (2013). https://doi.org/10.1109/ICGCE.2013.6823405
18. Lakshmikumar, K.R.: Analog PLL design with ring oscillators at low-gigahertz frequencies in nanometer CMOS: challenges and solutions. IEEE Trans. Circuits Syst. II Express Briefs **56**(5), 389–393 (2009). https://doi.org/10.1109/TCSII.2009.2019171

Unraveling the Dynamics of Classical Chaotic Systems and Strange Attractors: A Matlab-Based Exploration

R. Archana[1(✉)], Sebastian Treasa[1], and P. Sreenivas[2]

[1] Department of Electrical and Electronics Engineering, Federal Institute of Science and Technology, A P J Abdul Kalam Technological University, Thiruvananthapuram, Kerala, India
archana@fisat.ac.in

[2] Department of Mechanical Engineering, Government Model Engineering College, A P J Abdul Kalam Technological University, Thiruvananthapuram, Kerala, India
sreeni@mec.ac.in

Abstract. This paper explores classical chaotic systems, including the Lorenz system, Rössler system and Chen System. We systematically investigate the behavior of these chaotic systems by varying parameters. The Lorenz system, known for its butterfly-shaped attractor, showcases the intricate dynamics resulting from its sensitivity to initial conditions. The Rössler system, featuring a twisted double-scroll attractor, is analyzed for its chaotic behavior and parameter-dependent properties. The Matlab-based analysis allows for a comprehensive examination of strange attractors, providing visualizations that enhance our understanding of their complex structures. By systematically altering parameters, we observe the evolution of attractors, gaining insights into how changes in system conditions influence their dynamics. This exploration contributes to a deeper comprehension of chaos in nonlinear dynamical systems. The study underscores the significance of Matlab simulations in unraveling the complexities of classical chaotic systems, offering a valuable tool for researchers and practitioners seeking to analyze and understand chaotic dynamics in various scientific and engineering applications. Additionally, the paper presents a spectral analysis of Lorenz system utilizing FFT and wavelet decomposition techniques

Keywords: Strange attractors · Lorenz system · Rössler system · Deterministic aperiodic behavior · Matlab simulations · Parameter variations · Chaotic dynamics · Fractal structure · Nonlinear dynamical systems · Butterfly effect · Discrete-time dynamical system · Complex dynamics · Nonlinear modeling · Fast Fourier transform · wavelet transform

1 Introduction

Chaos, a phenomenon characterized by aperiodic and unpredictable behavior in deterministic systems, has captivated the attention of scientists and researchers

D. Gupta et al. (Eds.): PCEMS 2024, CCIS 2491, pp. 322–341, 2025.
https://doi.org/10.1007/978-3-031-90577-3_28

across various disciplines. The study of chaos provides insights into the inherent complexities of nonlinear dynamical systems, revealing patterns that defy conventional predictability. This paper embarks on a comprehensive exploration of classical chaotic systems, with a specific focus on the Lorenz system, Rössler system and Chen System as well as their associated strange attractors. Through the lens of Matlab simulations, we delve into the intricate dynamics and parameter-dependent properties of these systems, aiming to contribute to a deeper understanding of chaos in nonlinear dynamical systems.

Chaos, as observed in nonlinear dynamical systems, manifests as a unique form of aperiodic long-term behavior. This departure from traditional periodic behavior introduces an element of unpredictability, where even slight variations in initial conditions can lead to vastly different outcomes. The sensitivity to initial conditions, often referred to as the butterfly effect, underscores the profound impact that minute changes can have on the trajectory of a system. The butterfly effect, metaphorically illustrated by the idea that the flapping of a butterfly's wings could set off a chain of events resulting in a hurricane, exemplifies the delicate balance between deterministic systems and their inherent instability.

The exploration of chaotic systems begins with an acknowledgment of their deterministic nature. Unlike truly random processes, chaotic systems follow deterministic laws, where the evolution of the system is governed by well-defined equations. However, the sensitivity to initial conditions introduces an element of unpredictability, challenging traditional notions of determinism. This delicate interplay between determinism and chaos provides a fertile ground for scientific inquiry, prompting researchers to develop mathematical models that capture the essence of chaotic behavior.

1.1 Genesis of Chaotic Systems: A Brief Historical Perspective

Poincaré's work in the late 19th century indeed laid the foundation for our understanding of chaotic systems. While the study of irregular behavior in nonlinear systems dates back to earlier periods, the crystallization of chaos theory occurred in the latter half of the 20th century. The Lorenz system, characterized by its butterfly-shaped strange attractor, unveiled the complexity hidden within deterministic systems. The attractor's fractal structure became a hallmark of chaos, symbolizing the intricate and self-replicating patterns inherent in chaotic dynamics. Lorenz's work paved the way for the exploration of other chaotic systems, leading to the development of the Rössler system. Around the same time, Michel Hénon proposed the Hénon map, a discrete-time dynamical system that provided insights into the transition to chaos in dissipative systems. The Chua chaotic system originated from a remarkably common autonomous circuit proposed by Leon Chua at the University of California, Berkeley. Introduced by Chen in 1999, the Chen system is another chaotic system characterized by a set of ordinary differential equation. These chaotic systems have been studied in detail due to their significance in understanding nonlinear dynamics, chaos theory, and their applications in various fields including physics, engineering, and biology.

1.2 Strange Attractors and Fractal Structures

One of the defining features of chaotic systems is the presence of strange attractors, which are characterized by their intricate and often fractal structures. An attractor is considered strange when it exhibits a fractal geometry, indicating a self-similar and complex pattern. The term "strange" in this context does not imply an abnormality but rather denotes the unique and fascinating properties of these attractors.

The fractal nature of strange attractors, observed in systems like the Lorenz and Rössler systems, adds an additional layer of complexity to chaotic dynamics. Fractals are geometric shapes that exhibit self-similarity at different scales, and their presence in strange attractors highlights the richness of chaotic behavior. The study of these fractal structures provides a visual and mathematical framework for understanding the intricate patterns that emerge in chaotic systems.

1.3 Matlab Simulations: A Window into Chaotic Dynamics

Matlab, a powerful numerical computing environment, serves as an invaluable tool for the exploration and analysis of chaotic systems. The use of Matlab simulations enables researchers to visualize the dynamics of chaotic systems, offering a dynamic and interactive platform for studying strange attractors. The systematic variation of parameters in Matlab simulations allows for a comprehensive examination of how changes in the system conditions influence the behavior of chaotic systems.

The Lorenz system, Rössler system and Chen Systemare subjected to Matlab-based analyses to unravel the complexities of their strange attractors. Visualizations generated through these simulations provide a tangible representation of the evolving patterns, offering insights into the system's response to parameter variations. Matlab simulations not only facilitate a deeper understanding of chaotic dynamics but also serve as a bridge between theoretical models and real-world applications.

1.4 Spectral Analysis of Chaotic Systems

Fourier analysis is the traditional method in the frequency domain and is highly efficient in uncovering the periodicity of time series. Yet, chaotic time series display broad-band power spectra, and the Fourier spectrum fails to reveal their deterministic nature. An alternative method for time-frequency analysis is the Wavelet Transform. Both Fourier and wavelet analysis of Lorenz system is presented.

2 Literature Review

In his pioneering research, especially in his study of celestial mechanics, Poincaré discovered what is now known as deterministic chaos. He showed that even in deterministic systems governed by simple laws of motion, behavior could be highly sensitive to initial conditions.

One of his most significant contributions was in the context of the three-body problem in celestial mechanics. Poincaré demonstrated that the motion of three mutually gravitating bodies, like the Sun, Earth, and Moon, is inherently unpredictable over long periods. He found that even small changes in initial conditions could lead to dramatically different trajectories, making long-term predictions practically impossible.

This realization challenged the prevailing Newtonian view of a clockwork universe governed by deterministic laws. Poincaré's work paved the way for the development of chaos theory, which explores the behavior of nonlinear dynamical systems exhibiting sensitive dependence on initial conditions. It wasn't until the latter half of the 20th century that chaos theory saw significant growth, with the advent of computers enabling detailed numerical simulations and the discovery of chaotic systems like the Lorenz attractor.

Poincaré's insights remain fundamental to our understanding of chaos and complex systems, with his work influencing a wide range of disciplines beyond celestial mechanics, including physics, mathematics, biology, economics, and more [1]. Edward Lorenz's research on deterministic chaos and the butterfly effect has profoundly influenced our comprehension of intricate systems. His revelation, famously depicted in his essay "Deterministic Nonperiodic Flow" and later popularized in "The Essence of Chaos," fundamentally altered our understanding of predictability in dynamic systems. Lorenz's discoveries demonstrated that even minor fluctuations in initial conditions can lead to significantly different outcomes over time in certain nonlinear systems. This seminal work gave rise to the Lorenz system, a set of three coupled ordinary differential equations that would become an iconic representation of chaotic behavior. This responsiveness to initial conditions, commonly referred to as the butterfly effect, underscores the inherent unpredictability of complex systems despite their deterministic nature [2].

The Rössler system is a set of three nonlinear ordinary differential equations first introduced by Otto Rössler in 1976 [3]. It serves as a prototypical example of a chaotic dynamical system. The equations describe the evolution of three variables. The Rössler system exhibits chaotic behavior for certain parameter values, characterized by sensitive dependence on initial conditions and the presence of strange attractors. The system is named after Otto Rössler, a German biochemist and theoretical chemist, who studied it as a simplified model for chemical reactions. It has since become a widely studied example in the field of nonlinear dynamics and chaos theory.

The Chen system is another well-known example of a chaotic system, named after the Chinese mathematician Tien-Yien Li Chen who introduced it in 1999 [4]. Like the Lorenz system, the Chen system exhibits chaotic behavior under certain parameter regimes, characterized by its sensitive dependence on initial conditions and the presence of strange attractors. The Chen system has been studied extensively in the context of chaos theory and nonlinear dynamics. Its analysis has contributed to our understanding of chaotic systems and has practical applications in fields such as cryptography, secure communications, and chaos-based random number generation. Researchers continue to explore the dynamics and properties of the Chen system, as well as its connections to other chaotic systems and broader aspects of nonlinear dynamics [5,6].

In conservative systems, the volume of a given set remains constant in the finite-dimensional phase space over time, while in dissipative systems, the system state gradually contracts towards a compact set, ultimately converging towards an attractor set. Therefore, to establish sensitive dependence on initial conditions as a sufficient condition for chaos, it's imperative that the evolution of solutions is restricted within a certain region (global boundedness) [7].

The presence of sensitivity to initial conditions in a dynamic system is a prerequisite for chaos, but it's not solely sufficient. Additionally, it's crucial that the evolution of solutions remains confined within a certain region of the state space. This requirement implies that the orbits must be attracted to a specific set, known as an attractor. [8] This condition is always met in dissipative systems. The notion of the butterfly effect has since become a cornerstone in chaos theory, emphasizing the significance of comprehending nonlinear dynamics and the constraints of predictability in various natural and artificial systems [9].

Chaotic systems can be simulated with MATLAB by numerically integrating the differential equations or iterating the map equations that govern the system's dynamics. We can visualize the results using plots or animations to observe the system's chaotic behavior and analyze its properties [10,11].

3 Simulation and Analysis of Chaotic Systems

3.1 Lorenz System

Edward Lorenz made a groundbreaking contribution to the modern theoretical analysis of chaotic systems. Through repeated experiments with varying initial conditions, Lorenz confirmed the sensitivity to initial conditions, publishing his findings in 1965 to great acclaim in the scientific community. The system is encapsulated in three nonlinear differential equations featuring variables x, y, and z.

$$\dot{x} = \sigma(y - x) \tag{1}$$

$$\dot{y} = -xz + \gamma x - y \tag{2}$$

$$\dot{z} = xy - \beta z \tag{3}$$

Variables x, y and z are the three states of the system that can be separately plotted as three time series. Where σ, γ, and β are parameters that control the system's behavior.

In this context, the parameters are given as:

$\sigma = 8$ $\gamma = 26.5$ $\beta = 3$

The initial conditions for the simulation are:

$x_0 = 0$ $y_0 = 0.1$ $z_0 = 0$

These initial conditions, along with the parameter values, lead to chaotic behavior in the Lorenz system. This chaotic behavior is characterized by sensitive dependence on initial conditions, meaning that small changes in the initial state can lead to vastly different outcomes. In Figs. 1, 2 and 3, the variables x, y, and z are plotted over time to illustrate the dynamics of the Lorenz system. In these illustrations: x is represented as State 1 y is represented as State 2 and z is represented as State 3. Each of these states can be plotted as a time series to observe their evolution over time.

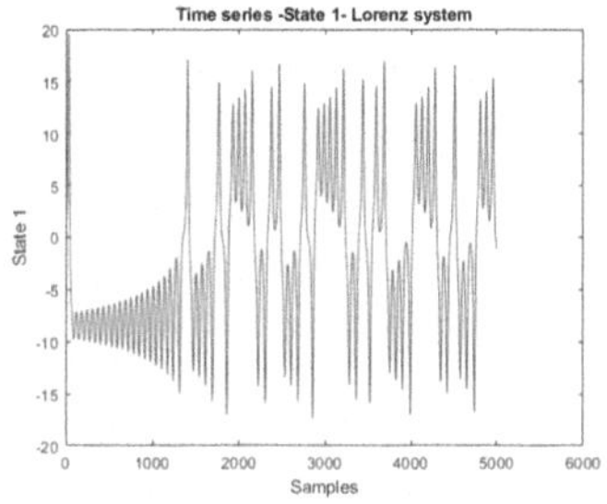

Fig. 1. Lorenz system time series-state 1

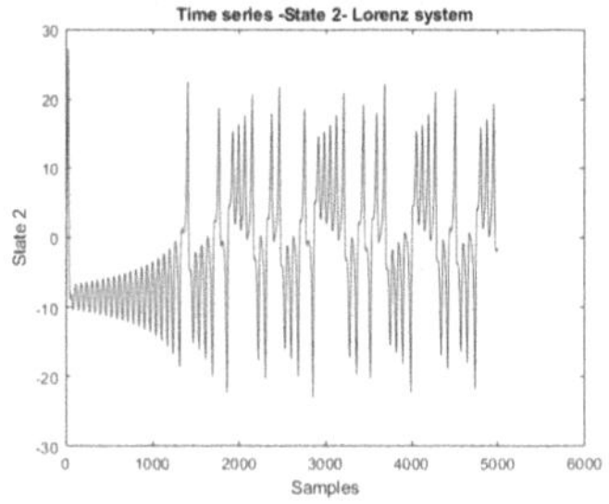

Fig. 2. Lorenz system time series-state 2

The Figs. 4, 5, 6, and 7 represents the Lorenz butterfly, which evolves in state space depicting the chaotic berhaviour of the system under consideration.

3.2 Rossler System

The Rössler system is a set of three nonlinear ordinary differential equations first introduced by Otto Rössler in 1976. is a classic example of a chaotic dynamical system, which consists of three coupled nonlinear ordinary differential equations.

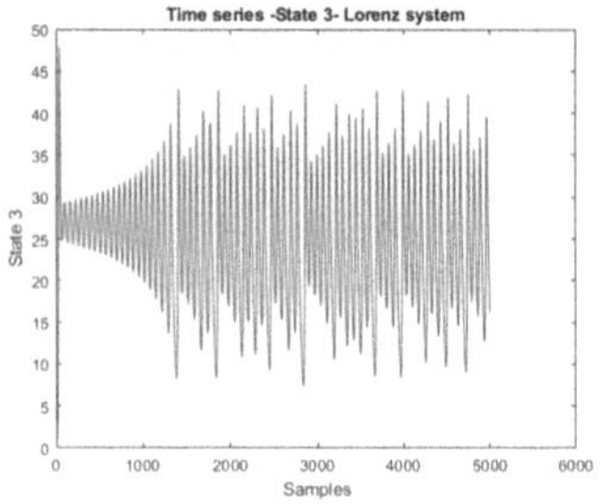

Fig. 3. Lorenz system time series-state 3 Figures 1, 2 and 3: Time series of x (State 1), Time series of y (State 2) and Time series of z (State 3). These plots will show how the value of x, y and z changes over time, revealing the chaotic oscillations typical of the Lorenz system

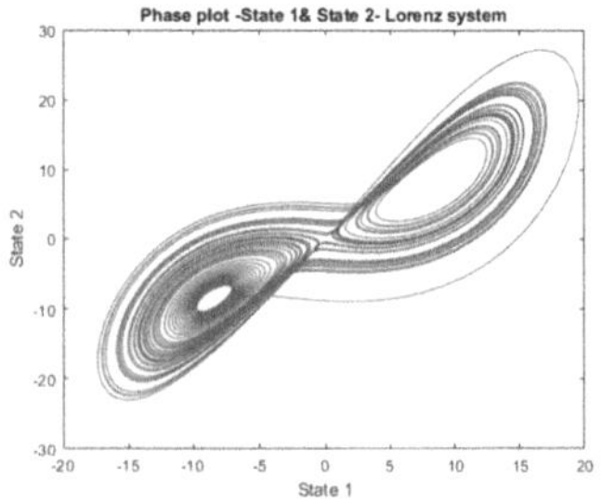

Fig. 4. Lorenz system Phase plot- state 1 and 2

The equations defining the Rössler system are:

$$\dot{x} = (-y - z) \tag{4}$$

$$\dot{y} = -x + ay \tag{5}$$

$$\dot{z} = b + z(x - c) \tag{6}$$

In this context, the parameters are given as:

a = 2 b = 0.2 c = 5.7

The initial conditions for the simulation are:

$x_0 = 0.1$ $y_0 = 0.1$ $z_0 = 0.1$ These initial conditions, along with the parameter values, lead to chaotic behavior in the Rossler system. This chaotic behavior is characterized by sensitive dependence on initial conditions, meaning that small changes in the initial state can lead to vastly different outcomes.

In Figs. 8, 9 and 10, the variables x, y, and z are plotted over time to illustrate the dynamics of the Rossler. In these illustrations: x is represented as State 1 y is represented as State 2 and z is represented as State 3 Each of these states can be plotted as a time series to observe their evolution over time.

Figures 11, 12, 13 and 14 represent the Rossler system, which evolves in state space depicting the chaotic behavior of the system under consideration. These pictures show the attractor.

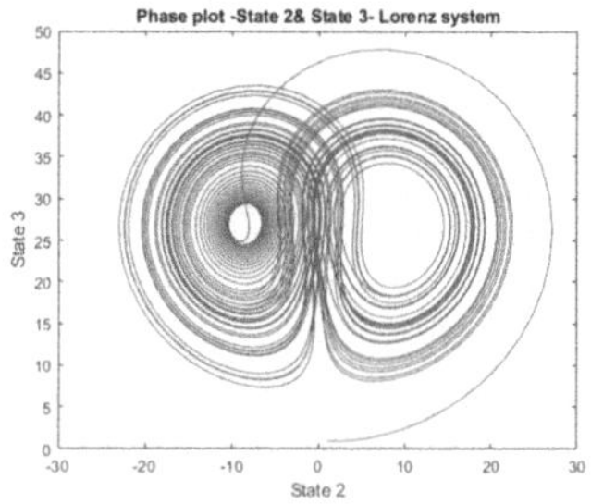

Fig. 5. Lorenz system Phase plot- state 2 and 3

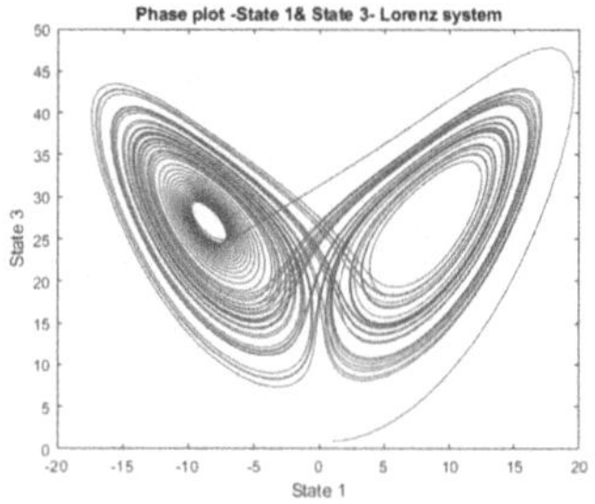

Fig. 6. Lorenz system Phase plot- state 1 and 3 Figures 4, 5 and 6: A 2D plot of the trajectory of the Lorenz system in the (x,y),(y,z) and (x,z) space. This plot is often referred to as the Lorenz attractor and visually represents the system's chaotic behavior as it evolves in two dimensions.

3.3 Chen System

Chen proposed a new chaotic system in 1999, described by the following three non-linear equations.

$$\dot{x} = a(y - x) \tag{7}$$

$$\dot{y} = -x + ay \tag{8}$$

$$\dot{z} = b + z(x - c) \tag{9}$$

The system exhibits chaotic behaviour with the values of a = 35, b = 3, c = 25. Chen systems are most suited for generating hyper chaos. The analysis of Chen system will contribute to a better understanding of a whole family of similar and closely related chaotic systems. The intrinsic dynamics of the Chen system deserves further investigation in the near future.

Figures 15, 16 and 17: Time series of x (State 1), Time series of y (State 2) and Time series of z (State 3). These plots will show how the value of x, y and z

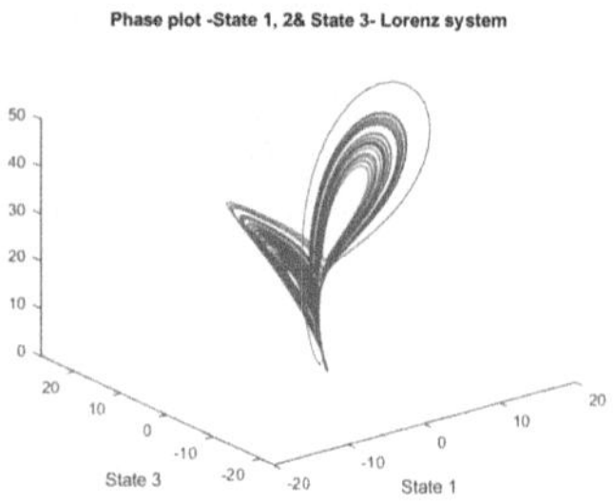

Fig. 7. Lorenz system Phase plot- state 1,2 and 3 Figure 7: A 3D plot of the trajectory of the Lorenz system in the (x,y,z),(y,z) and (x,z) space

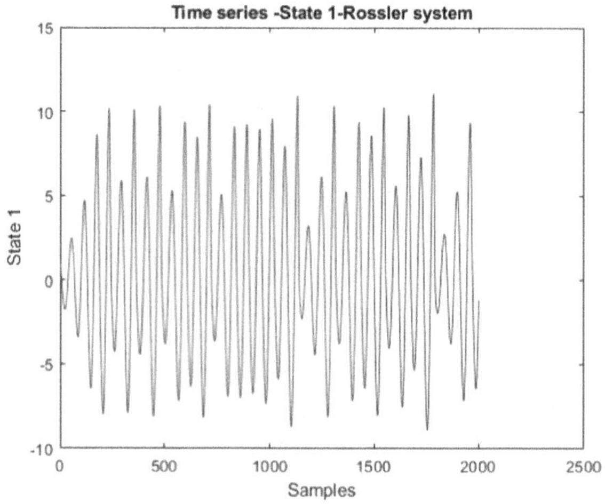

Fig. 8. Rossler system time series-state 1

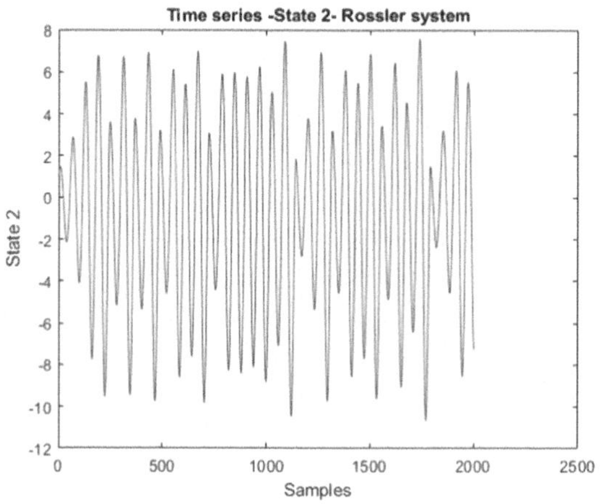

Fig. 9. Rossler system time series-state 2

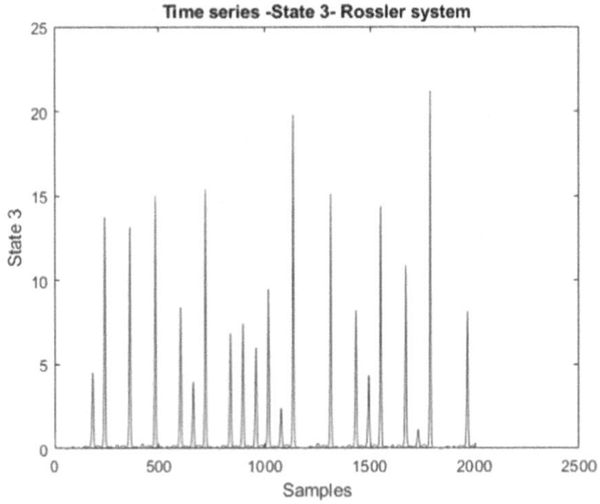

Fig. 10. Rossler system time series-state 3 Figures 8, 9 and 10: Time series of x (State 1), Time series of y (State 2) and Time series of z (State 3). These plots will show how the value of x, y and z changes over time, revealing the chaotic oscillations typical of the Rossler system

changes over time, revealing the chaotic oscillations typical of the Chen system (Fig. 18).

Figure 19: A 2D plot of the trajectory of the Chen system in the (x,y) space. This plot is often referred to as the Chen attractor and visually represents the system's chaotic behavior as it evolves in two dimensions.

Figure 20: A 2D plot of the trajectory of the Chen system in the (x,y) space. This plot is often referred to as the Chen attractor and visually represents the system's chaotic behavior as it evolves in two dimensions.

Figure 21: A 3D plot of the trajectory of the Chen system in the (x,y,z) space. This plot is often referred to as the Chen attractor and visually represents the system's chaotic behavior as it evolves in three dimensions.

4 Spectral Analysis of Chaotic Systems

Fourier analysis is the conventional approach in the frequency domain and is highly effective in uncovering the periodic patterns in time series. However, chaotic time series exhibit broad-band power spectra, and the Fourier spectrum does not indicate their deterministic origin [12,13]. The Fourier spectrum is not capable of representing the non-stationary nature of a chaotic signal. The Fourier spectrum cannot capture the non-stationary characteristics of a chaotic signal.

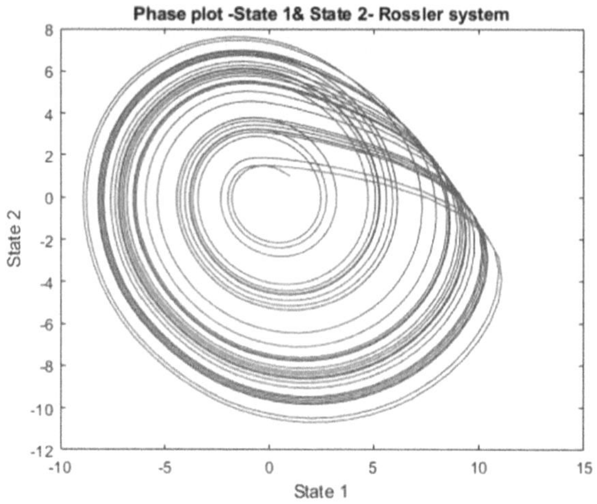

Fig. 11. Rossler system Phase plot- state 1 and 2

However, it can analyze the fundamental invariants of the underlying dynamical system using Fourier representation. Time-frequency distributions represent a robust suite of tools tailored for analyzing non-stationary signals, aiding in the exploration of time-frequency patterns inherent in nonlinear dynamics. A competing method for time-frequency analysis is the Wavelet Transforndamental invariants of the underlying dynamical system can be analysed using Fourier representation. Time-frequency distributions are powerful set of tools specifically designed for non-stationary signal analysis. It is helpful in the study of time-frequency pattern present in non- linear dynamics. One of the competing approaches for time-frequency analysis is Wavelet Transform [14,15].

Analysis Using Fourier Transform. The Lorenz system's behavior transitions from non-chaotic to chaotic based on variations in parameters. In the current analysis, the value of sigma is selected as 1 and 10. The system produces a non-chaotic time series for sigma is equal to 1, whereas it generates a chaotic time series for sigma 10. The behavior of the system in both scenarios under the Fourier domain is depicted in Figs. 22, 23, 24, 25 and 26 (Fig. 27).

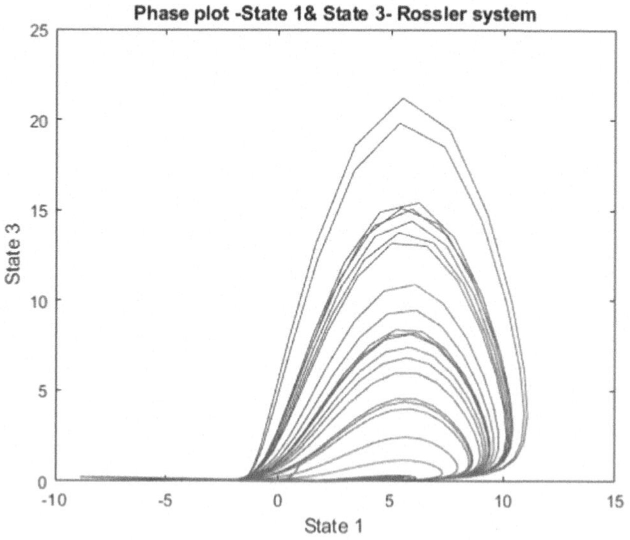

Fig. 12. Rossler system Phase plot- state 2 and 3

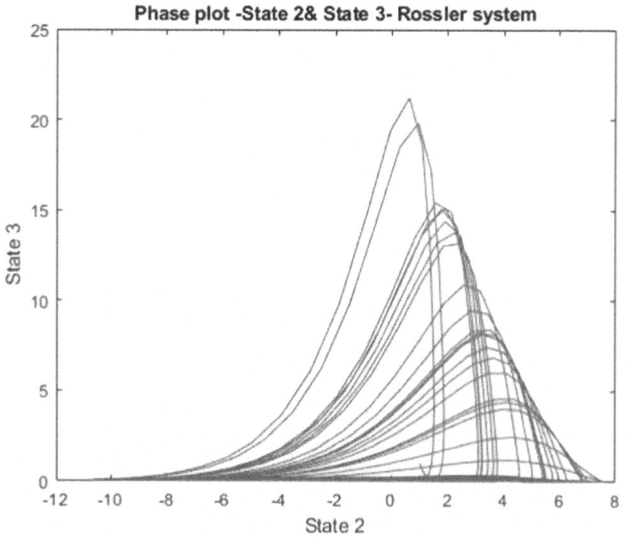

Fig. 13. Rossler system Phase plot- state 1 and 3 Figures 11, 12 and 13: 2D plots of the trajectory of the Rossler system in the (x,y), (y,z) and (x,z) space. This plot is often referred to as the Rossler attractor and visually represents the system's chaotic behavior as it evolves in two dimensions.

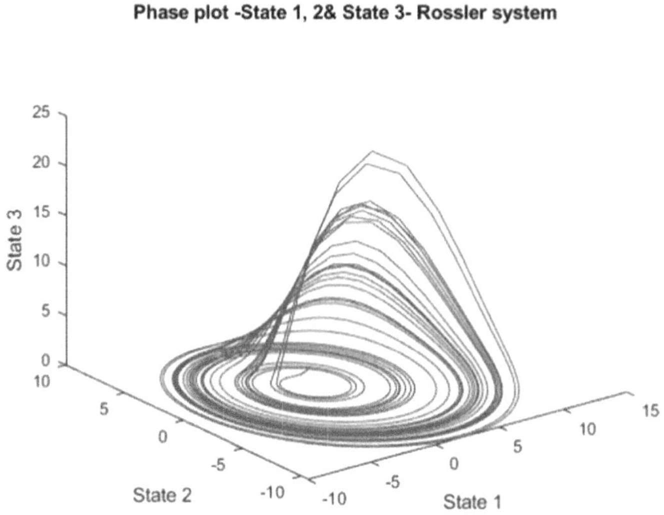

Fig. 14. Rossler system Phase plot- state 1,2 and 3 Figure 14: A 3D plot of the trajectory of the rossler system in the (x,y,z) space. This plot is often referred to as the Rossler attractor and visually represents the system's chaotic behavior as it evolves in two dimensions

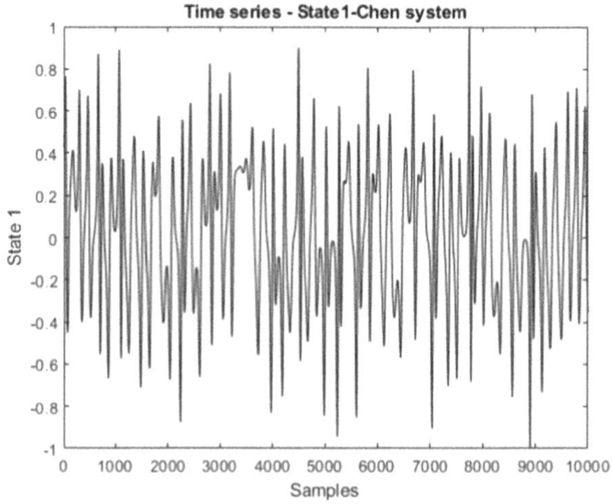

Fig. 15. Chen system time series-state 1

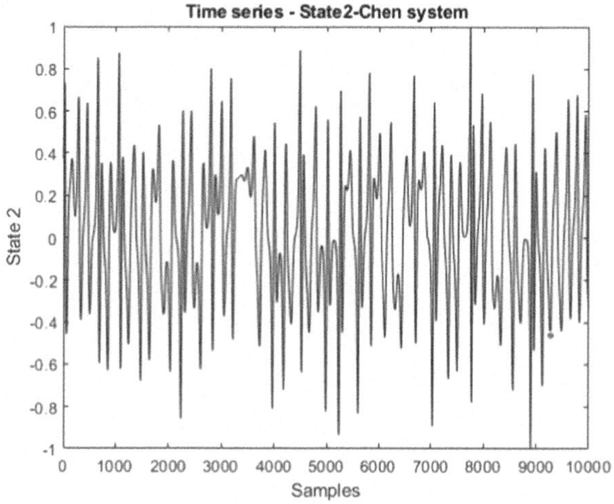

Fig. 16. Chen system time series-state 2

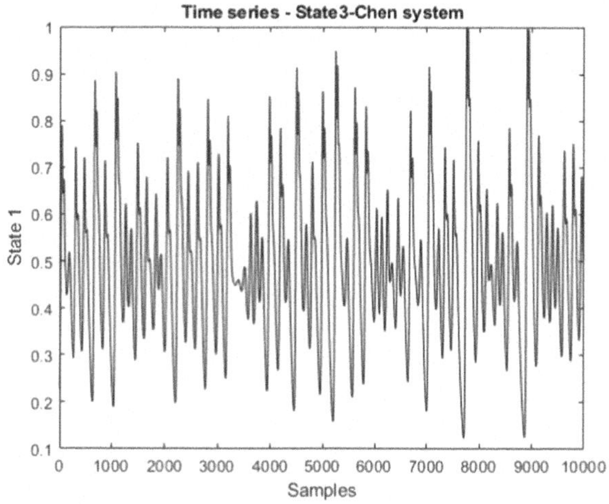

Fig. 17. Chen system time series-state 3

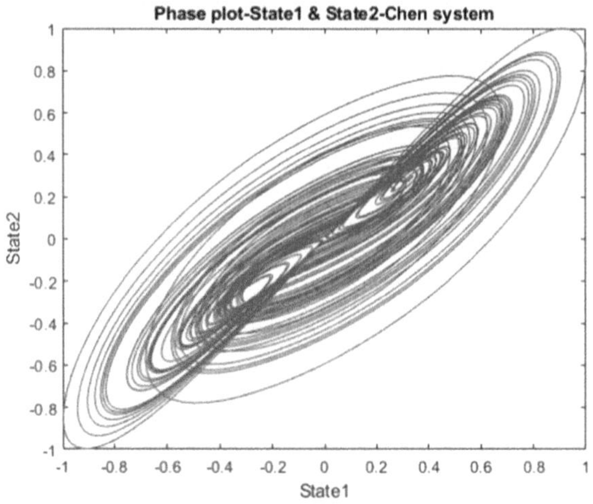

Fig. 18. Chen system Phase plot- state 1 and 2

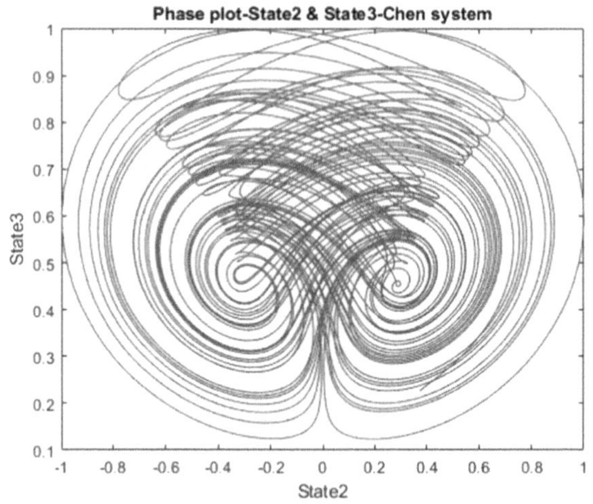

Fig. 19. Chen system Phase plot- state 2 and 3

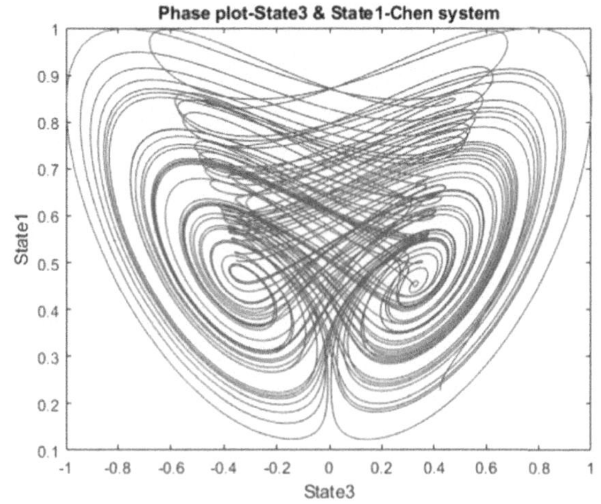

Fig. 20. Chen system Phase plot- state 1 and 3

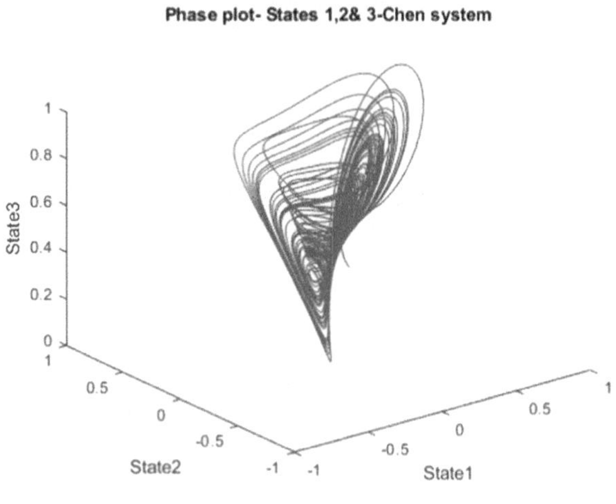

Fig. 21. Chen system Phase plot- state 1,2 and 3

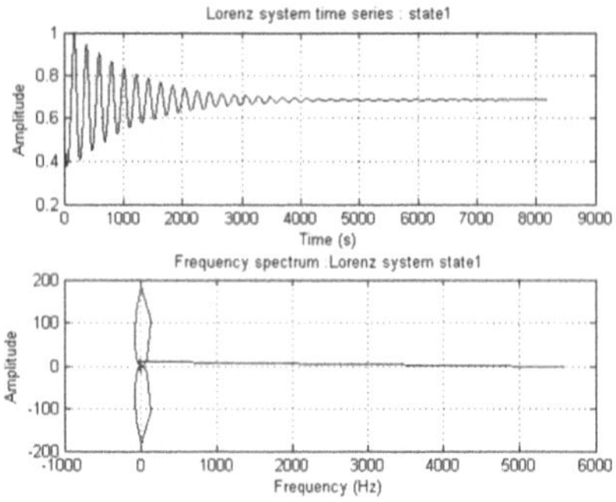

Fig. 22. Fourier coefficients

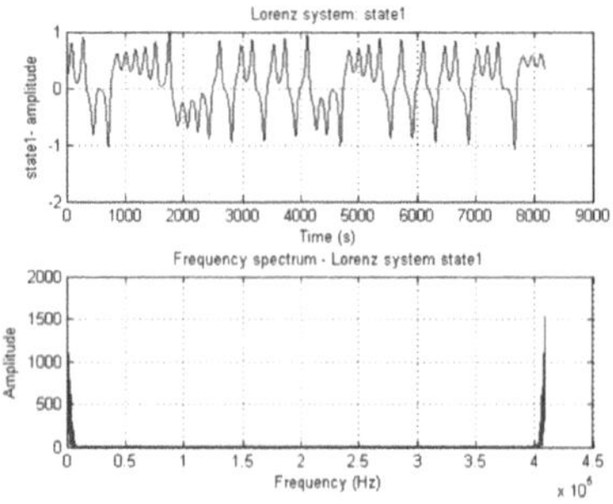

Fig. 23. Fourier coefficients

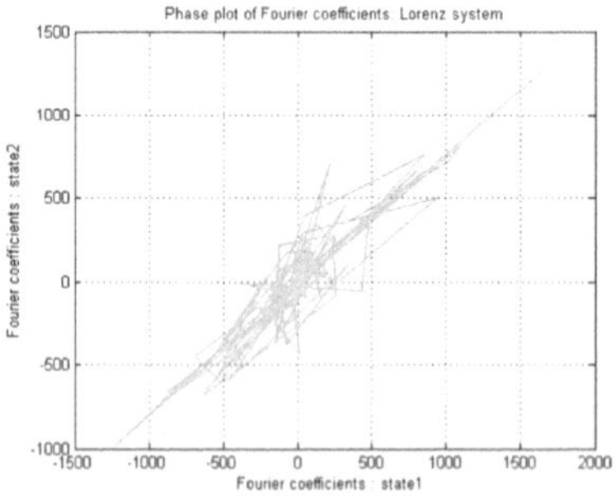

Fig. 24. Fourier coefficients

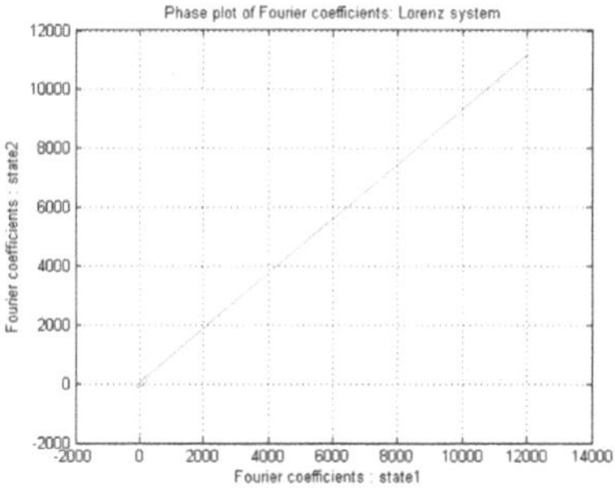

Fig. 25. Fourier coefficients

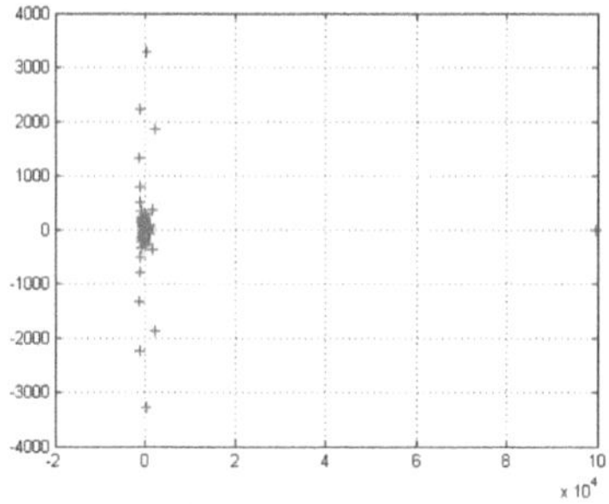

Fig. 26. Fourier coefficients of non chaotic state

4.1 Analysis Using Wavelet Transform

The Fourier method decomposes a signal into harmonically related complex exponential signals. Another approach to analyzing time series is the wavelet transform, which has been introduced and advanced for studying diverse phenomena.

The essential attributes of the wavelet transform include its ability to localize both in time (or space) and frequency. Different types of primary wavelets are documented in the literature, including Haar, Daubechies, and Mayer. Generally, Daubechies wavelets can be chosen for their maximal number of vanishing moments and minimal shift sensitivity. Wavelet analysis is used to examine both chaotic and non-chaotic states of the Lorenz time series. Chaotic time series are chiefly defined by strange attractors derived from phase plots. Hence, the examination of dominant coefficients at the highest energy level focuses on these phase plots. Subsequent analysis then elucidates the characteristics of wavelet coefficients in chaotic time series [16,17] (Figs. 28, 29 and 30).

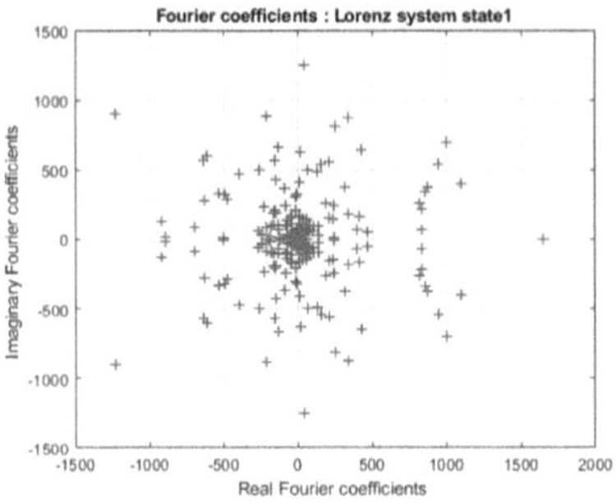

Fig. 27. Fourier coefficients of chaotic state

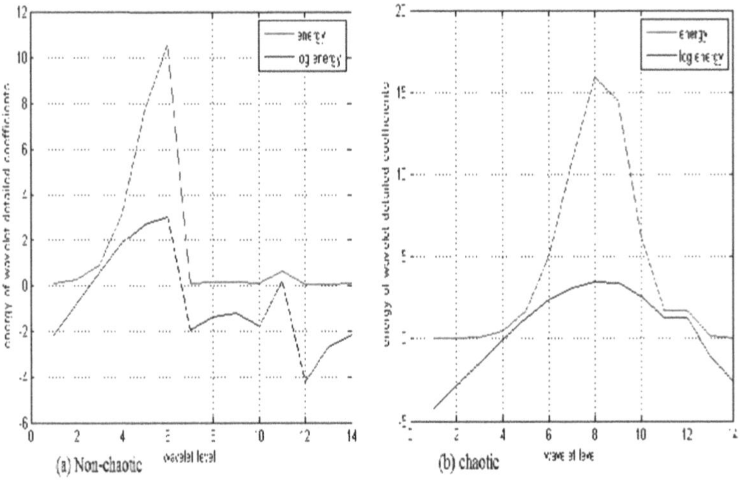

Fig. 28. Fourier coefficients of chaotic state

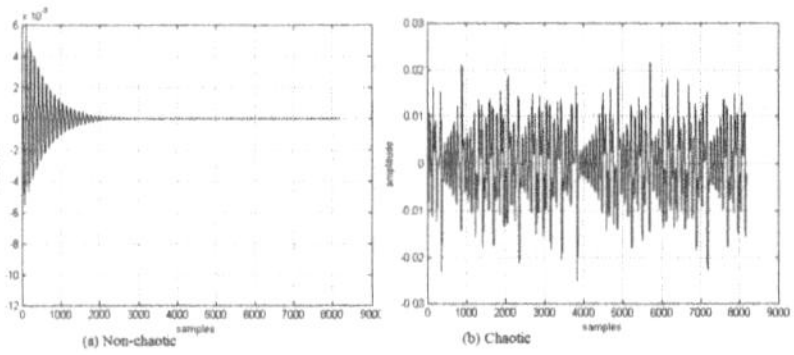

Fig. 29. Fourier coefficients of chaotic state

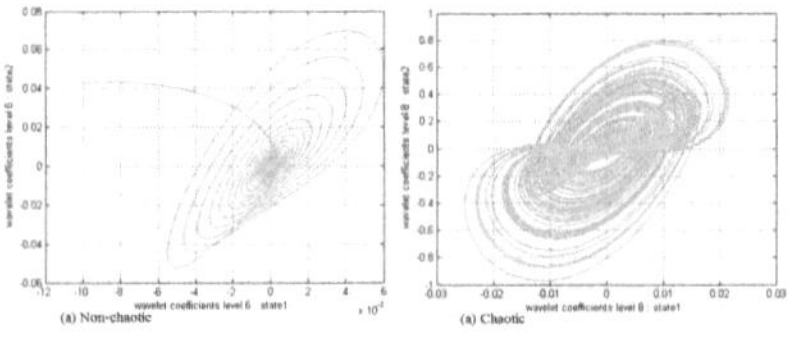

Fig. 30. Fourier coefficients of chaotic state

5 Conclusion

The Lorenz system, Rössler system, and Chen system are analyzed using Matlab to explore the intricacies of their strange attractors. Simulations in Matlab create visualizations that clearly depict the evolving patterns, helping to understand how these systems respond to changes in parameters. Matlab simulations enhance comprehension of chaotic dynamics and act as a connection between theoretical models and practical applications.

The spectral analysis of chaotic system is done using FFT and wavelet. The frequency spectrum is band limited for nonchaotic and is broad banded for chaotic systems. In a non-chaotic state, the amplitude and phase of Fourier coefficients are linear, while in a chaotic state, they become non-linear. The wavelet analysis demonstrated greater efficacy in many aspects.

References

1. The Three-Body Problem and the Equations of Dynamics. ASSL, vol. 443. Springer, Cham (2017). https://doi.org/10.1007/978-3-319-52899-1
2. Lorentz, E.: Deterministic non-periodic flow. J. Atmos. Sci. **20**, 130–141 (1963)
3. Rössler, O.E.: An equation for continuous chaos. Phys. Lett. A **57**(5), 397–398 (1976)
4. Chen, G., Ueta, T.: Yet another chaotic attractor. Int. J. Bifurcat. Chaos **9**(07), 1465–1466 (1999)
5. Takens, F.: On the numerical determination of the dimension of an attractor. Lect. Notes Math. **898**, 366–381 (1981)
6. Takens, F.: Detecting strange attractors in turbulence. In: Rand, D., Young, L.-S. (eds.) Dynamical Systems and Turbulence, Warwick 1980. LNM, vol. 898, pp. 366–381. Springer, Heidelberg (1981). https://doi.org/10.1007/BFb0091924
7. Banks, J., Brooks, J., Cairns, G., Davis, G., Stacey, P.: On devaney's definition of chaos. Am. Math. Mon. **99**(4), 332–334 (1992)
8. Ruelle, D.: Turbulence, Strange Attractors and Chaos, vol. 16. World Scientific (1995)
9. Abarbanel, H.: Analysis of Observed Chaotic Data. Springer, Heidelberg (2012)
10. Jafari, A., Hussain, I., Nazarimehr, F., Golpayegani, S., Jafari, S.: A simple guide for plotting a proper bifurcation diagram. Int. J. Bifurcat. Chaos **31**(01), 2150011 (2021)
11. Tlelo-Cuautle, E., Duarte-Villasenor, M.A., Garcia-Ortega, J.M.: Modeling and simulation of a chaotic oscillator by matlab. IEEE Latin Am. Trans. **5**(2), 94–97 (2007)
12. Cvitanovic, P., Feigenbaum, M.J., Pikovsky, A.: Recycling Fourier spectra of chaotic systems. In: Pennsylvania University, pp. 1–21 (2002)
13. Cooley, J.W., Tukey, J.W.: An algorithm for the machine calculation of complex fourier series. Math. Comput. **19**(90), 297–301 (1965)
14. Murguia, J.S., Campos-Cantón, E.: Wavelet analysis of chaotic time series. Revista mexicana de física **52**(2), 155–162 (2006)
15. Cattani, C., Bochicchio, I.: Wavelet analysis of chaotic systems. J. Interdisc. Math. **9**(3), 445–458 (2006)
16. Vela-Arevalo, L.V.: Time-frequency analysis based on wavelets for Hamiltonian systems. California Institute of Technology (2002)
17. Daubechies, I.: Ten lectures on wavelets. SIAM (1992)

Power-Efficient URLLC with RIS-Aided Rate Splitting Multiple Access

Mayur V. Katwe(✉)

National Institute of Technology Raipur, Chhattisgarh 492010, India
mvkatwe.ece@nitrr.ac.in

Abstract. This study investigates a novel resource allocation framework for a multi-user multiple-input multiple-output (MU-MIMO) downlink system that employs reconfigurable intelligent surfaces (RIS) within a rate-splitting multiple access (RSMA) scheme, specifically for ultra-reliable low-latency communication (URLLC). We aim to minimize the overall transmit power at the base-station while guaranteeing the rate quality of service of the downlink users in severely obstructed dead zones, where direct links between the base station (BS) and users are unavailable, with RIS serving as a relay to facilitate communication. To achieve this, we formulate a transmit power minimization problem that involves the joint design of active beamforming at the BS and passive beamforming at the RIS under given rate and URLLC requirements. Given the non-convex nature of the original problem, we propose an alternating optimization and successive convex approximation framework to solve it. Our proposed solution demonstrates rapid convergence within a few iterations, as validated by numerical simulations. The proposed RSMA scheme for the ARIS-aided downlink URLLC system outperforms conventional multiple access schemes in terms of power requirements.

Keywords: Rate-splitting multiple access (RSMA) · reconfigurable intelligent surface (RIS) · ultra-reliable low-latency communication (URLLC) · power minimization

1 Introduction

Sixth-generation (6G) wireless networks are set to offer services beyond the capabilities of fifth-generation (5G) networks [1]. According to the third-generation partnership project (3GPP) specifications, ultra-reliable low-latency communication (URLLC) relies on finite block-length (FBL) packet transmission to achieve end-to-end latency below 10ms and low packet error probability (PEP) below 10^{-5}. The performance analysis in URLLC systems can focus on [2]:

(a) Block-length minimization: reducing latency under given rate and PEP constraints,
(b) PEP minimization: enhancing reliability under given rate and block-length constraints,

D. Gupta et al. (Eds.): PCEMS 2024, CCIS 2491, pp. 342–356, 2025.
https://doi.org/10.1007/978-3-031-90577-3_29

(c) Rate maximization: increasing throughput under given block-length and PEP constraints.

These analyses support various wireless applications, such as industrial automation and autonomous driving, which require high reliability or low latency. In contrast, future cellular applications like data offloading and scheduling prioritize data rate maximization under reliability and latency constraints. Significant research has focused on determining the achievable transmission rate for finite block lengths [3].

Reconfigurable intelligent surface (RIS) technology has emerged as a key solution for future wireless networks due to its enhanced coverage, flexibility, low cost, easy deployment, and high spectral efficiency [4]. RIS comprises low-cost passive reflecting elements that create virtual paths between base stations and users by inducing desired phase shifts. This technology boosts signal strength and mitigates interference in multi-antenna, multi-user scenarios, promoting URLLC by establishing virtual line-of-sight links for cell-edge or dead-zone users. With the rising demand for high spectral and energy efficiency, unconventional multiple access techniques like space-division multiple access (SDMA) and non-orthogonal multiple access (NOMA) have been developed. These techniques simultaneously serve multiple users using spatial or power multiplexing, offering better power utilization than orthogonal multiple access (OMA). Rate-splitting multiple access (RSMA) has emerged as a promising approach for multi-antenna, multi-user communication, combining the benefits of SDMA and NOMA [5]. RSMA splits user messages into common and private parts, partially decoding interference and treating it as noise, thereby providing a robust framework for high data rates, low latency, and high reliability in URLLC applications.

1.1 Background

Numerous studies in the literature have investigated resource allocation strategies to improve the performance of reconfigurable intelligent surface (RIS)-aided wireless communication systems. These works explore the benefits of RIS deployment, including energy and spectral efficiency improvements, channel diversity enhancement, extended coverage, optimal RIS placement, and active beamforming design for multi-antenna systems (MIMO) [6]. Resource allocation in RIS-assisted networks for ultra-reliable low-latency communication (URLLC) services has emerged as a compelling research area [7–10]. These studies collectively demonstrate how RIS enhances coverage and reliability for URLLC by establishing robust indirect links between base stations, RIS, and users, particularly beneficial in challenging MIMO environments. For instance, Li et al. [7] proposed an aerial RIS (ARIS) system to minimize worst-case packet error probability (PEP) by optimizing ARIS placement, beamforming, and block length. Xie et al. [8] addressed the RIS-aided downlink URLLC system with user grouping and joint optimization of user grouping, block length, and passive beamforming. They employed convex approximations to achieve near-optimal latency reduction.

Recent studies, such as [2,5,11–15], have explored more reliable and efficient multiple access schemes, particularly RSMA, in RIS-aided networks. Most of these works, such as [11], consider infinite block length transmission. RSMA, when combined with URLLC, can achieve lower latency, higher reliability, and comparable transmission rates to NOMA and RSMA with shorter block lengths [5]. The low latency and high data-rate communication of RSMA with MU-MIMO FBL transmission offer significant advantages for machine-type communication (MTC). Like, [5] and [2,16] addressed resource allocation design for energy and spectral efficiency maximization under URLLC constraints. However, the beamforming design approximation in [16] is sub-optimal and does not fully utilize beyond-diagonal RIS.

1.2 Motivation and Contributions

Future wireless applications are set to provide extended ultra-reliable low latency communication (xURLLC), offering high data rates along with exceptional reliability and minimal latency that exceed the capabilities of current core services. The integration of these data rates and URLLC requirements in resource-limited systems demands a move from traditional architectures to more robust and advanced multiple access schemes [1]. Key to achieving high degrees of freedom (DoF) and efficient radio resource management in multiple access channels are sophisticated power control, intelligent interference management, and optimized spectrum usage. RSMA enhances spectral efficiency compared to conventional schemes by utilizing all available DoF for smart interference management and effective power control across multiple users. Moreover, RSMA's flexibility, adaptability to changing channel conditions, and ability to provide extremely low latency and high reliability through URLLC make it a promising option for future wireless networks [5]. RSMA and RIS techniques are leading candidates for 6G, addressing spectral and delay constraints in wireless applications. Their integration can open new possibilities for mBRLLC systems, improving spectral efficiency and reliability. By combining RSMA and URLLC with RISs, we aim to achieve high-performance communication, leveraging passive devices that enhance signal quality and coverage through controlled phase shifts.

Motivated by aforementioned challenges, this work investigate RIS-aided RSMA system to reduce the transmit power requirement to attain given rate QoS under URLLC. Unlike existing works [2,5,11,16,17], we investigates a resource allocation scheme for RIS -aided RSMA for MU-MIMO DL URLLC to improve the spectral efficiency of dead-zone users which are disconnected from the BS, i.e., a direct link between BS and user is unavailable due to the severe obstructions in an urban environment. We aim to maximize the spectral efficiency of the ARIS-aided RSMA systems within the constraints of short packet communication (SPC). However, this resource allocation design faces several challenges. Like, the non-convex nature of the sum throughput maximization problem requires advanced techniques for finding the global optimum. Joint optimization of active and passive beamforming, along with ARIS location, adds complexity due to multiple variables and constraints. Balancing spectral efficiency with URLLC requirements for data rate and packet error probability is difficult. Additionally, as the system grows in complexity, efficient and scalable algorithms are essential for real-time implementations. Uur performance evaluation provides empirical evidence of how well our framework performs under various conditions, ensuring it meets real-world requirements in terms of rate-reliability and latency. It is observed that RSMA with a shorter block length can achieve the same transmission rate as traditional long packet NOMA and SDMA schemes.

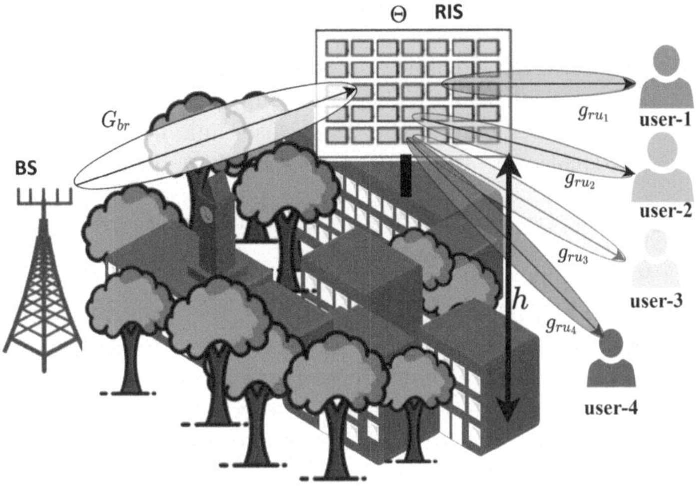

Fig. 1. RIS-aided MU-MIMO DL communication

2 System Model and Problem Formulation

Let us consider a MU-MIMO downlink multi-user URLLC system that consists of a set of U users, BS and ARIS. Assuming that the RIS is mounted on building and the users are found in a dead zone whose locations is known as priori. In other words, the direct ground link between the BS and the users is missing due to unfavorable propagation conditions and large obstacles in an urban setting, and RIS acts as a relay to the terrestrial base station as shown in Fig. 1. In other words, the BS with M antennas communicates with single-antenna users through RIS which consists of N passive reflecting elements i.e., phase-shifters. The RIS performs passive beamforming which facilitates smart propagation environment design with efficient interference management and high spectral efficiency for dead-zone users.

2.1 Channel and Signal Model

Let us denote $\boldsymbol{G}_{br} = \varrho_l \left(d_{br}\right)^{-\rho} \boldsymbol{A} \in \mathbb{C}^{M \times N}$ as the channel for the BS-RIS link and $\boldsymbol{g}_{r_u} = \varrho_l \left(d_{r_u}\right)^{-\rho} \boldsymbol{a} \in \mathbb{C}^{N \times 1}$ as the channel for the RIS- u^{th} user link such that

l_{br} and l_{r_u} are the large-scale path loss values and $\boldsymbol{A} \in \mathbb{C}^{M\times N}$ and $\boldsymbol{a}_u \in \mathbb{C}^{N\times 1}$ are the small-scale fading coefficients for BS-RIS and RIS-u^{th} user link, respectively. ϱ_L and ϱ_N denotes the attenuation constants for LOS and NLOS links, respectively and ρ denotes the path loss exponent (PLE). $d_{br} = ||\boldsymbol{r}_b - \boldsymbol{w}||_2$ and $d_{r_u} = ||\boldsymbol{w} - \boldsymbol{r}_u||_2$ denote the distance between BS-user and ARIS-user where $\boldsymbol{r}_b, \boldsymbol{w}$ and $\boldsymbol{r}_u$ are the positions of the BS, ARIS and u^{th} user, respectively.

Further, denoting $\boldsymbol{\Phi}$ as the passive beamforming matrix at the ARIS which is defined as $\boldsymbol{\Phi} = \mathbf{diag}\,(\boldsymbol{\phi})$, where $\boldsymbol{\phi} = [\phi_1, \ldots, \phi_N]^T, |\phi_n| = 1, \forall n \in \mathcal{N} = \{1, \ldots, N\}$, and $\boldsymbol{\phi}$ is referred as the phase-shifter of the ARIS. To this end, the optimal setting of $\boldsymbol{w}$ and $\boldsymbol{\phi}$ provide overall flexibility in the effective channel gain in terms of amplitude and phase coefficient which further improves the system performance. Now, the effective channel gain for the BS-u^{th} user link, i.e., $\boldsymbol{h}_u$ is explicit as the function of ARIS location $\boldsymbol{w}$ and ARIS phase-shifter $\boldsymbol{\phi}$ given as

$$\boldsymbol{h}_u(\boldsymbol{w}, \boldsymbol{\phi}) = \boldsymbol{G}_{br}\boldsymbol{\Phi}\boldsymbol{g}_{r_u} = \boldsymbol{G}_{br}\ \mathbf{diag}\left(\boldsymbol{g}_{r_u}\right)\boldsymbol{\phi}, \tag{1}$$

where $\boldsymbol{g}_u \in \mathbb{C}^{M\times 1}$, $u \in \mathcal{U} \triangleq \{1, \ldots, U\}$.

Assuming an RSMA scheme, the base station (BS) aims to transmit U messages to U users. In this scheme, each user's signal is divided into two parts: a common part and a private part. These parts are simultaneously transmitted at the same frequency and time slot. The common part of the users are encoded together into a common message s_0 whereas the private part of each user is encoded differently into U distinct private messages, $s_u, \forall u \in \mathcal{U}$ such that $\mathbb{E}\{|s_u|^2\} = 1, \forall u \in \{0\} \cup \mathcal{U}$. Now, the BS transmits a combined message vector $\boldsymbol{s} = [s_0, s_1, \ldots, s_u]^T$ using a precoder matrix $\boldsymbol{V} \triangleq [\boldsymbol{v}_0, \boldsymbol{v}_1, \ldots, \boldsymbol{v}_u] \in \mathbb{C}^{M\times(U+1)}$ where $v_0 \in \mathbb{C}^{M\times 1}$ and $v_u \in \mathbb{C}^{M\times 1}$ denotes precoder design for the common message and the uth private message. So, the total transmitted signal by BS is given as

$$\boldsymbol{x} = \boldsymbol{v}_0 s_0 + \sum\nolimits_{u\in\mathcal{U}} \boldsymbol{v}_u s_u,\ \boldsymbol{x} \in \mathbb{C}^{M\times 1}, \tag{2}$$

where, the maximum transmit power for the considered system is limited only up to $P^{\max}$, i.e., trace $\left(\boldsymbol{V}^H\boldsymbol{V}\right) \leq P^{\max}$.

Now, each uth user receives the transmitted signal through the ARIS and decodes the common message s_0 along with its private message s_u. The received signal at u^{th} user can be given as

$$y_u = \boldsymbol{h}_u^H\boldsymbol{x} + n_u, \tag{3}$$

where n_u representing the additive white Gaussian noise in the system, is modeled as a complex number following a circularly symmetric complex Gaussian distribution $\mathcal{CN}\left(0, \sigma_u^2\right)$. It has a mean of zero and a variance of σ_u^2.

At the receiver, each user decodes the common message and its own private message and treat other interfering messages as noise. So, the SINR of s_0 message after its decoding at the uth user can be given as

$$\gamma_{0,u}(\boldsymbol{w}, \boldsymbol{\phi}, \boldsymbol{V}) = \frac{\left|\boldsymbol{h}_u^H \boldsymbol{v}_0\right|^2}{\sum_{j\in\mathcal{U}} \left|\boldsymbol{h}_u^H \boldsymbol{v}_j\right|^2 + \sigma_u^2} = \frac{\left|\boldsymbol{\phi}^H \boldsymbol{g}_u^H \boldsymbol{v}_0\right|^2}{\sum_{j\in\mathcal{U}} \left|\boldsymbol{\phi}^H \boldsymbol{g}_u^H \boldsymbol{v}_j\right|^2 + \sigma_u^2}, \tag{4}$$

where $\boldsymbol{g}_u = \mathbf{diag}\left(\boldsymbol{g}_{r_u}^H\right) \boldsymbol{G}_{br}^H$. After successful decoding of the common message, the uth user decode its private message s_u while treating the private messages of other users as the interference. The SINR of s_u message at the uth user can be expressed as

$$\gamma_u(\boldsymbol{w}, \boldsymbol{\phi}, \boldsymbol{V}) = \frac{\left|\boldsymbol{h}_u^H \boldsymbol{v}_u\right|^2}{\sum_{j\in\mathcal{U}/u} \left|\boldsymbol{h}_u^H \boldsymbol{v}_j\right|^2 + \sigma_u^2} = \frac{\left|\boldsymbol{\phi}^H \boldsymbol{g}_u^H \boldsymbol{v}_u\right|^2}{\sum_{j\in\mathcal{U}/u} \left|\boldsymbol{\phi}^H \boldsymbol{g}_u^H \boldsymbol{v}_j\right|^2 + \sigma_u^2}. \tag{5}$$

To meet stringent delay requirements, the BS employs short packet communication. Specifically, the BS transmits the common message s_o and all private messages $s_u, \forall u$ using finite block length (FBL) with sizes l_o and l_u symbols, respectively. Given the latency constraints in the FBL regime, decoding errors are unavoidable, and the probability of error per symbol (PEP) is associated with each symbol. Furthermore, the maximum achievable data rate for private and common messages in the considered URLLC system cannot be provided by Shannon's capacity theorem. Notably, traditional infinite block length transmission systems follow Shannon's channel capacity theorem, which recommends the maximum achievable data rate for additive-white Gaussian noise channels. In other words, the Shannon's capacity does not characterize the maximal achievable data rate for URLLC applications as the block length is finite and the non-zero PEP for arbitrarily high signal to noise ratio (SNR) is trivial . Using Shanon's capacity theorem, the latency and reliability will be underestimated and overestimated, respectively. The approximate maximum achievable rate throughput for both the common message and the private message at the uth user can be expressed as

$$R_{0,u}(\boldsymbol{w}, \boldsymbol{\phi}, \boldsymbol{V}) = \log_2\left(1 + \gamma_{0,u}\right) - \frac{X_u\left(\gamma_{0,u}\right)}{\sqrt{l_o}} \frac{Q^{-1}\left(\varepsilon_u\right)}{\log_e 2}, \tag{6}$$

$$R_u(\boldsymbol{w}, \boldsymbol{\phi}, \boldsymbol{V}) = \log_2\left(1 + \gamma_u\right) - \frac{X_u\left(\gamma_u\right)}{\sqrt{l_u}} \frac{Q^{-1}\left(\varepsilon_u\right)}{\log_e 2}, \tag{7}$$

where ε_u is the PEP for u^{th} user, $Q^{-1}(\bullet)$ is inverse Gaussian Q-function defined as $Q^{-1}(x) = \frac{1}{\sqrt{2\pi}} \int_x^{\infty} \exp\left(\frac{-t^2}{2}\right)$ and $X_u(\gamma) \triangleq \sqrt{1 - (1+\gamma)^{-2}}$ is the channel dispersion function.

To ensure that the common message s_0 is successfully decoded by all the users, the achievable common rate shall not exceed $R_c \triangleq \min\{R_{0,1}, \ldots, R_{0,U}\}$. Given the common message rate R_c and the rate c_u allocated to u^{th} user, the constraint of each uth user data rate of receiving common message is given by $\sum_{u=1}^{U} c_u \leq \min_{u\in\mathcal{U}} R_{0,u}$, which indicates that the total data rates of all users receiving common message must be less than the rate of common message. Here,

c_u represents the portion of the common rate allocated to the uth user. In essence, the maximum achievable rate for a given user is the sum of their private rate and their assigned portion of the common rate. Consequently, the sum rate can be expressed as

$$R_{\text{sum}}(\boldsymbol{w}, \boldsymbol{\phi}, \boldsymbol{V}) = R_c + \sum_{u \in \mathcal{U}} R_u = \sum_{u \in \mathcal{U}} e_u + \sum_{u \in \mathcal{U}} R_u. \tag{8}$$

2.2 Problem Formulation

Our main objective is to minimize the overall transmit power in the considered RIS-aided DL RSMA system by jointly designing the precoder matrix at the base station $\boldsymbol{V}$ and the phase-shifter at the RIS $\boldsymbol{\phi}$. The optimization process takes into account the constraints related to FBL transmission, transmit power, and quality-of-service (QoS) under given PEP and block length for the messages which is formulated as

$$\min_{\boldsymbol{V}, \boldsymbol{\phi}, \boldsymbol{w}, \boldsymbol{e}} \sum_{u \in \mathcal{U}} \boldsymbol{v}_u^H \boldsymbol{v}_u$$

$$\text{s.t. } |\phi_n| = 1, n \in \mathcal{N}, \tag{9a}$$

$$c_u + \log_2(1+\nu_u) - \frac{X_u(\nu_u)}{\sqrt{l_u}} \frac{Q^{-1}(\varepsilon_u)}{\log_e 2} \geq r_u^{\min}, \ u \in \mathcal{U}, \tag{9b}$$

$$\sum_{u \in \mathcal{U}} c_u \leq \min_u \left\{ \log_2(1+\nu_{0,u}) - \frac{X_u(\nu_{0,u})}{\sqrt{l_o}} \frac{Q^{-1}(\varepsilon_u)}{\log_e 2} \right\}, \tag{9c}$$

$$c_u \geq 0, \ u \in \mathcal{U}, \tag{9d}$$

where $\boldsymbol{e} = [e_1, \ldots, c_U]^T$ and the constraint (9a) is the beamforming design constraint at the RIS. Furthermore, the QoS constraint (9b) guarantees that the minimum rate requirement $r_u^{\min}$ is met and the constraint (9c) ensures the successful decoding of the common message by each user. Notably, the formulated resource allocation design problem in (9) is a non-convex NP-hard problem that is intractable in its true form.

3 Proposed Solution

To address the non-convexity of the joint optimization problem presented in (9), we employ a decoupling strategy that splits the original problem into the design of precoding at the BS and the determination of phase shifts for RIS elements, and the determination of the ARIS location. Subsequently, we propose an alternating optimization framework that iteratively solves these sub-problems in a sequential manner which takes into account the available information on packet size and PEP.

Note that the objective function is a monotonic increasing function with the SINR v_u. Owing to this, the problem of precoder design can be alternatively written into the simplified form using auxiliary variables $\boldsymbol{\zeta} = \{\zeta_{0,u}, \zeta_u\}$ as

$$\min_{\boldsymbol{V},\boldsymbol{\phi},\zeta,\boldsymbol{e}} \sum_{u\in\mathcal{U}} \boldsymbol{v}_u^H \boldsymbol{v}_u$$

$$\text{s.t. (9a), (9d),}$$

$$\zeta_{0,u} \leq \frac{\left|\boldsymbol{\phi}^H \boldsymbol{g}_u^H \boldsymbol{v}_0\right|^2}{\sum_{j\in\mathcal{U}} \left|\boldsymbol{\phi}^H \boldsymbol{g}_u^H \boldsymbol{v}_j\right|^2 + \sigma_u^2}, \; u \in \mathcal{U}, \tag{10a}$$

$$\zeta_u \leq \frac{\left|\boldsymbol{\phi}^H \boldsymbol{g}_u^H \boldsymbol{v}_u\right|^2}{\sum_{j\in\mathcal{U}/u} \left|\boldsymbol{\phi}^H \boldsymbol{g}_u^H \boldsymbol{v}_j\right|^2 + \sigma_u^2}, \; u \in \mathcal{U}, \tag{10b}$$

$$c_u + \log_2(1+\zeta_u) - F_u(\zeta_u) \geq r_u^{\min}, \; u \in \mathcal{U}, \tag{10c}$$

$$\log_2(1+\zeta_{0,u}) - F_u(\zeta_{0,u}) \geq \sum_{u\in\mathcal{U}} c_u, u \in \mathcal{U}, \tag{10d}$$

where $\boldsymbol{\zeta} = \{\zeta_{0,u}, \zeta_u\}$ and $F_u(\zeta) = \frac{X_u(\zeta)}{\sqrt{l}} \frac{Q^{-1}(\varepsilon)}{\log_e 2}$ is the reduction factor in the achievable rate due to the consideration of FBL calculated under given values of PEP ε and packet size l for each user. Importantly, the introduced auxiliary variables and the constraints in (10a) and (10b) induce the maximization of the lower bound of the original objective function, instead of the original objective function.

Further, the SINR expressions in (10a) and (10b) can be further simplified using auxiliary variables $\varrho_{0,u}$ and ϱ_u as

$$\min_{\boldsymbol{V},\boldsymbol{\phi},\zeta,\boldsymbol{e}} \text{Trace}\left(\boldsymbol{V}^H \boldsymbol{V}\right)$$

$$\text{s.t. (9a), (9d), (10c), (10d)}$$

$$\sum_{j\in\mathcal{U}} \left|\boldsymbol{\phi}^H \boldsymbol{g}_u^H \boldsymbol{v}_j\right|^2 + \sigma_u^2 \leq \varrho_{0,u}, \; u \in \mathcal{U}, \tag{11a}$$

$$\sum_{j\in\mathcal{U}/u} \left|\boldsymbol{\phi}^H \boldsymbol{g}_u^H \boldsymbol{v}_j\right|^2 + \sigma_u^2 \leq \varrho_u, \; u \in \mathcal{U}, \tag{11b}$$

$$\chi_{0,u} = \frac{\left|\boldsymbol{\phi}^H \boldsymbol{g}_u^H \boldsymbol{v}_0\right|^2}{\varrho_{0,u}} \geq \zeta_{0,u}, \; \chi_u = \frac{\left|\boldsymbol{\phi}^H \boldsymbol{g}_u^H \boldsymbol{v}_u\right|^2}{\varrho_u} \geq \zeta_u, \; u \in \mathcal{U}. \tag{11c}$$

However, the problem of precoder design in (11) is still non-convex and does not provide closed-form solution due to non-convexity of the function $F_u(\zeta)$ and coupling of variables $\boldsymbol{V}$ and $\boldsymbol{\phi}$. To this end, a few inner approximations are utilized to convert the non-convex functions into affine form. Like, the function $F_u(\zeta)$ can be approximated using its upper bound as

$$F_u(\zeta) \leq \widetilde{F_u}(\zeta) = \frac{Q^{-1}(\varepsilon)}{\sqrt{l}\log_e 2} \widetilde{X}_u^{(a)}(\zeta) \tag{12}$$

where $\zeta^{(a)}$ is the value of ζ at a^{th} iteration and $\widetilde{Z}_u(\zeta)$ is the approximation of $F_u(\zeta)$ at the $(a+1)^{th}$ iteration and

$$\widetilde{X}_u^{(a)}(\zeta) = \sqrt{1-\left(1+\zeta^{(a)}\right)^{-2}} + \frac{\left(\zeta-\zeta^{(a)}\right)\left(1+\zeta^{(a)}\right)^{-3}}{\sqrt{1-\left(1+\zeta^{(a)}\right)^{-2}}}, \tag{13}$$

is the upper bound of $X_u(\zeta)$ at $(a+1)^{th}$ iteration.

Note that the expression in (11c) is non-convex and has strong coupling of variables. So, we adopt alternation optimization framework such that the precoder design is solved first with fixed beamforming design (sub-problem 1) and then passive beamforming design is solved with fixed precoder design (sub-problem 2).

So, the non-convex expressions in (11c) are relaxed for sub-problem 1, i.e., with fixed (given) passive beamforming design $\boldsymbol{\phi} = \overline{\boldsymbol{\phi}}$, at any $(a+1)^{th}$ iteration as

$$\widetilde{\chi}_{0,u}^{(a)}(\boldsymbol{V}) \triangleq \frac{2\mathrm{Re}\left(\overline{\boldsymbol{\phi}}^H \boldsymbol{g}_u^H \boldsymbol{v}_0^{(a)} \overline{\boldsymbol{\phi}}^H \boldsymbol{g}_u^H \boldsymbol{v}_0\right)}{\varrho_u^{(a)}} - \frac{\left|\overline{\boldsymbol{\phi}}^H \boldsymbol{g}_u^H \boldsymbol{v}_0^{(a)}\right|^2 \varrho_{0,u}}{\left(\varrho_{0,u}^{(a)}\right)^2} \geq \zeta_{0,u},\ u \in \mathcal{U}, \tag{14}$$

$$\widetilde{\chi}_u^{(a)}(\boldsymbol{V}) \triangleq \frac{2\mathrm{Re}\left(\boldsymbol{\phi}^H \boldsymbol{g}_u^H \boldsymbol{v}_u^{(a)} \boldsymbol{\phi}^H \boldsymbol{g}_u^H \boldsymbol{v}_u\right)}{\varrho_u^{(a)}} - \frac{\left|\overline{\boldsymbol{\phi}}^H \boldsymbol{g}_u^H \boldsymbol{v}_u^{(a)}\right|^2 \varrho_u}{\left(\varrho_u^{(a)}\right)^2} \geq \zeta_u,\ u \in \mathcal{U}, \tag{15}$$

respectively, where $\boldsymbol{\phi}^{(a)}$ are the passive beamforming design obtained at a^{th} iteration.

Similarly, the non-convex expressions in (11c) are relaxed for sub-problem 2, i.e., with fixed (given) precoder design $\boldsymbol{V} = \overline{\boldsymbol{V}}$, at any $(a+1)^{th}$ iteration as

$$\widetilde{\chi}_{0,u}^{(a)}(\boldsymbol{\phi}) \triangleq \frac{2\mathrm{Re}\left(\overline{\boldsymbol{v}_0}^H \boldsymbol{g}_u \boldsymbol{\phi}^{(a)} \overline{\boldsymbol{v}_0}^H \boldsymbol{g}_u \boldsymbol{\phi}\right)}{\varrho_u^{(a)}} - \frac{\left|\overline{\boldsymbol{v}}_0^H \boldsymbol{g}_u \boldsymbol{\phi}^{(a)}\right|^2 \varrho_{0,u}}{\left(\varrho_{0,u}^{(a)}\right)^2} \geq \zeta_{0,u},\ u \in \mathcal{U}, \tag{16}$$

$$\widetilde{\chi}_u^{(a)}(\boldsymbol{\phi}) \triangleq \frac{2\mathrm{Re}\left(\overline{\boldsymbol{v}_u}^H \boldsymbol{g}_u \boldsymbol{\phi}^{(a)} \overline{\boldsymbol{v}_u}^H \boldsymbol{g}_u \boldsymbol{\phi}\right)}{\varrho_u^{(a)}} - \frac{\left|\overline{\boldsymbol{v}}_u^H \boldsymbol{g}_u \boldsymbol{\phi}^{(a)}\right|^2 \varrho_u}{\left(\varrho_u^{(a)}\right)^2} \geq \zeta_u,\ u \in \mathcal{U}, \tag{17}$$

respectively, where $\{\boldsymbol{v}_0^{(a)}, \boldsymbol{v}_u^{(a)}\}$ are the precoder design obtained at a^{th} iterations.

Now, the problem in (11) is reformulated into two sub-problems as

$$\begin{aligned}
(\mathbf{P1}):\ &\min_{\boldsymbol{V},\boldsymbol{\zeta},\boldsymbol{e}}\ \mathrm{Trace}\left(\boldsymbol{V}^H \boldsymbol{V}\right)\\
&\text{s.t. (9a), (9d), (11a), (11b), (14), (15),}\\
&c_u + \log_2\left(1+\zeta_u\right) - \widetilde{Z}_u\left(\zeta_u\right) \geq r_u^{\min},\ \forall u,\\
&\log_2\left(1+\zeta_{0,u}\right) - \widetilde{Z}_u\left(\zeta_{0,u}\right) \geq \sum\nolimits_{u\in\mathcal{U}} c_u, \forall u,
\end{aligned} \tag{18}$$

Algorithm 1. Alternating Optimization for Unified Solution

Input: $\overline{V}, \overline{\phi}, A^{\max}, \epsilon_p$ and $l_o, \varepsilon_u, l_u, \varepsilon_u, \forall u$

1: **Initialize**: $a = 0, R_p^{(-1)} = 0$ and $V^{(0)}$

2: **while** $a \leq A^{\max}$ or $\left|R_p^{(a)} - R_p^{(a-1)}\right| \leq \epsilon_p$ **do**

3: Solve (P1) and (P2)

4: **end while**

Output: $\hat{R}_p = R_p^{(a)}, \hat{V} = V^{(a)}, \hat{\varrho} = \varrho^{(a)}$ and $\hat{\zeta} = \zeta^{(a)}$.

and

$$\begin{aligned}
\textbf{(P2)}: \max_{\phi,\zeta,e,\varrho} \; & \sum_{u\in\mathcal{U}} \left\{e_u + \log_2\left(1+\zeta_u\right) - \widetilde{Z}_u\left(\zeta_u\right)\right\} \\
& \text{s.t. (9a), (9d), (11a), (11b), (16), (17),} \\
& c_u + \log_2\left(1+\zeta_u\right) - \widetilde{Z}_u\left(\zeta_u\right) \geq r_u^{\min}, \forall u, \\
& \log_2\left(1+\zeta_{0,u}\right) - \widetilde{Z}_u\left(\zeta_{0,u}\right) \geq \sum\nolimits_{u\in\mathcal{U}} c_u, \forall u,
\end{aligned} \tag{19}$$

where $\varrho = \{\varrho_{0,u}, \varrho_u\}$. Without loss of generality, the power minimization with passive beamforming design becomes sum-rate maximization problem [18]. Finally, the optimization problem stated in (18) and (19) exhibits convexity in terms of all the optimization variables, concave objective and affine constraints. Therefore, the problem can be solved using the CVX toolbox [19]. The solution can be iteratively obtained until convergence is reached or the maximum number of iterations is reached, based on the provided information about the convergence factor ϵ_p and the finite-block length (FBL) transmission. Algorithm 1 provides a summary of the precoding design procedure considering these factors. Let $\hat{V}, \hat{\phi}, \hat{\zeta}, \hat{\varrho}$ represent the converged solution obtained from Algorithm 1. Further, we validate the convergence of Algorithm 1 in the next section through numerical simulations.

4 Numerical Simulations and Discussion

This section describes the performance analysis of the proposed ARIS-aided DL MIMO RSMA system with FBL transmission via extensive computer simulations and the results are averaged over 1000 Monte Carlo simulations.

4.1 General Parameter Settings

In our simulations, we consider a scenario where $U = 3$ DL users are deployed on a ground plane with dimensions 100×100 m^2, uniformly distributed. The base station (BS) is located at a fixed position of $[-50\text{m}, 50\text{m}, 0]^T$. We set the number of reflecting elements in the ARIS as $N = 40$. The path-loss channel model for all links, with parameters $\rho_L = 2.8, \rho_N = 3.3$, $\varrho_L = 11.4$ dB, and

$\varrho_N = 17.6$ dB. The initial points are generated as follows. Once the feasibility check is performed, the initial points $\boldsymbol{\phi}^{(0)}, V^{(0)}$ are selected randomly such that $|\phi_n| = 1, \forall n, ||v_k|| = P^{\max}/K, \forall k$ and $\boldsymbol{w}^{(0)} = (\boldsymbol{w}^{\min} + \boldsymbol{w}^{max})/2$, while ensuring that it satisfies problem constraints. Furthermore, we set the block length and PEP for all messages as $l_o = l_u = l = 50$ and $\varepsilon_u = \varepsilon = 10^{-5}$, respectively, where u represents the user index. The convergence factors and maximum number of iterations for the iterative algorithms are chosen as $\epsilon_a = \epsilon_b = \epsilon_c = 10^{-6}$, $A^{\max} = B^{\max} = 20$, and $C^{\max} = 5$, respectively. The maximum power allocated to the BS is set as $p^{\max} = 30$ dBm. The initial phase shift at the ARIS and precoder matrix at the BS are randomly selected within the constraints specified in equations. The minimum signal strength gap between the common and private signals is set as $\Omega = -120$ dB. The minimum QoS requirement for the downlink users is defined as $r_u^{\min} = 1$ for all $u \in \mathcal{U}$. Additionally, the noise power is set as $\sigma_u^2 = -130$ dB. For the PSO algorithm, the corresponding parameters are set as $N_p = 100$, $v^{\min} = 0$, $v^{\max} = 500$, $S^{\max} = 100$, $\psi = 0.6, \psi_1 = \psi_2 = 2$, $\boldsymbol{w}^{\min} = [0, 0, 20]$m, and $\boldsymbol{w}^{\max} = [100, 100, 200]$m.

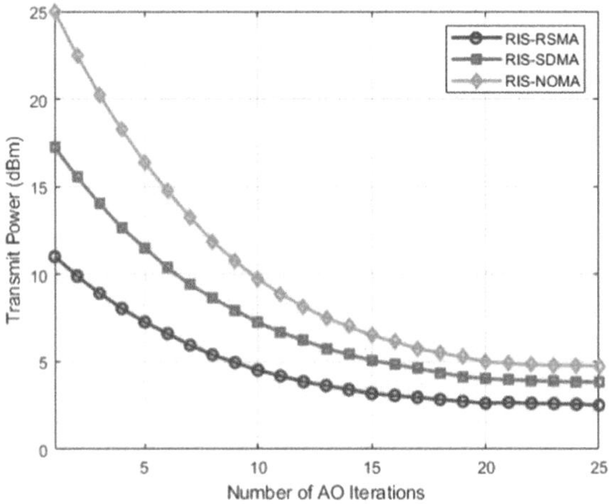

Fig. 2. Performance w.r.t. number of AO iterations

4.2 Convergence Analysis

Figure 2 illustrates the performance of three schemes-RIS-RSMA, RIS-SDMA, and RIS-NOMA - in terms of transmit power requirements over a series of AO iterations. As observed, all three schemes demonstrate a decrease in transmit power with an increasing number of AO iterations, indicating convergence within 15–20 iterations. Among the schemes, RIS-RSMA consistently exhibits the lowest transmit power requirements across all iterations. This superior performance is attributed to the rate-splitting multiple access technique's ability to efficiently manage interference and optimize resource allocation, thereby enhancing spectral efficiency. In comparison, RIS-SDMA shows a higher transmit power requirement than RIS-RSMA but still performs better than RIS-NOMA. The space-division multiple access approach leverages spatial multiplexing to serve multiple users, leading to better spectral efficiency and reduced power consumption compared to non-orthogonal multiple access. However, it does not match the efficiency of RIS-RSMA due to less effective interference management.

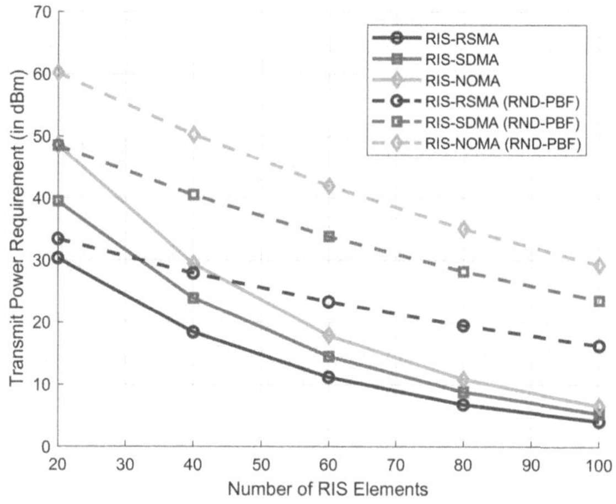

Fig. 3. Performance versus Number of RIS elements

4.3 Impact of a Varying Number of RIS Elements

Figure 3 elucidate the impact of a varying number of RIS elements by fixing number of BS anetnnas. Importantly, the increase in number of RIS elements decreases the transmit power requirement of the proposed system. The plot shows how the transmit power requirement evolves with an increasing number of RIS elements for three main schemes: RIS-RSMA, RIS-SDMA, and RIS-NOMA. Additionally, it contrasts these schemes with their counterparts employing random passive beamforming (RND-PBF). The results demonstrate an exponential decrease in transmit power for RIS-RSMA, RIS-SDMA, and RIS-NOMA, highlighting their superior efficiency compared to the RND-PBF variants, which exhibit a slower reduction in transmit power. This underscores the effectiveness of optimized beamforming strategies in achieving significant power savings and enhancing performance in RIS-enabled communication systems.

4.4 Impact of Varying FBL Transmission Parameters

Finally, we illustrate the system performance of proposed RSMA URLLC system w.r.t. varying FBL transmission parameter, i.e., block length size l as shown in

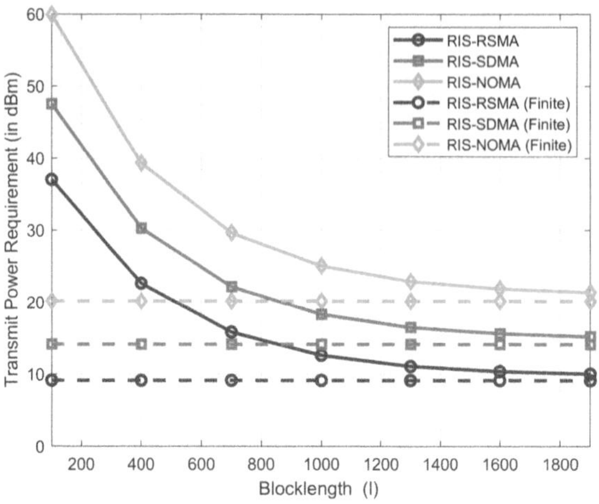

Fig. 4. Performance versus Blocklength

Fig. 4. The required transmit power for the proposed algorithms rises as the block length shortens. However, it stabilizes once the block length reaches $l \geq 200$. The performance gap between the rate-function based on normal approximation for URLLC systems [5] and the rate function derived from Shannon's capacity for infinite block length transmissions decreases as the block length increases. Since the complementary error function $Q^{-1}(\varepsilon)$ is a monotonically decreasing function of ε, the impact of channel dispersion in the normal approximation lessens. This reduction leads to an increase in the rate-throughput of URLLC. Considering all scenarios, RSMA gains exceptional performance as compared to NOMA and SDMA schemers even for stringent URLLC conditions i.e., at shorter block length and high reliability. In other words, RSMA, as compared to NOMA and SDMA, employs shorter block lengths and achieves lower latency while maintaining the same transmit power. This characteristic of RSMA makes it highly promising for reducing latency and ensuring high reliability in core services of future wireless networks.

5 Conclusion

This paper addressed the problem of minimizing transmit power at BS for dead zone users in an RIS-assisted MIMO-DL RSMA system while jointly optimizing the active beamforming at the BS and passive beamforming at the ARIS, considering the packet size and PEP for FBL transmission. To handle the non-convexity of the problem, an alternating algorithm was proposed, where the sub-problems of beamforming at the BS and ARIS are solved alternately. The active and passive beamforming sub-problems were solved using successive convex approximation. Simulation results demonstrated that the proposed algorithm converges within a few iterations. Additionally, the ARIS-aided RSMA system outperformed NOMA and SDMA schemes by approximately 40% and 60%, respectively. Conclusively, the optimal ARIS placement in RSMA brings potential benefits for future wireless networks, including low latency, high reliability, and improved coverage, particularly for core services in URLLC scenarios.

References

1. Saad, W., Bennis, M., Chen, M.: A vision of 6G wireless systems: applications, trends, technologies, and open research problems. IEEE Network **34**(3), 134–142 (2020)
2. Pala, S., Katwe, M., Singh, K., Clerckx, B., Li, C.-P.: Spectral-efficient RIS-aided RSMA URLLC: toward mobile broadband reliable low latency communication (mBRLLC) system. IEEE Trans. Wireless Commun. **23**(4), 3507–3524 (2024)
3. Popovski, P., et al.: Wireless access for ultra-reliable low-latency communication: principles and building blocks. IEEE Network **32**(2), 16–23 (2018)
4. Katwe, M., et al.: An Overview of Intelligent Meta-surfaces for 6G and Beyond: Opportunities, Trends, and Challenges arXiv preprint, pp. 1–7. arXiv:2405.03356
5. Katwe, M., Singh, K., Clerckx, B., Li, C.-P.: Rate splitting multiple access for energy efficient RIS-aided multi-user short-packet communications. In: IEEE Globecom Workshops (GC Wkshps), pp. 644–649 (2022)
6. Di, B., Zhang, H., Song, L., Li, Y., Han, Z., Poor, H.V.: Hybrid beamforming for reconfigurable intelligent surface based multi-user communications: achievable rates with limited discrete phase shifts. IEEE J. Sel. Areas Commun. **38**(8), 1809–1822 (2020)
7. Li, Y., Yin, C., Do-Duy, T., Masaracchia, A., Duong, T.Q.: Aerial reconfigurable intelligent surface-enabled URLLC UAV systems. IEEE Access **9**, 140248–140257 (2021)
8. Xie, H., Xu, J., Liu, Y.-F., Liu, L., Ng, D.: User grouping and reflective beamforming for IRS-aided URLLC. IEEE Wirel. Commun. Lett. **10**(11), 2533–2537 (2021)
9. Ranjha, A., Kaddoum, G.: URLLC facilitated by mobile UAV relay and RIS: a joint design of passive beamforming, blocklength, and UAV positioning. IEEE Internet Things J. **8**(6), 4618–4627 (2021)
10. Almekhlafi, M., Arfaoui, M.A., Elhattab, M., Assi, C., Ghrayeb, A.: Joint resource allocation and phase shift optimization for RIS-aided eMBB/URLLC traffic multiplexing. IEEE Trans. Commun. **70**(2), 1304–1319 (2022)
11. Katwe, M., Singh, K., Clerckx, B., Li, C.-P.: Rate-splitting multiple access and dynamic user clustering for sum-rate maximization in multiple RISs-aided uplink mmwave system. IEEE Trans. Commun. **70**(11), 7365–7383 (2022)
12. Katwe, M., Deshpande, R., Singh, K., Pan, C., Ghare, P.H., Duong, T.Q.: Spectrally-efficient beamforming design for STAR-RIS-aided URLLC NOMA systems. IEEE Trans. Commun. (Early Access Article)
13. Katwe, M., Singh, K., Clerckx, B., Li, C.-P.: Rate splitting multiple access for sum-rate maximization in IRS aided uplink communications. IEEE Trans. Wireless Commun. **22**(4), 2246–2261 (2023)
14. Katwe, M., Singh, K., Clerckx, B., Li, C.-P.: Improved spectral efficiency in STAR-RIS aided uplink communication using rate splitting multiple access. IEEE Trans. Wireless Commun. **22**(8), 5365–5382 (2023)

15. Pala, S., Katwe, M., Singh, K., Tsiftsis, T.A., Li, C.-P.: Robust design of RIS-aided full-duplex RSMA system for V2X communication: a DRL approach. In: GLOBECOM 2023–2023 IEEE Global Communications Conference, Kuala Lumpur, Malaysia, pp. 2420–2425 (2023)
16. Soleymani, M., Santamaria, I., Jorswieck, E.A., Clerckx, B.: Optimization of rate-splitting multiple access in beyond diagonal RIS-assisted URLLC systems. IEEE Trans. Wireless Commun. **23**(5), 5063–5078 (2024)
17. Yang, Z., Chen, M., Saad, W., Shikh-Bahaei, M.: Optimization of rate allocation and power control for rate splitting multiple access (RSMA). IEEE Trans. Commun. **69**(9), 5988–6002 (2021)
18. Pala, S., Singh, K., Katwe, M., Li, C.-P.: Joint optimization of URLLC parameters and beamforming design for Multi-RIS-aided MU-MISO URLLC system. IEEE Wirel. Commun. Lett. **12**(1), 148–152 (2023)
19. Grant, M., Boyd, S.: CVX: Matlab software for disciplined convex programming, version 1.21 (2010). http://cvxr.com/cvx

Data Transmission over x-Pol and y-Pol 1550 nm Carrier for Short Reach Optical Fiber Communications

S. Deepak[1(✉)], M. Naveenkumar[2,3], Neha Gupta[2], and E. S. Shivaleela[2]

[1] Cochin University of Science and Technology, Cochin, India
deepaks@pg.cusat.ac.in
[2] Indian Institute of Science, Bengaluru, India
{naveenkumarm,gneha}@iisc.ac, lila@iisc.ac.in
[3] Siddaganga Institute of Technology, Tumakuru, India
navinvcm@sit.ac.in

Abstract. Bandwidth hungry applications along with the increasing number of internet users are pushing the edge networks to provide higher bandwidths. Fiber optic cables have high bandwidth capacity and can carry large amounts of data, has lowest transmission loss in the c-band, and the number of available optical carriers have to be optimally used, to get highest spectral efficiency as well as capacity. For short-reach communications, intensity modulation/direct detection technique offers the simplest and lowest cost/bit solution.

There is scope of doubling the bandwidth capacity of the optical fiber, using polarization division multiplexing, on each wavelength, for short reach communications. The study focuses on the experiment of optical carrier with x and y polarization components carrying data independently. By exploiting both x and y polarizations as independent carriers of the 1550 nm wavelength, the capacity of transmission is doubled. We have generated linearly polarized and circularly polarized light using polarization controllers (PC) and transmitted over single mode fiber (SMF) of 50 km length, and also through polarization maintaining fiber (PMF) of length 10 m. The transmission loss of polarization maintaining fiber is significantly higher than SMF and hence only 10 m length is used. The comparison of the above two transmitted signals shows that, it is possible to carry independent x-pol and y-pol signals over SMF, provided the fiber doesnot go through physical disturbances (vibrations). Both x-pol and y-pol signals are successfully transmitted/received and observed on the Digital Storage Oscilloscope (DSO). In our future work, we plan to carryout transmission at higher bit rates.

Keywords: Polarization controller · Linearly polarized light · Circularly polarized light · Polarization division multiplexing

D. Gupta et al. (Eds.): PCEMS 2024, CCIS 2491, pp. 357–366, 2025.
https://doi.org/10.1007/978-3-031-90577-3_30

1 Introduction

The demand for high bandwidth and increased throughput is because of the increase in usage of internet services driven from high bandwidth applications. The optical communication enabled the growth in transmission capacity of communication system. The switching speed of the electronic circuits is unable to keep pace with the demand of higher signal processing rate required in high-speed networks. To meet this demand several techniques are used to increase the spectral efficiency/capacity. Manipulation of the state of polarization (SOP) of light, amplitude and the phase is done to encode information, exploiting different dimensions of the carrier, thereby the system capacity can be enhanced, like mode division multiplexing, time division multiplexing and wavelength division multiplexing [1]. Also, polarization division multiplexing is one of these techniques used for doubling the capacity, which has two linear and orthogonal state of polarizations of a wavelength, carrying independent data.

Higher order phase modulation like quadrature phase shift keying (QPSK) are used for long-haul transmission. While, higher order amplitude modulation like pulse amplitude modulated four-level can also be effectively used for short reach communications [1]. The real possibility of polarization division multiplexing (PDM) along with IM/DD technique, incorporating in the already existing fibers makes it a very significant method in increasing capacity. Thus, within a single fiber two independent transmission channels can be created of the same wavelength [2]. The PDM has the advantage of doubling the capacity of transmission [3,4], provided fiber doesnot go through any perturbation. x-pol and y-pol optical carriers are orthogonal to each other and hence can carry independent signals [5].

The polarization separation and twin-single side band (SSB) signal generation are implemented using a cascaded polarization beam splitter (PBS) and IQM [6], which is also helpful for the carrier suppression of the opposite polarization. The optical PDM technique is a promising method for spectral-efficient optical transmission, which can be used to double the system capacity and spectral efficiency [7–11]. The PDM silicon photonic integrated circuit (PIC) is proposed by [12], suitable for intra-chip optical interconnection.

The organization of the paper is as follows: in Sect. 2, the concept of generating linearly polarized and circularly polarized light, in our experimental setup is explained. Experimental results are analyzed in Sect. 3 and Sect. 4 is about the conclusions and future work.

2 Experimental Work

The passive optical network consists of consists of three main units optical line terminal (OLT), optical network unit (ONU) and optical distributed network (ODN). The architecture of the PON system is shown in the Fig. 1. Bandwidth is a vital factor in designing and maintaining a LAN. To meet the higher bandwidth demand, different methods are used. The schematic diagram of the experimental setup for carrying out polarization division multiplexing with distributed feedback (DFB) laser of wavelength 1550 nm is as shown in the Fig. 2. The DFB laser is connected to polarization controller (PC), and the output of PC to Polarization beam splitter (PBS). By varying the quarter/half wave plate of PC, the polarization level of optical signal can be controlled. Linearly polarized light is generated, with one polarization, x-pol, dominant and y-polarized light very low, as shown in the Digital Storage Oscilloscope (DSO) plots later. Both the outputs of PBS are detected by using photodetectors (PD) and the waveform is observed on the DSO. The schematic shown in the Fig. 2 is modified by adding the additional PC in between the first PC 1 and PBS as shown in the Fig. 3.

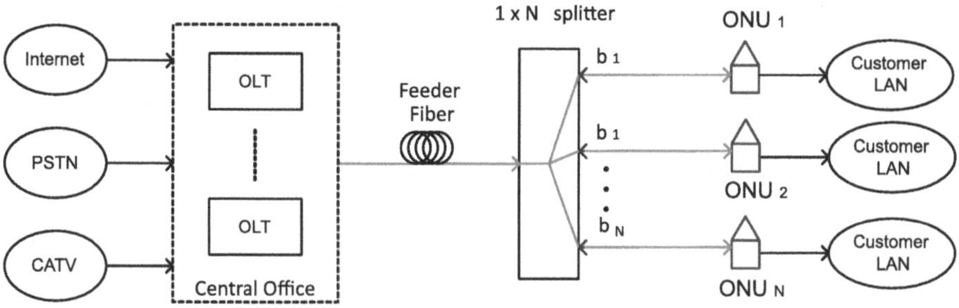

Fig. 1. Passive optical network architecture with fiber cabling distribution from the central office to customer premesis, providing various services

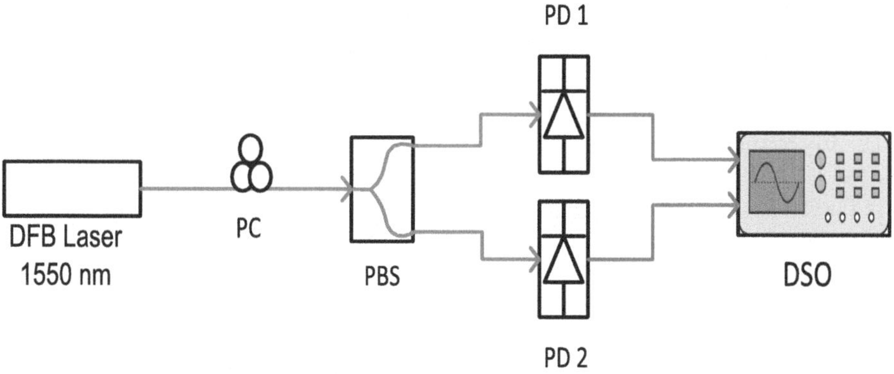

Fig. 2. Schematic for generating linearly polarized light from a DFB laser, using a polarization controller and the x-pol and y-pol light paths are split using polarization beam splitter and detected by two photoreceivers

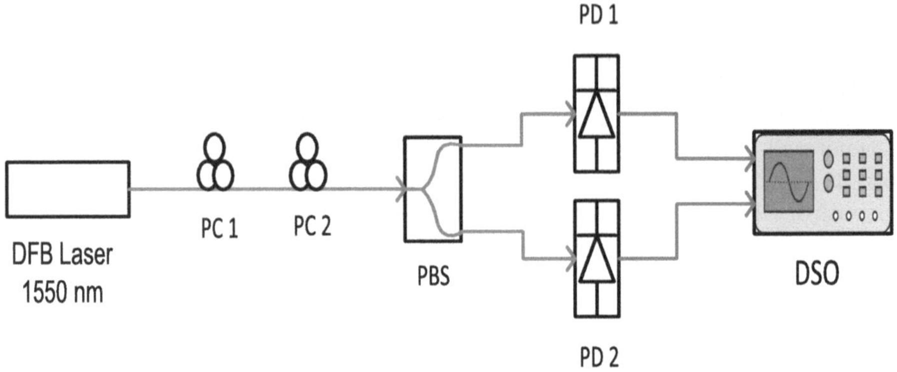

Fig. 3. Schematic for generating circularly polarized light from a DFB laser, using two polarization controllers, and the x-pol and y-pol light paths are split using polarization beam splitter and detected by two photoreceivers

The DFB laser is connected to a PC consists of quarter-wave plate, a half-wave plate, and a second quarter-wave plate, each of which can be rotated independently. The half-wave plate, which rotates the polarization direction of linearly polarized light, and the quarter-wave plate, which converts between circular polarizations. It is necessary to adjust three waveplates in order to transform any state of polarization to any other output covering the entire Poincare sphere [13].

By adjusting the waveplates of the first PC, linear polarization signal is generated, which is provided as input to second PC 2 for generating circularly polarized light as shown in the Fig. 3. A polarization beam splitter (PBS) is used for polarization splitting the optical signal and tested for equal x and y polarization amplitudes. The photo detector (PD) is connected to each output port of the PBS and optical to electrical conversion is done, and the outputs of PDs are connected to the digital storage oscilloscope to observe the electrical signal.

The circularly polarized light, with a modulation frequency of 2.4 MHz generated from the configuration of Fig. 3 is transmitted over a single mode fiber (SMF) spool of length 50 km. The schematic is shown in the Fig. 4. The transmission loss due to SMF is 0.2 dB/km, and polarization maintaining fiber is 0.2 dB/meter, at laser wavelength of 1550 nm.

Further, the SMF in the Fig. 4 is replaced with polarization maintaining fibers (PMFs) of 5 m and 10 m length, to transmit the circularly polarized light as shown in the Fig. 5. The DFB laser is modulated with same frequency of 2.4 MHz. The experimental set up for generating linearly and circularly polarized light is shown in the Fig. 6.

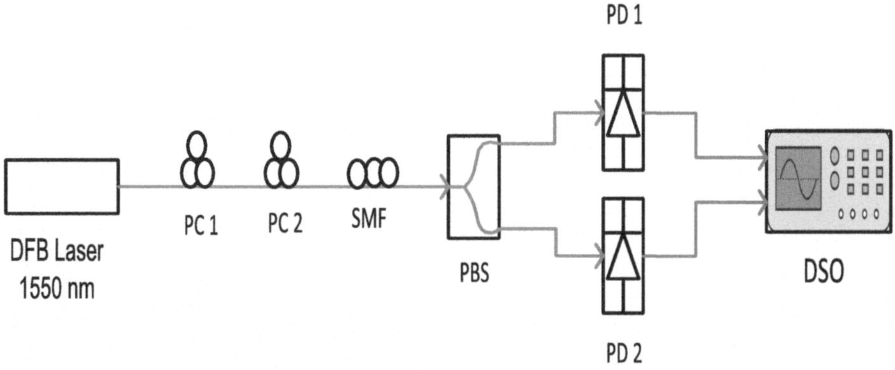

Fig. 4. Schematic of polarization division multiplexing of DFB laser for generating circularly polarized light, transmitted through single mode fiber

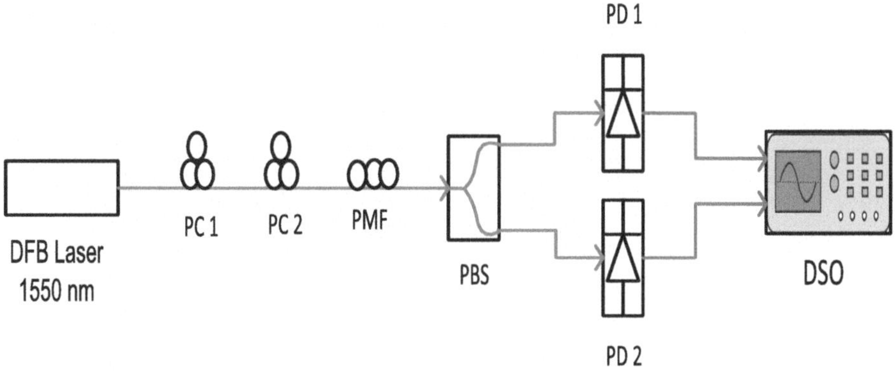

Fig. 5. Schematic of polarization division multiplexing of DFB laser for generating circularly polarized light, transmitted through polarization maintaining fiber (PMF) of length 10 m

3 Experimental Results

The linearly polarized and circularly polarized light are generated by using the schematic of Fig. 2 and 3. The DFB laser is modulated by frequency of 50 KHz and linearly polarized light is observed as shown in the Fig. 7.

Initially, using a frequency of 50 kHz a linearly polarized light was observed, as shown in Fig. 7. This was observed at the 0° angle configuration of the first and the third quarter wave plates and also for the half wave plate. By adjusting the third wave plate which is a quarter wave plate to an angle of 45°, a circularly polarized light is generated and the x and y polarization components have equal amplitude values, as shown in the Fig. 8. These two orthogonal components are obtained in the digital storage oscilloscope as yellow and violet signals. The modulating frequency for the DFB laser varied to 1.2 MHz and 2.4 MHz. The

Fig. 6. Experimental setup for generating circularly polarized light, transmitted through 50 km SMF or 10 m polarization maintaing fiber

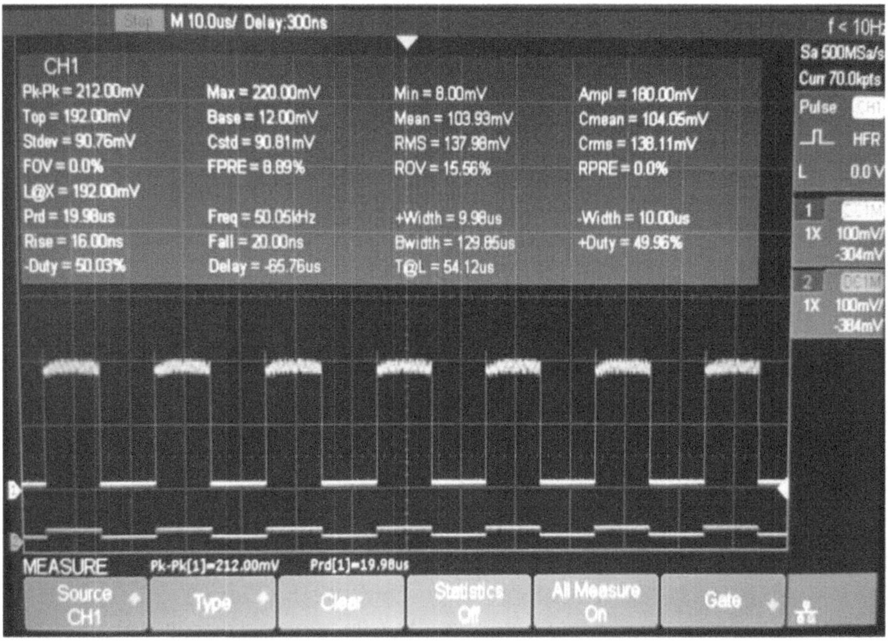

Fig. 7. Generation of linearly polarized light signal for DFB laser modulated with a frequency of 50 KHz.

same experimental configuration was maintained through out, and the outputs were measured.

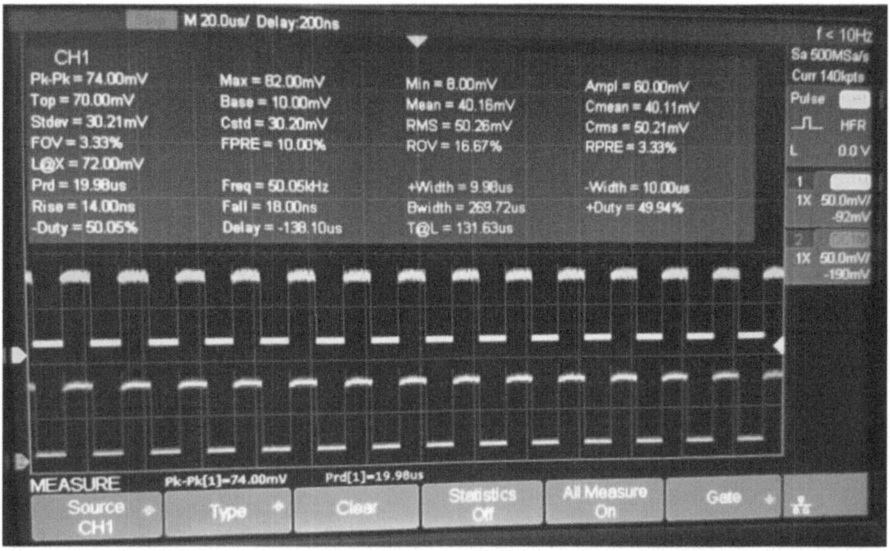

Fig. 8. Circularly polarized light signal generated and detected for a frequency of 50 kHz modulating signal

By rotating the third quarter waveplate, a signal with amplitude of 55 mV was obtained for the laser input power of 50% with modulating frequency of 50 kHz, and amplitude of 50 mV was obtained for the 70% laser power condition and modulating frequency of 1.2 MHz as shown in the Table 1.

Table 1. Results observed at DSO for the schematic of circularly polarized light without any external fiber connection

Frequency	Laser power	Output voltage
50 KHz	50%	55 mV
1.2 MHz	70%	50 mV

The single mode fiber of 2 km, 25 km, and 50 km are connected in the schematic of the Fig. 4, and the output pulse amplitudes of 50 mV, 20 mV and 5 mV are obtained. These results are measured for the for the 50% laser power modulated by 2.4 MHz frequency. The results are tabulated in th Table 2. The circularly polarized light signal generated using DFB laser modulated by frequency of 2.4 MHz with SMF of length 50 km connected in the schematic is shown in the Fig. 9.

The SMF connected in the schematic of the Fig. 4, are replaced by the PMF as shown in the Fig. 5. We have connected DFB laser with power 50%, modulated by frequency of 2.4 MHz. We observed amplitude of 50 mV and 8

Table 2. Results observed at DSO for the schematic of circularly polarized light with single mode fiber (SMF) connection

Fiber length	Frequency	Laser power	Output voltage
2 km	2.4 MHz	50%	50 mV
25 km	2.4 MHz	50%	20 mV
50 km	2.4 MHz	50%	5 mV

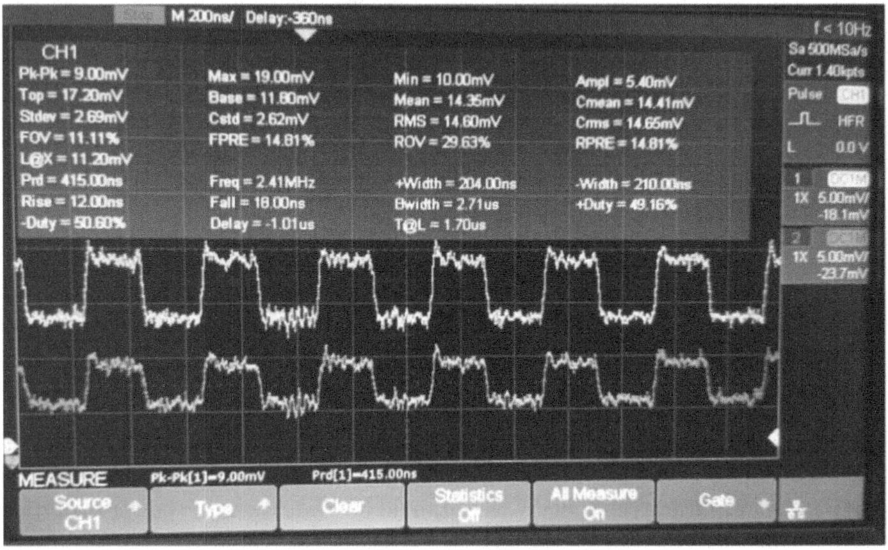

Fig. 9. Circularly polarized light signal generated and detected for a frequency of 2.4 MHz modulating signal transmitted over 50 km single mode fiber

mV for the PMF of length 5 m and 10 m respectively. By varying the laser power was changed to 100% from 50%, we observed change in the amplitude level at DSO, which is 32 mV. The results are tabulated in th Table 3. The different voltage levels in the Fig. 7 shows the plane polarized light, While, the same level of the output voltage observed in Fig. 8 indicates, it was circularly polarized light. The circularly polarized light signal generated using DFB laser modulated by frequency of 2.4 MHz with PMF of length 10 m connected in the schematic is shown in the Fig. 10.

Table 3. Results observed at DSO for the schematic of circularly polarized light using polarization maintaining fiber

Fiber length	Frequency	Laser power	Output voltage
5 m	2.4 MHz	50%	50 mV
10 m	2.4 MHz	50%	8 mV
10 m	2.4 MHz	100%	32 mV

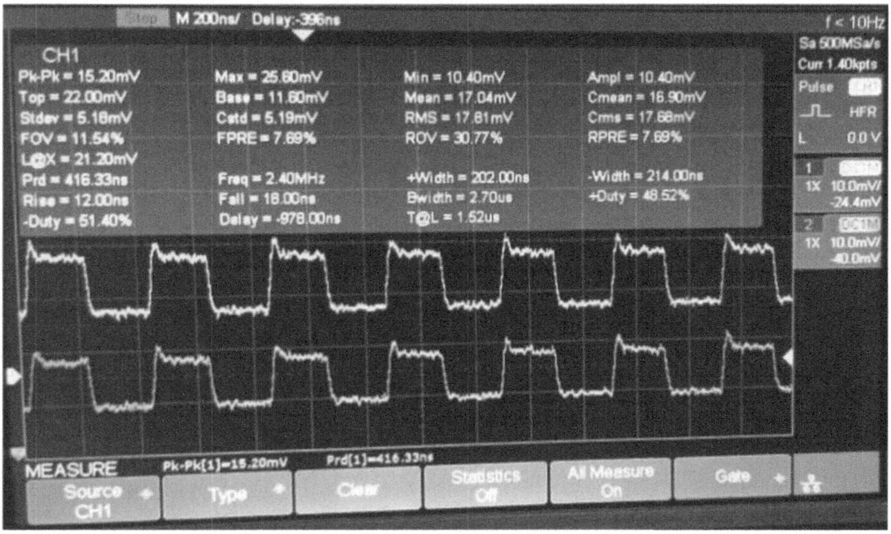

Fig. 10. Circularly polarized light signal generated and detected for a frequency of 2.4 MHz modulating signal transmitted over 10 m polarization maintaining fiber

4 Conclusions and Future Work

We have experimentally generated linearly and circularly polarized light using PC and PBS. By exploiting x and y polarization as independent carriers of the same wavelength, the capacity of transmission is doubled. The two separated polarization components x and y can be used for transmission along short-reach distances, if polarization is maintained throughout. Using the experimental technique used in the study, data rate can be doubled for each wavelength. Similarly for all N carriers, if polarization division multiplexing is implemented, the capacity of the network is increased $2N$ fold, with incoherent IM/DD transmission method, for short-reach communications.

References

1. Chen, Z.Y., Yan, L.S., Pan, Y., et al.: Use of polarization freedom beyond polarization-division multiplexing to support high-speed and spectral-efficient data transmission. Light Sci. Appl. **6**, e16207 (2017). https://doi.org/10.1038/lsa.2016.207
2. Kyselak, M., Dvorak, F., Vlcek, C.: Measurement techniques for determining the polarization division multiplexing tolerance field. In: 2019 International Conference on Advanced Technologies for Communications (ATC), Hanoi, Vietnam, pp. 269–272 (2019). https://doi.org/10.1109/ATC.2019.8924560.
3. Liu, X., Chraplyvy, A., Winzer, P., et al.: Phase-conjugated twin waves for communication beyond the Kerr nonlinearity limit. Nature Photon. **7**, 560–568 (2013). https://doi.org/10.1038/nphoton.2013.109
4. Zhang, J., Yu, J., Chi, N.: Transmission and full-band coherent detection of polarization-multiplexed all-optical Nyquist signals generated by Sinc-shaped Nyquist pulses. Sci. Rep. **5**, 13649 (2015). https://doi.org/10.1038/srep13649
5. Ao, X., Liu, L., Wosinski, L., He, S.: Polarization beam splitter based on a two-dimensional photonic crystal of pillar type. Appl. Phys. Lett. **89**(17), 171115 (2006). https://doi.org/10.1063/1.2360201
6. Zhou, Y., Ming, J., Wang, L., Wu, D., Zhao, L., Xiao, J.: Optical polarization division multiplexing transmission system based on simplified twin-SSB modulation. Sensors **22**(20), 7700 (2022). https://doi.org/10.3390/s22207700. PMID: 36298051; PMCID: PMC9612215
7. Xiao, J., Yu, J., Li, X.: 40-Gb/s PDM-QPSK signal transmission over 160-m wireless distance at W-band. Opt. Lett. **40**, 998–1001 (2015). https://doi.org/10.1364/OL.40.000998
8. Li, X., Yu, J., Xiao, J.: Fiber-wireless-fiber link for 128-Gb/s PDM-16QAM signal transmission at W-band. IEEE Photon. Technol. Lett. **26**, 1948–1951 (2014). https://doi.org/10.1109/LPT.2014.2342711
9. Li, F., Cao, Z., Li, X.: Fiber-wireless transmission system of PDM-MIMO-OFDM at 100 GHz frequency. J. Lightwave Technol. **31**, 2394–2399 (2013). https://doi.org/10.1109/JLT.2013.2266353
10. Zhang, J., Yu, J., Jia, Z.: 400 G transmission of super-nyquist-filtered signal based on single-carrier 110-GBaud PDM QPSK with 100-GHz grid. J. Lightwave Technol. **32**, 3239–3246 (2014). https://doi.org/10.1109/JLT.2014.2343016
11. Wang, W., Zou, D., Feng, X.: Optical polarization division multiplexing fiber-wireless integration system at Ka-band based on a low-cost dual-drive MZM. Opt. Exp. **29**, 37453–37463 (2021). https://doi.org/10.1364/OE.441081
12. Yu, Y., Chen, G., Sima, C., Zhang, X.: Intra-chip optical interconnection based on polarization division multiplexing photonic integrated circuit. Opt. Exp. **25**, 28330–28336 (2017)
13. Xiao-Guang, Z., Yuan, Z.: The number of least degrees of freedom required for a polarization controller to transform any state of polarization to any other output covering the entire Poincaré sphere. Chin. Phys. B **17**(7), 1–6 (2008). https://doi.org/10.1088/1674-1056/17/7/027
14. Yu, H., et al.: A 32-channel c-band hybrid wavelength/polarization division (de) multiplexer on silicon. IEEE Photon. J. **15**(3), 1–6 (2023)

An Approach for Extraction of Relevant Contextual Information from Exercise Logs Using Large Language Models

Anand Arumilli(✉), Amit Oberoi, and Ayesha Rifa

Oracle HCM, Oracle India Pvt. Ltd., Hyderabad, India
arumilli.anand@oracle.com, amit.oberoi@alumni.iitd.ac.in

Abstract. In this paper, we give out an approach towards utilisation of Large Language Models (LLMs) for extracting key and relevant information from exercise logs generated by people. We demonstrate the efficacy of six popular LLMs in handling exercise logs while classifying multiple dimensions of exercise type, intensity, duration, and feeling post-exercise. We compare the performance of LLMs with traditional models such as BERT classifiers and XLNet for few-shot classification tasks. We go on to suggest that we can use the power of LLMs to analyze exercise log data without the need for training. This can then be used for personalized recommendations related to fitness and also for marketing analysis while measuring the effectiveness of a coach/application. Our conceptual framework proposes LLM based classification and analysis resulting in an intelligent recommendation system.

Keywords: Large Language Models · Natural Language Processing · Semantic Analysis · Fitness Applications

1 Introduction

Exercise log is a record keeping tool that people use to document their physical activities. This log typically includes details such as the type of exercise, duration, intensity, and any relevant notes on how one felt before, during, and after the exercise. In the real world, the formal recording of exercise is not followed religiously by the population at large. Except for keen sportsmen and athletes, most people tend to comment or blog about their exercise. This may be accompanied by a photograph or a snapshot of running activity accompanied by a tweet or social media post.

These exercise logs on large-scale systems can be utilized fruitfully by using the power of Large Language Models (LLMs) to understand generic and application-specific context. For example, these may be matched with relevant feeling tags to track the post-exercise sentiments. Rather than using automated systems which can handle limited corpora, or using language models after resource hungry fine tuning, we can consider using foundational LLMs for classification of exercise logs to yield measurable and informative inputs. This can be

D. Gupta et al. (Eds.): PCEMS 2024, CCIS 2491, pp. 367–380, 2025.
https://doi.org/10.1007/978-3-031-90577-3_31

achieved with few-shot training or even prompt engineering [1,2] thus making the entire process very efficient. Also, the ability of LLMs to handle multiple languages or even multimodal data over a wide global context can be leveraged to understand logs across a wide population for yielding intelligible information. Similarly, LLMs can be effectively used in recommender systems for personalized inputs for trainers and users. The aim of this paper is therefore to present an approach towards beneficial use of human generated data relevant to fitness activities, while using applications, equipment or in gymnasiums or institutions.

2 Related Work

Recent studies have explored the use of deep learning architectures, including Convolutional Neural Networks (CNNs), Long Short-Term Memory Networks (LSTMs), and transformer-based models like BERT, RoBERTa, and XLNet, for sentiment analysis and emotion extraction in the fitness domain, without relying on large language models (LLMs). LLMs have found extensive applications in various domains in healthcare, including clinical natural language processing (NLP), electronic health records (EHR), social media analysis, medical imaging, biomolecular data, and biophysical signals [3]. However, there is a noticeable gap in the literature when it comes to the specific use of LLMs for sentiment analysis or emotion extraction from exercise-related logs. This study aims to address this gap by exploring the potential of LLMs to accurately capture and classify the sentiments associated with different exercise activities, thereby contributing to a novel application area within the healthcare and wellness domain. A comparison chart of the architecture and the strengths and limitations of each are presented in Table 1. The evolution of LLMs has seen various capabilities in text generation, finding similar context and question answering etc. [4]. The transformer architecture allowed for a deeper understanding of the context embedded within the text. Further understanding of Generative AI as given at [5] gives an insight into human-computer interaction using GPT. This enables us to understand the key features and solutions for conversational assistants [6]. Certain studies have explored performance using few shot learning with the dataset while introducing parameter-level approach to analyze the effect of various sizes of training samples.

Table 1. Comparison of CNN, LSTM, BERT, XLNet, and LLMs

Feature	CNN	LSTM	BERT	XLNet	LLMs
Architecture	Convolutional layers + pooling	RNN with gates	Transformer encoder (self-attention)	Transformer-XL (autoregressive + bidirectional)	Transformer decoder (self-attention)
Attention Mechanism	None	None	Self-attention (global)	Self-attention (global + autoregressive)	Self-attention (unidirectional)
Strengths	Efficient for local features	Captures long-term dependencies	Deep contextual understanding	Combines bidirectional + autoregressive features	General-purpose, scalable to many NLP tasks including text generation
Limitations	Poor for sequential data	Slow for long sequences	Not generative, requires masking	More complex and computationally expensive	Requires significant computational resources; unidirectional limitations for understanding tasks

Large language models have witnessed significant advancements in Natural Language Processing related tasks such as sentimental analysis incorporated in our work. Models such as BERT (Bidirectional Encoder Representations from Transformers) [7] based on the Transformer architecture, have demonstrated powerful representations learned through pre-training on massive text data, making them applicable to our Exercise Logging system. Context handling supports language models in generating relevant information through knowledge graphs constructed using GPT model [8].

This paper proposes a novel approach by utilizing large language models (LLMs), which build on the advances in transformer-based architectures while offering a more generalized framework for language understanding and generation. LLMs, trained on diverse datasets with billions of parameters, can perform well on few-shot and few-shot tasks, surpassing the capabilities of traditional architectures such as BERT, CNNs, LSTM, and XLNet in terms of scalability, versatility, and performance on extraction of contextual information.

The model during pre-training uses self-supervised learning methods with a multi-layer Transformer on a large corpus to predict the next tokens based on input and acquires the ability to capture information related to text structure. Data pruning method improves the model performance as discussed by Marion, Pozobon, Wang, Fadee and Hooker in [9]. Carefully pruning large training samples can be as effective as increasing data volume for enhanced outcomes. Prompt engineering with GenAI as discussed in [10] is a critical step that extends the capabilities of large language models (LLMs). It involves designing and optimizing various prompts in natural language to guide the model for different tasks, such as classification, generation, extraction, and categorization as discussed by Sahoo, Singh, Sahaans others in [11] This is a promising approach executed on a specific dataset, resulted in better accuracy on the target function by providing a detailed summary. NLP algorithms used in this such as [12] have further been improvised using large language models to build an interactive chat application that can generate a workout summary.

3 Conceptual Framework

We propose using a robust LLM with the capability of incorporating global context which will effectively capture context from user or subscriber comments/blogs or journals. These text-based expressions/utterances in natural language can be classified into domain-specific inputs about the activity and its effect on the user. Post this, relevant categorical data can be analysed in detail to evolve a recommender system [13] for users as well as fitness app/service providers. Corresponding conceptual framework is depicted at Fig. 1.

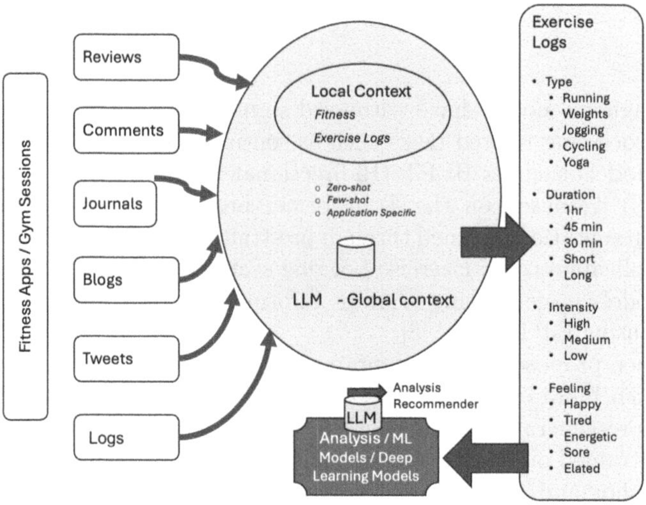

Fig. 1. Conceptual Framework: LLM Based Classification and Recommendation

The innovative concept of analysing exercise logs can be extended to global fitness trends by extracting contextual text data from reviews, comments, blogs, etc. illustrated in Fig. 1. The aim is to provide a deeper understanding of exercise habits and preferences. In the past, LLMs have been applied in healthcare contexts, our application specifically focuses on extracting nuances from exercise logs, and could be used for downstream recommendation using few-shot learning techniques.

4 Experimentation

4.1 Dataset

In this study, we utilized a dataset consisting of 3,000 samples [12]. This labeled dataset contained three primary columns

Log - Free-text descriptions of the exercise performed by individuals, typically written in natural language. This unstructured text field combined details about the activity, intensity, duration, and emotional response.

Exercise Tag - A structured description categorizing the type of exercise performed.

Feeling Tag - Semantic labels representing the emotional state or feeling experienced during or after the exercise.

To capture additional dimensions of contextual information, two more features, intensity and duration, were manually labeled to enhance the dataset.

Intensity - Categorized as low, medium, or high based on the type and exertion level of the exercise.

Duration - The total time taken to complete the exercise or workout session, typically recorded in minutes or hours.

The primary aim was to analyze the correlation between exercise metrics and post-exercise sentiments, using Large Language Models (LLMs) for personalized insights and marketing strategies (Table 2).

Table 2. Exercise Logs

Log	Exercise Tag	Feeling Tag	Intensity	Duration
I ran for 45 min on the treadmill. I felt so tired and out of shape	running	tired	high	45 min
I squatted for 20 min. I was exhausted afterwards	squats	exhausted	medium	20 min
I went bowling for 2 h. I felt happy	bowling	happy	low	2 h
I swam 50 laps in 30 min. I felt refreshed	swim	refreshed	high	30 min
I went to my barbell class for 25 min and felt so energized afterwards	weights	energised	medium	25 min

4.2 Data Preparation

To prepare the data for training and evaluation, we split the dataset into training and validation sets, adopting a few-shot learning approach. Specifically, 20% of the data (600 logs) is the test set, while the other 80% of the data (2,400 logs) is for training. The exercise log in the dataset describes different forms of physical activity or workout capturing various types and intensities as another class, and the post-exercise feelings are recorded in the feeling log that tells about the emotional and physical responses. The proposed method uses only 2% of samples to surpass the full data fine-tuning, reducing time costs by 97%. A remarkable accuracy is achieved through strategic selection and optimization techniques allowing various models to learn from the data that is present and making it useful to scale it to large learning task.

4.3 Models Used

In the initial phase of our research, we conducted a series of experiments using well-established transformer-based models, including BERT, RoBERTa, and XLNet, to extract sentiments and contextual information from exercise logs. The experimentation methodology involved fine-tuning these models on a dataset of annotated exercise descriptions to classify the emotional states and exercise types.

Model Initialization - The pre-trained versions of BERT (`bert-base-uncased`), RoBERTa(`roberta-base`), and XLNet (`xlnet-base-cased`) were loaded using the `transformers` library. The classification head was adapted to match the number of unique labels derived from the dataset.

Tokenization - Each model's respective tokenizer was employed to preprocess the input sentences. This involved tokenizing the sentences, adding special tokens, and padding or truncating to a fixed length (e.g., 128 tokens).

Model Training - The model was trained for 70 epochs, as this was deemed sufficient for comparison with the few-shot learning capabilities of large language models (LLMs) (Fig. 2).

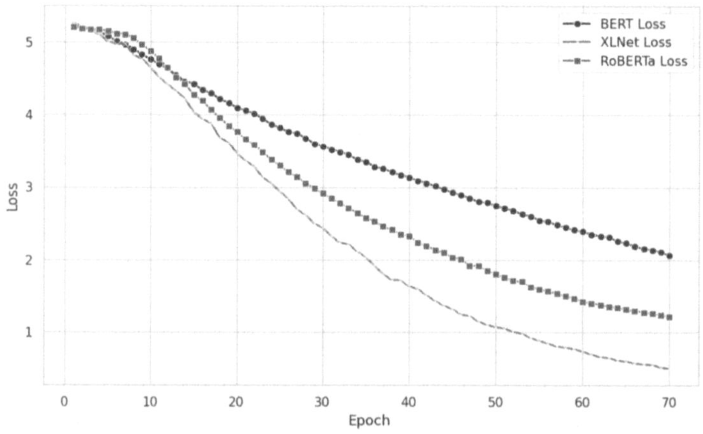

Fig. 2. Loss Curves for BERT, XLNet, and RoBERTa Over 70 Epochs

For each batch, the loss was computed, and the model weights were updated using the Adam optimizer.

Classification - The new exercise log entries were tokenized and processed through the trained model. This effectively segregated the log into distinct categories: exercise type, feeling, intensity, and duration (Fig. 3).

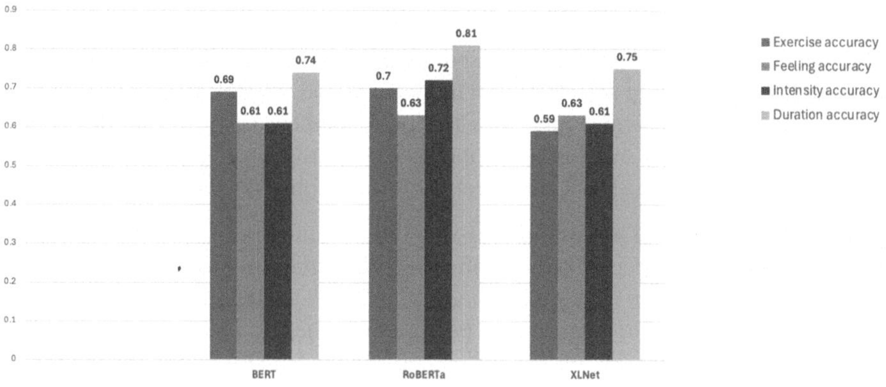

Fig. 3. Accuracy scores of three preliminary models

Despite extensive hyperparameter tuning and model optimization efforts, the performance metrics indicated sub-optimal results, with the models struggling to capture the nuanced emotional context effectively.

In the subsequent phase, we employed Large Language Models (LLMs) to address these challenges. We experimented with several prominent LLMs to classify the exercise, feeling tags, intensity and duration tags. Here, we provide an overview of key LLMs that were used in our experiments, their architectures, and their applications.

GPT 3.5 - The transformer architecture [14] employed in GPT represents a major leap forward compared to earlier NLP methodologies like RNNs and CNNs. The self-attention mechanism enables the model to take into account the context of the entire sentence when predicting the next word. This innovation has significantly enhanced the model's capability to comprehend and generate language. The decoder in the transformer is tasked with producing the output text based on the input representation.

Llama 2 - It is built upon the transformer architecture [14], optimizing for performance and efficiency in various language tasks. Designed to handle a broad spectrum of NLP applications with fewer parameters compared to GPT-3.5, it maintains competitive performance through architectural innovations. Building on the insights from the GPT-Neo-X project, LLaMA incorporates rotary positional embeddings (RoPE) at each layer, which significantly enhances its overall performance [15].

Gemma - The architecture is proprietary, but it likely incorporates elements of transformers with domain-specific enhancements. These enhancements are designed to optimize the model's performance for specialized tasks by leveraging advanced techniques such as transfer learning and domain adaptation. Applications are tailored for customized NLP tasks in specialized fields like healthcare and fitness where precision and domain-specific knowledge are crucial. It's design allows it to handle complex language processing tasks within these specialized areas effectively.

Mistral 7B - The architecture is based on the paper [16] that uses tokens from previous layers, allowing selection of window size and the chunk size to compute the attention over cache and measures the performances to ensure model's effective across a variety of use cases. Supports a wide range of applications in text generation, Translation, Data Interpretation, and context analysis.

Gemini 1.5 Pro - The architecture includes elements of transformers augmented with domain-specific enhancements and shares similarities with Gemma. These enhancements are intended to boost the model's capabilities in performing multitask learning and fine-tuning for particular domains. This model is designed for customized NLP tasks across various specialized fields. The model is built to excel in these environments by adapting to the specific requirements and challenges presented by each domain, providing robust and accurate language processing solutions.

FLAN-T5 - This unified approach simplifies the process of training and fine-tuning the model on a variety of tasks. FLAN (Fine-tuned Language Agnostic Networks) [17] represents an extension where models are fine-tuned on multiple tasks to improve generalization. The architecture utilizes a transformer-based, text-to-text paradigm.

4.4 Methodology

To build upon previous methods such as CNN, BERT classifier, and XLNet an advancement to large language models has been highlighted and demonstrated impressive capabilities in generating contextually relevant natural responses. Utilized a few-shot learning approach in the initial phase which provided limited interpretability. Hence, we narrowed down the approach to few-shot training with LLMs. Further, training the models to domain-specific with exercise samples and understanding to extract each tag from the sentence.

Few-Shot Learning - We have given 5 sample logs as examples for few shot learning and the corresponding labels for Exercise type, feeling, intensity and duration. This approach is applied on GPT-3.5, LLAMA-2, Gemma, Gemini 1.5 pro, FlanT5 to extract the associated tag such exercise, feeling tag and corresponding duration if present and the intensity. Our results indicate a slight improvement in performance with an increasing number of shots to 5. Additionally, the number that needs to be set depends on the data complexity and thus indicates challenges for small scale models.

Training and Validation - The training set consists of 2400 samples (80% of 3000), while the validation set consists of 600 samples (20% of 3000). The models are trained to classify the exercise and feeling tags based on the provided examples.

Model Evaluation - After training each model was evaluated on the test set. We used pre-trained embeddings from the BERT model (BertTokenizer and BertModel) to convert both the predicted and true labels into vector representations. These embeddings allowed us to measure the semantic closeness between the true and predicted labels using cosine similarity. This approach helped in cases where the model predicted tags that were close in meaning but not exact matches (e.g., "tired" vs. "fatigued"). A threshold of 0.7 was used for cosine similarity. Predictions with cosine similarity scores above this threshold were considered semantically correct, even if the exact words did not match the expected labels. The performance metrics focused on accuracy, measuring how well each model could correctly classify the exercise and feeling tags from the exercise logs [18] (Fig. 4).

5 Results

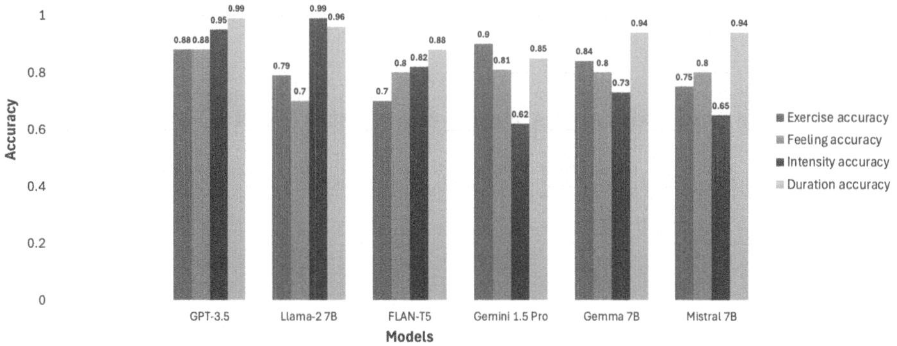

Fig. 4. Accuracy scores of six LLM models tested

In this study, we evaluated the performance of several large language models (LLMs) in predicting exercise-related outcomes using different accuracy metrics: Exercise Accuracy, Feeling Accuracy, Intensity Accuracy, and Duration Accuracy. The models compared include GPT-3.5, Llama-2 7B, FLAN-T5, Gemini 1.5 Pro, Gemma, and Mistral 7B.

We observed that the accuracy of the LLMs surpassed that of traditional models like BERT, RoBERTa, XLNet. GPT-3.5 demonstrated robust performance across various tasks, making it a strong candidate for applications requiring general high accuracy in NLP tasks. Llama-2 7B excelled in tasks requiring precision in intensity and duration, but improvements are needed in understanding and predicting feelings. FLAN-T5 is a well-rounded model but may require fine-tuning for specific tasks to achieve top-tier performance. Gemini 1.5 Pro shows potential in exercise-related tasks but needs improvement in intensity prediction to enhance its overall utility. Gemma and Mistral 7B are solid performers with strengths in predicting duration and exercise thus suitable for specialized domains (Table 3).

Table 3. Comparison of Accuracy Scores

Model	Exercise	Feeling	Intensity	Duration
BERT	0.69	0.61	0.61	0.74
RoBERTa	0.70	0.63	0.72	0.81
XLNet	0.59	0.63	0.61	0.75
GPT-3.5	0.88	0.88	0.95	0.99
Llama-2 7B	0.79	0.70	0.99	0.96
FLAN-T5	0.70	0.80	0.82	0.88
Gemini 1.5 Pro	0.90	0.81	0.62	0.85
Gemma 7B	0.84	0.80	0.73	0.94
Mistral 7B	0.75	0.80	0.65	0.94

5.1 Insights and Inferences Using the Best Model

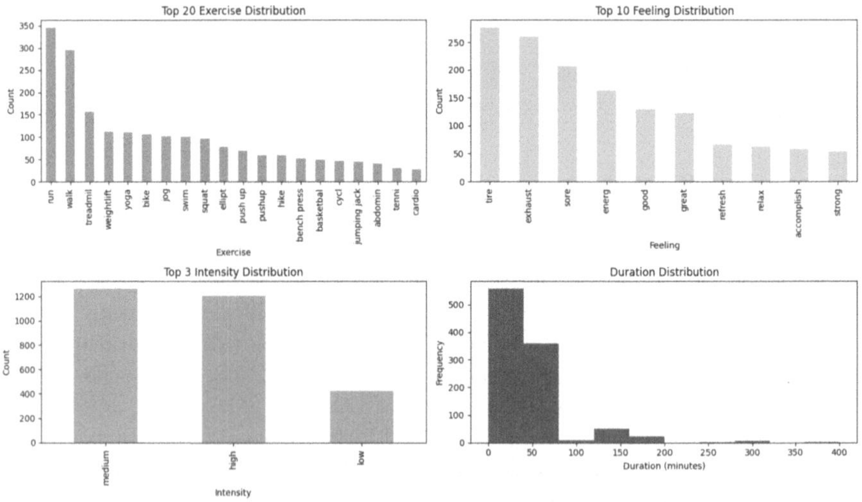

Fig. 5. Distribution of top Exercise, Intensity, Feeling and Duration

We use GPT - 3.5 to get further insights from the dataset in Fig. 5.

Top 20 Exercise Distribution - The bar plot illustrates the count of the top 20 exercises, showing a visual representation of which exercises are most popular.

Top 10 Feeling Distribution - The bar plot for the top 10 feelings displays the frequency of each feeling tag, offering insights into the emotional responses to different exercises.

Top 3 Intensity Distribution - The intensity distribution bar plot highlights the frequency of low, moderate, and high intensities, indicating the common levels of exercise intensity among participants.

Duration Distribution - The histogram of exercise duration, spread across 0-400 min, shows the frequency of different duration ranges, giving an overview of how long the exercise sessions typically last.

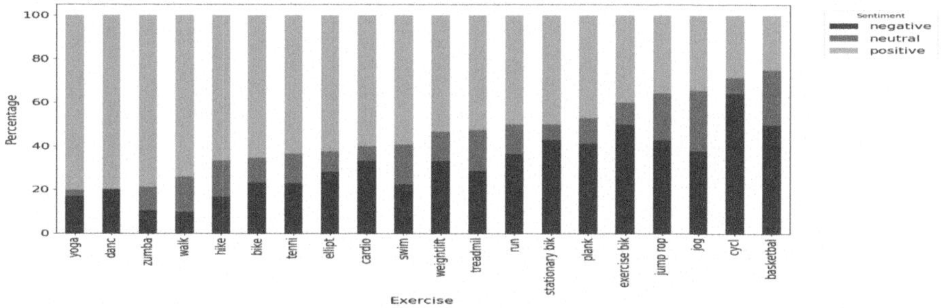

Fig. 6. Distribution of top 20 exercise types sorted by positive sentiment

Correlation Between Exercise Type and Sentiment - We used the LLM to further classify Feeling tag to positive, neutral and negative sentiment experienced by the subject post exercise. We display insights on the top 20 exercise types in Fig. 6.

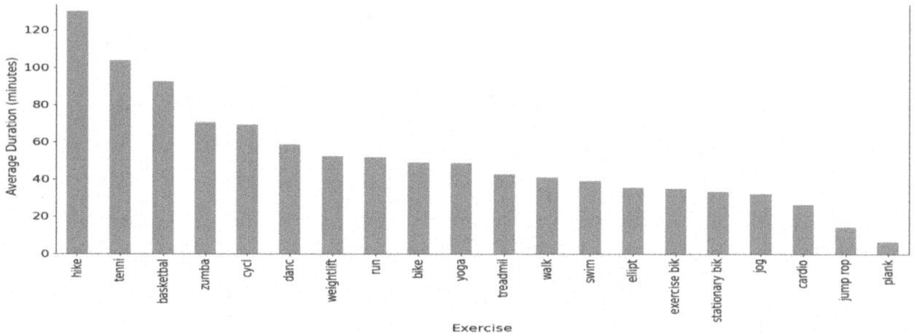

Fig. 7. Average Duration by exercise types

Average Duration by exercise Types - We used the LLM to provide the average duration spent on each of the top 20 exercise types as in Fig. 7.

5.2 Recommendation System

From the experimentation, GPT-3.5 emerges as the best-performing model with the highest overall average score and consistently high accuracy across all tasks. This indicates its strong generalization capabilities and effectiveness in understanding and generating accurate predictions. We use it for downstream sentiment analysis and inference tasks on the complete Exercise Log to provide recommendations to improve [19] through prompting as shown in Fig. 8.

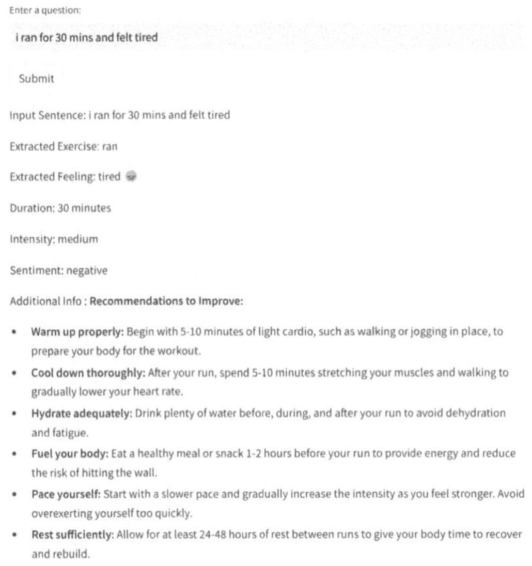

Fig. 8. Downstream usage in Recommendation engine

6 Conclusion

In this paper we explored the utilization of Large Language Models (LLMs) for extracting and analyzing key information from natural language texts generated by individuals post-exercise. We labeled available dataset [12] for feeling, intensity, duration and type of exercise. Our study evaluated six popular LLMs, namely GPT-3.5, Llama-2, FLAN-T5, Gemini 1.5 Pro, and Gemma 7B and Mistral 7B. These were evaluated for performance in handling exercise logs and classifying multiple dimensions such as exercise type, intensity, duration, and post-exercise feelings. Further we compared the performance of LLMs with the BERT and XLnet classifiers after training them on the same dataset.

We were able to extract valuable insights on the distributions of individual parameters and correlations between these. We were able to handle analysis such as, "Do certain exercises tend to make people feel more energized/positive while others make them feel exhausted/negative"?, and, "Which exercises do people spend more time on average?"

Our findings confirm the potential of LLM to effectively analyze exercise log data, providing valuable insights for personalized fitness recommendations, marketing analysis, and evaluating the effectiveness of coaches or applications. LLMs like GPT-3.5 performed 19% to 35% using few-shot approach better as compared to BERT and XLnet classifiers trained over 70 epochs. This highlights the significant advantages of LLMs in developing intelligent recommendation systems capable of handling complex datasets efficiently.

In future work, we plan to fine-tune our existing models and extend the experimentation to include multilingual as well as multi-modal support.

References

1. Gao, T., Fisch, A., Chen, D.: Making Pre-trained Language Models Better Few-shot Learners (2020)
2. Liu, P., Yuan, W., Fu, J., Jiang, Z., Hayashi, H., Neubig, G.: Pre-train, prompt, and predict: a systematic survey of prompting methods in natural language processing. ACM Comput. Surv. **55**(9), 195:1–195:35 (2023)
3. Nerella, S., et al.: Transformers in Healthcare: A Survey (2023). arXiv:2307.00067
4. Large Language Models (LLMs): Hypes and Realities
5. Mohammad, A.F., Clark, B., Hegde, R.: Large language model (LLM) & GPT, a monolithic study in generative AI. In: 2023 Congress in Computer Science, Computer Engineering, & Applied Computing (CSCE), pp. 383–388 (2023)
6. Dedhia, K., Konkar, M., Shah, D., Tawde, P.: A proposed algorithm to perform few shot learning with different sampling sizes. In: 2022 IEEE Fourth International Conference on Advances in Electronics, Computers and Communications (ICAECC), pp. 1–5 (2022). ISSN: 2642-6595
7. Devlin, J., Chang, M.W., Lee, K., Toutanova, K.: BERT: Pre-training of Deep Bidirectional Transformers for Language Understanding (2019). arXiv:1810.04805
8. Abu-Rasheed, H., Weber, C., Fathi, M.: Knowledge Graphs as Context Sources for LLM-Based Explanations of Learning Recommendations (2024). arXiv:2403.03008
9. Marion, M., Üstün, A., Pozzobon, L., Wang, A., Fadaee, M., Hooker, S.: When Less is More: Investigating Data Pruning for Pretraining LLMs at Scale (2023). arXiv:2309.04564
10. Effectiveness of Generative Artificial Intelligence for Scientific Content Analysis | IEEE Conference Publication | IEEE Xplore
11. Sahoo, P., Singh, A.K., Saha, S., Jain, V., Mondal, S., Chadha, A.: A Systematic Survey of Prompt Engineering in Large Language Models: Techniques and Applications (2024). arXiv:2402.07927
12. Epps, M., Uribe, J., Korpusik, M.: A new dataset for natural language understanding of exercise logs in a food and fitness spoken dialogue system. In: 2021 IEEE Spoken Language Technology Workshop (SLT), Shenzhen, China, pp. 1071–1078. IEEE (2021)

13. Zhao, Z., et al.: Recommender Systems in the Era of Large Language Models (LLMs) (2023)
14. Vaswani, A., et al.: Attention is all you need. In: Advances in Neural Information Processing Systems, vol. 30. Curran Associates, Inc. (2017)
15. Roumeliotis, K.I., Tselikas, N.D., Nasiopoulos, D.K.: Llama 2: Early Adopters' Utilization of Meta's New Open-Source Pretrained Model (2023)
16. Jiang, A.Q., et al.: Mistral 7B (2023). arXiv:2310.06825
17. Longpre, S., et al.: The flan collection: designing data and methods for effective instruction tuning. In: Proceedings of the 40th International Conference on Machine Learning, pp. 22631–22648. PMLR (2023). ISSN: 2640-3498
18. anandaru. code (2024)
19. Xu, L., et al.: Prompting Large Language Models for Recommender Systems: A Comprehensive Framework and Empirical Analysis (2024)

A New MIMO Antenna with High Gain for 5G mm-Wave Application

Swati Verma[1](✉), Vandana Thakare[1], and Sweety Jain[2]

[1] Madhav Institute of Technology and Science, Gwalior, India
swativerma01694@gmail.com, vandana@mitsgwalior.in
[2] Samrat Ashok Technological Institute, Vidisha, India

Abstract. The paper provides an in-depth examination of the development & features of 5G spectrum allocations, along with MIMO antenna design. It gives an overview of massive MIMO technology & its uses to base station applications. The paper's primary objective is to highlight the frequencies reserved for 5G and to demonstrate the impacts of MIMO designs. An array antenna for 5G applications has been developed, where we originally designed an array of rectangular patch antennas functioning as a Multi-Input-Multi-Output (MIMO) antenna operating at 28 GHz. This antenna is compact, measuring 14.89 mm × 42.90 mm. The antenna's two-dimensional radiation patterns exhibit circularly polarized radiation pattern due to the array configuration. The proposed antenna is designed and simulated, the simulation result showing return loss of less than −10 dB with isolation, a gain of 6.7 dB at 28 GHz, and an 8.07 dB directivity at the operating frequency. This makes it a strong candidate for MIMO applications at 28 GHz band, allowing 5G cellular communications.

Keywords: MIMO Antenna · 2D Radiation Pattern · 5G Millimeter wave (mm-wave) Frequency Band · Microwave studio (CST)

1 Introduction

The advent of 5G technology heralds a new era in wireless communications, promising unprecedented data rates, ultra-low latency, and massive connectivity. As articulated by Andrews et al. [1], 5G technology is poised to revolutionize various sectors by providing enhanced mobile broadband, ultra-reliable low latency communications, and massive machine-type communications. This transformation necessitates innovations in multiple domains, including non-orthogonal multiple access [2], advanced antenna designs, and millimeter-wave (mm Wave) communications. The design of efficient antennas is a cornerstone for the successful deployment of 5G networks. For instance, the enhancement of return loss through defected ground planes has been demonstrated to significantly improve the performance of microstrip patch antennas [3]. Wideband high-gain antenna subarrays [4] and millimeter-wave microstrip patch antennas [5] are also crucial for meeting the stringent requirements of 5G applications. These advancements are underscored by industry and academic collaborations that highlight the challenges and research priorities in 5G [6].

D. Gupta et al. (Eds.): PCEMS 2024, CCIS 2491, pp. 381–391, 2025.
https://doi.org/10.1007/978-3-031-90577-3_32

Empirical studies, such as those conducted on rain attenuation at 26 GHz in Malaysia [7], provide valuable insights into the real-world performance of 5G systems under various environmental conditions. Moreover, the investigation of space diversity in 28/38 GHz MIMO antennas [8] and directional radio propagation path loss models [9] are essential for optimizing 5G network design and deployment. The development of 5G millimeter-wave antenna arrays [10] and high isolation MIMO antennas [11] further underscores the importance of advanced antenna technology in achieving the desired performance metrics.

Recent research has also focused on novel antenna designs for 5G smartphones [12], massive MIMO systems [13], and performance enhancement techniques for 5G antennas [14]. The design of high-gain wideband MIMO antennas [15] and meta surface-based circularly polarized MIMO antennas [16] exemplifies the innovative approaches being taken to meet the diverse requirements of 5G applications. Comparative studies on different metamaterial surfaces [17] and dual-band antenna arrays [18] provide a comprehensive understanding of the design and performance trade-offs involved.

Furthermore, the integration of LTE and millimeter-wave 5G MIMO antenna systems [20], the use of defected ground structures [21], and the development of antennas for 28-GHz MIMO applications [23] highlight the ongoing efforts to enhance the capabilities of 5G networks. These advancements are complemented by studies on mobile-phone antenna arrays with diamond-ring slot elements [24], multi-band MIMO systems [25], and integrated antenna systems for future handheld devices [26].

Research on mutual coupling reduction in MIMO antennas using parasitic elements [28], EBG structures [29], and MNG metamaterials [30] is pivotal for improving the isolation and performance of 5G antennas. The design of compact super-wideband antennas [32] and tri-band stacked patch antennas [33], as well as wideband dielectric resonator antennas [34], further illustrates the breadth of innovations in antenna technology.

The research efforts highlighted in this introduction demonstrate a wide range of approaches and solutions aimed at overcoming the challenges associated with 5G communications, paving the way for a connected future.

1.1 Multiple Input Multiple Output (MIMO)

Congestion, multipath fading, and radiation losses are all risks associated with wireless transmission [14] systems. Furthermore, at longer wavelengths, these issues become more intense. When dealing with all these problems, the use of MIMO antennas becomes essential as they increase the communication range while simultaneously reducing signal strength. Consequently, MIMO architecture may be utilized in 5G networks to provide low latency, bandwidth, and high performance while maintaining low power consumption. With MIMO, multiple signals can be transmitted efficiently by utilizing several antennas, resulting in a substantial increase in available bandwidth.

The use of multiband antennas, which offer coverage for a variety of wireless systems, is one way of reducing the multiple antennas required in MIMO systems. MIMO antennas may be divided into two categories based on their frequency band: wideband antennas & multiband antennas [14].

Numerous mm-wave antennas have been described in the literature for use in mm-wave applications. According to research [15], a 5G antenna with a MIMO design with a response range of 23–40 GHz has been developed. The suggested design has an area of 80 × 80 mm^2, according to the specifications. Even though the suggested MIMO architecture achieves a significant amount of bandwidth, the size of the bandwidth is found to be very big [16] presents a rounded, polarizing MIMO antenna having 4 ports that are identical to the earlier one. The placement of an antenna element inside the center of big number of dependent components has been intended to enhance radioactivity efficiency; however, the use of passive components increases the difficulty of the suggested MIMO system. In another example [17], researchers present a single antenna component with total size of 106 mm^2 and maximum gain of 6.59 dB. The application of electromagnetic band-gap structures (EBG) causes significant coupling problems or make manufacturing of antenna element more challenging, even though the gain obtained by antenna element is adequate. Similarly, an array antenna operating at 2 frequency bands—namely, 28 & 38 GHz—is suggested [18] for study. This approach is interesting because of the higher gain of a stated antenna; however, it does not have the MIMO capability. The authors [19] suggest an 8 x 8 MIMO system by an overall size of 31.2 × 31.2 × 1.57 mm^3. The antenna structure radiation response suggests a resonance at 25.2 GHz with a maximum gain of 8.7 dB, as well as a substantial percentage of backside lobe is detected in the antenna design radiation responses. Another study presents [20], an antenna for 4G & 5G applications with a resonance frequency of 28 GHz and overall size of the antenna is 75 × 85 mm^2. While the antenna demonstrates a gain of 5.13 dB for the millimeter wave band, its radiation pattern exhibits a high number of back & side lobes, making it unsuitable for usage in mm-wave communications networks. The authors [21] suggest a mm-wave antenna designed with a T- shaped configuration that covers a bandwidth rate of nearly 25.1 to 37.5 GHz, while authors [22, 23] suggest a 4-component antenna with just an size of 30 × 35 mm^2 with a peak gain of 8.3 dB which operates at 28 GHz frequency band. The 28 GHz mm-wave band has been identified as a promising candidate for upcoming mm-wave phones after the release of 5G sub-6 GHz & millimeter-wave prospective frequencies. Numerous studies [24–27] had indicated memory buffer for mm-wave antennas, and most of these analyses have incorporated mm-wave architecture. A multi-band 4-port antenna system with total size of 158 × 78 mm^2 and a gain of 5.1 dB at 28 GHz band is described in Reference [25]. Mm-wave antenna produced 3 distinct resonances at 28, 37 & 39 GHz frequencies. However, because of their unique size, each of the 4 antennas can only support a maximum of 4-port on the system. In Reference [26], the new T- slot antenna design consisting of two element-feed systems is described, resonating at two different bands: one under sub-6 GHz & another in the mm-wave spectrum from 24 GHz to 29 GHz with a peak gain of 5 dB at 28 GHz. Despite fact that research revealed consistent resonant frequencies & optimization techniques, the ground plane assemblage and large dimensions of radiating components prohibit the construction from being utilized by a greater collection of components. Lastly, [27] describes an 8-port tapered slot antenna system with total size of 104 × 104 × 0.501 mm^3.

2 Design of the Proposed Antenna System

The layout of the proposed four-element MIMO antenna system is shown in Fig. 1. The antenna is designed and optimized using the computer simulated software (CST). The antenna system consisted of a ground structure of dimension $W_g \times L_g$ (14.89 mm ×

42.90 mm), on the bottom side of the FR4 substrate with a dielectric constant of 4.2 and thickness of 3.10 mm. Four patch elements were printed on the top side of the substrate to achieve a four-element antenna array system. The antenna dimensions are given in Fig. 2. The overall size of the antenna system is 14.89 mm (W) × 42.90 mm (L) × 3.10 mm (H). Whole antenna system is fed via a 50-Ω microstrip line, with the dipole radiating in the y-direction. The radiation direction varies depending on which antenna is excited, by an array of meta material unit cells.

Each antenna is positioned at a certain distance. Mutual coupling between the ports, produced by surface waves is minimized by inserting a stub in the ground plane among the two antennas. The stub creates resonance when surface current builds around the stub, enhancing isolation between ports. For a stub length of LSB = 4.45 mm, isolation corresponds to a frequency of 28 GHz. Every antenna element has equal slots. The slots are introduced into each element to reduce the size of the element and transform it into a multi-band antenna element. The sizes of the patch elements and lengths of the slots were optimized using parametric analysis to achieve a compact multi-band MIMO antenna system with a high diversity gain performance (Fig. 3).

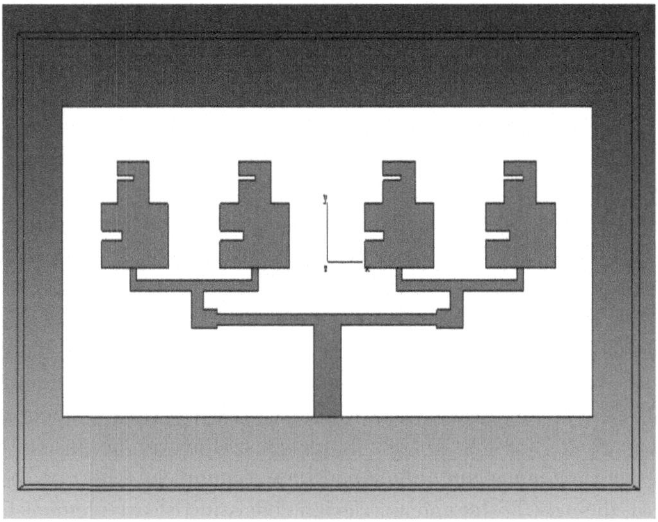

Fig. 1. Front view of Designed four-port MIMO antenna

The proposed antenna operates over a range of 25 GHz to 45 GHz. It achieves a steady two-dimensional radiation pattern in such a broad range by using an asymmetrical structure. On this basis, MIMO antennas are suggested and investigated. Because of its unique structure, when applied to large-scale arrays, the strong isolation among antenna elements is maintained.

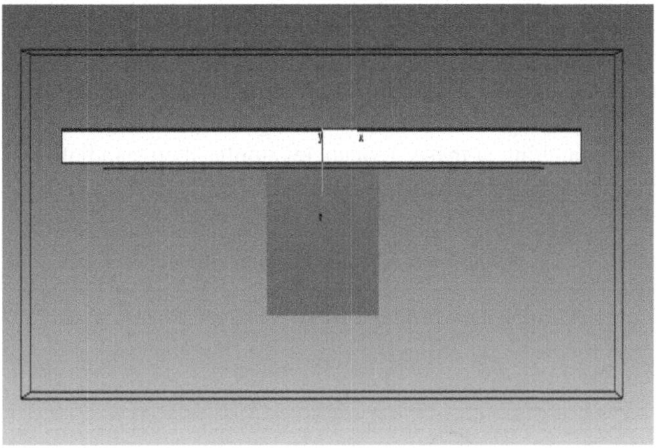

Fig. 2. Top view of Designed four-port MIMO antenna

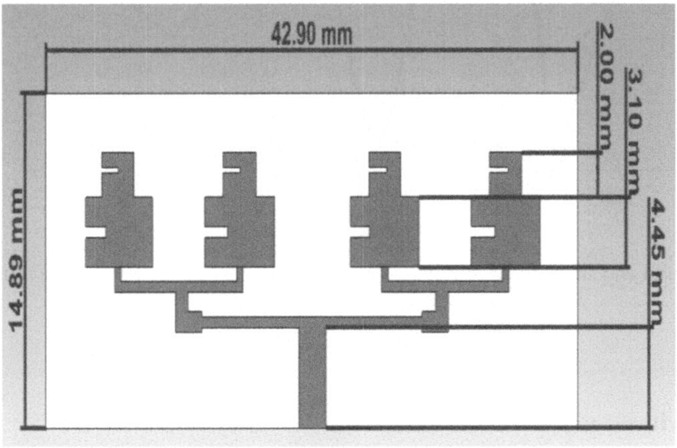

Fig. 3. Dimensions of Designed four-port MIMO antenna

3 Experimental Outcomes

The proposed 4-port antenna is manufactured in-house using the lithography technique for MIMO applications operating at the 28 GHz frequency range.

3.1 Return Loss

This graph shows the S-parameter Snn, also known as the reflection coefficient, plotted in terms of magnitude (in dB) versus frequency (in GHz). The frequency range on the x-axis spans from 25 GHz to 45 GHz, covering a broad band in the millimeter-wave spectrum. The y-axis shows the magnitude of Snn in dB, ranging from −20 dB to −5 db. Lower Snn values which is −10 dB indicate better impedance matching and less reflection, which implies more power is being delivered to the antenna and less is reflected back to the source. In this graph, significant dips in the S11 magnitude can be seen around 28 GHz, 32 GHz, 38 GHz frequencies. The multiple dips at different frequencies suggest that this antenna is designed for operation across multiple bands 28 GHz: A slight dip below −10 dB, which might correspond to a targeted resonance in your design. The strong impedance matching at these frequencies implies better radiation efficiency and power transfer at these points (Fig. 4).

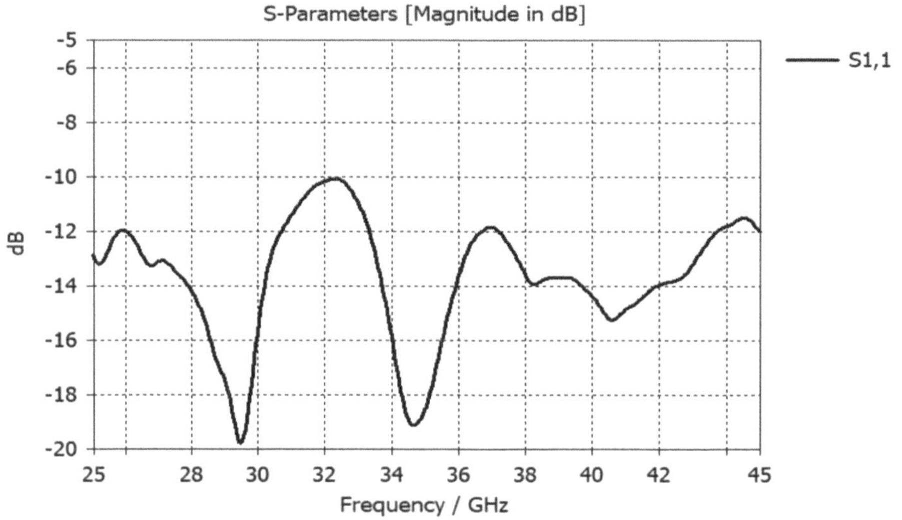

Fig. 4. Measured and simulated return loss showing value less than - 10 dB for a range of 25 GHz to 45 GHz.

3.2 Directivity

Figure 5 shows the 3D far-field radiation pattern that represents the directivity of the antenna at a frequency of 28 GHz. Directivity is a measure of how directional the antenna is, with a higher value indicating more focused radiation. The color scale on the right shows the directivity values in dB, with red and yellow areas indicating regions of high directivity and green/blue regions showing lower directivity. The pattern appears somewhat spherical with slight lobes, suggesting the antenna has a nearly omnidirectional radiation pattern. The directivity of the antenna is shown as 8.070 dB, which indicates the antenna focuses energy in certain directions rather than radiating it equally which is beneficial for applications needing concentrated signal strength. The radiation efficiency is noted as −1.358 dB A value of −1.358 dB The total efficiency is listed as −1.525 dB, meaning approximately 71% of the input power is radiated.

3.3 Gain

Figure 6 shows the 3D far-field radiation pattern that represents the gainy of the antenna at a frequency of 28 GHz. It represents a gain of 6.713 dB, which indicates the directional

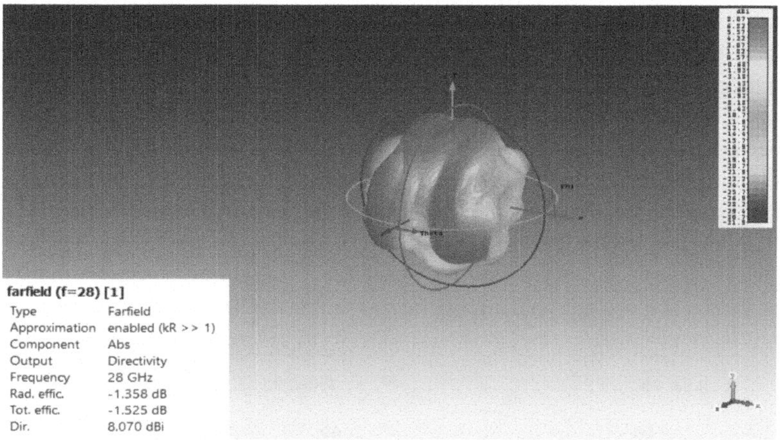

Fig. 5. Measured & simulated directivity of 8.07dB

amplification of the signal. This is a relatively high gain, suggesting the antenna is directional in this configuration. The radiation efficiency is −1.358 dB shows the efficiency of converting input power into radiated power, accounting for losses like dielectric and conductor losses.

Also, it has the Total Efficiency of −1.525 dB, considering all sources of loss, including matching loss. This value tells how effectively the input power is radiated.

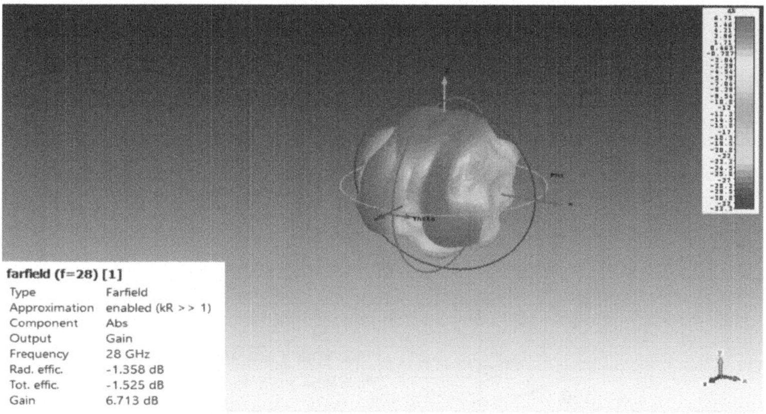

Fig. 6. Measured & simulated gain of 6.7dB

3.4 Radiation Pattern

The measured 2D radiation pattern of a 4-port antenna at 28 GHz is shown in Fig. 7. It represents the theta (degrees) versus dB at 28 GHz, demonstrating phi = 90 and phi = 270 in the far-field, which have complementary radiation patterns highlighting the presence of pattern diversity and therefore diversity gain. The radiation patterns demonstrate that the proposed MIMO antenna emits nearly omnidirectional radiation. Additionally, variation in the radiation patterns is minimal across a broad operating frequency band. It has been determined that the suggested antenna's maximum directivity reaches up to 8.070 dB and gain reach up to a limit 6.7 dB at various frequencies. The magnitude and direction of the major lobe are 3.83 dB & 61.0°, respectively, while the angular width of 92.4°. It has a sidelobe level of −3.4 db.

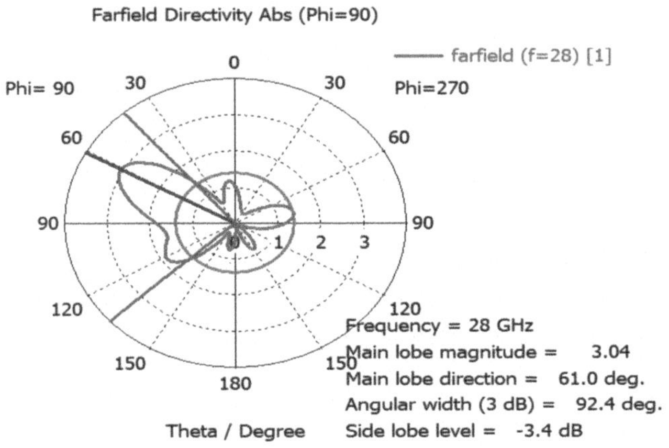

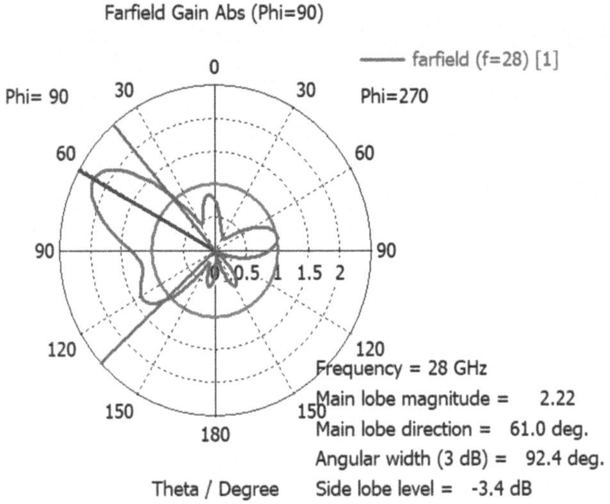

Fig. 7. 2D Radiation Patterns of Directivity and Gain

4 Conclusion

The proposed antenna design delivers high gain, excellent isolation, and robust radiation patterns within a compact structure. The observed and simulated findings indicate the suggested antenna structure may be configured into a variety of MIMO antenna structures as required. The suggested antenna's distinctive shape results in a steady radiation pattern and high isolation. The experimental findings highlight several benefits of the proposed antenna, including its distinct shape, steady radiation pattern, high isolation, and potential for higher gain. However, it's essential to acknowledge certain limitations and areas for future research.

Future work could focus on optimizing the antenna design to further enhance its performance characteristics, such as reducing sidelobe levels, reducing its size and improving angular coverage. Additionally, investigating the antenna's performance in real-world scenarios and addressing potential challenges, such as environmental interference and signal degradation, would be valuable for practical deployment in next-generation communication systems. Moreover, exploring the integration of advanced signal processing techniques with the MIMO antenna system could unlock new opportunities for improving overall system efficiency and reliability.

References

1. Andrews, J.G., et al.: What will 5G be? IEEE J. Sel. Areas Commun. **32**(6), 1065–1082 (2014)
2. Liu, Y., et al.: Non-orthogonal multiple access for 5G and beyond. Proc. IEEE **105**(12), 2347–2381 (2017)
3. Hakanoglu, B.G., Sen, O., Turkmen, M.: A square microstrip patch antenna with enhanced return loss through defected ground plane for 5G wireless networks. In: 2018 2nd URSI Atlantic Radio Science Meeting (ATRASC), Meloneras, pp. 1–4 (2018)
4. Ershadi, S., Keshtkar, A., Abdelrahman, A.H., Xin, H.: Wideband high gain antenna subarray for 5G applications. Prog. Electromagnet. Res. C **78**, 33–46 (2017)
5. Misbah, M.A.: Design and analysis of millimeter-wave microstrip patch antenna for 5G applications. In: International Conference on Technical Sciences, pp. 137–142 (2019)
6. NetWorld2020 ETP: 5G: Challenges, Research Priorities, and Recommendations; Joint White Paper (2014). https://networld2020.eu/wpcontent/uploads/2015/01/Joint-Whitepaper-V12-clean-after-consultation.pdf. Accessed 6 Jan 2021
7. Shayea, I., Rahman, T.A., Azmi, M.H., Islam, M.R.: Real measurement study for rain rate and rain attenuation conducted over 26 GHz microwave 5G link system in Malaysia. IEEE Access **6**, 19044–19064 (2018)
8. Aliakbari, H., Abdipour, A., Costanzo, A., Masotti, D., Mirzavand, R., Mousavi, P.: Performance investigation of space diversity for a 28/38 GHz MIMO antenna (applicable to mm-wave mobile network). In: Proceedings of the 2016 Fourth International Conference on Millimeter-Wave and Terahertz Technologies (MMWaTT), Tehran, Iran, 20–22 December 2016, pp. 41–44 (2016)
9. Sulyman, A.I., Alwarafy, A., MacCartney, G.R., Rappaport, T.S., Alsanie, A.: Directional radio propagation path loss models for millimeter-wave wireless networks in the 28-, 60-, and 73-GHz bands. IEEE Trans. Wireless Commun. **15**, 6939–6947 (2016)
10. Zhang, J., Ge, X., Li, Q., Guizani, M., Zhang, Y.: 5G millimeter-wave antenna array: design and challenges. IEEE Wirel. Commun. **24**, 106–112 (2016)

11. Wang, F., Duan, Z., Wang, X., Zhou, Q., Gong, Y.: High isolation millimeter-wave wideband MIMO antenna for 5G communication. Int. J. Antennas Propag. **2019**, Article ID 4283010 (2019)
12. Guo, J., Cui, L., Li, C., Sun, B.: Side-edge frame printed eight-port dual-band antenna array for 5G smartphone applications. IEEE Trans. Antennas Propag. **66**, 7412–7417 (2018)
13. Yang, B., Yu, Z., Dong, Y., Zhou, J., Hong, W.: Compact tapered slot antenna array for 5G millimeter-wave massive MIMO systems. IEEE Trans. Antennas Propag. **65**, 6721–6727 (2017)
14. Kumar, S., Dixit, A.S., Malekar, R.R., Raut, H.D., Shevada, L.K.: Fifth generation antennas: a comprehensive review of design and performance enhancement techniques. IEEE Access **8**, 163568–163593 (2020). https://doi.org/10.1109/ACCESS.2020.3020952
15. Sehrai, D.A., et al.: A novel high gain wideband MIMO antenna for 5G millimeter wave applications. Electronics **9**, 1031 (2020)
16. Hussain, N., Jeong, M.J., Abbas, A., Kim, N.: Metasurface-based single-layer wideband circularly polarized MIMO antenna for 5G millimeter-wave systems. IEEE Access **8**, 130293–130304 (2020)
17. Khan, J., et al.: Design and performance comparison of rotated Y-shaped antenna using different metamaterial surfaces for 5G mobile devices. CMC-Comput. Mater. Contin. **60**, 409–420 (2019)
18. Khan, J., Sehrai, D.A., Ali, U.: Design of dual band 5G antenna array with SAR analysis for future mobile handsets. J. Electr. Eng. Technol. **14**, 809–816 (2019)
19. Shoaib, N., Shoaib, S., Khattak, R.Y., Shoaib, I., Chen, X., Perwaiz, A.: MIMO antennas for smart 5G devices. IEEE Access **6**, 77014–77021 (2018)
20. Naqvi, S.I., et al.: Integrated LTE and millimeter-wave 5G MIMO antenna system for 4G/5G wireless terminals. Sensors **20**, 3926 (2020)
21. Jilani, S.F., Alomainy, A.: Millimetre-wave T-shaped antenna with defected ground structures for 5G wireless networks. In: Proceedings of the 2016 Loughborough Antennas & Propagation Conference (LAPC), Loughborough, UK, 14–16 November 2016, pp. 1–3 (2016)
22. Khalid, M., et al.: 4-port MIMO antenna with defected ground structure for 5G millimeter wave applications. Electronics **9**, 71 (2020)
23. Wani, Z., Abegaonkar, M.P., Koul, S.K.: A 28-GHz antenna for 5G MIMO applications. Prog. Electromagnet. Res. **78**, 73–79 (2018)
24. Parchin, N.O., et al.: Mobile-phone antenna array with diamond-ring slot elements for 5G massive MIMO systems. Electronics **8**, 521 (2019)
25. Al Abbas, E., Ikram, M., Mobashsher, A.T., Abbosh, A.: MIMO antenna system for multi-band millimeter-wave 5G and wideband 4G mobile communications. IEEE Access **7**, 181916–181923 (2019)
26. Naqvi, S.I., et al.: An integrated antenna system for 4G and millimeter-wave 5G future handheld devices. IEEE Access **7**, 116555–116566 (2019)
27. Ikram, M., Nguyen-Trong, N., Abbosh, A.: Multiband MIMO microwave and millimeter antenna system employing dual-function tapered slot structure. IEEE Trans. Antennas Propag. **67**(8), 5705–5710 (2019)
28. Li, Z.Y., Du, Z.W., Takahashi, M., Saito, K., Ito, K.: Reducing mutual coupling of MIMO antennas with parasitic elements for mobile terminals. IEEE Trans. Antennas Propag. **60**(2), 473–481 (2012)
29. Assimonis, S., Yioultsis, T., Antonopoulus, C.: Design and optimization of uniplanar EBG structures for low profile antenna applications and mutual coupling reduction. IEEE Trans. Antennas Propag. **60**(10), 4944–4949 (2012)
30. Lee, Y., Ga, D., Choi, J.: Design of a MIMO antenna with improved isolation using MNG metamaterial. Int. J. Antennas Propag. **2012**, Article ID 864306, 7 p (2012)

31. Kiem, N.K., Phuong, H.N.B., Hieu, Q.N., Chien, D.N.: A novel metamaterial MIMO antenna with high isolation for WLAN applications. Int. J. Antennas Propag. **2015**, Article ID 851904, 9 p (2015)
32. Liu, J., Esselle, K.P., Hay, S.G., Sun, Z., Zhong, S.: A compact super-wideband antenna pair with polarization diversity. IEEE Antennas Wirel. Propag. Lett. **12**, 1472–1475 (2013)
33. Suyan, N., Lohar, F.L., Solunke, Y., Dhote, C.: Design of compact size tri-band stacked patch antenna for GPS and IRNSS applications. In: Intelligent Learning for Computer Vision: Proceedings of Congress on Intelligent Systems 2020, pp. 501–509. Springer Singapore (2021)
34. Patel, P., Solunke, Y.: New wideband vertical tetraskelion-shaped dielectric resonator antenna. In: 2020 IEEE 17th India Council International Conference (INDICON), pp. 1–6. IEEE (2020)
35. Lohar, F.L., Dhote, C., Solunke, Y., Suyan, N.K.: Design of circularly polarized IRNSS receiver antenna using characteristic mode analysis. In: 2019 IEEE Indian Conference on Antennas and Propagation (InCAP), pp. 1–5. IEEE (2019)
36. Nitin Kumar, S., Kumar, K.V.P., Karthikeyan, S.S.: Compact dual-wideband bandpass filter for wireless applications. AEU-Int. J. Electron. Commun. **95**, 69–72 (2018)
37. Solunke, Y., Kothari, A., Abegaonkar, M.P.: Design of dual-band pass FSS with stable oblique incidence using characteristics mode analysis. In: 2022 IEEE Microwaves, Antennas, and Propagation Conference (MAPCON), pp. 1018–1022. IEEE (2022)
38. Dewangan, L., Patinavalasa, M.S., Acharjee, J., Solunke, Y., Ghosh, S., Mishra, N.K.: Broadband metamaterial absorber for stealth applications at K-band. AEU-Int. J. Electron. Commun. **170**, 154828 (2023)
39. Solunke, Y., Kothari, A.: An ultra-thin, low-RCS, dual-bandpass novel fractal-FSS for planar/conformal C&X bands applications. AEU-Int. J. Electron. Commun. 155073 (2023)
40. Solunke, Y., Patanvariya, D.G., Kothari, A., Sharma, K., Singh, D.: Design of dual-band pass polarization insensitive and transparent frequency selective surface for wireless applications. In: 2023 International Conference on Artificial Intelligence and Smart Communication (AISC), Greater Noida, India, pp. 1246–1250 (2023). https://doi.org/10.1109/AISC56616.2023.10085319

Dual-Band Antenna for Seamless 900 MHz and 2.4 GHz Connectivity

Divyam Verma[1], Shrivishal Tripathi[2](✉), and D. Venkata Siva Prasad[2](✉)

[1] Shri Shankaracharya Technical Campus, Bhilai, Chhattisgarh, India
divyam030903@gmail.com
[2] IIIT Naya Raipur, Atal Nagar-Nava Raipur, Chhattisgarh, India
{shrivishal,venkata}@iiitnr.edu.in
https://www.iiitnr.ac.in/

Abstract. The rapid proliferation of Internet of Things(IoT) devices increases the demand for a strong communication system. Since microstrip patch antennas are compact in size, low profile, and easily integratable, they become a preferred choice. However, conventional designs generally work on a single frequency, restricting their applicability in IoT applications where multi-band operations are necessary. In this paper, a dual-band microstrip patch antenna is proposed for IoT and Wi-Fi applications. This antenna operates on a frequency of 900 MHz for IoT and a frequency of 2.4 GHz for Wi-Fi applications, which facilitates device communication on both bands simultaneously for enhanced performance and functions. The antenna is designed using an FR4 substrate of 1.5 mm in thickness, having a dielectric constant of 4.4. These simulation and analysis results further validate the antenna performance metrics in terms of resonant frequency, bandwidth, and radiation pattern. The results obtained clearly confirm the efficiency of the antenna for IoT and Wi-Fi applications.

Keywords: Dual-band microstrip patch antenna (DBMPA) · IoT Connectivity · Multi-Frequency Operation

1 Introduction

In recent years, together with the rapid proliferation of IoT devices, the need has also been pressing for communication systems to be more efficient and reliable. Among the integral components of such systems, antennas play an absolutely necessary role in the transmission and reception of signals. Microstrip patch antennas are particularly preferred, as they offer compact size with a low profile, thus easily being integrable into many devices. However, the traditionally designed microstrip antennas are normally single-frequency operating, noticeably reducing their applicability to IoT devices where multi-band functionality is required.

Supported by IIIT Naya Raipur, Chhattisgarh.

D. Gupta et al. (Eds.): PCEMS 2024, CCIS 2491, pp. 392–401, 2025.
https://doi.org/10.1007/978-3-031-90577-3_33

In this paper, a dual-band microstrip patch antenna design and analysis have been proposed that can support 900 MHz for IoT applications and 2.4 GHz for Wi-Fi connectivity. Due to the dual-band design, concurrent communications across these frequency bands can be executed, which improves overall performance and functionality of IoT devices [2]. In domains like energy efficiency, environment, urbanization, logistics, and industrialization, IoT technology caters to a wide array of challenges, improving productivity, safety, health, and quality of life. These range from applications concerning smart farming, like remote crop monitoring, to remote patient monitoring in healthcare systems, waste management in smart cities, and control of machinery in manufacturing and lighting, heating, and security systems in smart homes. Microstrip patch antennas are lightweight, easily fabricated using modern printed circuit technology, mechanically robust, and can be conveniently attached to planar and non-planar surfaces. Although these are the benefits, compact MPA designing at the operating frequencies of 900 MHz and 2.4 GHz in the ISM band for IoT applications is quite challenging. In this sense, critical needs that require an antenna include covering both frequency bands using a single element effectively so as to make it possible for its simultaneous use without the inefficiencies associated with traditional single-band whip or monopole antennas. This dual-band MPA design is of benefit, mostly in device connectivity lines related to IoT devices within multiple frequency ranges. It simply means that the compact MPA developed will be of dual band, resonating at 900 MHz and 2.4 GHz to give flexibility in the design of IoT devices with respect to their size and performance. It leads to cost-effective and space-saving solutions for IoT applications requiring connectivity within the frequency range of 900 MHz to 2.4 GHz.

For dual-band operation, slots of different dimensions are created on the patch. The created slots will definitely distort the current distribution on the surface of the antenna, creating dual resonances. A parametric study on the length and width of the respective slots will be conducted in pursuit of the optimized values that will generate the resonances at 900 MHz and 2.4 GHz [6]. With these optimized parameters, the performance of the antenna will be simulated on an electromagnetic simulation software tool. The simulation results of the designed antenna demonstrate the attractive impedance matching and bandwidth characteristics at dual-band frequencies, 900 MHz and 2.4 GHz. The current distribution analysis shows that edges of the slots and the patch contribute to dual resonances. The radiation patterns are bidirectional in the E-plane and omnidirectional in the H-plane, with gains of 2.33 dBi at 900 MHz and 6.35 dBi at 2.4 GHz. Compared with the previously reported antennas, the proposed design has the following advantages: it is compact and dual-band, with dimensions of $70.3 \times 65 \times 1.6\,\text{mm}^3$, thus allowing integration into IoT devices. The technique for the slots in creating the dual resonances works effectively and might be further optimized to reach resonances at 900 MHz and 2.4 GHz frequencies for IoT and Wi-Fi applications. This work proposes a compact design for a dual-band microstrip patch antenna that can perform at 900 MHz for IoT applications and 2.4 GHz for Wi-Fi connectivity. In this paper, dual-band operation has been achieved by introducing slots on the patch to alter the current distribution that creates resonances at the desired frequencies. Simulation results show that the proposed antenna could support IoT and Wi-Fi communications simultaneously within a single compact module. This approach provides an economical and space-efficient solution for IoT applications requiring connectivity across multiple frequency bands within the 900 MHz to 2.4 GHz range.

Table 1. IoT operating frequencies

Region	Frequency (MHz)
EU	863–870
US	902–928
AU	915–928
CN	779–787 and 470–510

2 Design Procedure of the Antenna

To achieve a improved performance, a rectangular MPA as shown in Fig. 1 was designed on an FR-4 substrate at 900 MHz and 2.4 GHz. The substrate backed by a full ground plane made up of copper has a thickness of 0.035 mm. The calculated and the optimized parameters of the antenna at 2.4 GHz are shown in Table 2.

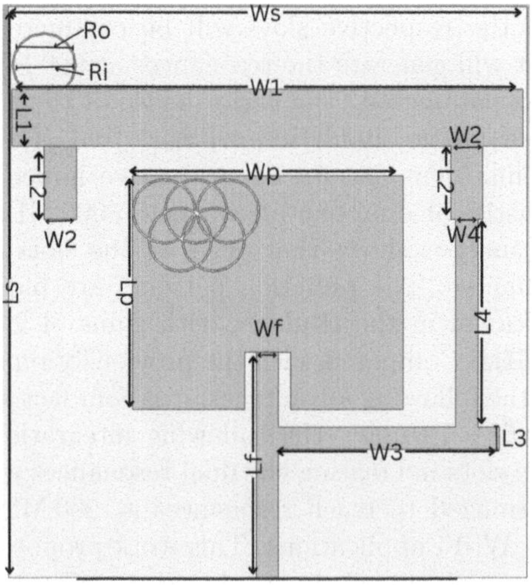

Fig. 1. Structure of the 900 MHz and 2.4 GHz Antenna Patch

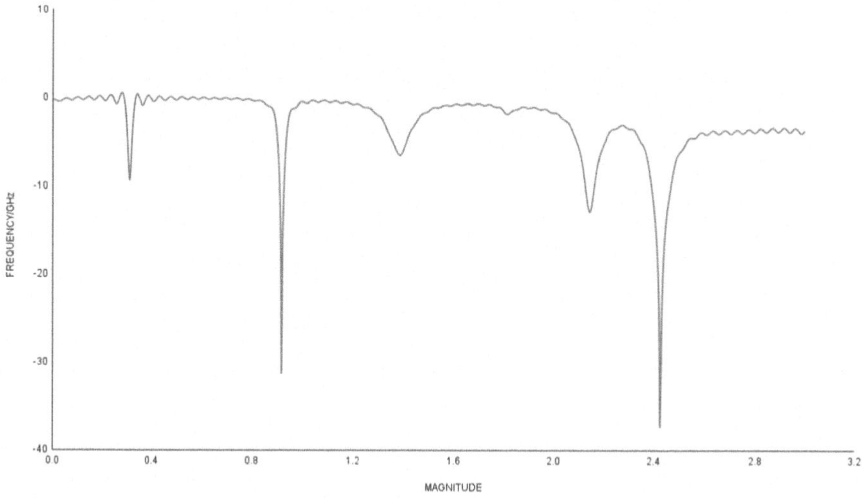

Fig. 2. S11 of the Antenna at 0.9 GHz and 2.4 GHz

As shown in Fig. 2, the return loss of the antenna is represented over 2.4 and 0.9 GHz. For the 900 MHz band, it has S11 less than –10 dB with a magnitude of –31.225 dB and impedance bandwidth of 20 MHz from 900 MHz to 920 MHz, while for the 2.4 GHz band, the magnitude is –37.115 dB with an impedance and bandwidth of 80 MHz from 2.38 GHz to 2.46 GHz.

3 Evolution of Antenna Design

3.1 Initial Design of the 2.4 GHz Antenna

The evolution of antenna design for dual-band applications began from a basic design of the 2.4 GHz microstrip patch antenna. It provided a base model of wireless communication that worked on a single frequency of 2.4 GHz but still gave a practical solution in many applications. This single-frequency antenna could serve quite well, but it still showed poor versatility in modern IoT applications, hence accelerating the requirement for multiple frequencies to ensure smooth operation. Other capabilities were missing, reducing its applicability and therefore calling for more improvements in design to include the diverse connectivity needs of IoT devices as shown in Fig. 3.1 for patch design and Fig. 4 for S parameter.

3.2 Ring Enhanced Design

The next antenna design evolution step was to modify the basic antenna structure by removing rings in the patch to achieve higher performance. Basically, two major issues with respect to bandwidth and efficiency for the frequency of

Table 2. Parameters of the Proposed Antenna

Parameter	Calculated (mm)	Optimized (mm)	Description
Ls	70	70.3	Length of Substrate
Ws	70	65	Width of Substrate
Wp	58.14	28.8	Length of Patch
Lp	45.28	34	Width of Patch
Wf	3.06	2.8	Width of Feed
Lf	-(ls/2-lp/2)	-(ls/2-lp/2)	Length of Feed
Wi	3	1.5	Width of Inset
Li	7	7	Length of Inset
MetT	0.035	0.035	Metal Thickness
SubT	1.57	1.57	Substrate Thickness
L1	64	64	Lenght of Stub
W1	7	7	Width of Stub
L2	9	9	Lenght of Stub
W2	4	4	Width of Stub
L3	13.75	13.75	Lenght of Stub
W3	2.8	2.8	Width of Stub
L4	2.8	2.8	Lenght of Stub
W4	10.4	10.4	Width of Stub
Ro	4	4	Outer Ring Radius
Ri	3.5	3.5	Inner Ring Radius

interest of 2.4 GHz were improved by this modification. The removed rings further improved impedance matching and increased the effective radiating area, making the antenna more appropriate for a wider range of applications. This improved design greatly differed from the previous model because some of the disadvantages of single-frequency antennas were improved and the performance characteristics were enhanced, enhancing the possibilities for application in several scenarios of wireless communication(see Fig. 3.2 for the patch design and Fig. 4 for the S-parameter).

3.3 Dual-Band Function with Slots

The final evolution in the design of the antenna was the introduction of slots into the patch such that the antenna operated effectively at 900 MHz and 2.4 GHz. Apart from achieving this dual-band functionality, this approach influenced the current distribution across the patch for the enhancement of overall performance of the antenna. Introduction of slots in the antenna increased its effective electrical length and supported resonance at a lower frequency of 900 MHz without

(3.1) (3.2) (3.3)

Fig. 3. (3.1) 2.4 GHz Antenna Patch, (3.2) 2.4 GHz Antenna With Rings, (3.3) 900 MHz and 2.4 GHz Antenna Patch

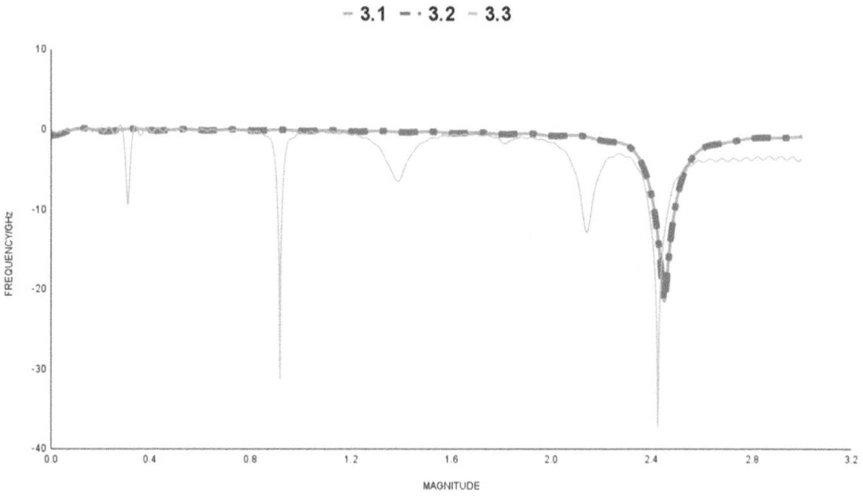

(b)S-parameter of the 2.4GHz Antenna Patch

Fig. 4. S21 of the Antenna at 2.4 GHz

degradation of its performance at 2.4 GHz. It did, however, result in the evolution of the antenna design from some simple model of single-frequency operation to a sophisticated dual-band system capable of meeting diversity in connectivity needs within modern IoT devices. It provides an effective cost and space solution for many applications that require simultaneous operations across different frequency bands. To make the antenna operable at two bands, different dimension slots were added on the patch, as shown in Fig. 3. These slots will badly distort the current distribution at the surface of the antenna and hence easily create the dual resonance. Some parametric studies have been done in order to determine the exact length and width for each slot to maintain its final

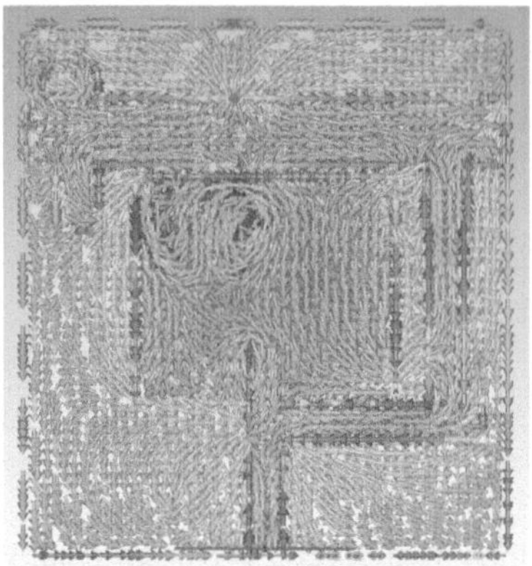

(a)Current Distribution at 2.4 GHz

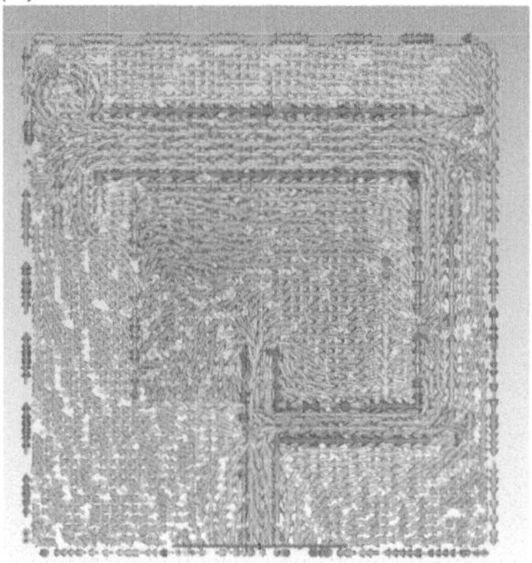

(b)Current Distribution at 0.9 MHz

Fig. 5. Antenna's Current Distribution (a) 2.4 GHz (b) 900 MHz

design resonating at both frequencies: 900 MHz and 2.4 GHz. These dimensions, for these slots, are very critical values to have the desired characteristics and are summarized in Table 1. Such an optimization procedure has led to systematic variations of the slot dimensions; later, simulations were made to see the effect on antenna performance. These studies approved the fact that the chosen dimension could provide effective dual-band operation, so it would be useful in antenna functioning for IoT applications at 900 MHz and also for Wi-Fi connectivity at 2.4 GHz. Particularly, this is very useful in modern IoT devices. In bulk, they call for connectivity covering a number of frequency ranges. This satisfies regional regulations and standards for their operation. It improves functionality, compactness, and cost for IoT devices by efficient operation at both frequency bands in a single antenna element. The patch design is shown in Fig. 3.3, with the S-parameter shown in Fig. 4.

4 Results and Analysis

The S11 of the antenna with the optimized parameters depicted in Table 1 is plotted in Fig. 2. It is clearly shown that it has achieved dual resonances: one at 900 MHz for IoT and another at 2.4 GHz for Wi-Fi connectivity.

The current distribution on the surface of the dual-band antenna is shown in Fig. 5. As can be seen from Fig. 5(a). On the other hand Fig. 5(b), the edges of slot and the edges of the patch produced the resonance at 900 MHz. Due to the presence of the slots, the current has to follow edges of the slot since it can't pass through it. This action increases the path length through which the current flows on the antenna surface thereby creating dual bands. Table 3 compares the performance of the designed antenna with some available literature. The proposed antenna has a dimension of 70.3 × 65 × 1.6 MM" with a comparative high directivity, making it quite a viable candidate for IoT applications and WiFi.

Table 3. Comparison of Proposed Antenna

Ref.	No. of Bands	Frequency (MHz)	Substrate	Dielectric	Conductor	Gain	Size (mm)
[1]	2	865.1	FR-4	4.4	Copper	1.0201 dB	102×81×1.6
		935.8				-5.4037 dB	
[2]	4	868	FR-4	4.3	Copper	3.36 dBi	300×30×0.8
		1210				3.05 dBi	
		1570				3.05 dBi	
		2400				4.17 dBi	
[7]	3	433	FR-4	4.4	Copper	2 dBi	80 × 50
		868				1.9 dBi	
		915				1.9 dBi	
[8]	2	400	FR-4	NA	Copper	-5 dBi	78 × 88
		900.2					
[9]	2	868	FR-4	4.4	Copper	1.92 dBi	100 X 40
		2400				4.2 dBi	
[10]	1	848-950-	FR-4	4.4	Copper	2.1 dBi	40 × 26
This Work	2	900	FR-4	4.4	Copper	2.33 dBi	70.3x65×1.6
		2400				6.35 dBi	

5 Conclusion

In this paper, the design and analysis of a dual-band microstrip patch antenna that could work at 900 MHz for IoT applications and 2.4 GHz for Wi-Fi connectivity were presented.

Key highlights of the proposed antenna design can be summarized as follows:-

- Good impedance matching and bandwidth performance at both frequency bands
- Broadside radiation patterns with acceptable gain levels
- Compact in size and low-profile, it is suitable for integration in IoT devices.

It is observed from the result of the simulation that the dual-band microstrip patch antenna has the possibility of realizing the IoT and Wi-Fi communication functionalities simultaneously within one compact module. This design approach offers a cost-effective and space-efficient solution to IoT applications that require connectivity across multiple frequency bands.

References

1. Boursianis, A.D., Goudos, S.K., Yioultsis, T.V., Siakavara, K.: Low-cost dual-band E-shaped patch antenna for energy harvesting applications using grey Wolf optimizer. In: 2019 13th European Conference on Antennas and Propagation (EuCAP), pp. 1–5. IEEE (2019)
2. Wang, Y., Santamaria, L., Ferrero, F., Lizzi, L.: Design of a multi-antenna portable IoT terminal. In: 2021 IEEE Conference on Antenna Measurements and Applications (CAMA), pp. 597–599. IEEE (2021)
3. Yahya, M.S., et al.: A compact dual band microstrip patch antenna for LoRa IoT applications. In: IEEE International RF and Microwave Conference (RFM). Kuala Lumpur, Malaysia, pp. 1–4 (2022). https://doi.org/10.1109/RFM56185.2022.10065043
4. Ferrero, F., Toure, M.B.: Dual-band LoRa antenna: design and experiments. In: 2019 IEEE Conference on Antenna Measurements and Applications (CAMA), Kuta, Bali, Indonesia, pp. 243–246 (2019). https://doi.org/10.1109/CAMA47423.2019.8959760.
5. Parthiban, P.: IoT antennas for industry 4.0 - design and manufacturing with an example. In:: IEEE International IoT, Electronics and Mechatronics Conference (IEMTRONICS). Vancouver, BC, Canada, pp. 1–5 (2020). https://doi.org/10.1109/IEMTRONICS51293.2020.9216349
6. Taouzari, M., Aoufi, J.E., Mouhsen, A., Nasraoui, H., Mrabat, O.E.: 900 MHz and 2.45 GHz compact dual-band circularly-polarized patch antenna for RFID application. In: 2015 Conference on Microwave Techniques (COMITE), Pardubice, Czech Republic, p. 1–4 (2015). https://doi.org/10.1109/COMITE.2015.7120226.
7. Yahya, M.S., et al.: Triple-band reconfigurable monopole antenna for long-range IoT applications. Sensors **23**(12), 5359 (2023)
8. Krishna, M.V., Raju, G.: Triangle shaped antenna design for IoT-based lorawan applications. SAMRIDDHIJ. Phys. Sci. Eng. Technol. **13**, 8–11 (2021)
9. Reha, A., Tarbouch, M., El Amri, A.: A dual band compact PIFA antenna for internet of things networks sigfox, lorawan and zigbee. In: Proceedings of the Colloque sur les Objets et systémes Connectès; Institut Universitaire de Technologie d'Aix-Marseille, Casablanca, Morocco (2019)
10. Tarbouch, M., Reha, A., El Amri, A., Mejdoub, Y.: A compact PIFA antenna for internet of things network LORAWAN at 900 Mhz band. In: Colloque sur les Objets et Systémes Connectès; Institut Universitaire de Technologie d'Aix-Marseille, Casablanca, Morocco (2019)

11. Lizzi, L., Ferrero, F., Monin, P., Danchesi, C., Boudaud, S.: Design of miniature antennas for IoT applications. In: 2016 IEEE Sixth International Conference on Communications and Electronics (ICCE), Ha-Long, Vietnam, pp. 234–237 (2016). https://doi.org/10.1109/CCE.2016.7562642.
12. Mohta, M., Rajawat, A., Gupta, S.H.: Design of an inset- fed slotted triple band antenna. In: 3rd International Conference On Internet of Things: Smart Innovation and Usages (IoT-SIU). Bhimtal, India, pp. 1–5 (2018). https://doi.org/10.1109/IoT-SIU.2018.8519937

Metaheuristic Optimization for 3D UAV Placement

M. Nithish Kumar, Dikshant Mendhe, Shaik Rafee, Praveen Pawar(✉), and Arvind Kumar

Department of Electronics and Communication Engineering, Visvesvaraya National Institute of Technology, Nagpur, India
{praveenpawar,arvindkumar}@ece.vnit.ac.in

Abstract. Unmanned aerial vehicles (UAVs), commonly known as drones, have emerged as versatile platforms with applications spanning various industries. From surveillance and monitoring to disaster response and infrastructure inspection. However, maintaining reliable communication between drones and ground stations is crucial. Path loss is a major challenge, affecting communication effectiveness. This paper explores reducing path loss using Particle Swarm Optimization (PSO). By utilizing the Rayleigh probability density function, the PSO algorithm, and Line of Sight Probability, we conducted an analysis of air-to-ground propagation by varying environmental and programming parameters. This comprehensive approach yielded a diverse range of results for the project. We provide insights into optimal drone placement across different environments. Looking ahead, future research can explore complex scenarios involving multiple UAVs, dynamic UAV networks, and coordinated UAV operations to further enhance the resilience, flexibility, and effectiveness of UAV-based communication systems in real-world environments. As technology continues to advance and regulations evolve, the potential for UAVs to transform various industries and aspects of daily life remains promising.

Keywords: UAV · PSO · Path loss · Optimization

1 Introduction

Unmanned aerial vehicles (UAVs), commonly known as drones, are autonomous or remotely piloted aircraft that do not require any onboard human pilot. UAVs have rapidly evolved and found diverse applications across various industries because of their versatility, maneuverability, and advancements in technology. UAVs are growing rapidly across many application domains, such as real-time monitoring, providing wireless coverage, remote sensing, search and rescue, delivery of goods, security, and surveillance, precision agriculture, civil infrastructure inspection, military, construction, medical, farming, wireless communication, and aerial surveillance [7], and many more. Smart UAVs are the next big

D. Gupta et al. (Eds.): PCEMS 2024, CCIS 2491, pp. 402–413, 2025.
https://doi.org/10.1007/978-3-031-90577-3_34

revolution in UAV technology promising to provide new opportunities in different applications [10] across different sectors. The UAV-enabled communication platform is recognized as a promising way for secure transmission in the air-to-air and air-to-ground(ATG) networks [13]. Looking toward the future, the applications of UAVs are poised to expand further. One notable trend is that when UAVs are used as flying, aerial base stations, they can support the connectivity of existing terrestrial wireless networks such as cellular and broadband networks.

Effective communication between UAVs and ground base stations is crucial for a wide range of applications, from surveillance to disaster response. Reducing path loss offers several benefits. It enhances communication reliability by minimizing signal degradation and disruptions. Also, it extends coverage range, enabling UAVs to maintain connectivity over vast areas required for expansive surveillance. It improves data transmission rates, providing real-time transmission of large datasets for enhanced decision-making. Additionally, minimizing path loss optimizes resource utilization, conserving energy and bandwidth for prolonged missions. Lastly, it enables autonomous UAV operations by ensuring reliable communication with ground control stations, crucial for navigation and mission execution. Reducing the path loss between UAVs and base stations is crucial for ensuring reliable, efficient, and flexible communication in UAV systems. By mitigating the effects of signal attenuation and optimizing communication links, UAVs can operate more effectively across a wide range of applications.

The Particle Swarm Optimisation (PSO) technique was proposed and initially developed by the electrical engineer Russell C. Eberhart and the social psychologist James Kennedy [5]. PSO is a population-based stochastic optimization algorithm motivated by intelligent collective behavior of some animals such as flocks of birds or schools of fish. Since presented, it has experienced a multitude of enhancements [12]. It operates with a population of particles navigating a search space to find optimal solutions, with each particle maintaining position and velocity, updated iteratively based on personal and collective experiences. The particles adjust their positions by considering their own best-known position (personal best) and the best-known positions found by any particle in the entire swarm (global best). Essential parameters include population size, inertia weight, acceleration constants, maximum velocity, and termination criteria, influencing the algorithm's exploration-exploitation balance and efficiency.

A study investigated optimizing drone basestation(DBS) deployment for next-generation Internet-of-Things(NG-IoT) applications using swarm intelligence. Comparative analysis among algorithms(CS, EHO, GWO, MBO, SSA, and PSO) revealed GWO as most effective in determining optimal DBS positions in 3D space, enhancing network performance [9]. Another study observes path loss for UAVs connected to cellular networks in rural areas, revealing height-related signal degradation. Factors, like expanded radio horizon, line-of-sight clearance, and reduced obstruction in the first Fresnel zone, were identified. It emphasized the need for improved quantification of these elements for future research, particularly in developing a generic path loss model for interfering cells

[4]. Additionally, a paper provided insights into the radio propagation characteristics of the cellular-to-UAV channel in suburban environments. It derived a statistical path-loss model considering depression angle and terrestrial coverage beneath the UAV, aiding in simple and accurate prediction of path loss [2]. Moreover, a study introduced innovative path loss models for precise distance estimation in wireless sensor networks (WSNs), particularly in indoor settings. The hybrid PE-PSO model, integrating PSO with Polynomial Equation (PE), outperformed conventional models significantly, enhancing distance estimation accuracy by 85% [11].

In this paper, we address the optimization of path loss between UAVs and base stations using PSO algorithm. Our study begins with an introduction section that provides insights into UAV applications, PSO algorithm, and relevant literature. We then look into the system model section, where we implement the PSO algorithm across various environmental and simulation parameters, aiming to identify optimal UAV placement strategies. Through rigorous experimentation, in the result section, we analyze the impact of different settings on path loss in UAV communications. Our results showcase the effectiveness of PSO-based optimization in mitigating path loss and provide valuable insights into enhancing the reliability and efficiency of UAV-based communication systems. Finally, we conclude by outlining future research directions, including the exploration of alternative optimization algorithms and the integration of dynamic environmental factors, to further enhance the performance of UAV communication systems.

2 System Model

Propagation Model: ATG propagation involves transmitting signals between an airborne transmitter, like a UAV or aircraft, and a ground-based receiver, such as a base station or ground terminal. It involves both line-of-sight (LOS) and non-line-of-sight (NLOS) paths, influenced by factors like distance, obstacles, and atmospheric conditions. This mode of communication poses distinct challenges compared to traditional ground-to-ground communication (Fig. 1).

In the above mentioned model, there are two regions named A and B. A represents Free Space Path loss and B region represents Additional Pathloss due to reflection , shadowing etc. θ represents angle of elevation between UAV and User Equipment.

The Mean ATG pathloss(expressed in dB) can be expressed as:

$$PL = FSPL + \eta_{\xi} \tag{1}$$

where η_{ξ} is mean value of the exessive pathloss due to urban environment and FSPL stands for free space path loss. It is computed using the following formula:

$$\text{FSPL(dB)} = 20\log_{10}\left(\frac{d}{d_0}\right) + 20\log_{10}(f) + 20\log_{10}(4\pi) \tag{2}$$

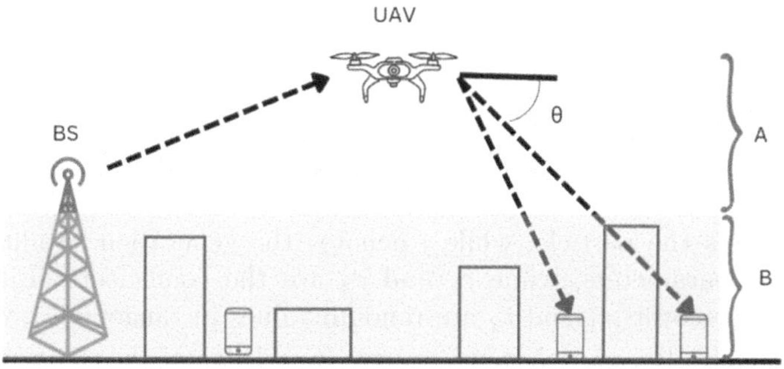

Fig. 1. ATG model

here, d is the distance between the transmitter and receiver and d_0 is the reference distance whose value is usually taken as 1.

$$\eta_\xi \; = P(\mathrm{LoS}) \times N_{\mathrm{LoS}} + P(\mathrm{NLoS}) \times N_{\mathrm{NLoS}} \tag{3}$$

P(LoS) is likelihood of LoS which depends on the surroundings and angle of elevation θ. $N_L oS$ denotes the number of LoS communications. P(NLoS) is likelihood of NLoS whose value can be found by 1-P(Los). $N_N LoS$ denotes the number of non-line of sight communications between transmitter and receiver.

The Rayleigh probability density function (PDF) is a mathematical description used in wireless communication to model the distribution of signal amplitudes in a Rayleigh fading channel. This function characterizes the different signal amplitude values occurring in scenarios where there's no dominant LoS path, and signals undergo multiple reflections and scattering, leading to variations in signal strength. The function of Rayleigh communication channel is given by [3]:

$$f(H) = (H/\gamma^2)^{(-H^2/2\gamma^2)} \tag{4}$$

where H is the building height in meters and γ indicates environmental elements like weather, atmospheric interference, and the density of barriers like buildings or trees that might affect the chance of LoS.

LoS communication will play a major role in the next generation of wireless technologies; from UAV-to-UAV networks, to land mobile satellite services, and mmWave cellular coverage [1]. LoS denotes a direct path between a transmitter and receiver in absence of obstructions, crucial for maintaining strong signal strength in wireless communication. It's vital across various technologies like cellular networks and Wi-Fi, especially for long-range communication. Parameters such as P(LoS, θ), a (land area ratio), b (building density), and θ (angle/distance) characterize LoS probability and the environment's obstruction level, impacting signal reliability and strength.

PSO algorithm: Particles begin with random solutions and velocities. Through iterations, each particle's fitness is assessed, updating its personal best and the global best. By adjusting positions and velocities iteratively, PSO converges towards the optimal solution. The position of each particle is obtained according to the following equation [6]:

$$x_i^{j+1} = x_i^j + wv_i^j + c_1 r_1 (x_{\text{best}}^i - x_i^j) + c_2 r_2 (x_{\text{best}}^{\text{global}} - x_i^j) \tag{5}$$

where, i denotes the particle, while j denotes the generation. Additionally, w is the inertia parameters, while c_1 and c_2 are the coefficients adjusting the exploration. Moreover ,r_1 and r_2 are random values in range [0, 1], while x_{best}^j and x_{best}^{global} denote the particles and swarms knowledge of the solution location, respectively.
The velocity of each particle is obtained according to the following equation [6]:

$$v_i^{j+1} = wv_i^j + c_1 r_1 (x_{\text{best}}^i - x_i^j) + c_2 r_2 (x_{\text{best}}^{\text{global}} - x_i^j) \tag{6}$$

where w is the inertia weight . learning coefficient in current direction, c_1 and c_2 are the acceleration coefficients. c_1 is personal learning coefficient, and c_2 is global learning coefficient.

As described earlier we use the ATG communication channel by considering Rayleigh distribution to develop a LoS and NLoS probability model, the final equation for FSPL and mean additional losses is given by,

$$FSPL_{\text{total}} = 20 \cdot \log_{10} \left(\frac{4\pi f_c \left((h^2 + d^2)^{0.5} + (h_1^2 + d_1^2)^{0.5} \right)}{c} \right) \tag{7}$$

$$\begin{aligned} \eta_\xi^{\text{total}} &= P_{\text{LoS}} \cdot n_{\text{LoS}} + (1 - P_{\text{LoS}}) \cdot n_{\text{NLoS}} \\ &+ P_{\text{LoS1}} \cdot n_{\text{LoS1}} + (1 - P_{\text{LoS1}}) \cdot n_{\text{NLoS1}} \end{aligned} \tag{8}$$

where, h and d are the height and distance difference between the base station antenna position and the UAV position, and h1 and d1 are the height and distance difference between the UAV position and the user position. P_{LoS}, η_{LoS}, P_{NLoS}, η_{NLoS} are for the channel between antenna and UAV and, P_{LoS1} , η_{LoS1} ,P_{NLoS1}, η_{NLoS1} are for the channel between UAV and user.

The final Cost Function that the PSO Algorithm is used for optimisation is,

$$PL_{total} = FSPL_{total} + \eta_\xi^{total} \tag{9}$$

One crucial aspect is the impact of altitude (h) on the LOS path, which plays a significant role in determining the probability of LOS transmission (P(LoS)). From the equation used in the LoS path, high values of altitude cause the series in the equation to smooth out. As a result, P(LoS) may be considered a continuous function of θ (angle of elevation) and the environmental factors. This observation is supported by the trend closely resembles a modified sigmoid function (S-curve) with parameters α, β, and γ. The modified Sigmoid function is represented as follows:

$$P(\text{LoS}, \theta) = \frac{1}{1 + a \exp(-b(\theta - a))} \tag{10}$$

where a and b are S-curve parameters. This approximation greatly simplifies the computation of the LoS probability and allows the analytical method. To generalize the solution, we have directly connected the S-curve parameters a and b to the environment variables α, β, andγ. Surface fitting with two variables was used to conduct this connection. The first variable is considered to be (α β), and the second to be (γ) [8].

Here we consider various environmental parameters representing urban, sub-urban, dense urban, and high-rise urban settings, as summarized in the Table 1, by exploring these distinct environmental contexts, we aim to analyze the impact of varying urban landscapes on path loss in UAV communications and evaluate the effectiveness of the PSO algorithm in optimizing UAV placement strategies across different scenarios.

Table 1. Environmental Parameters

Environment	a	b	N_{LoS}	N_{NLoS}
Suburban	4.88	0.43	0.1	21
Urban	9.61	0.16	1	20
Dense urban	12.08	0.11	1.6	34
High-rise urban	27.23	0.08	2.3	34

PSO algorithm from (6), we consider several parameters that influence the behavior and convergence characteristics of the swarm. These parameters include the inertia weight (w, learning coefficients (c_1 and c_2), and acceleration coefficients with the following considered values as shown in Table 2:

Table 2. PSO Parameters

Fitting parameters	Value
w	1
c_1	1.5
c_1	2
w_{damp}	0.99

Here, w_{damp} plays the role of decreasing the inertia weight after every iteration. After every iteration, the search space gradually decreases depending on all these parameters.

In our experimental setup, we take into account essential simulation parameters for PSO algorithm implementation and air-to-ground communication simulation. These parameters encompass dimensions (representing optimization variables like antenna positions), user and base station counts impacting task complexity, search agents shaping PSO behavior, iteration cycles, communication frequency, and environmental factors (urban variations affecting pathloss). This holistic approach enables thorough system modeling and evaluation across diverse scenarios, fostering comprehensive analysis and optimization (Table 3).

Table 3. Programming Parameters

Parameter	Values
Dimensions	500 m × 500 m × 500 m
Number of Users	[1, 10, 20, 30, 40, 50]
Number of Base Stations	5
Search agents	[20, 40, 50, 60, 80, 100]
Temporary search agents	50
Iterations	[100, 200, 300, 400, 500]
Temporary iterations	100
Frequency	2 Ghz
Environments	urban, sub-urban, dense urban, high-rise urban

3 Results

3.1 UAV Placement for Multiple Environments

In this scenario, we investigate the optimal placement of a UAV hovering over the mentioned area, serving a set of 30 randomly distributed users with 5 base stations. The objective is to minimize path loss by optimizing the UAV's position in various environments. The number of search agents (population) and maximum generations are set to 50 and 100, respectively.

The Table 4 depicts the optimal 3D placement of the UAV in different environments:

Table 4. Result

Environment	Uav Position	Uav altitude	Pathloss(in dB)
Suburban	(262.25,324.35)	68.74	85.62
Urban	(210.83,312.25)	128.34	91.70
Dense urban	(266.02,330.13)	185.79	97.16
High-rise urban	(309.31,376.61)	251.64	122.30

The Figs. 2a, 2b, 2c and 2d plot base stations (blue stars), user positions (green dots), and UAV positions (red diamond shapes). In each diagram, base stations are distributed across the area, with user positions scattered throughout. The UAV position is marked in red to signify the optimal 3D placement to minimize path loss.

From Table 4, we observe a trend where UAV height increases from suburban to high-rise urban areas, correlating with population density. As urban density rises, LoS communication probability decreases, necessitating higher UAV altitudes for coverage. This underscores the need for optimized UAV placement to counter increased path loss and maintain reliable communication across varied urban landscapes.

3.2 Average Pathloss for Multiple Numbers of Users

In this scenario, we assess the average pathloss in multiple environments while varying the number of randomly distributed users. The UAV hovers over the provided area, serving users with 5 base stations. The number of search agents (population) and maximum generations are set to 50 and 100, respectively.

The Fig. 3 illustrates the average pathloss for five scenarios, each varying the number of users in multiple environments. These scenarios provide insights into how pathloss varies with user density and environmental factors, highlighting the impact of urbanization on communication performance. However, slight changes in pathloss can be observed in other environments such as urban, suburban, and high-rise urban.

3.3 Comparison of Average Pathloss in Direct Link and UAV Link

In this scenario, we compare the average pathloss in two different communication links: the direct terrestrial link between the base station and user (Direct Link), and the UAV-assisted relay link between Base Station-UAV-User (UAV Link). The evaluation is conducted in multiple environments where a UAV hovers over the considered area, with 5 base stations and one user. The number of search agents (population) and maximum generations are set to 50 and 100, respectively.

Figure 4 illustrates the average pathloss comparison between UAV-assisted communication and conventional terrestrial communication over the considered area in different environments. The figure reveals a notable change in pathloss between the two communication links. As the environmental density increases, terrestrial pathloss experiences a heavy increase. In contrast, the pathloss in UAV-assisted communication remains relatively constant, indicating that the placement of the UAV adapts to the environment, maintaining optimal pathloss.

Furthermore, we conducted another scenario to evaluate the average pathloss in both communication links concerning the distance of the user from the base station. We introduced a scenario where line of sight is completely removed at a distance of 300 m from a single base station located at a fixed position (0,0) in a 1000 m $\times$ 1000 m area, with all other parameters remaining the same as in the previous scenario.

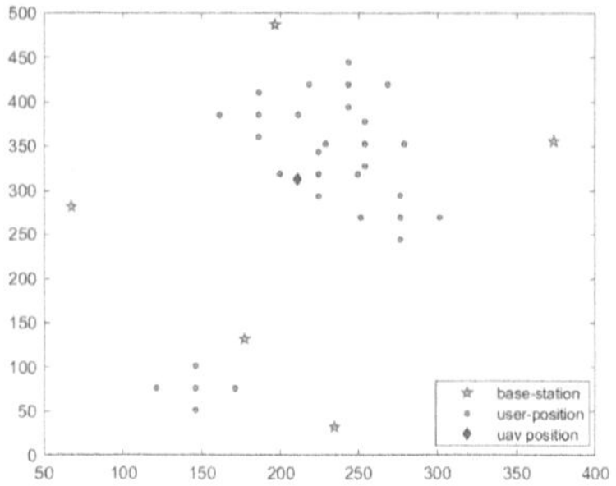

(a). Urban environment

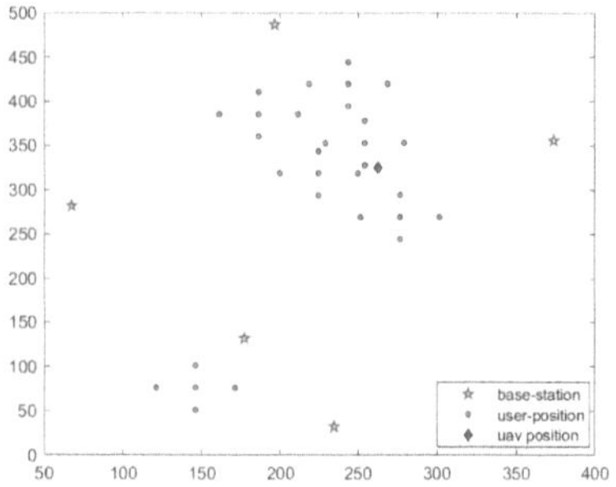

(b). Suburban environment

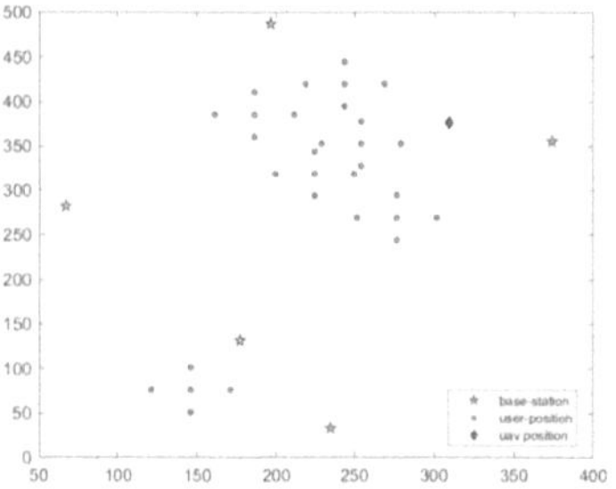

(c). High-rise urban environment

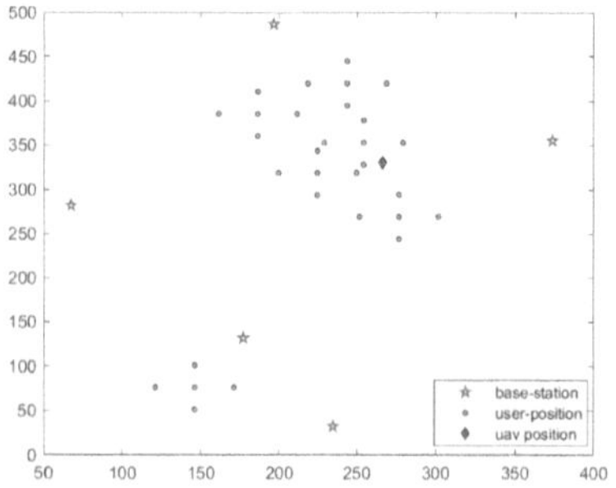

(d). Dense urban environment

Fig. 2. UAV placement in different environments. (a) Urban environment. (b) Suburban environment. (c) High-rise urban environment. (d) Dense urban environment

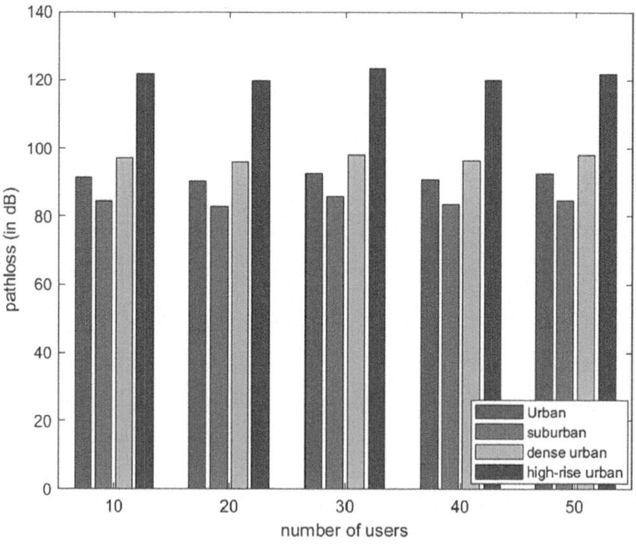

Fig. 3. Pathloss vs Number of users

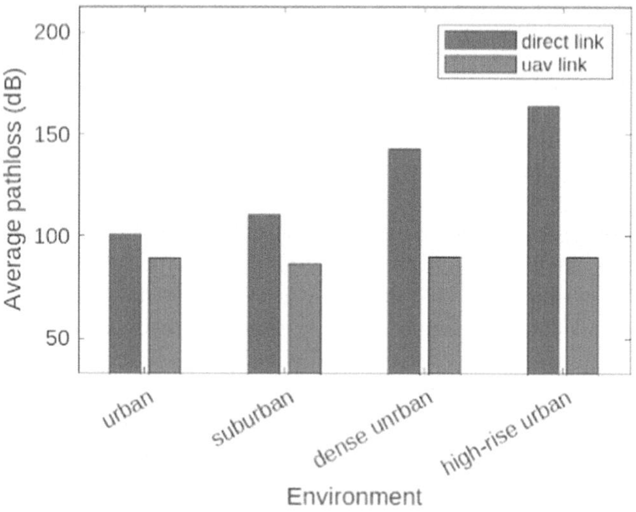

Fig. 4. Average pathloss comparison

Figure 5 depicts the average pathloss comparison between UAV-assisted communication and conventional terrestrial communication over a 1000 m × 1000 m area with respect to the user distance from the base station. The figure demonstrates a notable change in pathloss between the two communication links after line of sight is removed. While terrestrial communication performs better during line of sight propagation, UAV-assisted communication outperforms terrestrial

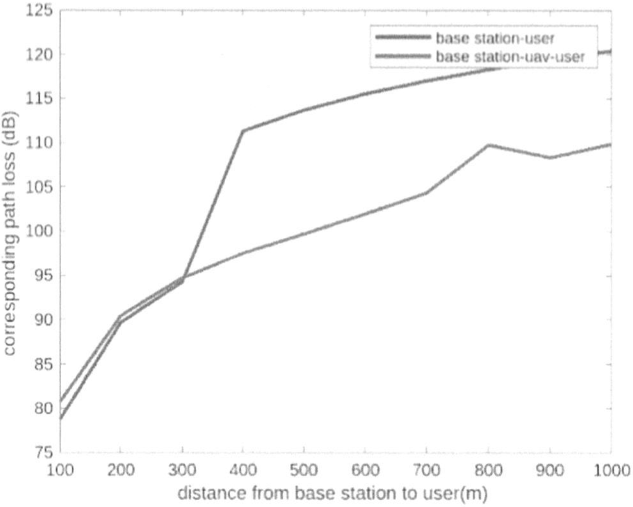

Fig. 5. Pathloss vs Distance from Base Station

communication significantly after line of sight is removed. Additionally, as the user moves away from the base station, the pathloss in terrestrial communication increases notably compared to UAV-assisted communication.

4 Conclusion and Future Work

This paper focused on improving UAV-assisted communication systems to address pathloss issues, which are essential for reliable network operation. We used the PSO metaheuristic to place UAVs in a way that minimizes signal loss in a variety of situations. Higher altitude placements mitigated building interference in densely populated urban regions, while mid-altitude locations helped suburban areas. We found that UAV-assisted communication outperforms direct lines in terms of pathloss reduction and signal quality enhancement, particularly in metropolitan areas. We provide insights into UAV placement tactics, which are essential for improving ground user coverage, by investigating metaheuristic optimization. This work establishes the foundation for more dependable UAV communication networks by demonstrating the effectiveness of PSO in addressing pathloss issues. Subsequent investigations ought to explore more optimization strategies and practical implementation situations, confirming and improving the practical applicability of UAV-based communication systems.

Future research in UAV-assisted communication systems could delve into complex scenarios involving multiple UAVs, aiming to mitigate co-channel interference through sophisticated algorithms incorporating dynamic power regulation, interference cancellation, and frequency reuse. Additionally, exploring dynamic UAV networks, where UAVs adjust positions and altitudes in real-time, could optimize coverage while minimizing path loss and preventing interference, particularly in urban or disaster-prone areas. Coordinated UAV networks could further enhance communication effectiveness by developing algorithms for cooperative path planning, resource sharing, and workload distribution among UAVs, ultimately improving the resilience and flexibility of UAV-based communication systems across diverse contexts.

References

1. Al-Hourani, A.: On the probability of line-of-sight in urban environments. IEEE Wirel. Commun. Lett. **9**(8), 1178–1181 (2020)
2. Al-Hourani, A., Gomez, K.: Modeling cellular-to-uav path-loss for suburban environments. IEEE Wirel. Commun. Lett. **7**(1), 82–85 (2017)
3. Al-Hourani, A., Kandeepan, S., Lardner, S.: Optimal lap altitude for maximum coverage. IEEE Wirel. Commun. Lett. **3**(6), 569–572 (2014)
4. Amorim, R., Mogensen, P., Sorensen, T., Kovács, I.Z., Wigard, J.: Pathloss measurements and modeling for uavs connected to cellular networks. In: 2017 IEEE 85th Vehicular Technology Conference (VTC Spring), pp. 1–6. IEEE (2017)
5. Freitas, D., Lopes, L.G., Morgado-Dias, F.: Particle swarm optimisation: a historical review up to the current developments. Entropy **22**(3), 362 (2020)
6. Hao, W., Wu, C., Lin, S.: Research on uav path planning based on improved particle swarm algorithm with inertia weight. In: 2023 IEEE International Conference on Control, Electronics and Computer Technology (ICCECT), pp. 738–741. IEEE (2023)
7. Mohsan, S., Khan, M.A., Noor, F., Ullah, I., Alsharif, M.H.: Towards the unmanned aerial vehicles (UAVs): a comprehensive review. Drones **6**(6), 147 (2022)
8. Moon, I., Dung, L.T., Kim, T.: Optimal 3d placement of uav-bs for maximum coverage subject to user priorities and distributions. Electronics **11**(7), 1036 (2022)
9. Pliatsios, D., Goudos, S.K., Lagkas, T., Argyriou, V., Boulogeorgos, A., Sarigiannidis, P.: Drone-base-station for next-generation internet-of-things: a comparison of swarm intelligence approaches. IEEE Open J. Antennas Propagat. **3**, 32–47 (2021)
10. Shakhatreh, H., et al.: Unmanned aerial vehicles (UAVs): a survey on civil applications and key research challenges. IEEE Access **7**, 48572–48634 (2019)
11. Sotiroudis, S.P., Athanasiadou, G., Tsoulos, G., Sarigiannidis, P., Christodoulou, C.G., Goudos, S.K.: Evolutionary ensemble learning pathloss prediction for 4g and 5g flying base stations with uavs. IEEE Trans. Antennas Propag. **71**(7), 5994–6005 (2023)
12. Wang, D., Tan, D., Liu, L.: Particle swarm optimization algorithm: an overview. Soft. Comput. **22**, 387–408 (2018)
13. Wu, H., Wen, Y., Zhang, J., Wei, Z., Zhang, N., Tao, X.: Energy-efficient and secure air-to-ground communication with jittering UAV. IEEE Trans. Veh. Technol. **69**(4), 3954–3967 (2020)

Multi-feature Reinforcement Learning for Financial Market Decision Making

Jitendra Kumar Awasti[1(✉)], Ashish Tiwari[1], M. M. Dhabu[1], Rishi Chhabra[2], and Aditya Bhagat[2]

[1] Department of Computer Science and Engineering, Visvesvaraya National Institute of Technology, Nagpur, India
dt23cse026@students.vnit.ac.in, {at,meeradhabu}@cse.vnit.ac.in

[2] Department of Computer Science and Engineering, Visvesvaraya National Institute of Technology, Nagpur, India
{chhabrarishi2001,adityavbhagat77}@gmail.com

Abstract. The financial industry grapples with the rising complexity and dynamic nature of stock markets, predicting trends and risk management are key components to maximize profits and reduce losses. Artificial Intelligence have been utilized in many industries successfully, numerous efforts have been made to integrate AI into algorithmic trading. Notably, Reinforcement Learning, which can adapt to dynamic decision making. This study utilizes a deep reinforcement learning model–Deep Q-Network and assess their potential for stock market movement estimation and utilizing prediction probabilities as signal strengths and risk management. 61 candlestick patterns as well as percent change of six features open, high, low, close, volume, and adjusted close are incorporated into this process. Together with the near and volume measurements, we use integrated gradients to identify the top 10 features that have the highest integrated gradient values, yielding a total of 12 features. These are fed into a GRU (Gated Recurrent Unit) model, and the output is then mixed with holding and gain percentage information from a dense layer. Following concatenation, the output is run through more dense layers. Using a softmax function to produce probabilities for the buy, sell, and hold actions. The reward is computed using the difference between the starting amount and the total value of holdings plus the remaining balance, and actions are carried out according to the highest probability. Our method, which has been validated on American and Indian stocks, shows that the DQN model produces trading signals that are both profitable and highly effective.

Keywords: Reinforcement learning (RL) · Gated Recurrent Unit (GRU) · Deep Q-Network (DQN) · Integrated Gradient (IG) · Financial market analysis · Algorithmic trading · Technical analysis · Stock market trends · Decision-making · Profit maximization

R. Chhabra and A. Ahagat—Research Intern

D. Gupta et al. (Eds.): PCEMS 2024, CCIS 2491, pp. 414–426, 2025.
https://doi.org/10.1007/978-3-031-90577-3_35

1 Introduction

Gaining maximum profits requires making astute decisions in the financial markets. High profit margins with low risk can be achieved by having the capacity to predict market changes and make well-informed trading decisions. Recently, there has been a surge in the use of artificial intelligence and machine learning in the financial markets, providing new approaches to analysis and trend prediction. this study employs a cutting-edge reinforcement learning technique in order to examine the Deep Q-Network (DQN) model's potential for stock market movement estimation and decision improvement.

Financial markets, robotics, and gaming are just a few of the areas where reinforcement learning–and especially the DQN model–has shown promise. The DQN model is able to learn complex decision policies and handle high-dimensional data since it integrates Q-learning and deep neural networks. Our objective is to create a trading strategy that is resilient to changes in the market and produces profitable trading signals by utilizing the capabilities of reinforcement learning.

Technical analysis techniques like candlestick patterns are used to forecast asset price movements by analyzing historical price changes. In order to identify possible market trends and reversals, traders and technical analysts frequently employ candlestick patterns, which we have incorporated into our research. These trends are crucial for making wise trade selections since they offer insightful information about market behavior. Six basic measures are also included: open, high, low, close, volume, and adjusted close. For precise trend prediction, these measures are essential as they provide a perspective of market activity.

Integrated gradients [41], a method for attributing a neural network's output to its input characteristics, is utilized for feature selection. Together with the near and volume metrics, we determine the top 10 features with the highest relevance scores by computing the integrated gradients. This approach interprets how different attributes contribute to the model's predictions. This is especially crucial in the financial markets, since knowing how various variables affect trade can help us make better feature selections. We can keep the most pertinent data while simplifying the model by focusing on the most crucial traits.

Holdings and percent gain information are merged with the output of the GRU model in a dense layer. This combination provides a thorough depiction of the state of the market by capturing both the fundamental and temporal features of the market data. After processing the concatenated output through a second dense layer, a sigmoid function is used to create probabilities for the buy, sell, and hold actions.

GRU models are useful in financial time-series analysis because of their ability to handle sequential data and retain long-term dependencies. This enables the algorithm to comprehend how market trends evolve over time, resulting in more precise forecasts. We may also take into consideration the effect that past actions have had on the state of the market by integrating holding and percent gain information.

Our empirical investigations underline the quite robust performance of DQN in our model, outperforming more traditional techniques at generating trading signals reliably and profitably. In this work, we combine multi-feature RL methodologies; hence, this system is continuously increasing profit margins, therefore showing the resilience and adaptability of our system across different market scenarios. These results not only confirm the validity of RL in improving financial decisions but also stress its potential to revolutionize automated trading systems with adaptive strategies in accordance with changing market conditions.

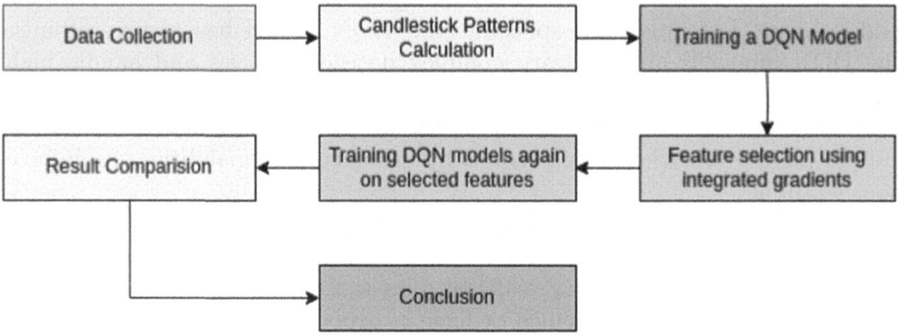

Fig. 1. Study flow chart.

2 Related Work

Reinforcement learning methods in algorithmic trading have made huge development, propelled by enhancements in computational techniques and the requirement for adaptive trading strategies. In this regard, this section reviews a few seminal studies and current innovations for an RL-based trading system, highlighting how they contribute to the methodology of automating and optimizing financial markets.

Early work on the application of RL to trading was concentrated on some core concepts, including direct RL methods. For example, Moody and Saffell [12] gave direct RL methodologies for the optimization of trading systems, which focus on performance functions aiming at strengthening decision-making. Such work laid the groundwork for further research into the applicability of RL techniques across scenarios of financial markets.

Deep reinforcement learning has evolved into a strong paradigm for dealing with the intricacies of financial data and market dynamics. Ansari et al. [3] proposed a DRL-based decision support system tailored for the purpose of making timely decisions pertaining to automated stock market trading. The system showed better scalability in handling large-scale data and dynamic optimization of trading strategy, hence a major advancement of adaptive financial decision-making.

The integration of directional change detection into RL frameworks enhanced the adaptiveness and predictive efficiency of trading systems. Aloud and Alkhamees [2] pioneered an intelligent algorithmic trading strategy that integrates principles of RL with methodologies of directional change. Being able to react at a quicker pace to changes occurring in the market, this enabled the enhancement of trading outcomes without exposure to market-driven risks.

Building on this insight, Lei et al. [7] went a step further in the application by proposing a jointly deep time-driven and feature-aware RL approach. Their model aimed at enriching the representations of financial signals and improving the efficiency of algorithmic trading by infusing temporal features into the learning process. This enhanced the robustness of trading algorithms besides increasing their responsiveness to market regime changes.

Portfolio optimization remains one of the most challenging tasks in an RL-based trading system. Almahdi and Yang [14] proposed an adaptive, recurrent RL-based approach for the construction of a trading portfolio. In the approach described, risk-return optimization was preferred with competent management of the maximum drawdown. It showed that RL could very well handle complex portfolio dynamics to optimize returns in volatile market scenarios.

Studies that have made comparisons have always emphasized how much better RL is than conventional trading tactics. A thorough analysis explaining RL's flexibility and improved performance in dynamic market situations over static trading rules was given by Bertoluzzo and Corazza [11]. These kinds of discoveries play a crucial role in promoting the use of RL as a revolutionary instrument in contemporary finance decision-making.

Applications in the real world have confirmed RL's applicability in trading platforms. By using adaptive reinforcement learning techniques to construct an automated FX trading system, Dempster and Leemans [20] achieved substantial profitability and demonstrated the effectiveness of RL in real-time trading scenarios. These uses highlight RL's ability to transform trading procedures by automating choices and maximizing results under erratic market circumstances.

Kuo et al.'s [21] generative adversarial market model has advanced efforts to generalize RL-based trading methods. By reducing the biases present in RL frameworks, this approach hopes to improve trade robustness in a variety of market scenarios and open the door for scalable RL structures appropriate for expansive financial markets.

New developments include multi-agent reinforcement learning models specifically designed for stock market simulation. Collaborative multi-agent reinforcement learning frameworks, intended to capture intricate market interactions and improve trading strategies jointly, were investigated by Lussange et al. [22]. These methods demonstrate how cooperative learning paradigms might improve financial decision-making procedures in the face of changing market conditions.

Even with these developments, there are still issues that need to be researched, such as computational complexity and model interpretability. More advancements in the industry and the possibility of more complex automated trading systems could be made possible by future research focused at improving the scalability and effectiveness of RL algorithms in practical trading applications.

The latest studies on reinforcement learning with regard to algorithmic trading concern the robustness and adaptability of a model in a dynamic market. For example, Fu et al. proposed a multi-feature supervised learning-based RL for enhancing stock trading performance. The main emphasis of the study was to harness diverse sources of information to enhance an individual's decision-making ability, thereby solving the challenge of data heterogeneity across financial markets.

In the realm of analyzing financial data and making decisions, DRL application significantly mitigated the complexities involved. Deng et al. [17] have put forward deep direct reinforcement learning methods for financial signal representation and trading. It realized significant improvements in predictive accuracy and trading efficiency within their methodology. Their proposed method, based on deep neural networks, captured more sophisticated patterns of market data to execute more rational trading decisions regarding the unknown market.

Furthermore, collaborative frameworks have also been developed besides the single-agent RL models, so as to better approximate the market dynamics. Lussange et al. [22] discussed multi-agent reinforcement learning strategies tailored for the simulations regarding the stock market. Their work highlighted that use of more than one agent will lead to powerful cooperative learning, so that adaptive trading strategies are constructed that handle the intricate dynamics within financial markets better. A fresh approach, leveraging collective intelligence to make improved decisions within competitive trading environments, the interest in which has marked a prominent departure from traditional single-agent RL.

In addition, work continues on how to deal with central problems, such as model interpretability and computational efficiency, for RL-based trading systems. For example, Bakhach et al. [30] presented an approach to be able to modify an RL model for financial timeseries analysis in such a way that it made it possible to detect key direction/reversal events in the markets, and consequently, enable such methods to increase transparency in the decision-making process–an extremely important aspect, since stakeholders might require explanations and confirmations on the automated trading strategies.

Looking forward, other breakthroughs in this area are bound to be capable of making paradigm changes within branches of financial trading, from risk management to portfolio optimization. Gao and Laiwan [24] explore Q-learning techniques and the process of portfolio management as an attempt to optimize strategies of investment with maximum risk-adjusted returns. Their work brings into focus how RL stands to automatize not just trading but also portfolio allocation optimization dynamically.

3 Methodology

Problem Statement. The principal aim of this study is to apply the Deep Q-Network (DQN) model to improve financial market trend forecasting and decision-making. The DQN model can learn and make predictions based on a rich blend of past price movements and candlestick patterns. This study integrates these disparate data sets to provide a comprehensive picture of market behavior. We ensure that our model is capable of identifying complex patterns and making well-informed decisions in dynamic market environments by integrating key metrics like volume and adjusted close prices with candlestick patterns, which are well-known for their capacity to visually represent price action over specific time frames.

3.1 Datasets

We use open source stock data that we can access from Yahoo Finance. This data includes all of the important characteristics, including Date, Open, High, Low, Close, Adjusted Close, and Volume. This dataset offers a thorough picture of market behavior across time and forms the basis for our research of American and Indian stocks. For the purpose of training and assessing our Deep Q-Network (DQN) model, the data retrieval procedure guarantees consistency and reliability to capture the past price movements and trade volumes.

We calculated wide range of candlestick patterns. By using these patterns, which include well-known patterns like "Doji," "Two Crows," and "Engulfing Pattern," we can recognize important market signals. Every pattern adds valuable information to our dataset.

Our dataset gains more granularity when these candlestick patterns are combined with basic features like volume and adjusted close prices. The amalgamation of these factors not only enhances our prognostic models but also facilitates a refined comprehension of market dynamics, endowing the DQN model with more precision and dependability in learning and forecasting.

3.2 DQN Architecture

Deep Q-Networks (DQNs) represent a significant advancement in reinforcement learning (RL), combining the traditional Q-learning algorithm with deep neural networks to handle high-dimensional state spaces. In the context of stock trading, DQNs are particularly useful as they enable the agent to learn optimal trading strategies based on historical market data. The key components of DQN include:

Neural Network Architecture:

Input Layer: The input to the DQN is a representation of the state, which, in the case of stock trading, can include OHLCV data, technical indicators, and position information. The data is often structured in sequences, and in our case, we use an GRU (Gated Recurrent Unit) to capture the temporal dependencies inherent in financial time series data.

Hidden Layers: These can consist of fully connected layers or GRU cells to process the input and learn patterns in the data.

Output Layer: The output layer predicts the Q-values for each possible action (buy, sell, hold). Each action is associated with a different Q-value, representing the expected return for taking that action in the given state.

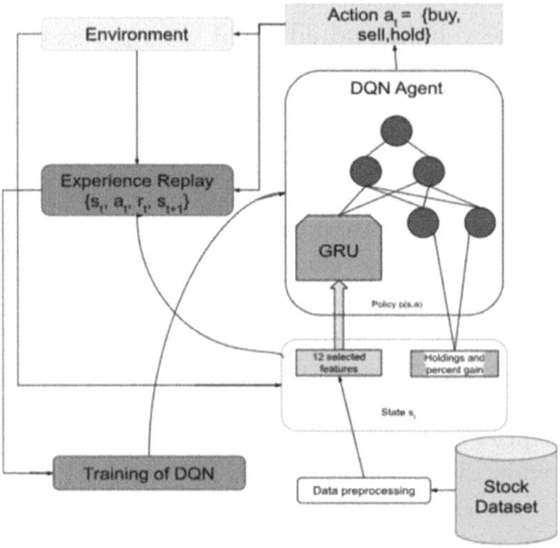

Fig. 2. DQN Architecture.

Action Space: The output from the GRU model is concatenated with additional information from a dense layer that includes holding and percent gain data:

$$h_{concat} = [\text{GRU output}, \text{Holding}, \text{Percent Gain}]$$

This concatenated output h_{concat} is then passed through another dense layer equipped with a softmax function:

$$\text{Prob}(a) = \sigma(W_{\text{action}} h_{concat} + b_{\text{action}})$$

where σ denotes the softmax function. The output probabilities $\text{Prob}(a)$ correspond to potential actions a (buy, sell, or hold), with actions chosen based on the highest probability determined by the model. in Addition to utilizing Prob(a) for action selection, it was also utilized to determine strength of the signal, e.g. if action calculated by model is "buy" then we subtract probability of sell from probability of buy and utilize this fraction as a percent value to buy new holdings.

$$holdings_{new} = (prob_{buy} - prob_{sell}) * margin$$

if signal is sell then $prob_{buy}$ is subtracted from $prob_{sell}$ and based on this fraction we sell stocks and book profit.

$$holdings_{sell} = (prob_{sell} - prob_{buy}) * holdings_{current}$$

this technique ensures that agent is able to manage risk based on current state of the environment. i.e. if agent predicts that stock prices will go down in next few days then it sells a fraction of stocks from current holdings and if stock price to go up in next few days it buys new holdings with a fraction of current available margin.

Reward Calculation: When the agent performs an action, a reward in the form of numerical points is provided to the agent to indicate how good the action is at a given stage. For the DQN model, the reward function is calculated using the Bellman equation:

$$R_t = r_t + \gamma \max_{a'} Q(s_{t+1}, a'; \theta^-) - Q(s_t, a_t; \theta)$$

where R_t represents the reward at time t, r_t is the immediate reward from the action a_t, γ is the discount factor, Q is the action-value function, and θ denotes the model parameters.

In this study the reward is calculated based on net worth of the portfolio i.e. by subtracting the initial capital from the total value of holdings plus available margin.

$$reward = (holdings * price_{current} + margin) - capital_{initial}$$

3.3 Feature Selection Using Integrated Gradients

Integrated Gradients: Integrated Gradients is an attribution method used in deep learning to interpret and explain the predictions of a model, particularly neural networks. Integrated Gradients provides a way to understand which parts of the input (or which features) are driving the model's predictions, which is especially important in fields like healthcare or finance where transparency is crucial.

$$IG_i = (x_i - x_{baseline}) \times \int_{\alpha=0}^{1} \frac{\partial F(x_{baseline} + \alpha \times (x - x_{baseline}))}{\partial x_i} d\alpha$$

here x_i represents each feature, $x_{baseline}$ is the baseline input (typically zero), and F is the output of the GRU model.

In this study initially we trained a base DQN model on AAPL stock data on all 67 features, after training and evaluation we used integrated gradients for feature selection. This method identifies the top 10 most important features (with the highest integrated gradient values), alongside current holdings and total gain, resulting in a curated set of 12 features for further training of DQN models on different stocks. The basic idea behind this feature selection is to reduce the size of state space by eliminating less important features, which further results in reduced duration of training DQN agents for further study.

4 Results and Discussion

To demonstrate that DQN agents action probabilities can be used as signal to buy, sell or hold and same probabilities can be used as signal strength for risk management we conducted experiments on AAPL, NFLX, TATAMOTORS and INFY. all these stocks are from different sectors as well as different stock markets, this fact shows that our strategy works on given OHLCV data independent of sectors.

Table 1. Cumulative percent gain per year.

Stock	Period	Gain (Cumulative per year)
AAPL	1-Jan-2005 to 31-Dec-2014	28.29%
AAPL	1-Jan-2015 to 29-Dec-2023	25.5%
NFLX	1-Jan-2005 to 31-Dec-2014	30.21%
NFLX	1-Jan-2015 to 29-Dec-2023	17.35%
INFY	1-Jan-2005 to 31-Dec-2014	11.88%
INFY	1-Jan-2015 to 29-Dec-2023	11.73%
TATAMOTORS	1-Jan-2005 to 31-Dec-2014	16.08%
TATAMOTORS	1-Jan-2015 to 29-Dec-2023	9.12%

The DQN agent is trained on different stockes from US and India. First data was splited into train and test sets. Data from 1-Jan-2005 to 31-Dec-2014 is used as train set and from 1-Jan-2015 to 29-Dec-2023 is used as test set. The DQN agent is trained on train set and test set is used for validation purposes.

In the following figures (from 3, 4, 5 to 6), it is shown that how Net worth is changing over time, the x-axis represents time and y axis represents net worth. In the Table 1, we have shown Cumulative percent gain of each stock for train and test sets.

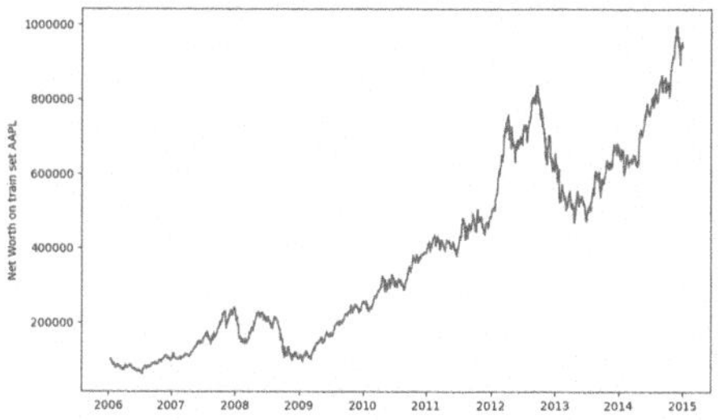

(a) Net worth over train set

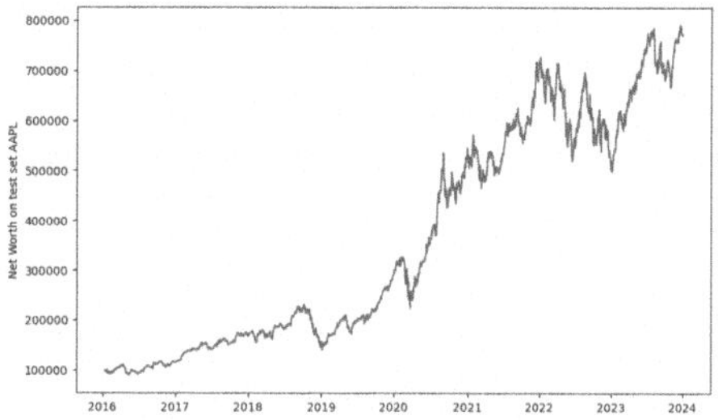

(b) Net worth over test set

Fig. 3. AAPL

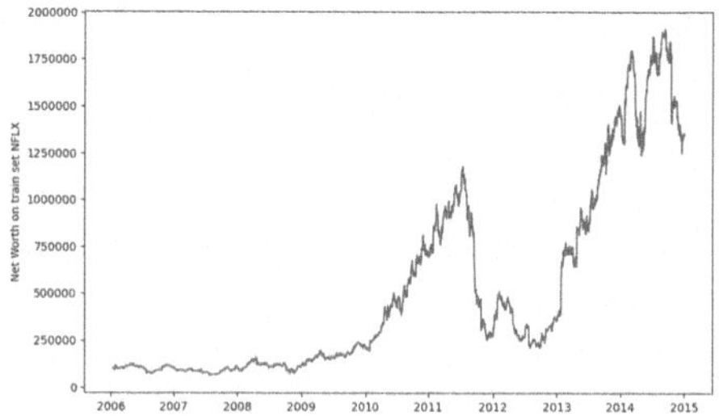

(a) Net worth over train set

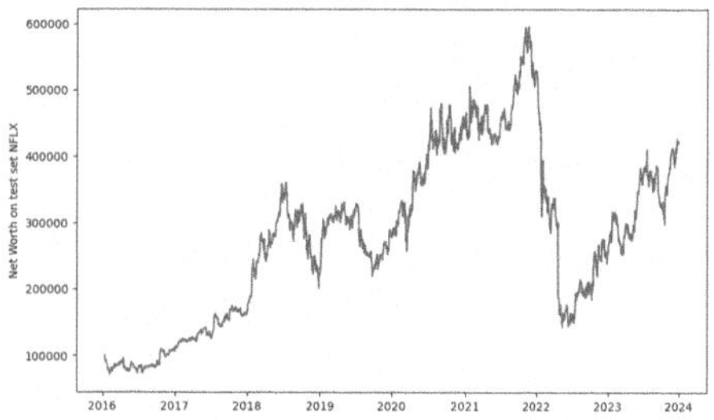

(b) Net worth over test set

Fig. 4. NFLX

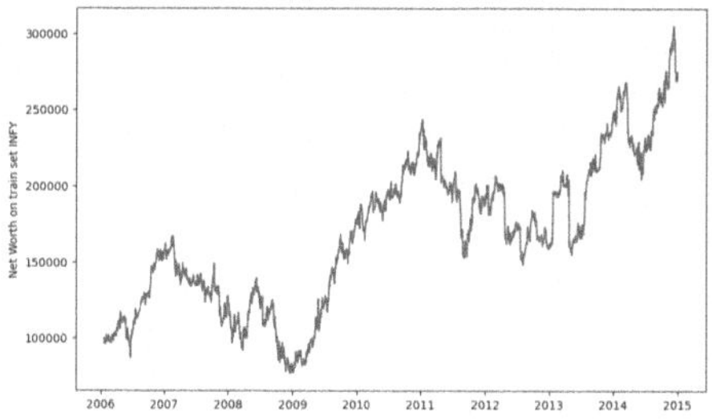

(a) Net worth over train set

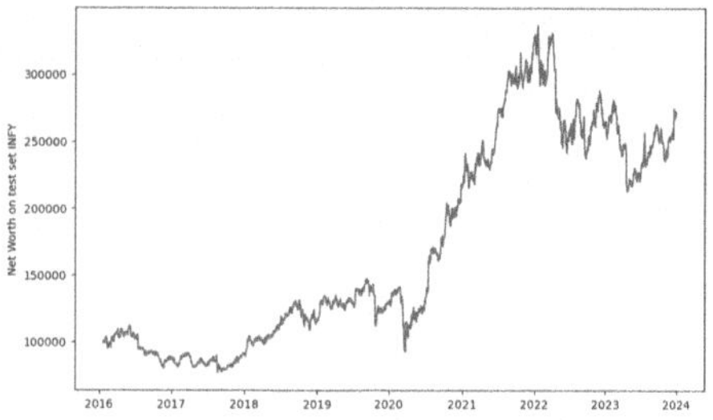

(b) Net worth over test set

Fig. 5. INFY

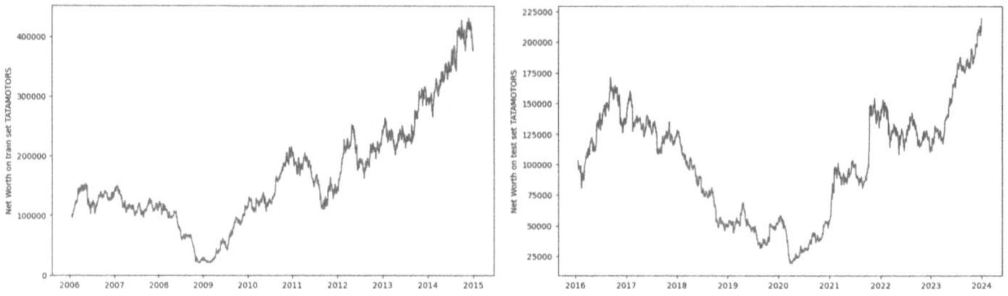

(a) Net worth over train set

(b) Net worth over test set

Fig. 6. TATAMOTORS

5 Discussion

5.1 Interpretation of Results

After analyzing the experiment results we can conclude that using signal probabilities of DQN agent as signal strength and risk management can create effective trading strategies in stock markets. However, using multi-agent setup too can provide better performance. In multi-agent setup 1 agent can provide signals to buy, sell or hold and second agent can perform as a risk manager. But using multi-agent setup will require higher computational resources and increased duration of training.

5.2 Challenges and Limitations

Despite their potential, RL-based trading systems face several challenges, including data sparsity, non-stationarity, and overfitting. Additionally, the implementation of RL algorithms in real-world trading environments requires careful consideration of transaction costs, market impact, and regulatory constraints.

6 Conclusion

In conclusion, this study provides a comprehensive analysis of reinforcement learning techniques for stock trading, focusing on the signal generation as well as risk management. Our findings contribute to the growing body of literature on automated trading systems and offer valuable insights into the practical application of reinforcement learning in financial markets.

By addressing key challenges and limitations, we pave the way for future research aimed at developing more effective and robust trading algorithms. This expanded elaboration provides a deeper understanding of the research methodology, results, and implications discussed in the paper, enriching the overall content and contributing to the advancement of knowledge in the field of reinforcement learning based stock trading.

References

1. Fu, K., Yu, Y., Li, B.: Multi-feature supervised reinforcement learning for stock trading. IEEE Access **9**, 77854–77867 (2023)
2. Aloud, M.E., Alkhamees, N.: Intelligent algorithmic trading strategy using reinforcement learning and directional change. IEEE Access **9**, 114670–114684 (2021)
3. Ansari, Y., et al.: A deep reinforcement learning-based decision support system for automated stock market trading. IEEE Access **10**, 50738–50754 (2022)
4. Treleaven, P., Galas, M., Lalchand, V.: Algorithmic trading review. Commun. ACM **56**, 76–85 (2013)
5. Bruce, B.: Trading Algorithms. 2nd edn. Student-Managed Investment Funds, Cambridge, U.K.: Academic,pp. 285–315 (2020)
6. Fama, E., Blume, M.: Filter rules and stock market trading profits. J. Bus. **39**, 226–241 (1966)
7. Lei, K., Zhang, B., Li, Y., Yang, M., and Shen, Y.: Time-driven feature-aware jointly deep reinforcement learning for financial signal representation and algorithmic trading. Expert Syst. Appl. 140, Art. no. 112872 (2020)
8. Sutton, R., Barto, A.: Reinforcement learning: An introduction. MIT Press, Cambridge, MA, USA (1998)
9. Chinchali, S., et al.: Cellular network traffic scheduling with deep reinforcement learning. In: Proceedings of the AAAI Conference on Artificial Intelligence, pp. 766–774 (2018)
10. Alkhamees, N., and Aloud, M.: DCRL: Approach to identify financial events from time series using directional change and reinforcement learning. Int. J. Adv. Comput. Sci. Appl. 12 (8) (2020)
11. Bertoluzzo, F., and Corazza, M.: Reinforcement learning for automatic financial trading: Introduction and some applications. Dept. Econ., Ca' Foscari Univ. Venice, Venice, Italy, Work. Paper 2012:33 (2012)
12. Moody, J., Saffell, M.: Learning to trade via direct reinforcement. IEEE Trans. Neural Netw. **12**(4), 875–889 (2001)
13. Moody, J., Wu, L., Liao, Y., Saffell, M.: Performance functions and reinforcement learning for trading systems and portfolios. J. Forecasting **17**(5–6), 441–470 (1998)
14. Almahdi, S., Yang, S.: An adaptive portfolio trading system: A risk-return portfolio optimization using recurrent reinforcement learning with expected maximum drawdown. Expert Syst. Appl. **87**, 267–279 (2017)
15. Glattfelder, J.B., Dupuis, A., Olsen, R.B.: Patterns in high-frequency FX data: Discovery of 12 empirical scaling laws. Quant. Finance **11**(4), 599–614 (2011)
16. Alkhamees, N., and Fasli, M.: Event detection from time-series streams using directional change and dynamic thresholds. In:Proceedings IEEE international conference on big data (Big Data)), Boston, MA, USA, pp. 1882–1891 (2017)
17. Deng, Y., et al.: Deep direct reinforcement learning for financial signal representation and trading. IEEE Trans. Neural Netw. Learn. Syst. **28**(3), 653–664 (2017)
18. Pendharkar, P.C., Cusatis, P.: Trading financial indices with reinforcement learning agents. Expert Syst. Appl. **103**, 1–13 (2018)
19. Nevmyvaka, Y., Feng, Y., and Kearns, M.: Reinforcement learning for optimized trade execution. In: Proceedings of the 23rd international conference on Machine learning (ICML), pp. 1–8 (2006)
20. Dempster, M., Leemans, V.: An automated FX trading system using adaptive reinforcement learning. Expert Syst. Appl. **30**, 543–552 (2006)

21. Kuo, C.-H., et al.: Improving generalization in reinforcement learning-based trading by using a generative adversarial market model. IEEE Access **9**, 50738–50754 (2021)
22. Lussange, J., et al.: Modelling stock markets by multi-agent reinforcement learning. Comput. Econ. **57**(1), 113–147 (2021)
23. Bertsimas, D., Lo, A.W.: Optimal control of execution costs. J. Financ. Markets **1**(1), 1–50 (1998)
24. Gao, X., and Laiwan, C.: An algorithm for trading and portfolio management using Q-learning and Sharpe ratio maximization. In: Proceedings of the international conference on neural information processing, pp. 832–837 (2000)
25. Zhang, J., and Maringer, D.: Indicator selection for daily equity trading with recurrent reinforcement learning. In: Proceedings of the 15th annual conference companion on Genetic and evolutionary computation, pp. 1757–1758 (2013)
26. Neuneier, R.: Optimal asset allocation using adaptive dynamic programming. In: Proceedings Adv. Neural Inf. Process. Syst., Cambridge, MA, USA: MIT Press, pp. 952–958 (1996)
27. Neuneier, R.: Enhancing Q-learning for optimal asset allocation. In: Proceedings Advances in neural information processing systems,pp. 936–942 (1998)
28. Carapuço, J., Neves, R., Horta, N.: Reinforcement learning applied to Forex trading. Appl. Soft Comput. **73**, 783–794 (2018)
29. Aloud, M., et al.: A directional-change events approach for studying financial time series. Econ. Open Access Open Assess. E-J. **6**, 1–18 (2012)
30. Bakhach, A., et al.: Forecasting directional changes in the FX markets. In: Proceedings IEEE Symposium Series on Computational Intelligence (SSCI), pp. 1–8 (2016)
31. Tsang, E.P.K., et al.: Profiling high-frequency equity price movements in directional changes. Quant. Finance **17**(2), 217–225 (2017)
32. Ao, H., and Tsang, E.: Trading algorithms built with directional changes. In: Proc. IEEE Conference on Computational Intelligence for Financial Engineering and Economics (CIFEr), pp. 1–7 (2019)
33. Bakhach, A. M., et al.: TSFDC: A trading strategy based on forecasting directional change. Intell. Syst. Accounting, Finance Manage. 25 (3), 105–123 (2018)
34. Alkhamees, N., and Fasli, M.: An exploration of the directional change based trading strategy with dynamic thresholds on variable frequency data streams. In: Proceedings International Conference on the Frontiers and Advances in Data Science (FADS), pp. 108–113 (2017)
35. Sani, A.H.S., et al.: Deep reinforcement learning: A selective overview of literature and applications. IEEE Trans. Neural Netw. Learn. Syst. **29**(6), 2290–2303 (2018)
36. Li, B.: The impact of directional changes on volatility forecasting. J. Banking Finance **87**, 214–228 (2018)
37. Aloud, M., and Alkhamees, N.: A hybrid intelligent trading system for financial market prediction. In: Proc. 12th Int. Conf. Hybrid Intell. Syst. (HIS), pp. 328–333 (2012)
38. Nguyen, D., Arjun, N.: A deep reinforcement learning model for forex trading. Comput. Econ. **53**(1), 1–22 (2019)
39. Das, S.: Financial prediction using machine learning and directional change algorithm. In: Proc. IEEE Int. Conf. Big Data (Big Data), pp. 3986–3991 (2019)
40. Fama, E.F., Blume, M.E.: Filter rules and stock market trading profits. J. Business **39**, 226–241 (1966)
41. Sundararajan, M., Taly, A., Yan, Q.: Axiomatic attribution for deep networks. International conference on machine learning, PMLR (2017)

Xception-Based Face Anti-Spoofing Approach for Secure Biometric Authentication

K. Vannurswamy[1](✉), B. H. Shekar[2], Bharathi Pilar[3], Kotegar A. Karunakar[4], and Frank Jiang[1,2,3,4]

[1] Department of Computer Science, Mangalore University, Konaje, Mangalore, Karnataka, India
vanurmucs@gmail.com
[2] Department of Computer Science, University College, Mangalore, Karnataka, India
[3] Department of Data Science, MAHE, Manipal, India
[4] School of Information Technology, Deakin University, Melbourne, Australia
karunakar.ak@manipal.edu, frank.jiang@deakin.edu.au

Abstract. In this paper, we explore a Convolutional Neural Network (CNN) binary classifier that leverages the Transfer Learning Technique, specifically Xception, integrated within the TensorFlow framework. Transfer Learning harnesses knowledge acquired from one task to enhance the performance of a related task. In this context, we exploit the pre-existing knowledge within the extensive dataset learned by Xception to enhance the binary classifier's performance. A global average pooling layer is integrated into the Xception model architecture to further reduce the spatial dimensions, focusing on the most relevant and significant information from each feature map. A sigmoid activation function is employed to facilitate binary classification. Extensive experimentation on both the CASIA-FASD and LCC FASD datasets demonstrates that when classifying images as real or fake, the model achieves an F1-score of 99.64%. In addition, we conducted a comparative analysis with several state-of-the-art techniques. The results indicate that the proposed method outperforms these well-known techniques in face anti-spoofing.

Keywords: Convolutional Neural Network · Pre-trained Models · Face Anti-Spoofing · Biometrics · Face and Iris · Presentation Attacks

1 Introduction

Facial spoofing involves the creation of fake facial images or videos to deceive the face, usually in order to gain unauthorized access. These fake faces can be used in many ways, including photo prints, paper cutouts, video replays, 3D masks, or artificially created face replacement videos. To ensure the security and reliability of facial recognition technology, identifying and preventing facial spoofing is crucial. Deep learning-based learning features as well as intrinsic features are learned through training data and are capable of performing well.

D. Gupta et al. (Eds.): PCEMS 2024, CCIS 2491, pp. 427–437, 2025.
https://doi.org/10.1007/978-3-031-90577-3_36

Most anti-spoofing programs focus on analyzing facial patterns. This analysis involves observing signs such as eye blinking, lip movements, and head movements that are important in determining whether a face seen in a painting or a dragon is a momentary pattern or an illusory representation. But these methods often create features specific to a particular fraud or dataset and may not be good for those that go undetected. Facial anti-spoofing technology plays an important role in protecting facial recognition from malicious attacks. Therefore, designing a face protection method with accurate detection and focusing ability is important to increase the security of face recognition and identification [1].

Many algorithms have been developed to detect fraud according to various methods. Many challenges have been created to take the necessary measures to prevent attacks that will support the development of this field. Another technique that uses hand-generated features to detect facial spoofing is based on image quality and motion [2].

In light of these challenges, we propose utilizing the Xception model as the foundation for our facial spoofing detection system. The Xception model is particularly suited for this task due to its unique architecture, which employs depthwise separable convolutions. This design enables the model to efficiently capture intricate features while significantly reducing the number of parameters compared to traditional convolutional networks. Consequently, the Xception model demonstrates high accuracy and robustness, making it effective in distinguishing subtle differences between genuine and spoofed faces.

Moreover, the Xception model has consistently outperformed other state-of-the-art architectures, such as VGG16 and ResNet, in various image classification benchmarks. Its ability to generalize across diverse datasets enhances its applicability in real-world scenarios where spoofing attempts may vary widely. By leveraging the strengths of the Xception model, we aim to develop a facial anti-spoofing method that combines deep learning-based feature extraction with intrinsic feature analysis, thereby increasing the accuracy and reliability of facial recognition systems in the face of malicious attacks (Fig. 1).

In Sect. 2, we provide a survey of recent advancements in deep learning approaches for face spoofing and anti-spoofing, summarizing existing literature. Section 3 outlines the methodologies employed in our experiments. Section 4 details our experimental setup, results, and comprehensive evaluation. Finally, Sect. 5 summarizes our findings and draws conclusions from our research study.

2 Related Work

Attacks on automatic face recognition (AFR) systems are generally classified into two main types: digital manipulation and physical presentation attacks. Digital manipulation involves subtly altering images in the digital domain to deceive the face recognition system, as described by Tolosana [3]. On the other hand, physical presentation attacks attempt to trick AFR systems by presenting images or replicas of faces on physical media in front of imaging sensors, as discussed by Liu [4].

(a) Examples of Real Images

(b) Example of Spoofed Images

Fig. 1. Real and Spoofed Images

Face presentation attacks (FPAs) [5] can be classified based on attackers' intentions into two main types: impersonation and obfuscation. Impersonation involves using spoof techniques to mimic a genuine user's facial features, often employing photos, electronic screens, or 3D masks for deception. On the other hand, obfuscation focuses on concealing the attacker's true identity by utilizing methods like glasses, makeup, wigs, or disguises to evade detection by facial recognition systems. These tactics underscore the dual strategies of assuming another's identity or masking one's own to exploit vulnerabilities in security systems reliant on facial recognition.

The deep learning method maximizes the potential of unstructured data extraction and uses CNN-LSTM architecture to capture physical features. Results [6].

Face Anti-Spoofing (FAS) is essential to protect facial recognition from attacks. This research reframes FAS as a challenge of data analysis, using binary communication networks (BCNs) to capture specific patterns. BCN integrates multi-level data, multi-level optimization (MFRM) and multi-head tracking to improve learning. This new method achieved a significant improvement of $11.3 \pm 9.5\%$ EER in the competitive evaluation of six parameters of the SiW-M dataset, surpassing the existing method [7].

Recently, the focus of research in deep-learning-based face anti-spoofing has shifted towards CNNs. For example, CNN-based face anti-spoofing using ResNet and Inception architecture [8] combines deep learning with multi-level local binary model (ELBP) texture features to accurately perform face spoofing detection [9]. This method has been tested on the Replay-Attack, CASIA FASD, and MSU-MFSD databases, demonstrating its effectiveness in both intra- and inter-dataset situations [10]. The deep convolutional neural network (CNN) based approach [11] demonstrates supervised learning of discriminative features. When subjected to testing across two challenging datasets, CASIA and REPLAY-ATTACK, it underscores CNN's capability to extract features with exceptional generalization properties, resulting in a significant reduction in the Half Total Error Rate (HTER).

Presentation attacks (PA) are divided into 2D and 3D categories based on their geometric characteristics. 2D PA [12] involves displaying facial attributes from images or videos, including flat or wrapped images, eye/mouth cutouts, and animated film. Advances in 3D printing have made 3D masks a threat to automatic facial recognition (AFR); 3D masks are more accurate in color, texture, Softness and geometry compared to 2D PA [13]. These 3D masks are made from a variety of materials: Rigid masks can be made from paper, resin, plaster, or plastic, while flexible masks usually use silicone or latex.

Facial Anti-Spoofing (FAS) is necessary to protect facial recognition against different attacks. This work presents PatchNet, which redefines FAS as a detailed patch type definition challenge. It improves data variability and feature discernment by identifying combinations of capturing devices and presentation materials from undistorted face image patches. Our approach, incorporating Asymmetric Margin-based Classification Loss and Self-supervised Similarity Loss, significantly improves generalization, outperforming existing methods on multiple benchmarks and proving robust against unseen spoof types [14].

Convolution operations are employed to capture local and densely informative features. These methods are specifically designed to detect various forms of spoofing, such as printed photos, paper cutouts, replayed videos, and 3D masks, often tailored to specific spoofing types or datasets [15].

One common issue is the lack of diversity in the datasets, as they often consist of samples with uniform backgrounds and invariant lighting conditions due to data collection constraints. This limitation hinders models from effectively learning and generalizing patterns for distinguishing between live and spoofed samples. To address these challenges, creating a more extensive dataset with a wider range of samples and developing a model capable of detecting various types of spoofing becomes crucial in the current scenario.

3 Proposed Methodology

The proposed approach incorporates a CNN binary classifier that makes use of the Transfer Learning Technique, specifically Xception, integrated within the TensorFlow framework. Transfer learning uses knowledge gained from one task to improve the performance of another task. The Xception model, well-known for its remarkable performance in image classification tasks, stands as a powerful deep convolutional neural network. It is specifically designed to handle color images, typically represented as three-dimensional arrays of pixel values, with red, green, and blue color channels included in each pixel. It comprises multiple blocks with diverse convolutional layers and skip connections. In this context, Xception, which has already acquired knowledge from a comprehensive dataset, is harnessed to enhance the performance of the binary classifier.

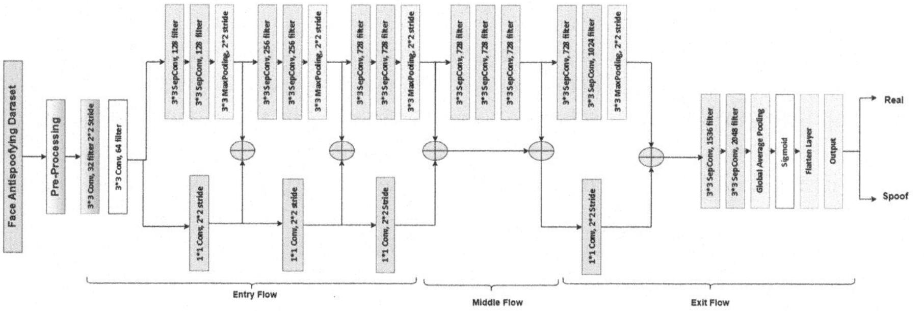

Fig. 2. Architecture of Xception model [16]

The proposed approach is structured around three principal phases: pre-processing, feature extraction, and model training. To prepare the input image for the Xception model and normalize pixel values, a pre-processing layer is applied. Throughout the experimentation, a model with a batch size of 16 and image dimensions of 229×229 was employed. This approach involves adjusting the pixel range from [0, 255] to [0, 1] for each input image. Techniques for data normalization and augmentation encompass rotation, adjustments to height and width, random zoom, and horizontal flipping.

The input image is fed into the Xception model, which has been selected due to its innovative use of depthwise separable convolutions. This architecture allows for efficient feature extraction by separating the spatial filtering and feature combination processes, significantly reducing the number of parameters compared to traditional convolutional layers. Xception's design enhances its capability to capture complex patterns, shapes, and textures across different spatial scales while maintaining computational efficiency. This model has demonstrated superior performance on benchmark datasets such as ImageNet, outperforming traditional architectures like VGG16 and ResNet, particularly in tasks requiring fine-grained classification.

To extract features from the image, a number of separable convolution layers are applied. Within these layers, the image undergoes processing with a variety of filters, enabling the detection of intricate details. To preserve essential features and enhance gradient flow during training, skip connections facilitate the integration of information from early layers into deeper ones. Following the convolutional layer traversal, a global average pooling (GAP) layer is introduced. This layer diminishes the spatial dimensions of the feature maps, effectively summarizing the most pertinent information from each feature map.

The selection of GAP over traditional pooling methods is justified by its ability to reduce the risk of overfitting while maintaining critical global information. Unlike max pooling, which focuses on the most dominant features and can lead to loss of spatial context, GAP averages the feature maps, ensuring a more holistic representation of the input. This is particularly beneficial in scenarios where the overall context of the image is more informative than local details.

The feature vector produced by the global average pooling layer is fed into the fully connected output layer, which generates a single value. Within this layer, the elements of the feature vector undergo weighted summation to determine the ultimate result. Apply the sigmoid function to the output of the exact concatenation to convert it to binary. The sigmoid function indicates the probability of the image falling into the good category by compressing the output between 0 and 1.

The steps involved in the proposed approach are visually depicted in the block diagram 2, and these steps are explained below.

1. **Entry Flow:** The model begins with a conventional convolution layer, succeeded by batch normalization and activation functions (ReLU). Subsequently, it proceeds with another convolution layer, along with batch normalization and ReLU activation. This entry flow is responsible for extracting fundamental features from the input image
2. **Middle Flow:** The middle flow encompasses a sequence of repetitive blocks, each comprising three depth-wise separable convolution layers coupled with batch normalization and ReLU activation. To preserve gradient flow during training and enable the model to effectively learn both low-level and high-level features, residual connections are established between the output of a block and its input. The middle flow is iterated multiple times (typically eight times in the original Xception architecture) to extract more intricate features.
3. **Exit Flow:** The exit flow further enhances the features by introducing additional depth-wise separable convolution layers, accompanied by max-pooling operations and batch normalization. The output of the exit flow undergoes global average pooling to reduce spatial dimensions.
4. **Fully Connected Layer:** Xception ends with all layers connected by a sigmoid activation function (Dense). In the classification function, this layer produces values ranging from 0 to 1, indicating the probability that the input image is a real face or a fake face, respectively.

4 Experiments Results and Discussion

In this section, we evaluate our framework using two general dataset. We will briefly describe this data and then analyze the performance of our examples by comparing it with different designs.

4.1 Dataset

To demonstrate the effectiveness of the proposed method, experiments were conducted on two competing datasets: LCC-FASD and CASIA-FASD.

LCC-FASD: The experiments utilized the LCC-FASD dataset, which includes three subsets: training, development, and evaluation. This dataset consists of 1,942 genuine images and 16,885 spoofing images sourced from platforms such as YouTube, Amazon Mechanical Turk, and Yandex Toloka. A notable strength of this database is its wide variety of spoofing techniques and recording devices. The dataset is divided into training and testing sets with an 80/20 split. Detailed statistics on subjects and images are provided in Table 1.

Table 1. Statistic of subjects and images for LCC FASD

Part	Identities	Genuine Images	Spoofing Images
Training	118	1223	7076
Development	25	405	2543
Evaluation	100	314	7266
TOTAL	243	1942	16885

CASIA-FASD: The CASIA-FASD [17] database contains 12,657 images from 50 subjects, including real and fake images. These include 5,123 real images and 7,534 fake images, divided into different quality levels: low, medium and high. Fake footage replicates scenes such as cropped footage, corrupted footage, and film-based spoofs. This information is useful for testing and improving anti-spoofing algorithms designed to improve the security and reliability of facial recognition in various practical applications.

4.2 Evaluation Metrics

The evaluation metrics for the proposed approach include precision, recall, and accuracy. Recall assesses the system's proficiency in accurately identifying positive samples, especially faked faces, among all of the dataset's real positive samples. Precision evaluates the proportion of correctly identified positive samples (spoofed faces) relative to all samples classified as positive by the system. It gauges the system's accuracy specifically in identifying positives. Accuracy, on the other hand, reflects the overall correctness of the system's predictions, encompassing both true positives (correctly detected spoofed faces) and true negatives (accurately detected genuine faces).

$$\text{Accuracy} = \frac{\text{TP} + \text{TN}}{\text{TP} + \text{TN} + \text{FP} + \text{FN}} \tag{1}$$

The F1 score for face spoofing is a metric that combines and inverts to provide an equal measure of ability to identify real and fake faces. It includes the following models: This metric is especially useful in situations where striking a balance between accuracy and recall is important to the performance of a facial spoofing detection system.

$$F1\text{-score} = \frac{2 \times \text{Precision} \times \text{Recall}}{\text{Precision} + \text{Recall}} \tag{2}$$

4.3 Results

In this study, we focus on facial aggression and validate our model by comparing the CASIA-FASD and LCC-FASD data. For CASIA-FASD data, previous work [18] used symbols such as F1-Score, yield, accuracy, and correctness, as shown in a Table 2.

Table 2. The evaluation metrics of the proposed Xception approach, along with the performance of the ResNet50 and Inception architectures on CASIA-FASD Dataset

Model	Accuracy	Precesion	Recall	F1-Score
Resnet	92.07%	92.47%	92.22	92.71%
Inception	96.45%	96.47%	96.76%	97.58%
Xception	99.45%	99.09%	99.22%	99.54%

For the LCC-FASD dataset, which includes 1,942 genuine images and 16,885 spoofing images from diverse sources like YouTube and Amazon Mechanical Turk, we evaluated using similar metrics. Our proposed Xception-based approach achieved outstanding results, as summarized in Table 3.

The experimental results for all 3 learning models are summarized in Table 2 and Table 3. The experimental outcomes demonstrate a noteworthy f1-score rate of *99.64*% in successfully detecting facial spoofing attacks, outperforming previously published methods in the literature. Comparative analysis reveals that the Xception model consistently outperforms the ResNet50 and Inception architectures on both datasets. In the CASIA-FASD dataset, the Xception model achieved a remarkable accuracy of 99.45% and an F1-Score of 99.54%, significantly surpassing ResNet and Inception. Similarly, on the LCC-FASD dataset, the Xception model demonstrated superior performance with an accuracy of 99.53% and an F1-Score of 99.64%, highlighting its effectiveness in detecting facial spoofing attacks.

Table 3. The evaluation metrics of the proposed Xception approach, along with the performance of the ResNet50 and Inception architectures on LCC-FASD Dataset

Model	Accuracy	Precesion	Recall	F1-Score
Resnet	93.05%	93.40%	93.10	93.61%
Inception	97.45%	98.50%	97.46%	98.35%
Xception	99.53%	99.19%	99.42%	99.64%

Table 4 compares our method with other state-of-the-art techniques on the LCC-FASD dataset, showcasing our approach's outstanding F1-Score of 99.64%. This achievement surpasses previous methods such as ELBP+Modified Xception [10], CNN+Stereo Matching Network [19], SVM+LBP [20], and others.

Table 4 presents an overview of the performance metrics of the proposed deep-learning architectures, focusing on recall, precision, and accuracy. The F1-Score assesses the overall accuracy of the model's predictions, while recall measures the ratio of true positive results to all positive predictions. Precision indicates the proportion of the model's positive predictions that were correct, specifically identifying accurately detected true positives by the model.

Table 4. Comparison of the proposed method with other state-of-the-art

Model	Accuracy	Precision	Recall	F1-Score
ELBP+ Modifed Xception [10]	80.12%	60.45%	80.50%	68.53%
CNN+Stereo Matching Network [19]	92.68%	90.65%	90.50%	92.53%
SVM+LBP [20]	98.07%	98.37%	97.22%	95.71%
CNN [1]	97.00%	96.47%	95.66%	97.48%
CNN+ MTCNN [21]	99.32%	96.27%	94.76%	96.58%
Proposed Approach	**99.53%**	**99.19%**	**99.42%**	**99.64%**

This classifier applies the Xception model as a basis for bias. The top layer was removed and a simple Dense layer with one neuron and sigmoid activation was used in its place. We have computed accuracy, precision, recall, and F1-Score to evaluate the performance of the Xception model. The experimental results on LCC FASD dataset give F1-Score of 99.64%, as shown in Table 4. This proposed method was evaluated against several state-of-the-art models, highlighting its superior performance across various metrics. The ELBP combined with Modified Xception achieved an accuracy of 80.12%, with a precision of 60.45%, a recall of 80.50%, and an F1-score of 68.53%. The CNN paired with a Stereo Matching Network performed significantly better, yielding an accuracy of 92.68%, precision of 90.65%, recall of 90.50%, and an F1-score of 92.53%. The SVM integrated with LBP demonstrated impressive results, reaching an accuracy of 98.07%, with precision at 98.37%, recall at 97.22%, and an F1-score of 95.71%. Similarly, the CNN approach reported an accuracy of 97.00%, precision of 96.47%, recall of 95.66%, and an F1-score of 97.48%. The CNN combined with MTCNN achieved an accuracy of 99.32%, along with a precision of 96.27%, recall of 94.76%, and an F1-score of 96.58%. In comparison, the proposed approach outperformed all others, achieving an impressive accuracy of 99.53%, with a precision of 99.19%, recall of 99.42%, and an F1-score of 99.64%. This comparison underscores the effectiveness of the proposed method in enhancing performance metrics in the relevant tasks.

5 Conclusion

In this paper, we present an effective approach for detecting manipulated or spoofed face images, thereby enhancing the security and reliability of facial recognition systems. Our method utilizes a CNN classifier and leverages Transfer Learning, specifically integrating Xception within the TensorFlow framework. This integration results in a significant improvement in face anti-spoofing accuracy. Our approach focuses on reducing spatial dimensions via a global average pooling layer while preserving crucial image features. The use of a sigmoid activation function enables precise binary classification, ensuring dependable detection of spoofed face images. Experimental evaluations on both CASIA-FASD and LCC FASD datasets, particularly LCC FASD dataset, show promising results, achieving an impressive F1-Score of 99.64%. Furthermore, comparative analyses against state-of-the-art techniques demonstrate the effectiveness of our proposed method in anti-spoofing face detection. This research paves the way for future advancements in the field and offers a robust solution for safeguarding facial recognition systems against fraudulent activities.

Acknowledgement. This work is supported by the Scheme for Promotion of Academic and Research Collaboration (SPARC), Ministry of Education, Govt. of India, for collaborative research work with Deakin University, Australia under Grant No. SPARC/2019-2020/P1695/P1695/SL.

References

1. Chaudhari, P., Kulkarni, N., More, P.: Antispoofing for facial recognition through separable convolution. In: 2021 IEEE Pune Section International Conference (PuneCon), pp. 1–4. IEEE (2021)
2. Feng, L., et al.: Integration of image quality and motion cues for face anti-spoofing: a neural network approach. J. Visual Commun. Image Represent. **38**, 451–460 (2016)
3. Tolosana, R., Vera-Rodriguez, R., Fierrez, J., Morales, A., Ortega-Garcia, J.: Deepfakes and beyond: a survey of face manipulation and fake detection. Inf. Fusion **64**, 131–148 (2020)
4. Liu, A., et al.: Cross-ethnicity face anti-spoofing recognition challenge: a review. IET Biometr. **10**(1), 24–43 (2021)

5. Marcel, S., Nixon, M.S., Fierrez, J., Evans, N.: Handbook of Biometric Anti-Spoofing: Presentation Attack Detection, vol. 2. Springer, Heidelberg (2019)
6. Zhang, M., Zeng, K., Wang, J.: A survey on face anti-spoofing algorithms. J. Inf. Hiding Priv. Protect. **2**(1), 21 (2020)
7. Yu, Z., Li, X., Niu, X., Shi, J., Zhao, G.: Face anti-spoofing with human material perception. In: Vedaldi, A., Bischof, H., Brox, T., Frahm, J.-M. (eds.) ECCV 2020. LNCS, vol. 12352, pp. 557–575. Springer, Cham (2020). https://doi.org/10.1007/978-3-030-58571-6_33
8. Nagpal, C., Dubey, S.R.: A performance evaluation of convolutional neural networks for face anti spoofing. In: 2019 International Joint Conference on Neural Networks (IJCNN), pp. 1–8. IEEE (2019)
9. Pei, M., Yan, B., Hao, H., Zhao, M.: Person-specific face spoofing detection based on a siamese network. Pattern Recogn. **135**, 109148 (2023)
10. Antil, A., Dhiman, C.: A two stream face anti-spoofing framework using multi-level deep features and elbp features. Multimedia Syst. 1–16 (2023)
11. Sun, Y., Xiong, H., Yiu, S.M.: Understanding deep face anti-spoofing: from the perspective of data. Visual Comput. **37**(5), 1015–1028 (2021)
12. Venkatesh, S., Ramachandra, R., Raja, K., Busch, C.: Face morphing attack generation and detection: a comprehensive survey. IEEE Trans. Technol. Soc. **2**(3), 128–145 (2021)
13. Jia, S., Guo, G., Zhengquan, X.: A survey on 3d mask presentation attack detection and countermeasures. Pattern Recogn. **98**, 107032 (2020)
14. Wang, C.Y., Lu, Y.D., Yang, S.T., Lai, S.H.: Patchnet: a simple face anti-spoofing framework via fine-grained patch recognition. In: Proceedings of the IEEE/CVF Conference on Computer Vision and Pattern Recognition, pp. 20281–20290 (2022)
15. Liu, Y., Stehouwer, J., Jourabloo, A., Liu, X.: Deep tree learning for zero-shot face anti-spoofing. In: Proceedings of the IEEE/CVF Conference on Computer Vision and Pattern Recognition, pp. 4680–4689 (2019)
16. Dhillon, A., Verma, G.K.: Convolutional neural network: a review of models, methodologies and applications to object detection. Prog. Artif. Intell. **9**(2), 85–112 (2020)
17. Zhang, Z., Yan, J., Liu, S., Lei, Z., Yi, D., Li, S.Z.: A face antispoofing database with diverse attacks. In: 2012 5th IAPR International Conference on Biometrics (ICB), pp. 26–31. IEEE (2012)
18. Chen, F.-M., Wen, C., Xie, K., Wen, F.-Q., Sheng, G.-Q., Tang, X.-G.: Face liveness detection: fusing colour texture feature and deep feature. IET Biometr. **8**(6), 369–377 (2019)
19. Li, Z., Yuan, J., Jia, B., He, Y., Xie, L.: An effective face anti-spoofing method via stereo matching. IEEE Signal Process. Lett. **28**, 847–851 (2021)
20. Khurshid, A., Tamayo, S.C., Fernandes, E., Gadelha, M.R., Teofilo, M.: A robust and real-time face anti-spoofing method based on texture feature analysis. In: Stephanidis, C. (ed.) HCII 2019. LNCS, vol. 11786, pp. 484–496. Springer, Cham (2019). https://doi.org/10.1007/978-3-030-30033-3_37
21. Cai, P., Quan, H.: Face anti-spoofing algorithm combined with cnn and brightness equalization. J. Central South Univ. **28**(1), 194–204 (2021)

Gesture Prediction Using Surface-EMG Signals

Sibani Panigrahi[1(✉)], Sohham Seal[1], Shyam Lal[2], and Ganesh Naik[3]

[1] Dept. of Mathematical and Computational Sciences, National Institute of Technology Karnataka, Surathkal, India
sibani.pngrh@gmail.com, sohhamofficial@gmail.com

[2] Dept. of Electronics and Communication Engineering, National Institute of Technology Karnataka, Surathkal, India
shyamfec@nitk.edu.in

[3] Department of Computer Science and IT, Flinders University, Adelaide, Australia
ganesh.naik@flinders.edu.au

Abstract. Gesture prediction plays a crucial role in enhancing human-computer interaction by enabling intuitive and natural control methods, thereby reducing reliance on traditional input devices. It significantly improves accessibility for individuals with physical disabilities by providing alternative means of communication and control. Moreover, gesture prediction has broad applications in fields such as robotics, virtual reality, and prosthetics, enhancing both the functionality and user experience of these technologies. This study presents the design and development of an Electromyogram (EMG) signal-based gesture recognition system utilizing recent Deep Learning (DL) techniques. The Hyser EMG dataset was used for experimentation, and its data was pre-processed and analyzed using both sliding window and a combination of sliding window and Fourier transform methods. The performance of the EMG signal-based gesture recognition system was evaluated and compared across different DL models. The results demonstrate that RCCGNet-based gesture prediction outperforms other models.

Keywords: Gesture Identification · Fourier Transform · Sliding Window Method · ResNet · VGGNet · AlexNet · RCCGNet

1 Introduction

Gestures have long been a fundamental part of non-verbal communication, serving as a means to express intent, emotions, and information without the use of speech. From primitive human interactions to modern-day technology, gestures continue to be a natural and intuitive extension of human expression. Over recent decades, gesture recognition has gained considerable attention as a key area of research, with applications ranging from Human-Computer Interaction (HCI) and robotics to gaming, assistive technologies and rehabilitation.[7].

D. Gupta et al. (Eds.): PCEMS 2024, CCIS 2491, pp. 438–449, 2025.
https://doi.org/10.1007/978-3-031-90577-3_37

A promising technique in gesture recognition involves the use of electromyographic (EMG) signals, which are electrical signals generated by muscle contractions. First recorded by Emil du Bois-Reymond in 1849, EMG signals have played a critical role in biomedical engineering, enabling detailed studies of muscle function and providing the foundation for digitizing human motion for computational analysis [10]. These signals are typically characterized by amplitude, frequency, and phase, but interpreting them can be challenging due to various factors such as muscle anatomy, physiological variations, and signal noise.

Historically, EMG has been employed for studying neuromuscular disorders using invasive methods such as needle electrodes. These techniques, while effective for diagnosing motor unit action potentials (MUAP), have significant limitations when it comes to capturing overall muscle activity, making them unsuitable for gesture prediction. Furthermore, traditional EMG methods often utilize unsupervised pattern recognition, clustering, and neuro-fuzzy models, which are designed for disorder classification rather than recognizing generalized muscle movements [12–15]

In contrast, the development of non-invasive surface EMG (sEMG) technology has opened new possibilities for gesture recognition. sEMG utilizes electrodes placed on the skin to capture muscle activity, offering advantages such as ease of application, cost-effectiveness, and the ability to monitor muscle behavior without causing discomfort. These attributes make sEMG ideal for real-time gesture recognition, which requires both high accuracy and responsiveness. Applications of sEMG include human-robot interaction, assistive technologies for individuals with physical disabilities, and Extended Reality (XR) systems such as Virtual Reality (VR) and Augmented Reality (AR) [8].

Numerous studies have investigated the potential of sEMG for gesture classification, exploring various preprocessing techniques and machine learning (ML) algorithms. For instance, [11] applied discrete wavelet transform (DWT) for preprocessing sEMG signals and classified four gestures (up, down, right, and left) using an artificial neural network (ANN), achieving an accuracy of 88.4%. Similarly, [16] employed a Radial Basis Function (RBF) for preprocessing and reported 81.69% accuracy for classifying three hand gestures. Additionally, studies such as [17] have demonstrated improvements in accuracy using classical ML models like Decision Trees (DT), Support Vector Machines (SVM), and k-Nearest Neighbor (kNN), with feature selection methods boosting accuracy from 70.04% to 89.75%. More recently, deep learning approaches have also been explored, such as the Slow Fusion and Inception model architecture introduced by [18], which achieved an accuracy of 83.97%. A comparative analysis by [19] highlighted the importance of preprocessing in improving gesture recognition accuracy, noting that distortions in EMG data could cause variations in accuracy of up to 26.5%.

Moreover, research has shown that combining sEMG with other sensor modalities, such as gyroscopic data, can further enhance classification accuracy. For example, [20] reported a high classification accuracy of up to 96.3% for finger gestures using sEMG and gyroscopic signals from the Myo armband. While their results are impressive, their study relies on a dataset that is not publicly benchmarked, making direct comparison across studies difficult. Additionally, the use of gyroscopic signals adds complexity, which may not always be feasible in simpler setups. Their dataset also includes only 10 participants, which may limit the generalizability of the results.

This paper aims to contribute to the field of sEMG-based gesture recognition by leveraging the Hyser dataset and exploring two signal preprocessing techniques: the sliding window method and a combination of sliding window and Fourier transform. The sliding window method segments continuous sEMG signals into manageable portions for analysis, while the Fourier transform extracts frequency domain features, revealing underlying signal patterns. Multiple deep learning models were evaluated to assess their effectiveness in recognizing gestures from these processed sEMG signals. By comparing the results of these techniques, this study seeks to provide valuable insights into improving the efficiency and accuracy of gesture prediction systems, with potential applications in HCI, rehabilitation, and beyond.

2 Methodology

The methodology adopted in this research is structured into three key stages: preprocessing of EMG Dataset, model development and evaluation. This approach emphasizes a quantitative analysis leveraging deep learning(DL)techniques for surface-EMG signal classification. The primary dataset utilized is sourced from Krilova et al. [1], discussed in details in Sect. 3.1, which forms the basis for the analysis.

In the preprocessing stage, techniques such as the sliding window method and Fourier transformation are applied to refine the dataset, ensuring its compatibility with deep learning models. These preprocessing steps are crucial in enhancing the dataset's features, making it more suitable for model training. Given that most of the DL models employed are designed for 2D data, and the dataset consists of 1D signals, we adapt the architectural structure of the models by modifying each layer to operate effectively on 1D signals. This adaptation allows us to leverage the benefits of convolutional layers, traditionally used in 2D models, for the analysis of 1D surface-EMG signals.

For model development we follows a sequential progression, where each subsequent model aims to improve the accuracy of the analysis, as detailed in the following sections. The final model, depicted in Fig. 1, achieves the highest accuracy, demonstrating the effectiveness of this methodical approach.

Sliding Window: Sliding Window: Electromyographic (EMG) signals corresponding to different hand movements exhibit unique properties that must be analyzed to uncover specific movement patterns. Given the stochastic nature of EMG signals, they are processed in discrete segments, commonly referred to as sliding windows. These windows may overlap or be non-overlapping, allowing for the analysis of the signal in manageable time intervals. As EMG signals are influenced by external factors, subsequent filtering is necessary to retain critical information, reduce dimensionality, and extract relevant waveform features within a defined time frame. The sliding window technique involves partitioning the data into segments and applying statistical measures such as averaging or determining the maximum value for each feature within the windowed segment [9].

Fourier Transform: EMG signals are composed of multiple frequency components, each associated with different physiological activities, such as muscle contractions. The Fourier transform is employed to decompose the EMG signal into its frequency-domain components, facilitating the identification of frequency patterns relevant to gesture classification. Certain movements may exhibit distinct frequency characteristics in the EMG signal, and Fourier analysis enables the extraction of these features for classification purposes. Additionally, the Fourier transform aids in signal denoising by filtering out noise or interference within specific frequency bands.

2.1 Convolutional Neural Network

Convolutional Neural Networks (CNNs) are foundational in extracting spatial and temporal features from raw EMG signals. For EMG data, which typically involves multiple channels capturing muscle activity over time, CNNs are adept at recognizing patterns and dependencies within this data. The CNN architecture typically consists of three key layers: convolutional layers for feature extraction, pooling layers for dimensionality reduction, and fully connected layers for classification [2]. These layers are structured to learn hierarchical features, starting from simple patterns in the initial layers to more complex patterns in deeper layers, making CNNs particularly effective for the task of gesture classification based on EMG signals. The convolutional layers can capture local features in the signal, such as muscle activations, while the pooling layers help in making the model invariant to small variations in signal amplitude.

2.2 AlexNet

AlexNet builds upon the basic CNN architecture by introducing more depth and complexity, which is crucial for handling the intricacies of EMG signal data. The network consists of five convolutional layers and three fully connected layers [6], with the Rectified Linear Unit (ReLU) used as the activation function to mitigate the vanishing gradient problem-a common issue in deep networks. For EMG signal classification, the deeper architecture of AlexNet allows for the extraction of more sophisticated features, improving the model's ability to distinguish between subtle variations in muscle activations associated with different gestures. The use of larger filters in the initial layers enables the capture of broader signal patterns, which are crucial for accurately classifying gestures that involve multiple muscles.

2.3 VGGNet

As EMG data can be noisy and complex, VGGNet's design philosophy of using smaller convolutional filters (3×3) across its layers enhances the network's ability to capture fine-grained details in the signal. VGG-16, with its 13 convolutional layers followed by 3 fully connected layers, reduces the number of trainable parameters through its uniform filter size, thus improving training efficiency and generalization on small datasets, which is often the case with EMG datasets. The progressive increase in the number of filters-from 64 to 512-across the layers allows the model to capture increasingly complex features [4], which is critical for distinguishing between subtle gesture differences that might be encoded in the EMG signals.

2.4 ZFNet

ZFNet offers an improvement over AlexNet by employing smaller initial filters (7×7 instead of 11×11), which is particularly beneficial for EMG signal classification as it allows for the preservation of more detailed features in the early stages of the network [3]. This refinement is crucial for EMG signals, where subtle differences in muscle activity patterns need to be captured accurately for reliable gesture classification. Additionally, the deeper architecture of ZFNet, combined with its use of ReLU activations and batch stochastic gradient descent, enhances the network's ability to learn from EMG data, leading to better generalization and improved classification accuracy.

2.5 ResNet

Residual Network(ResNet),a revolutionized deep learning by addressing the challenge of training very deep neural networks [5]. A core problem with increasing the depth of a network is the vanishing gradient problem, where gradients become too small to effectively train the network. This often leads to a degradation in performance as the network grows deeper, which is counterintuitive. ResNet solves this by introducing residual connections-shortcut connections which skip one or more layers. These connections allow the network to learn residual functions, or the difference between the desired output and the input, rather than trying to learn the full transformation directly. The skip connections in ResNet work by allowing the input of a layer to bypass one or more layers and be added directly to the output. Mathematically, if the transformations (typically a series of convolutions, batch normalization, and activation) applied to the input x in a block is represented as $F(x)$, then the output of the residual block becomes $y = F(x) + x$. This addition of the input x directly to the output means that even if the weights in the layers are not optimal or the gradients are small, the model can still pass along useful information from earlier layers to later ones.

For sEMG signals, which often contain subtle and intricate patterns related to muscle activations, preserving these features across layers is vital. The skip connections in ResNet ensure that important features are not lost as the data passes through multiple layers of convolution and pooling. This preservation of information allows the network to learn both shallow and deep features effectively. Shallow features might capture basic signal characteristics, while deeper layers can capture more complex and abstract patterns [21]. By maintaining a direct path for gradients to flow back through the network, ResNet also allows for much deeper architectures-enabling the network to model complex dependencies and nuances in the sEMG data without suffering from the vanishing gradient problem. This leads to more accurate and robust gesture classification, as the network can be trained to a greater depth without losing critical information, ensuring that even the smallest and most complex features are learnt effectively.

2.6 RCCGNet

Renal Cell Carcinoma Grading Network (RCCGNet), although originally designed for renal cell carcinoma grading, offers a highly innovative architecture that is particularly well-suited for the classification of multi-channel sEMG signals [22]. The centerpiece of this architecture is the Shared Channel Residual (SCR) block, which plays a critical role in enhancing the model's ability to capture and integrate complex inter-channel relationships. In the context of sEMG signal classification, where each channel represents the activity of a different muscle group, the SCR block is invaluable.

The SCR block is designed to connect and share information across different data channels, allowing the network to learn not only from individual channels but also from the relationships between them. This is particularly important in sEMG data because gestures often involve coordinated activity across multiple muscles. For example, a single gesture like grasping can involve the activation of several muscles in the forearm, each contributing to the overall movement. The SCR block enables the network to capture these coordinated patterns by ensuring that information is shared across channels in a structured manner.

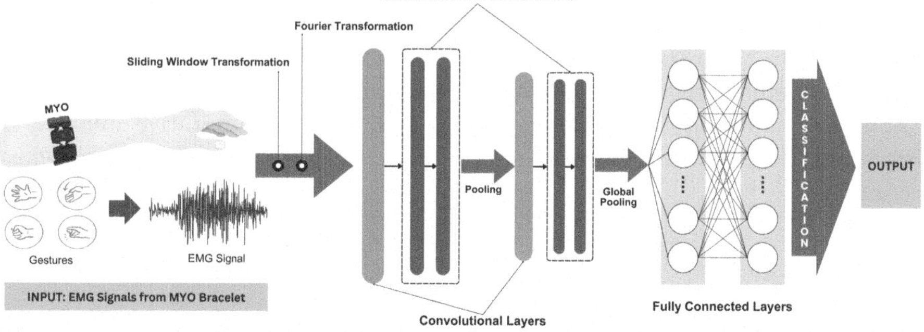

Fig. 1. Flowchart of the final model pipeline incorporating sliding window and Fourier transform as preprocessing steps, followed by the application of a modified RCCGNET

The block operates by using residual connections to link different channels, much like how ResNet uses skip connections to link layers. These residual connections within the SCR block allow the model to preserve and combine features from different channels effectively. By sharing information across channels, the SCR block ensures that the model does not miss any critical inter-channel dependencies, which are essential for accurate gesture recognition. Additionally, the SCR block incorporates gating mechanisms that control the flow of information between channels, allowing the model to focus on the most relevant features and discard irrelevant noise. This selective information sharing enhances the model's ability to differentiate between similar gestures by considering the full context of muscle activity.

In the final grading (or classification) phase, the features processed by the SCR blocks are further refined and combined in fully connected layers to produce the final classification output. Because the SCR blocks have effectively integrated information from all channels, the classifier has access to a comprehensive and unified representation of the gesture. This results in higher accuracy and robustness in sEMG signal classification, as the model can better account for the complex and often subtle patterns of muscle activation that define different gestures.

3 Simulation Results and Discussions

The EMG dataset undergoes preprocessing through two distinct approaches: first using the sliding window method, and second by augmenting this with Fourier Transformation. The sliding window technique effectively reduces dimensionality while extracting essential features from the signals, facilitating improved model performance.

For the implementation of the EMG-based gesture recognition system, we employ a variety of deep learning models, including AlexNet, VGGNet, ZFNet, ResNet18, ResNet34, and RCCGNet. Among these, we select the models that demonstrate the highest accuracies for further analysis. To ensure robustness and adaptability to the dataset, we perform a 3-fold cross-validation. This method helps mitigate the bias associated with random sampling of the training and testing sets, enhancing the reliability of our results.

Detailed information regarding the dataset and a comparative analysis of the various deep learning models' performance are presented in the subsequent sections.

3.1 Dataset

This study utilizes the Hyser dataset [1], which contains raw s-EMG data recorded by MYO Thalmic armbands. These armbands are equipped with eight sensors, evenly spaced around the forearm, which capture myographic signals corresponding to a range of static hand gestures. The dataset is organized as time-series data, where each column represents data from one of the eight sensors, labeled as Channel 1 through Channel 8.

Data was collected from 36 subjects who performed two series of hand gestures, with each gesture held for 3 s and a 3-s pause between gestures. The gestures include hand at rest, clenched fist, wrist flexion, wrist extension, radial deviation, ulnar deviation and extended palm gestures. Each subject's session includes recordings across all eight sensors, resulting in approximately 30,000 instances within the dataset.

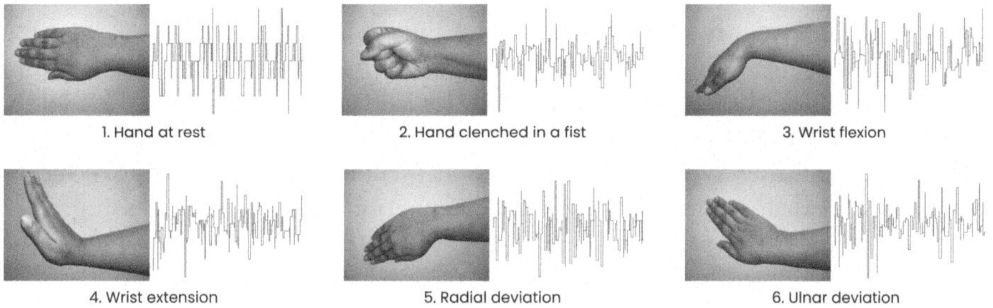

Fig. 2. Visualization of the dataset showing various gestures alongside their corresponding time series data from Channel 1

The MYO Thalmic armband is designed for versatile use, supporting touch-free control through gesture recognition and connecting to Mac, Windows, iOS and Android devices via Bluetooth Smart for data storage and analysis. However, variability in the dataset arose because the extended palm gesture was not consistently performed by all subjects. This inconsistency necessitated the removal of the extended palm gesture during the data preprocessing phase, leaving six gesture types as illustrated in Fig. 2 for subsequent classification tasks. Despite minor variability in dataset size, with each column containing between 40,000 to 50,000 recordings, the absence of missing values ensures a robust dataset suitable for gesture recognition research.

3.2 Evaluation Metrics

The performance of Electromyogram signal-based gesture recognition system is evaluated using 2 commonly employed metrics: Accuracy and F1 score. These metrics provide valuable insights for the effectiveness of predictive models in multi-class classification tasks.

Accuracy : Accuracy score is the metrics that is defined as the ratio of true positives (TP) and true negatives (TN) to all positive (TP+FP) and negative observations (TN+FN). It tells how often model will correctly predict an outcome out of the total number of times it made predictions.

$$Accuracy = \frac{TP + TN}{TP + FN + TN + FP} \tag{1}$$

F1-score: Also known as balanced F-score or F-measure is the harmonic mean of precision and recall score. It gives equal weight to both the Precision and Recall for measuring its performance in terms of accuracy.

$$F1score = 2 \times \frac{Precision \times Recall}{Precision + Recall} \tag{2}$$

3.3 State-of-the-Art Comparison Models

In this section, we present a comparative analysis of various deep learning models to evaluate their performance across different metrics. Table 1 summarizes the accuracy and F-measure metrics for six models: AlexNet, ResNet18, ResNet34, VGG16, ZFNet, and RCCGNet. The analysis considers two scenarios: using the Sliding Window technique and a combination of Sliding Window and Fourier Transform for dimensionality reduction. These preprocessing techniques are applied to optimize the model's ability to generalize by reducing the size of the dataset and mitigating the risk of overfitting.

Table 1. Metrics comparison of different DL Models

Models	Sliding Window				Sliding Window and Fourier Transform			
	Training		Testing		Training		Testing	
	Accuracy	F1-Score	Accuracy	F1-Score	Accuracy	F1-Score	Accuracy	F1-Score
AlexNet	71.62	71.70	70.09	70.24	90.30	90.32	88.95	89.03
ResNet18	72.53	72.43	71.11	71.04	92.22	92.24	90.49	90.52
ResNet34	61.42	58.13	61.08	57.52	91.60	91.61	90.16	90.17
VGG16	45.86	43.20	39.55	39.30	88.84	88.83	88.37	88.34
ZFNet	82.01	81.80	80.74	80.53	91.60	91.57	90.11	90.07
RCCGNet	99.28	99.28	95.88	99.87	95.84	92.02	95.09	92.03

Upon comparison, we observe that ResNet models and RCCGNet exhibit the highest performance, with RCCGNet outperforming all other models in terms of accuracy. This is likely due to its architectural advantages, including optimized convolutional layers and superior parameter tuning.

To further refine these models, we conduct a hyperparameter tuning process on the original RCCGNet model. This involves adjusting learning rates, number of epochs, and selectively removing layers such as SCR blocks (used for spatial-channel reduction), which may contribute to overfitting. By employing a combination of sliding window techniques, we achieved a notable testing accuracy of 95.88%.

To validate the robustness of the best-performing models, we applied k-fold cross-validation (k = 3) on ResNet18, ResNet34 and RCCGNet. This technique divides the dataset into k subsets and iteratively trains the model on k-1 subsets while testing on the remaining one. The results for k-fold validation, displayed in Table 2, further confirm the superior performance of RCCGNet. This model consistently achieves the highest accuracy and F-measure values in both the training and testing phases across all datasets.

Table 2. 3-fold Cross-Validation model performance comparison

Methods\Models	fold no.	Metrics	Resnet-18	RCCGNET
Sliding Window	1	Accuracy	92.20	92.43
		F1-score	92.24	92.40
	2	Accuracy	92.00	92.70
		F1-score	91.99	92.74
	3	Accuracy	92.60	92.06
		F1-score	92.67	92.00
	Average	Accuracy	92.31	92.40
		F1-score	92.30	92.42
Sliding Window & Fourier Transform	1	Accuracy	91.46	92.49
		F1-score	91.48	92.51
	2	Accuracy	92.73	92.75
		F1-score	92.70	92.73
	3	Accuracy	92.32	92.90
		F1-score	92.32	92.90
	Average	Accuracy	92.17	92.71
		F1-score	92.16	92.71

4 Conclusion

This paper highlighted comprehensive simulation study and performance analysis of an EMG-based gesture recognition system utilizing advanced deep learning techniques. The sEMG dataset was preprocessed using both the sliding window method and a combination of sliding window with Fourier transformation, leading to improved performance metrics for the gesture recognition system. Various models, including ResNet18, ResNet34, VGGNet, AlexNet, ZFNet, and RCCGNet, were implemented to analyze 1D surface EMG signals for gesture recognition task. The results indicated that RCCGNet model achieved the highest accuracy and F1 score. This is attributed to its use of skip connections, which bypass layers where performance degrades, thereby enhancing information sharing between layers. By incorporating data from previous layers, the network can improve local semantic feature representation without adding computational complexity.

Future work can explore a variety of EMG signal preprocessing techniques to further enhance model performances. Common approaches include filtering methods (low-pass, high-pass, band-pass, and notch) to remove noise, rectification to highlight muscle activity, and normalization against a reference like the Maximum Voluntary Contraction (MVC) for standardization. Smoothing techniques such as moving averages and RMS reduce signal fluctuations, while wavelet and Fourier transforms facilitate time-frequency analysis. Additionally, feature extraction methods like PCA and ICA, along with artifact removal techniques, can isolate relevant signal components, collectively improving signal clarity and gesture recognition performance. Further more, recording sEMG signals can also be optimized by employing various methods beyond the Myo bracelet. Alternatives include traditional wet and dry electrodes, wireless systems like Shimmer EMG and Delsys Trigno, and wearable garments with integrated sensors. High-density electrode arrays and innovative textile electrodes enhance signal quality and comfort, while custom solutions such as 3D-printed electrodes cater to specific clinical and fitness applications. By utilizing diverse capture methods, the accuracy of gesture prediction improves through enhanced signal quality and optimal electrode placement. Wireless systems and wearable technologies enable more natural movement during data collection, resulting in richer datasets and more effective gesture recognition algorithms.

Furthermore, future work could explore comparisons between various preprocessing techniques, as well as the application of advanced methods such as Discrete Wavelet Transform (DWT) and Radial Basis Function (RBF). These approaches have been effectively demonstrated in previous studies.

References

1. Krilova, N., Kastalskiy, I., Kazantsev, V., Makarov, V.A., Lobov, S.: EMG data for gestures. UCI Machine Learning Repository (2019). https://doi.org/10.24432/C5ZP5C
2. Yamashita, R., Nishio, M., Do, R., Togashi, K.: Convolutional neural networks: an overview and application in radiology. Insights Imaging **9**(4), 611–629 (2018). https://doi.org/10.1007/s13244-018-0639-9
3. Zeiler, M.D., Fergus, R.: Visualizing and understanding convolutional networks. In: Fleet, D., Pajdla, T., Schiele, B., Tuytelaars, T. (eds.) ECCV 2014. LNCS, vol. 8689, pp. 818–833. Springer, Cham (2014). https://doi.org/10.1007/978-3-319-10590-1_53
4. Simonyan, K., Zisserman, A.: Very deep convolutional networks for large-scale image recognition. arXiv preprint arXiv:1409.1556 (2014)
5. He, K., Zhang, X., Ren, S., Sun, J.: Deep residual learning for image recognition. In: Proceedings of the IEEE Conference on Computer Vision and Pattern Recognition, pp. 770–778 (2016)
6. Krizhevsky, A., Sutskever, I.: Hinton., GE: AlexNet. Adv. Neural Inf. Process. Syst. (2012)
7. McNeill, D.: Gesture and Thought. University of Chicago Press (2005)
8. Jiang, N., Rehbaum, H., Vujaklija, I., Graimann, B., Farina, D.: Intuitive, online, simultaneous, and proportional myoelectric control over two degrees-of-freedom in upper limb amputees. IEEE Trans. Neural Syst. Rehabil. Eng. **22**(3), 501–510 (2013)

9. Sultana, A., Ahmed, F., Alam, M.S.: A systematic review on surface electromyography-based classification system for identifying hand and finger movements. Healthcare Anal. **3**, 100126 (2023). https://doi.org/10.1016/j.health.2022.100126
10. Finkelstein, G.: Mechanical neuroscience: Emil du Bois-Reymond's innovations in theory and practice. Front. Syst. Neurosci. **9** (2015). https://doi.org/10.3389/fnsys.2015.00133
11. Ayachi, M., Seddik, H.: Overview of EMG signal preprocessing and classification for bionic hand control. In: 2022 IEEE Information Technologies & Smart Industrial Systems (ITSIS), pp. 1–6 (2022). https://doi.org/10.1109/ITSIS56166.2022.10118387
12. Subasi, A.: Classification of EMG signals using combined features and soft computing techniques. Appl. Soft Comput. **12**(8), 2188–2198 (2012). https://doi.org/10.1016/j.asoc.2012.03.035
13. Subasi, A., Yilmaz, M., Ozcalik, H.R.: Classification of EMG signals using wavelet neural network. J. Neurosci. Methods **156**(1), 360–367 (2006). https://doi.org/10.1016/j.jneumeth.2006.03.004
14. Christodoulou, C.I., Pattichis, C.S.: Unsupervised pattern recognition for the classification of EMG signals. IEEE Trans. Biomed. Eng. **46**(2), 169–178 (1999). https://doi.org/10.1109/10.740879
15. Pattichis, C.S., Schizas, C.N., Middleton, L.T.: Neural network models in EMG diagnosis. IEEE Trans. Biomed. Eng. **42**(5), 486–496 (1995). https://doi.org/10.1109/10.376153
16. Shilaskar, S., Bhatlawande, S., Chavare, R., Ingale, A., Joshi, R., Vaishale, A.: Human hand movement classification based on EMG signal using different feature extractors. Biomed. Pharmacol. J. **17**(1), 71–82 (2024). https://doi.org/10.13005/bpj/2835
17. Jena, A., Baberwal, K., Gehlot, N., Kumar, R.: Impact of feature selection on sEMG signal classification. In: 14th International Conference on Computing Communication and Networking Technologies (ICCCNT), pp. 1–6 (2023). https://doi.org/10.1109/ICCCNT56998.2023.10306464
18. Tutak Erözen, A.: A new CNN approach for hand gesture classification using sEMG data. J. Innov. Sci. Eng. **4**(1), 44–55 (2020). https://doi.org/10.38088/jise.730957
19. Nazari, F., Mohajer, N., Nahavandi, D., Khosravi, A.: Comparison of gait phase detection using traditional machine learning and deep learning techniques. In: IEEE International Conference on Systems, Man, and Cybernetics (SMC), pp. 403–408. IEEE (2022). https://doi.org/10.1109/SMC53654.2022.9945397
20. Tepe, C., Erdim, M.: Classification of surface electromyography and gyroscopic signals of finger gestures acquired by Myo armband using machine learning methods. Biomed. Signal Process. Control **75**, 103588 (2022). https://doi.org/10.1016/j.bspc.2022.103588
21. Al-Moosawi, N.M.A.M., Khudeyer, R.S.: RESNET-34/DR: a residual convolutional neural network for the diagnosis of diabetic retinopathy. Informatica. **45**(7) (2021). https://doi.org/10.31449/inf.v45i7.3774
22. Chanchal, A.K., Lal, S., Kumar, R., Kwak, J.T., Kini, J.: A novel dataset and efficient deep learning framework for automated grading of renal cell carcinoma from kidney histopathology images. Sci. Rep. **13**(1), 5728 (2023). https://doi.org/10.1038/s41598-023-31275-7

Author Index

D. Gupta et al. (Eds.): PCEMS 2024, CCIS 2491, pp. 451–454, 2025.
https://doi.org/10.1007/978-3-031-90577-3

The manufacturer's authorised representative in the EU is Springer Nature Customer Service Centre GmbH, Europaplatz 3, 69115 Heidelberg, Germany. If you have any concerns regarding our products, please contact ProductSafety@springernature.com

Printed and bound by CPI Group (UK) Ltd, Croydon, CR0 4YY
30/09/2026
02225585-0012